Introduction to Magnetic Recording

OTHER IEEE PRESS BOOKS

Insights Into Personal Computers, *Edited by A. Gupta and H. D. Toong*
The Space Station: An Idea Whose Time Has Come, *By T. R. Simpson*
Marketing Technical Ideas and Products Successfully! *Edited by L. K. Moore and D. L. Plung*
The Making of a Profession: A Century of Electrical Engineering in America, *By A. M. McMahon*
Power Transistors: Device Design and Applications, *Edited by B. J. Baliga and D. Y. Chen*
VLSI: Technology Design, *Edited by O. G. Folberth and W. D. Grobman*
General and Industrial Management, *By H. Fayol; revised by I. Gray*
A Century of Honors, *an IEEE Centennial Directory*
MOS Switched-Capacitor Filters: Analysis and Design, *Edited by G. S. Moschytz*
Distributed Computing: Concepts and Implementations, *Edited by P. L. McEntire, J. G. O'Reilly, and R. E. Larson*
Engineers and Electrons, *By J. D. Ryder and D. G. Fink*
Land-Mobile Communications Engineering, *Edited by D. Bodson, G. F. McClure, and S. R. McConoughey*
Frequency Stability: Fundamentals and Measurement, *Edited by V. F. Kroupa*
Electronic Displays, *Edited by H. I. Refioglu*
Spread-Spectrum Communications, *Edited by C. E. Cook, F. W. Ellersick, L. B. Milstein, and D. L. Schilling*
Color Television, *Edited by T. Rzeszewski*
Advanced Microprocessors, *Edited by A. Gupta and H. D. Toong*
Biological Effects of Electromagnetic Radiation, *Edited by J. M. Osepchuk*
Engineering Contributions to Biophysical Electrocardiography, *Edited by T. C. Pilkington and R. Plonsey*
The World of Large Scale Systems, *Edited by J. D. Palmer and R. Saeks*
Electronic Switching: Digital Central Systems of the World, *Edited by A. E. Joel, Jr.*
A Guide for Writing Better Technical Papers, *Edited by C. Harkins and D. L. Plung*
Low-Noise Microwave Transistors and Amplifiers, *Edited by H. Fukui*
Digital MOS Integrated Circuits, *Edited by M. I. Elmasray*
Geometric Theory of Diffraction, *Edited by R. C. Hansen*
Modern Active Filter Design, *Edited by R. Schaumann, M. A. Soderstrand, and K. B. Laker*
Adjustable Speed AC Drive Systems, *Edited by B. K. Bose*
Optical Fiber Technology, II, *Edited by C. K. Kao*
Protective Relaying for Power Systems, *Edited by S. H. Horowitz*
Analog MOS Integrated Circuits, *Edited by P. R. Gray, D. A. Hodges, and R. W. Brodersen*
Interference Analysis of Communication Systems, *Edited by P. Stavroulakis*
Integrated Injection Logic, *Edited by J. E. Smith*
Sensory Aids for the Hearing Impaired, *Edited by H. Levitt, J. M. Pickett, and R. A. Houde*
Data Conversion Integrated Circuits, *Edited by D. J. Dooley*
Semiconductor Injection Lasers, *Edited by J. K. Butler*
Satellite Communications, *Edited by H. L. Van Trees*
Frequency-Response Methods in Control Systems, *Edited by A. G. J. MacFarlane*
Programs for Digital Signal Processing, *Edited by the Digital Signal Processing Committee, IEEE*
Automatic Speech & Speaker Recognition, *Edited by N. R. Dixon and T. B. Martin*
Speech Analysis, *Edited by R. W. Schafer and J. D. Markel*
The Engineer in Transition to Management, *By I. Gray*
Multidimensional Systems: Theory & Applications, *Edited by N. K. Bose*
Analog Integrated Circuits, *Edited by A. B. Grebene*
Integrated-Circuit Operational Amplifiers, *Edited by R. G. Meyer*
Modern Spectrum Analysis, *Edited by D. G. Childers*
Digital Image Processing for Remote Sensing, *Edited by R. Bernstein*
Reflector Antennas, *Edited by A. W. Love*
Phase-Locked Loops & Their Application, *Edited by W. C. Lindsey and M. K. Simon*
Digital Signal Computers and Processors, *Edited by A. C. Salazar*
Systems Engineering: Methodology and Applications, *Edited by A. P. Sage*
Modern Crystal and Mechanical Filters, *Edited by D. F. Sheahan and R. A. Johnson*
Electrical Noise: Fundamentals and Sources, *Edited by M. S. Gupta*
Computer Methods in Image Analysis, *Edited by J. K. Aggarwal, R. O. Duda, and A. Rosenfeld*
Microprocessors: Fundamentals and Applications, *Edited by W. C. Lin*
Machine Recognition of Patterns, *Edited by A. K. Agrawala*
Turning Points in American Electrical History, *Edited by J. E. Brittain*

Introduction to Magnetic Recording

Edited by

Robert M. White

Vice President, Engineering and Technology
Control Data Corporation

Published under the sponsorship of the IEEE Magnetics Society.

The Institute of Electrical and Electronics Engineers, Inc., New York

PRINTED IN THE UNITED STATES OF AMERICA

IEEE Order Number: PC01784

Library of Congress Cataloging in Publication Data

Main entry under title:

Introduction to magnetic recording.

Includes indexes.
1. Magnetic recorders and recording.
I. White, Robert M., 1938–
TK7881.6.I58 1985 621.389'324 85-104
ISBN 0-87942-184-3

Contents

Reprint Papers

Preface

MAGNETIC recording is presently a $20 billion industry in the United States. It spans audio, video, and digital applications in the form of tapes, cassettes, and disks. This industry is expected to grow to $100 billion by 1990. This growth will be accompanied by dramatic improvements in the technology. It is generally believed, for example, that magnetic recording densities can be improved by at least an order of magnitude! Yet, ironically, there are no more than four or five universities in the U.S. where students can learn the fundamentals of this subject. The Japanese, on the other hand, have identified magnetic recording as a critical industry and an elaborate university–industry collaboration has been established. This has alarmed U.S. industry to the extent that in the past year the major manufacturers of recording equipment such as IBM, Control Data, MMM, etc., have pledged over $10 million to establish recording centers at the University of California at San Diego and Carnegie Mellon. However, at an NSF-sponsored workshop on Magnetic Information Technology (MINT), held in the spring of 1983, it was estimated that at least two more major centers and numerous smaller ones are necessary to provide the trained students required by this growing industry.

Not surprisingly, there also exists very little pedagogical material on this subject. This book represents an effort to fill that void. It was developed from a seminar course given at Stanford University in the fall of 1983. The goal of this seminar was to introduce graduate students, with a wide range of backgrounds, to the fundamental concepts of magnetic recording. One of the stimulating aspects of this technology is that it encompasses a wide range of disciplines, including solid-state physics, electromagnetics, fluid flow, mechanical design, and information theory. This also makes it a difficult subject to present. Consequently, more effort was made at breadth than at depth. An attempt was made to focus on the origin of concepts used throughout the recording literature, such as the Stoner–Wohlfarth model, reciprocity, etc.

Included in this book is a selection of reprinted papers on recording. References to these papers, as well as others, appear in the specially written material. Several criteria were used in selecting the papers to be reprinted. Some, such as the Westmijze and Karlqvist papers, were chosen for their historical value to the subject, and the fact that they are not readily available. Most, however, were chosen for their impact on recording, either through the presentation of new results, such as Iwasaki's introduction of perpendicular media, or through the presentation of new techniques, such as Lean and Wexley's boundary element method. In instances where decisions were difficult I chose those papers which were more self-contained and provided the necessary background material.

I would like to thank Dr. D. Bloomburg of Xerox and Prof. K. Parvin of San Jose State University for reading the entire manuscript and making many helpful suggestions. I would also like to thank Dr. I. Beardsley of IBM and Dr. N. Bertram of Ampex for numerous helpful discussions.

The material for this book was prepared while I was at Xerox Palo Alto Research Center. I would like to thank F. Mannes and F. Boderman for their assistance with the manuscript and S. Wallgren for his excellent work on the figures.

ROBERT M. WHITE
Editor

Chapter 1
Introduction

1.1 History of Magnetic Recording

THE first demonstration that a magnetic medium could be used to record information was provided by Valdemar Poulsen in 1898. In particular, Poulsen recorded and reproduced sound with an invention called the "telegraphone" which was exhibited at the Paris Exposition of 1900. This was 23 years after Thomas Edison invented the phonograph. The telegraphone consisted of a steel piano wire which was wound on a spiral groove around the surface of a drum. An electromagnet made contact with the wire and was free to slide along a rod positioned parallel to the drum. When the drum was rotated the electromagnet was pulled down the rod. When current from a microphone passed through the electromagnet the wire was magnetized in proportion to the current. Although the telegraphone won the grand prize at the Exposition, the recorded signal was very weak, requiring listening with headphones.

With the development of the vacuum tube amplifier in 1920 it became possible to reproduce sound from a steel wire as loud as a listener wanted. However, the noise was also amplified! In 1921 W. L. Carlson and G. W. Carpenter invented the concept of ac biasing which endows magnetic recording with the high performance it has today. The Carlson–Carpenter discovery went unrecognized at the time, with the result that the development of magnetic recording progressed slowly.

Prior to 1927 all recording was done on wire or steel tape. In that year, J. A. O'Neill received a patent for a paper tape which had been coated with a magnetic liquid and then dried. In Germany, Fritz Pfleumer received a similar patent involving iron powder. Pfleumer's tape was later employed in a German recorder called the "magnetophone." While experimenting with a magnetophone in 1940, H. J. von Braunmuhl and W. Weber rediscovered the concept of ac biasing. This was employed successfully by the Germans during World War II. At the end of the war John T. Mullin brought two magnetophones back to the U.S. Mullin's application of magnetic recording to broadcast radio with the Bing Crosby show in 1947 did much to stimulate recording in the U.S. The 3M Company developed oxide tapes in 1947 and Ampex started shipping audio recorders in 1948. Video recording was demonstrated in 1951.

Meanwhile, in the digital world, the need was developing for a data storage system that offered "random access," that is, rapid access to data at any location in a file. This problem caught the attention of Rey Johnson, an IBM inventor, who had been sent to San Jose in 1952 to establish a West Coast research center. Johnson decided to try to develop a device that would allow information to be recorded and retrieved in any order, rather than in the batches required by magnetic tape. After some investigation, the IBM group decided that a disk offered the best choice. In 1955 IBM announced the first magnetic disk drive, the IBM 305, or RAMAC.[1] This first drive consisted of fifty 24-in-diameter disks mounted on a vertical shaft rotating at 1200 rpm (see Fig. 1.1). The recording density involved 100 bits/in and 20 tracks/in, giving a total drive capacity of 5 Mbytes. Since then IBM has dominated the disk drive industry. The revenues from the captive market provided by its computer systems far exceed those of its closest competitor. This dominance is based on IBM's continual improvements in the recording technology. The state-of-the-art at any given time is measured by the areal density of IBM's latest products (see Fig. 1.2).

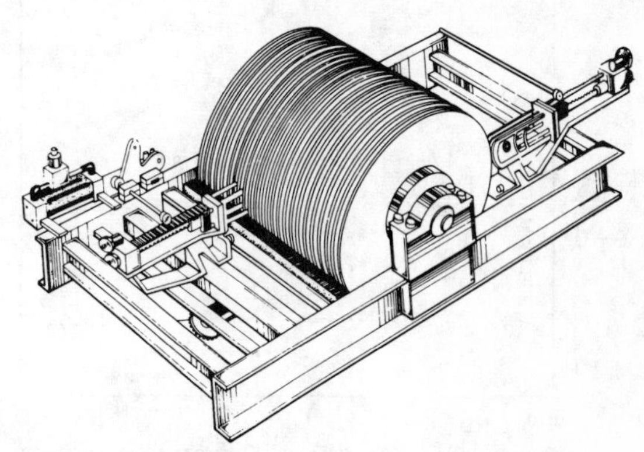

Fig. 1.1: Artist's conception of the IBM RAMAC file.

1.2 Storage Hierarchy

In audio and video applications, magnetic tapes serve an independent function. In digital data processing systems, however, tapes and disks often appear in complementary roles that utilize their unique features. Thus, disk drives with their random access

[1]The acronym originally referred to the complete 305 system and was coined from Random access memory Accounting MAChine. In later usage the word became generic and was applied to any machine with a disk file. IBM sales literature changed the meaning to the more general Random Access Method of Accounting and Control.

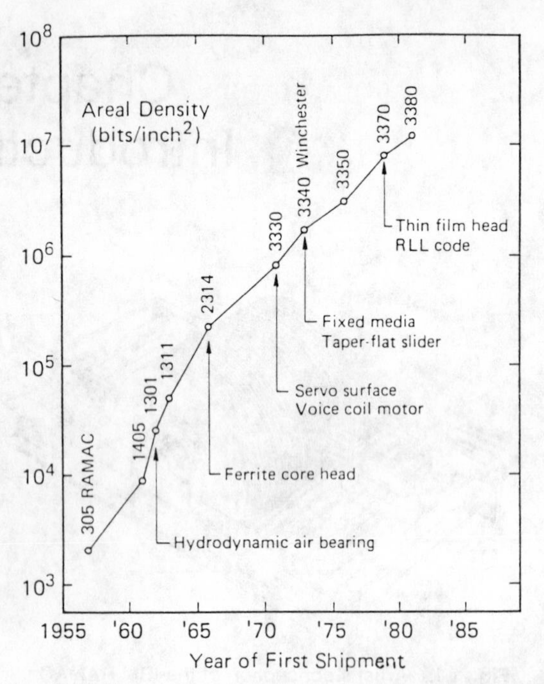

Fig. 1.2: Evolution of IBM Direct Access Storage Devices (DASD's).

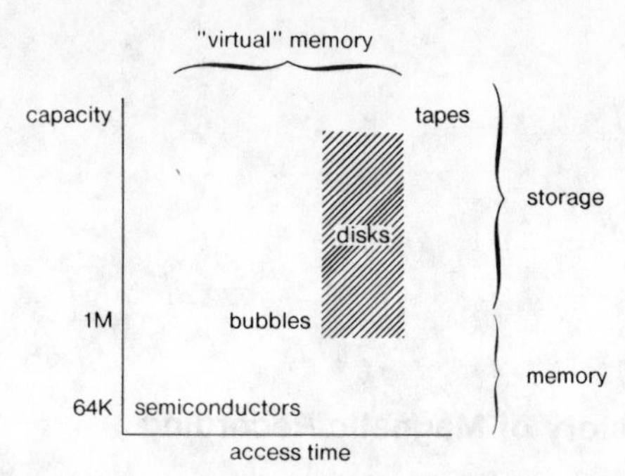

Fig. 1.3: Capacity-access time description of the memory hierarchy.

have access times much shorter than tape drives. However, being mechanical, they are still slower than semiconductors. On the other hand, since a magnetic "data cell" does not require its own individual design, magnetic disk storage is less expensive than semiconductor storage on a per-bit basis. Thus we have the "hierarchy" of storage illustrated by Fig. 1.3.

Notice that disk drives offer a wide variety of capacities. This reflects the ubiquitous nature of disk drives. Large computer systems require large amounts of on-line storage. Such storage requirements are satisfied by having numerous large disks on one spindle. For example, IBM's 3380, which offers 2.5 Gbytes, consists of two spindles each having eight 14 in disks. On the other hand, a personal work station may only require tens of megabytes which can be provided by a compact 5 1/4 in disk drive. The interchangeability of data among drives is accomplished by removable media. These may be either rigid disks or flexible, i.e., "floppy" disks.

Chapter 2
Magnetic Media

MAGNETISM has fascinated man since the discovery of the magical properties of lodestone even before reliable written records existed. The first application of this magnetic material appears to have been the compass. Our understanding of magnetism has only come in relatively recent times. Although Oersted discovered the relation between magnetism and electrical currents in 1820, the real explanation of magnetic moments and their interactions came only with the development of quantum mechanics.

2.1 Intrinsic Magnetic Properties

2.1.1 Magnetic Moments

Particles, like light, exhibit both particle-like and wave-like characteristics. In particular, when an electron is confined to the dimensions of an atom its wave-like nature governs its behavior: its energy and certain other physical properties can take on only discrete values. Furthermore, the exact location of the electron cannot be specified, merely its probability of being at some distance from the nucleus. This probability is related to the eigenfunctions and the quantized energies to the eigenvalues of the Schrödinger equation

$$\mathscr{H}\psi = E\psi. \tag{2.1}$$

Here $\mathscr{H}$ is the Hamiltonian, or energy, expressed as an operator by writing the momentum as the differential operator,

$$\boldsymbol{p} = -i\hbar\nabla.$$

For an electron in a spherical potential

$$V(\boldsymbol{r}) = V(r)$$

and the Hamiltonian is simply

$$\mathscr{H} = p^2/2m + V(r) = \hbar^2\nabla^2/2m + V(r).$$

The eigenfunctions associated with this operator are products of the spherical harmonics, $Y_l^m(\theta,\varphi,)$, and a function of r that depends upon the particular form of $V(r)$. The eigenfunctions ψ_{nlm} are therefore labeled by a radial quantum number n $(=0,1,2,\cdots)$ and the angular quantum numbers l $(=0,1,2,\cdots)$ and m $(=0,\pm 1,\cdots,\pm l)$. An electron in a state with $l=0$, 1, 2, 3, etc., is referred to as an *s, p, d, f*, etc. electron.

If we apply a magnetic field to this electron the quantum mechanical description requires that we replace the momentum $\boldsymbol{p}$ by the canonical momentum

$$\boldsymbol{p} \to \boldsymbol{p} - (e/c)\boldsymbol{A}$$

where $\boldsymbol{A}$ is the vector potential giving the magnetic field, i.e.,

$$\boldsymbol{H} = \nabla \times \boldsymbol{A}.$$

If $\boldsymbol{H}$ is uniform, then a convenient form for $\boldsymbol{A}$ is

$$\boldsymbol{A} = \tfrac{1}{2}\boldsymbol{H} \times \boldsymbol{r}.$$

The Hamiltonian then reduces to

$$\begin{aligned}\mathscr{H} &= (\boldsymbol{p} - (e/c)\boldsymbol{A})^2/2m + V(r)\\ &= p^2/2m + V(r) - \frac{e}{2mc}(\boldsymbol{r}\times\boldsymbol{p})\cdot\boldsymbol{H}\end{aligned}$$

plus a term of order H^2 which is responsible for diamagnetism. This quantity $\boldsymbol{r}\times\boldsymbol{p}$ is the angular momentum, l. it is easy to show that the spherical harmonics are eigenfunctions of this angular momentum operator, i.e.,

$$(\boldsymbol{r}\times\boldsymbol{p})_z Y_l^m(\theta,\varphi) = m\hbar Y_l^m(\theta,\varphi).$$

The extra energy associated with the magnetic field is therefore

$$E_{zl} = -m\mu_B H \tag{2.2}$$

where μ_B is the Bohr magneton,

$$\mu_B = 0.927\times 10^{-20}\ \text{erg/G}.$$

This enables us to define an orbital magnetic moment

$$\mu_l = -\mu_B l \tag{2.3}$$

where l is now in units of $\hbar$.

Thus, if our electron were in a *d*-orbital, i.e., $l=2$, the ground state in a magnetic field would correspond to $m=-2$.

At finite temperatures the magnetic moment is obtained by averaging over the allowed states. Thus, for a *d* electron,

$$\mu = \frac{\sum_{m=-2}^{2}(-\mu_B m)\exp(-m\mu_B H/k_B T)}{\sum_{m=-2}^{2}\exp(-m\mu_B H/k_B T)}. \tag{2.4}$$

For small values of the argument of the exponential

we find

$$\mu = \frac{2\mu_B^2 H}{k_B T}. \tag{2.5}$$

This inverse dependence of the moment on the temperature is much more general than this simple example would suggest. It was observed in many materials by P. Curie in the early 1900's and is a characteristic feature of *paramagnetic* materials.

Let us now progressively add more complications to the problem. One aspect that must be considered is the fact that in a crystalline solid an ion does not see a spherical symmetry, but rather the symmetry of the crystalline environment. The electric fields from such an environment have the effect of mixing the angular momentum states to form new eigenstates. For example, in a cubic environment the states of the *d*-electron become

$$\psi_e = \begin{cases} \psi_0 \\ \frac{1}{\sqrt{2}}(\psi_2 + \psi_{-2}) \end{cases} \psi_{t_2} = \begin{cases} \psi_1 \\ \psi_{-1} \\ \frac{1}{\sqrt{2}}(\psi_2 - \psi_{-2}) \end{cases}. \tag{2.6}$$

The letters *e* and t_2 are a labeling notation which has its basis in early group-theoretical treatments. An interesting consequence of this mixing is that the orbital magnetic moment is reduced. In particular,

$$\langle \psi_e | l | \psi_e \rangle = 0. \tag{2.7}$$

Thus, if the state *e* is the lowest state (this depends on the nature of the cubic coordination) the low temperature moment will be completely "quenched."

But the magnetic moment also has a *spin* origin. From relativistic quantum mechanics we find that the energy of an electron spin in a magnetic field is

$$E_{Zs} = \sigma \mu_B H \tag{2.8}$$

where $\sigma = \pm 1$.

Most systems, of course, involve more than one electron. As soon as we have more than one electron we must also include the Coulomb interaction between these electrons. Suppose we have two electrons in the same *orbital*. Since the Pauli exclusion principle requires that no two electrons be in the same state, these two electrons must be in opposite *spin* states. Thus, for example, there would be no net magnetic moment associated with two electrons in an *e* orbital. Actually, there will be a small moment, induced by an applied field in the direction *opposite* to the field. This *diamagnetism* arises from Lenz's law and is very small.

Now consider two electrons in different orbitals. In the absence of the Coulomb interaction the wave function will involve products of the individual wave functions. For example, if $\varphi(1)\alpha(1)$ denotes an "up" spin located at r_1 in an orbital α, such a product might be

$$\varphi_a(1)\alpha(1)\varphi_b(2)\beta(2). \tag{2.9}$$

But since the electrons are indistinguishable, the product

$$\varphi_b(1)\beta(1)\varphi_a(2)\alpha(2) \tag{2.10}$$

is equally valid. Thus, any linear combination of (2.9) and (2.10) is a possible solution. If we had *n* electrons and *n* states there would be *n*! ways of permuting the electrons. Just as the spherical harmonics are the basis functions for a system with rotational symmetry, one can construct basis functions for this permutation symmetry. What is remarkable, however, is that of all these possible functions nature allows only the one which is completely antisymmetric! This is a mathematical statement of the Pauli exclusion principle. What it means is that only the combination

$$\varphi_a(1)\alpha(1)\varphi_b(2)\beta(2) - \varphi_b(1)\beta(1)\varphi_a(2)\alpha(2) \tag{2.11}$$

is a valid starting basis.

Since there are three other ways in which we can orient the two spins we have a basis of four functions with the same energy. When the interaction is "turned on," these four states mix and the degeneracy is lifted. In particular, we obtain a triplet and a singlet whose energies differ by an amount $2J_{12}$ where

$$J_{12} = \int\int d^3r_1\, d^3r_2\, \varphi_a^*(1)\varphi_b^*(2)(e^2/r_{12})\varphi_a(2)\varphi_b(1) \tag{2.12}$$

is called the exchange integral. The magnitude of the total spin, $|\mathbf{S}|^2 = |\mathbf{S}_1 + \mathbf{S}_2|^2$, associated with the triplet is 2 while that of the singlet is 0. This is consistent with a vector model for the addition of the spin where $\mathbf{S} = 1$, i.e., $|\mathbf{S}|^2 = S(S+1)$. In fact, Dirac showed that this manifestation of the exclusion principle through the Coulomb interaction could be represented as a scalar spin–spin interaction

$$-2J_{12}\mathbf{S}_1 \cdot \mathbf{S}_2. \tag{2.13}$$

When the orbitals are orthogonal, as they are on one site, then the sign of the exchange is positive, i.e., the triplet state lies lowest. This is the origin of Hund's rule which says that electrons interact in such a manner as to maximize their total spin. Thus Fe^{3+}, which has five 3*d* electrons, has a total spin of 5/2, giving a spin magnetic moment of $5\mu_B$.

This description works very well for situations where the "magnetic electrons" are localized. However, if we dissolve a magnetic ion in a metallic host, or if we bring many magnetic ions together to form a metal, then the "magnetic electrons" can hop onto neighbor-

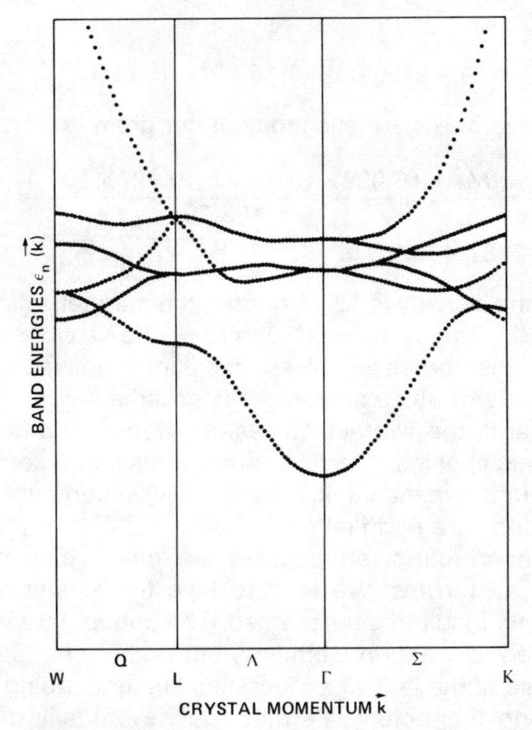

Fig. 2.1: Representative transition-metal energy-band structure, showing the extent to which the five relatively flat *d* bands can be distinguished from the more dispersive *sp* band. In "strong" magnets, the relatively flat *d* bands for majority-spin electrons lie entirely below ε_F, thereby contributing five electrons to $n_\uparrow$. These bands were calculated for paramagnetic Ni.

ing sites. The electron eigenstates are no longer localized but extended throughout the metal, and the energies spread into bands. Fig. 2.1 shows a representative energy-band structure for a transition metal. The electrons fill these states up to the Fermi energy which is determined by the number of electrons. The effect of exchange on this description is to split the "spin-up" and "spin-down" bands. The moment is then

$$\mu = (n_\uparrow - n_\downarrow)\mu_B. \tag{2.14}$$

But the total number of electrons involved is the chemical valence Z,

$$Z = n_\uparrow + n_\downarrow. \tag{2.15}$$

Iron, for example, has a chemical valence of 26–18 or 8. It turns out that elements to the right of Fe in the periodic table possess completely filled up-spin *d* bands, i.e., $n_{d\uparrow} = 5$. The exchange splitting of the *sp* bands is much smaller. Calculations show that $n_{sp\uparrow}$ is about 0.3 for the elements to the right of Fe. Thus, the total moment in units of μ_B is

$$\mu = 2(n_{d\uparrow} + n_{sp\uparrow}) - Z = 10.6 - Z. \tag{2.16}$$

The chemical valence of Co is 9. So, $\mu = 1.6$. For Ni we find 0.6. Fe itself does not quite have filled up-spin *d* bands. Therefore, its moment is not 2.6 but is slightly smaller, 2.2. For alloys of Fe, Co, Ni, or Cu the *d* bands

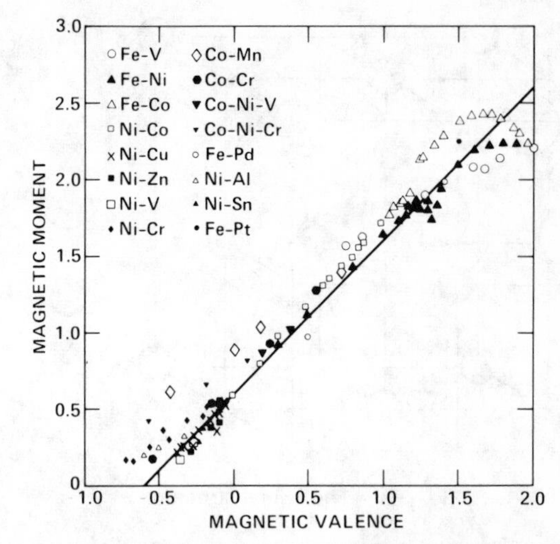

Fig. 2.2: Generalized Slater–Pauling curve [1]. Average alloy magnetization per alloy atom versus average magnetic valence. Magnetic valence is an integer for each column of the periodic table. It is the negative of the usual chemical valence, except for the Fe, Co, and Ni columns, where it is 10 minus the chemical valence.

are complicated. But the up-spin bands still lie below the Fermi level. Therefore the simple expression given above still applies with Z being the average valence [1]. Thus the moment varies linearly with concentration as shown in Fig. 2.2.

2.1.2 Long Range Order

In the previous section we have described how exchange is responsible for the formation of magnetic moments. At zero temperature the entropy associated with a system of such moments goes to zero, and the moments "freeze" into a specific configuration. The nature of this long range order depends upon the interactions among the moments. These interactions are also exchange interactions. If the wave functions of the electrons on the different sites overlap then there will be an intersite interaction of the Dirac form involving the individual electrons. However, the sign of the exchange may be either positive or negative. Also, under some conditions, it is possible to rewrite this interaction as an exchange interaction between the *total* spins of each site, i.e.,

$$-2J\mathbf{S}_A \cdot \mathbf{S}_B \tag{2.17}$$

where $\mathbf{S}_A = \Sigma \mathbf{S}_i$. This interaction forms the basis for what is called the Heisenberg model of ferromagnetism. In general, however, the origin of exchange in real materials is more subtle. In many magnetic insulators, for example, there are anions such as oxygen, fluorine, etc., intervening between the magnetic ions. In this case the exchange arises through a mechanism known as superexchange. P. W. Anderson has shown that the sign of this effective exchange is negative, that is, it tends to favor antiparallel alignment of the moments. In insulators with relatively simple crystal structures this leads to *antiferromagnetism* at low tem-

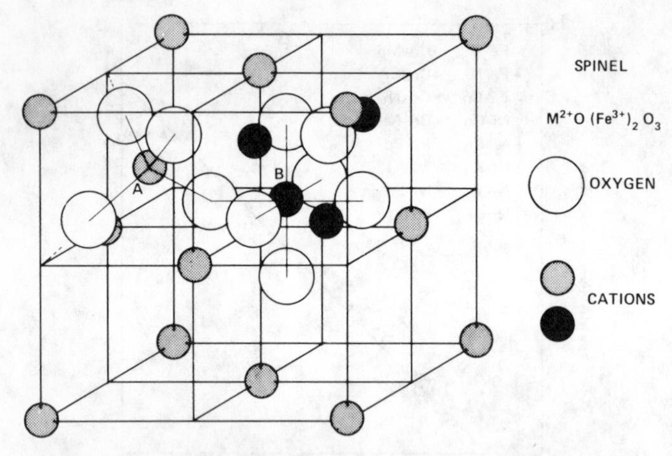

Fig. 2.3: The spinel structure.

peratures. In materials with more complicated structures the magnetic structures may also be complex.

As an example of a complex magnetic structure let us consider the basic component of magnetic media, $\gamma-Fe_2O_3$, or maghemite.

This material has the so-called *spinel* crystal structure. This structure is shown in Fig. 2.3. The smallest cubic unit cell contains 32 oxygen ions. The cations occupy interstitial positions in this cubic oxygen lattice, of which there are two types.

In one, called the *A* site, the cation is surrounded by four oxygen ions located at the corners of a tetrahedron. In the other, called the *B* site, the cation is surrounded by six oxygen ions at the vertices of an octahedron. Eight *A* sites and 16 *B* sites are occupied per unit cell. If these *A* sites are occupied by divalent ions and the 16 *B* sites by trivalent ions the structure is called a *normal* spinel. If, however, the *B* sites are occupied half by divalent and half by trivalent ions distributed at random, with the other trivalent ions in the *A* sites, the structure is called an *inverse* spinel. In the case of $\gamma-Fe_2O_3$, which has the inverse structure, there are no divalent ions; two-thirds of the octahedral *B* sites that would normally be occupied by divalent ions are occupied by Fe^{3+} ions, the other one-third remaining vacant. Thus, the formula for the unit cell may be written

$$\underset{A\ \text{sites}}{Fe_8}\left[\underset{\substack{\text{divalent}\\ B\ \text{sites}}}{Fe_{16/3}\square_{8/3}}\right]\left[\underset{\substack{\text{trivalent}\\ B\ \text{sites}}}{Fe_8}\right]O_{32}$$

where $\square$ stands for a vacancy.

The exchange interactions in $\gamma-Fe_2O_3$ are such as to align the *A* sites antiparallel to the *B* sites. Since the moment of Fe^{3+} is $5\mu_B$, the net magnetic moment of a unit cell is

$$\mu = 8(5\mu_B) - [16/3(5\mu_B) + 8(5\mu_B)] = (80/3)\mu_B$$

or $(80/3)/(32/3) = 2.5\mu_B$ per molecular unit. The number of molecules per mole is Avogadro's number *N*, and the number of grams per mole is the molecular weight

$$A = 2(55.85) + 3(16.00) = 159.70.$$

Therefore, the magnetic moment per gram is

$$\sigma = \frac{N\mu}{A} = \frac{(6.022\times10^{23})(2.5)(0.927\times10^{20})}{159.70}$$
$$= 87.4\ \text{EMU/g}.$$

Since the density is $5.074\ \text{g/cm}^3$, the magnetization is $443\ \text{EMU/cm}^3$. A material such as $\gamma-Fe_2O_3$ whose net moment is the result of an imbalance between two oppositely oriented sublattices is called a *ferrimagnet*. In general, the two sublattices could involve different magnetic species. The rare earth garnets, for example, which underlie magnetic bubble devices, contain Fe^{2+}, Fe^{3+}, and rare earth ions.

Nonconducting ferrimagnets are often referred to simply as ferrites. We tend to take the existence of magnetic insulators for granted. The concept was first patented in 1909 (in Germany) but nothing happened because of the lack of a theoretical understanding and a driving technology. Ferrites were eventually developed largely as a result of Snoek's work at Philips during the early 1940's on spinels and Néel's 1948 theory of ferrimagnetism (for which he shared the 1970 Nobel Prize in Physics).

If we have an ordered net magnetic moment, as in a ferromagnet or ferrimagnet, then it will cost exchange energy to introduce any spatial variation in this magnetization. This is contained in the Heisenberg exchange. However, it is often convenient to express this energy in a more macroscopic form. In particular, it must be proportional to the spatial variation in *M*. It can be shown [2] that if we consider only the magnetic energy the exchange energy density takes the form

$$E_x = A\left(|\nabla M_x|^2 + |\nabla M_y|^2 + |\nabla M_z|^2\right) \qquad (2.18)$$

where the exchange parameter *A* is related to the microscopic Heisenberg exchange *J*.

2.1.3 Transition Temperature

Thermal fluctuations tend to destroy the long range magnetic order. The temperature at which this occurs is of the order of the "intermoment" exchange. In the case of the insulating magnets this exchange is much less than the "intramoment" exchange which is responsible for the moment formation. Thus, above a certain temperature, called the Curie temperature for ferro- or ferrimagnets and the Neel temperature for antiferromagnets, one has disordered magnetic moments whose amplitudes are not appreciably affected by the temperature.

In the case of metallic magnetic materials the spatial extent of the magnetic moment may be quite large. Consequently, temperature may actually destroy the long range order by destroying the moment itself. This

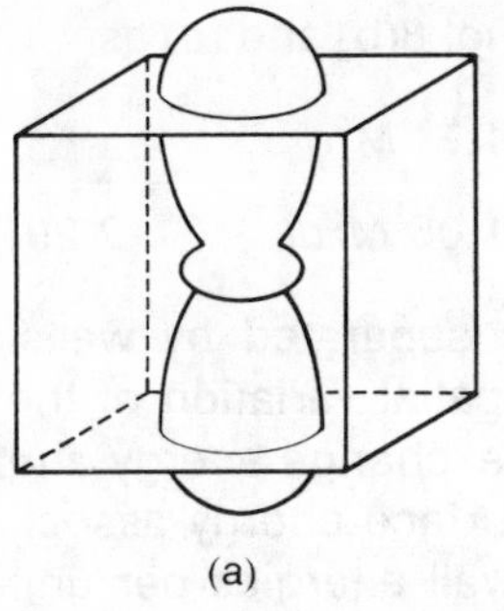

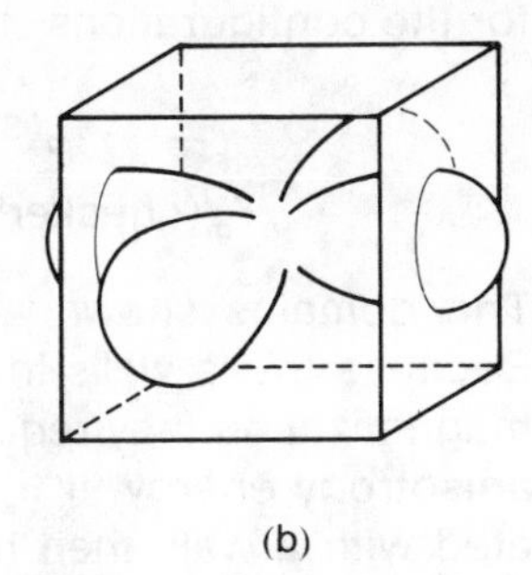

(a) (b)

Fig. 2.4: Charge densities associated with (a) one of the *e* orbitals and (b) one of the t_2 orbitals.

is what is generally referred to as itinerant magnetism. The fact that Fe and Ni show a Curie-law response above their transition temperatures indicates that they behave very much like localized moment systems. Cr is perhaps the best example of a truly itinerant system.

2.1.4 Crystalline Anisotropy

Fe and Ni both have a cubic crystalline structure (Fe is bcc, Ni fcc). Experimentally, it is found that in the ordered state the magnetic moment of Fe prefers to point along the [100] axis, while that of Ni along the [111] axis. We speak of these directions as their easy axes. This magnetic anisotropy arises from the coupling between the electronic spin, which is largely responsible for the magnetic moment, and the electronic charge distribution. This charge distribution itself is influenced by the crystalline environment as we mentioned. The *e* state and t_2 states, for example, have the charge distributions shown in Fig. 2.4. The total charge distribution of the ion or the magnetic cell will contain contours that will determine its position in the environment. The magnetic moment is coupled to this charge and will therefore point in a specific direction. Notice that if the [111] axis is an easy axis there will be four equivalent easy axes in the crystal.

The largest crystalline anisotropies are found in crystals of $RECo_2$ and RE_2Co_{17}. This is presumably connected with the very anisotropic charge distributions of the rare earths (RE).

Phenomenologically, the anisotropy energy associated with an axial anisotropy may be written

$$E_A = K\sin^2\theta \qquad (2.19)$$

where θ is the angle the magnetization makes with the axis. When $K>0$ we refer to this as an easy axis; when $K<0$ it becomes a hard axis. Although iron and nickel are cubic, when Ni–Fe films are evaporated in the presence of a magnetic field an easy axis is induced in the direction of this field due to the directional ordering of Fe–Fe pairs in the Ni matrix.

If the anisotropy does reflect a cubic symmetry, as in iron or nickel, the phenomenological anisotropy energy density is written

$$E_A = K_1\left(\alpha_1^2\alpha_2^2+\alpha_2^2\alpha_3^2+\alpha_3^2\alpha_1^2\right)+K_2\alpha_1^2\alpha_2^2\alpha_3^2 \qquad (2.20)$$

where the α_i are the direction cosines of the magnetization referred to the cube edges. The sign of K_1 for Ni is opposite that for Fe:

$$K_1\,(\mathrm{Fe}) = 4.6\times10^5\ \mathrm{erg/cm^3}$$
$$K_1\,(\mathrm{Ni}) = -0.56\times10^5\ \mathrm{erg/cm^3}.$$

As a result, the anisotropy of Ni–Fe alloys passes through zero at 70 percent Ni.

2.1.5 Magnetostriction

When a material is magnetized there is also a change in its length. The fractional change $\lambda = \delta l/l$ associated with a change in magnetization from zero to saturation is called *magnetostriction*. This coupling between the magnetization and the strain is due to the strain dependence of the anisotropy energy. If we denote the strain components by e_{ij}, the magnetoelastic coupling energy density in a cubic material must have, by symmetry, the form

$$E_{me} = B_1\left(\alpha_1^2e_{xx}+\alpha_2^2e_{yy}+\alpha_3^2e_{zz}\right) + B_2\left(\alpha_1\alpha_2e_{xy}+\alpha_2\alpha_3e_{yz}+\alpha_3\alpha_1e_{zx}\right). \qquad (2.21)$$

The elastic energy density has the form

$$E_e = \tfrac{1}{2}c_{11}\left(e_{xx}^2+e_{yy}^2+e_{zz}^2\right)+\tfrac{1}{2}c_{44}\left(e_{xy}^2+e_{yz}^2+e_{zx}^2\right) + c_{12}\left(e_{yy}e_{zz}+e_{xx}e_{zz}+e_{xx}e_{yy}\right) \qquad (2.22)$$

where the c's are the elastic moduli. Minimizing $E_{me}+E_e$ with respect to the strains we find that the elongation in the direction $(\beta_1,\beta_2,\beta_3)$ is given by

$$\delta l/l = (3/2)\lambda_{100}\left(\alpha_1^2\beta_1^2+\alpha_2^2\beta_2^2+\alpha_3^2\beta_3^2-1/3\right) + 3\lambda_{111}\left(\alpha_1\alpha_2\beta_1\beta_2+\alpha_2\alpha_3\beta_2\beta_3+\alpha_1\alpha_3\beta_1\beta_3\right) \qquad (2.23)$$

where

$$\lambda_{100} = -\frac{2}{3}\frac{B_1}{c_{11}-c_{12}} \qquad (2.24)$$

and

$$\lambda_{111} = -\frac{1}{3}\frac{B_2}{c_{44}} \qquad (2.25)$$

are the magnetostriction coefficients. For iron and nickel

$$\lambda_{100}(\mathrm{Fe}) = 20.7\times10^{-6}$$
$$\lambda_{111}(\mathrm{Fe}) = -21.2\times10^{-6}$$
$$\lambda_{100}(\mathrm{Ni}) = -45.9\times10^{-6}$$
$$\lambda_{111}(\mathrm{Ni}) = -24.3\times10^{-6}.$$

For a polycrystalline sample with crystallites oriented at random, the appropriate magnetostriction coeffi-

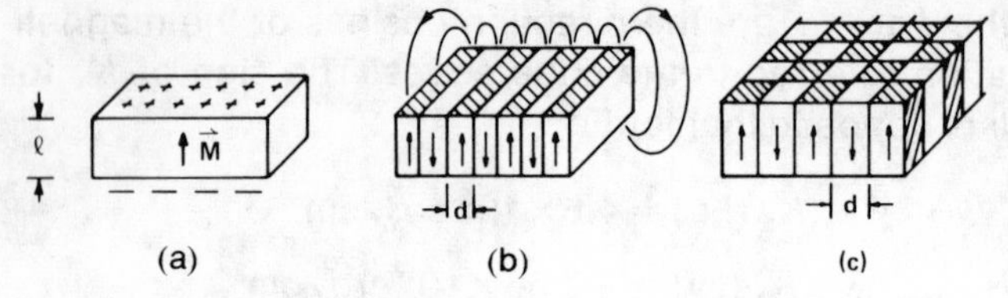

Fig. 2.5: Examples of domain configurations.

cient is

$$\lambda_s = (2/5)\lambda_{100} + (3/5)\lambda_{111}. \quad (2.26)$$

In Ni–Fe alloys this goes through zero at 80 percent Ni. Because of its very small anisotropy and magnetostriction, the 78 percent alloy, called *permalloy*, has very useful extrinsic properties, as we shall see.

Since stress and strain are linearly related through the elastic constants, the magnetoelastic energy may be expressed in terms of an applied stress. In particular, suppose we have an isotropic situation where $\lambda_{100} = \lambda_{111} = \lambda$. Then if a stress σ makes an angle θ with the magnetization, the energy density becomes

$$E_{me} = -(3/2)\lambda\sigma\cos^2\theta. \quad (2.27)$$

Note that this has the consequence of a uniaxial anisotropy.

2.1.6 Magnetic Domains

Another concept that is intrinsic to magnetic materials is that of magnetic domains. This concept has its basis in the classical *magnetostatic* interaction. A magnetic moment, according to Maxwell's equation, gives rise to a magnetic field. The interaction between this field and the magnetic moment producing it gives the magnetostatic energy.

Consider first a uniformly magnetized film of thickness l as shown in Fig. 2.5. The energy associated with the magnetostatic fields can be expressed in terms of the surface magnetic pole density or, equivalently, the demagnetization field arising from these poles. The demagnetizing field in this case is $\boldsymbol{H}_D = -4\pi\boldsymbol{M}$. Therefore, the magnetostatic energy per unit area of surface is

$$\epsilon_m = -(1/2A)\int d^3r\,\boldsymbol{M}\cdot\boldsymbol{H}_D = 2\pi M^2 l. \quad (2.28)$$

Kittel[1] [3] has calculated the corresponding energies for the configurations shown in Fig. 8(b) and (c) as

$$\epsilon_m(\text{stripes}) = 1.71\,M^2 d$$
$$\epsilon_m(\text{checkerboard}) = 1.06\,M^2 d. \quad (2.29)$$

The domains shown above are separated by walls. Because these walls involve a spatial variation of the magnetization they require both exchange energy and anisotropy energy. If σ_w is the surface energy associated with a wall, then the total wall energies per unit (sample) surface area for these configurations are

$$\epsilon_w(\text{stripes}) = \sigma_w l/d$$
$$\epsilon_w(\text{checkerboard}) = 2\sigma_w l/d. \quad (2.30)$$

Adding this wall energy to the magnetostatic energy and minimizing with respect to the domain size d leads to an optimum stripe domain with a width of

$$d = \sqrt{\sigma_w l/1.71M^2}. \quad (2.31)$$

Wall energies are typically of the order of a few erg/cm^2. Therefore, the domain size is of the order of $[\sqrt{l(\text{cm})}/M(\text{G})]$cm. The energy at this minimum for the stripe pattern is

$$\epsilon_{min}(\text{stripe}) = 2M\sqrt{1.71\sigma_w l} = 3.42\,M^2 d. \quad (2.32)$$

The ratio between this energy and that of the uniformly magnetized film is approximately d/l, which, by our original assumption, is much less than one. Similar arguments apply to other geometries. Thus, most "magnetic" materials will break into domains greatly reducing the observable magnetic moment.

The wall between domains can have a very complex structure. In Fig. 2.6 we illustrate the two simplest types of walls, the Bloch wall and the Néel wall. In the Bloch wall the magnetization rotates about an axis normal to the plane of the wall. Therefore, $\nabla\cdot\boldsymbol{M} = 0$ which means there are no *volume* demagnetizing fields. At the center of the wall, however, as can be seen in the figure, the magnetization is normal to the surface which generates a *surface* demagnetizing field. The Néel wall does generate a volume demagnetizing field. In order to determine these fields, and hence the magnetostatic energy, we must know the magnetization distribution. This becomes a complicated variational problem. Néel, and later Middelhoek [4] calculated the energies associated with Bloch and Néel walls. These energies are also plotted in Fig. 2.6 as functions of film thickness for numbers appropriate to permalloy. As the film becomes very thin, the Néel wall becomes preferred, but notice that its width far exceeds the thickness of the film.

Domain walls are readily observed by applying a ferrofluid to the surface of the magnetic film. Such a fluid consists of a colloidal suspension of magnetic particles of the order of 100 Å to 150 Å which migrate into the field gradients associated with the walls. This

[1] In a plate of infinite extent the demagnitization field is $-4\pi M$ everywhere. Therefore, the B field inside is zero and since $B_\perp$ must be continuous, the B field outside will also be zero. That is, there are no field lines outside! Therefore, domain formation does not occur in order to reduce the field energy outside the sample as is commonly taught. Rather, it is the internal demagnetization field that is reduced by domain formation. If we have a domain of characteristic dimension d in a film of thickness l, the demagnetization factor is reduced from unity to something of the order of the aspect ratio d/l. This is the origin of the appearance of the dimension d in these energies.

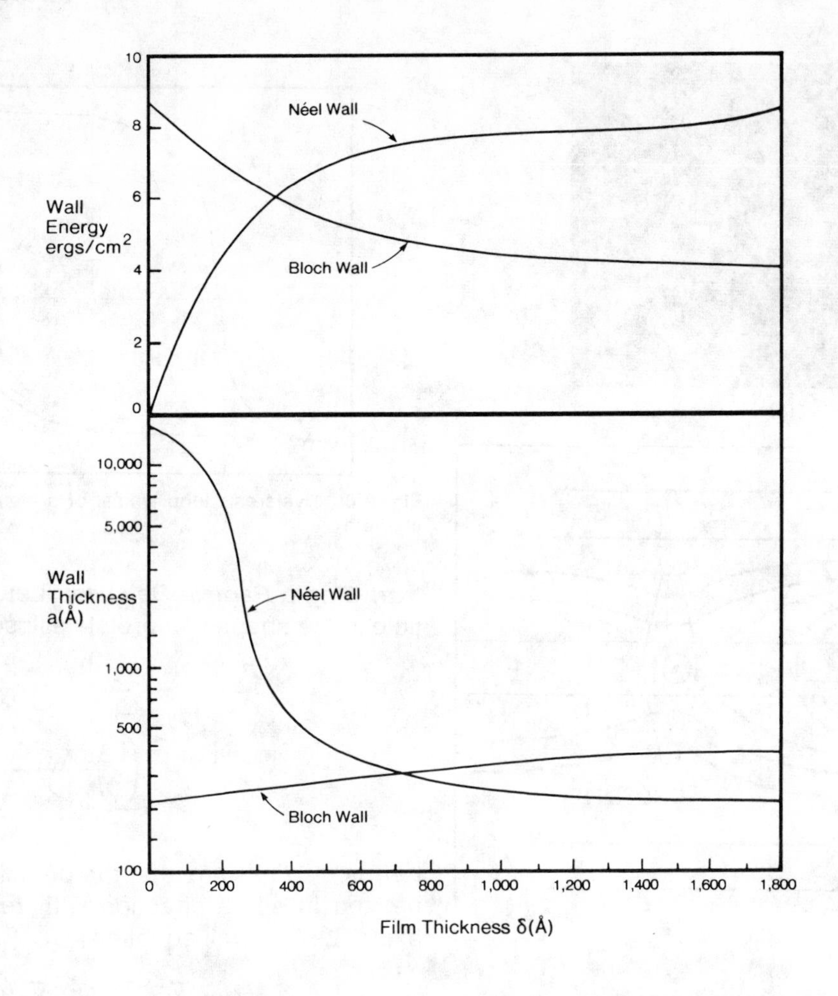

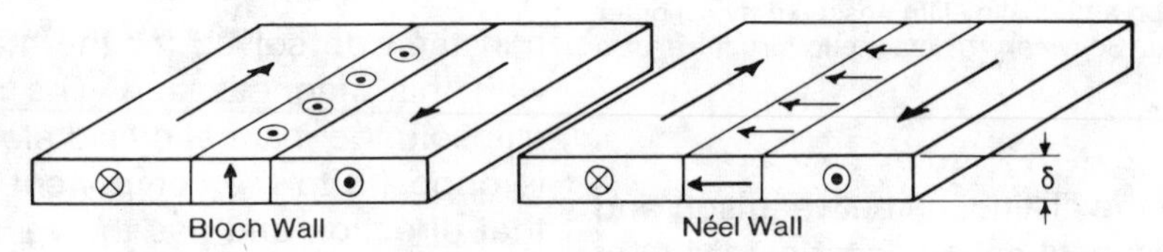

Fig. 2.6: Bloch and Néel domain wall energies and wall thicknesses as functions of film thickness.

technique was first tried by Bitter in 1931. He observed patterns complicated by surface domains. In 1949, Williams, Bozorth, and Shockley [5] observed bulk domains in carefully electropolished samples. Since thin films have very smooth surfaces they may be studied by the Bitter technique without being polished. When Huber, Smith, and Goodenough [6] investigated permalloy films in the thickness range of several hundred angstroms, they observed the "crosstie" patterns shown in Fig. 2.7. The crosstie wall is associated with the fact that a Néel wall may rotate in either of two directions. Let us denote these two types of Néel walls as N^+ and N^-. Then where an N^+ and N^- segment meet we have a magnetic configuration called a *Bloch line*. The crosstie wall basically consists of an array of Bloch lines. However, half of the Bloch lines will have the "wrong" sense relative to the directions of the magnetizations on either side of the wall. These produce the crossties as shown in Fig. 2.7.

2.2 Extrinsic Magnetic Properties

By extrinsic properties we mean those properties associated with macroscopic media. As such these properties are greatly influenced by the composition of the medium. These properties are reflected in the response of the medium to an applied magnetic field. This response takes the form of the magnetic hysterisis loop shown in Fig. 2.8.

There are four parameters that characterize such a loop. The first is the value of the magnetization at very high fields, the saturation magnetization M_s. This should correspond to the magnetic moment per unit volume. The second parameter is the remanent magnetization which depends upon the domain structure in zero field. The third parameter is the value of the applied field required to drive the magnetization to zero from its saturated state. This is the coercive field, or coercivity. The coercivity varies over a wide range

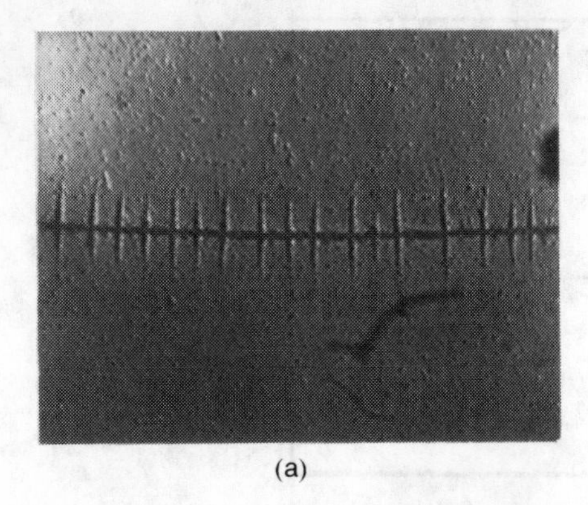

(a)

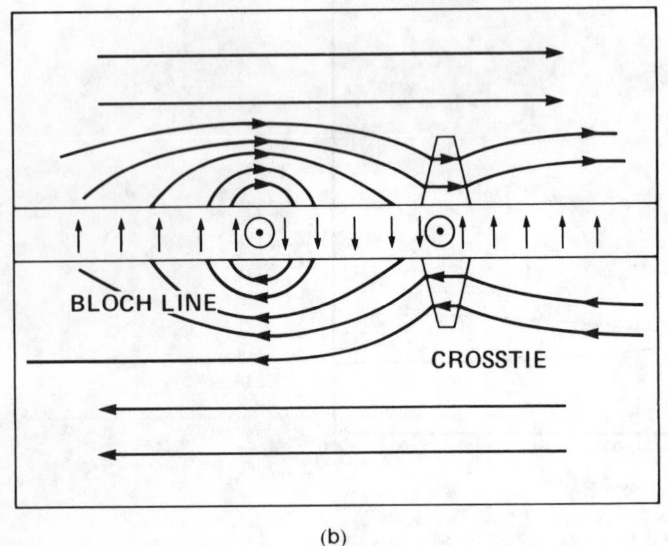

(b)

Fig. 2.7: Crosstie pattern on a permalloy film observed by (a) bitter technique (courtesy of L. J. Schwee); (b) magnetic structure of a crosstie.

of values and leads to the distinction between soft and hard magnets. Soft magnets have a low coercivity, of the order of 0.05 Oe. Such magnets, as we shall see, are desirable for magnetic "head" media. Hard magnets have large coercivities. For permanent magnets we want the largest coercivities possible. For recording media, however, if the coercivities are too high the medium cannot be written upon. Typical coercivities are in the vicinity of 500 Oe. The coercivity is a strong function of chemical and structural defects and, therefore, of the preparation process. The fourth parameter characterizing the loop is its squareness. We shall define this quantitatively in Chapter 4.

2.2.1 Particulate Media

The feature of magnetic media that makes it attractive for storing information is that it can be produced in large areas. There are basically two ways of doing this. One is to dissolve magnetic particles in some binder and coat them onto a substrate. This has been the approach for tapes and disks for the past 30 years. The other is to evaporate or plate magnetic films. In this section we shall consider the properties of magnetic particles.

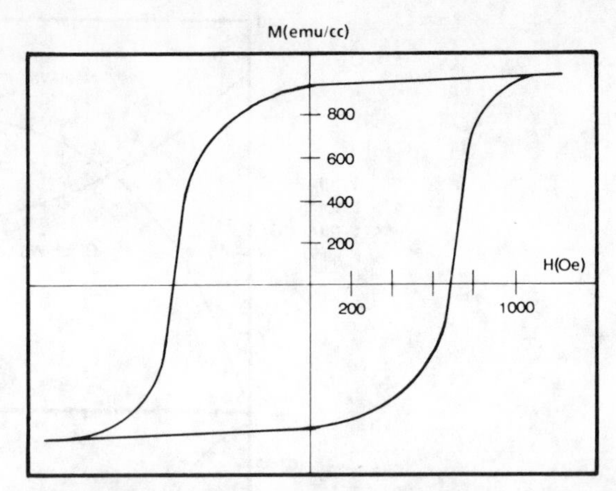

Fig. 2.8: Hysteresis loop typical of a metallic magnetic recording medium.

a) Single Particle Behavior: Let us consider a particle in the shape of a prolate ellipsoid:

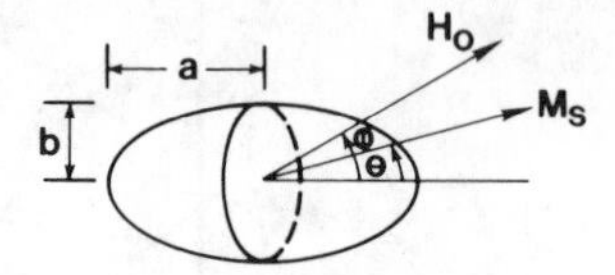

Let us also assume that the particle is uniformly magnetized in some direction with respect to the major axis. The Maxwell equation

$$\nabla \cdot \boldsymbol{H} = -4\pi \nabla \cdot \boldsymbol{M}$$

can then be solved for the magnetic field associated with this magnetization. The result is that inside the ellipsoid the magnetic field along the ith principle axis is opposite to the component of the magnetization in that direction and has the value $-N_i M_i$, where $N_i (i = a, b)$ are the "demagnetizing" factors characterizing the ellipsoid

$$N_a = \frac{4\pi}{a^2/b^2 - 1} \left\{ \frac{(a/b)\ln\left[a/b + (a^2/b^2 - 1)^{1/2}\right]}{(a^2/b^2 - 1)^{1/2}} - 1 \right\}. \tag{2.33}$$

The corresponding demagnetization energy density has the form

$$-\tfrac{1}{2}\boldsymbol{M} \cdot \boldsymbol{H}_D = -\tfrac{1}{2} N_a M_a^2 - \tfrac{1}{2} N_b M_b^2 = (N_a - N_b) M_s^2 \sin^2\theta.$$

This is referred to as *shape* anisotropy and is written as $K \sin^2\theta$. Since $N_a + 2N_b = 4\pi$, K may be expressed in terms of N_a,

$$K = \left(\pi - \tfrac{3}{4} N_a\right) M_s^2. \tag{2.34}$$

If our external field is applied to the particle at some angle φ the magnetization will be pulled into an angle θ giving a total energy

$$E = -M_s H_a \cos\theta - M_s H_b \sin\theta + K \sin^2\theta. \tag{2.35}$$

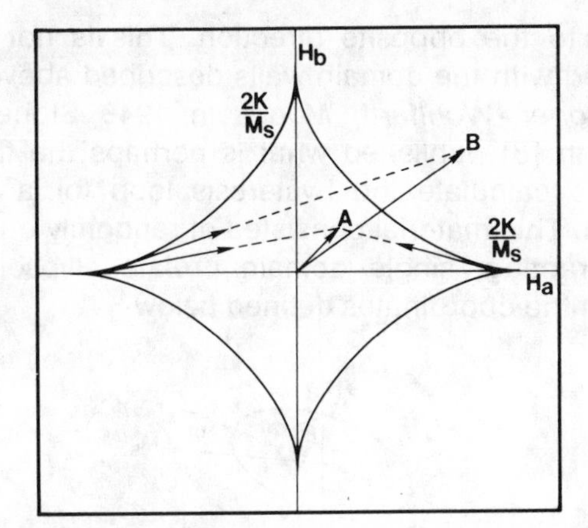

Fig. 2.9: Asteroid construction for determining the equilibrium directions of the magnetization.

The equilibrium magnetization direction is determined by

$$\partial E/\partial \theta = 0. \tag{2.36}$$

This will be a stable minimum if

$$\partial^2 E/\partial \theta^2 > 0. \tag{2.37}$$

Setting the second derivative to zero gives the equations which have the solution

$$H_a = -(2K/M_s)\cos^3\theta$$
$$H_b = (2K/M_s)\sin^3\theta. \tag{2.38}$$

Therefore the curve

$$H_a^{2/3} + H_b^{2/3} = (2K/M_s)^{2/3}, \tag{2.39}$$

which is the equation for an *asteroid*, separates the region where the system has two stable minima (inside) from that where it has only one stable minimum (outside). (See Fig. 2.9.) This construction is very useful, for it can be shown that the stable magnetization directions for any given field (H_a, H_b) are obtained by drawing tangents from this point to the upper half of the asteroid. Consider, for example, a field of magnitude and direction characterized by point *A*. Then there are two possible orientations for the magnetization of the particle:

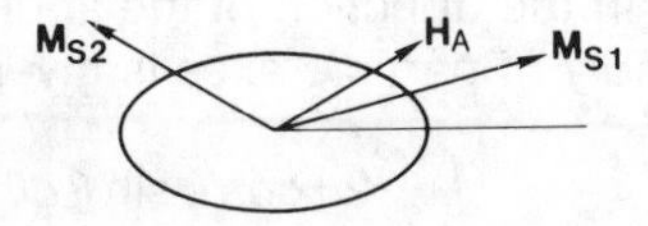

A field whose magnitude places it outside the asteroid, at point *B* for example, produces a magnetization in only one direction:

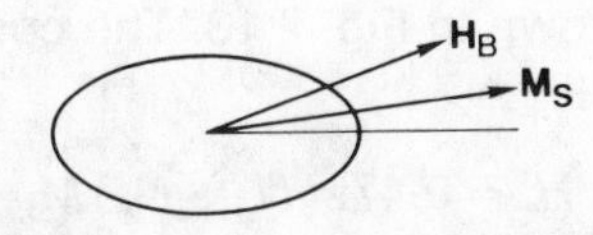

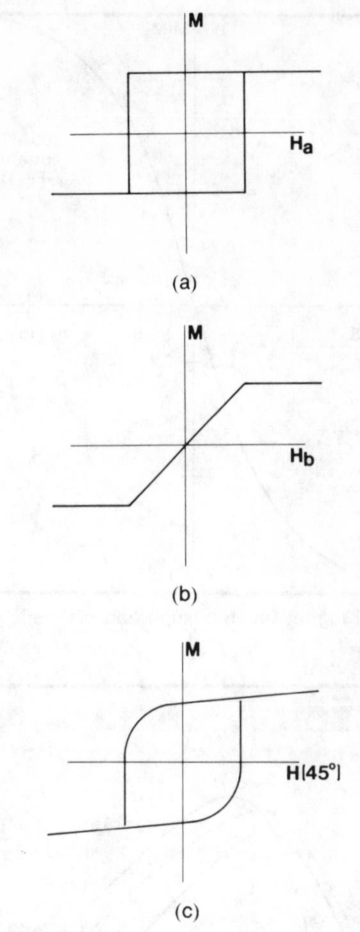

Fig. 2.10: Hysteresis loops for different directions of the field relative to the particulate axis.

Thus, suppose the field is parallel to the major axis. Then M is either parallel or antiparallel to H for $H < 2K/M_s$. When H exceeds $2K/M_s$, then M is only parallel to H. This gives the single-particle hysteresis loop shown in Fig. 2.10(a). Two other cases, (b) and (c), are also shown.

b) Size Effects: In the previous discussion we assumed that the magnetization remained uniform in a direction determined by the applied field. This neglects the possibility of spontaneous domain formation and the effect of temperature. Above some critical size, domains will form. This will occur when the magnetostatic energy of a single domain, $M_s^2 ab^2$, becomes comparable to a wall energy $\sigma_w ab$. Thus,

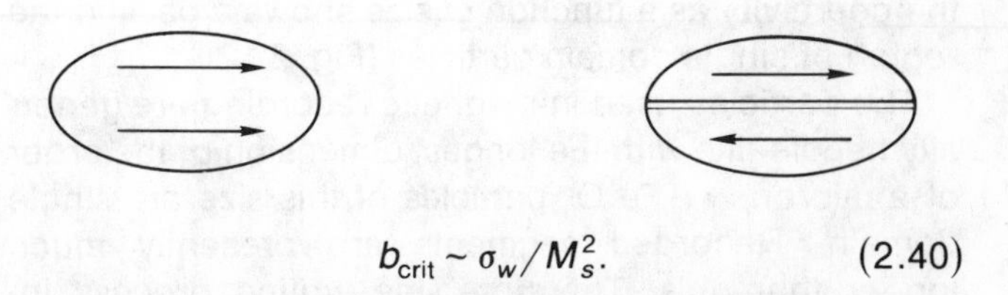

$$b_{crit} \sim \sigma_w / M_s^2. \tag{2.40}$$

On the other hand, if the sample becomes too small, thermal fluctuations will flip the magnetization. The shape demagnetization field provides an energy barrier $(N_b - N_a)M^2ab^2$. When the thermal energy K_BT is

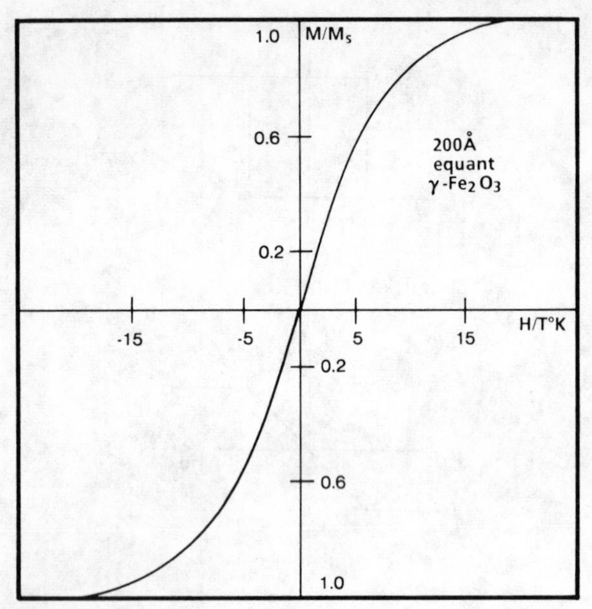

Fig. 2.11: Magnetization of superparamagnetic particles.

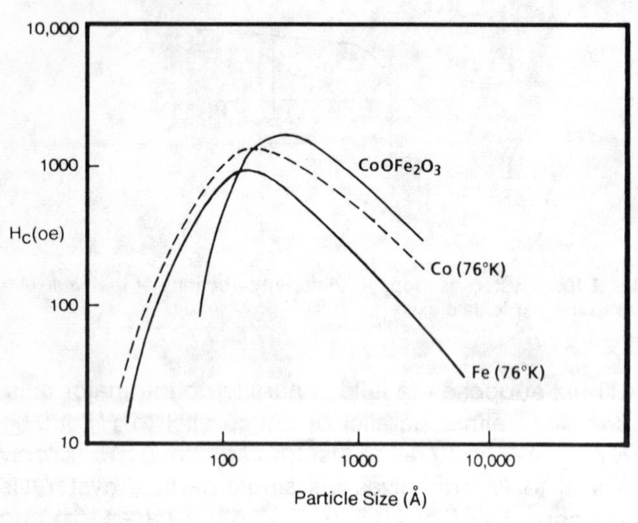

Fig. 2.12: Relations between coercivity and particle diameter for several materials [7].

comparable to, or greater than this barrier, the net magnetic moment of the particle responds with a Curie-law. This situation is referred to as *superparamagnetism*. Fig. 2.11 shows the magnetization of a 200 Å particle of $\gamma - Fe_2O_3$. Since the coercivity of a superparamagnet is zero while the coercivity of a multidomain particle is small because of domain wall motion, the coercivity as a function of size shows a peak in the region of single domain particles (Fig. 2.12).

The particles used in magnetic recording are generally needle-like with the longest dimension of the order of a micron. $\gamma - Fe_2O_3$ particles of this size are single domain. Recorded segments are presently much longer than this. Therefore, the writing process involves the reversal of large clusters of single domain particles. The transition region between two oppositely magnetized segments involves a gradual shift in the distribution of particles magnetized one direction relative to the opposite direction. This is not to be confused with the domain walls described above.

c) Stoner–Wohlfarth Model: In 1948 Stoner and Wohlfarth [8] published what is perhaps the first attempt to calculate the hysteresis loop for a "real" material. This material consisted of randomly oriented, noninteracting, single domain prolate ellipsoids. In terms of the coordinates defined below

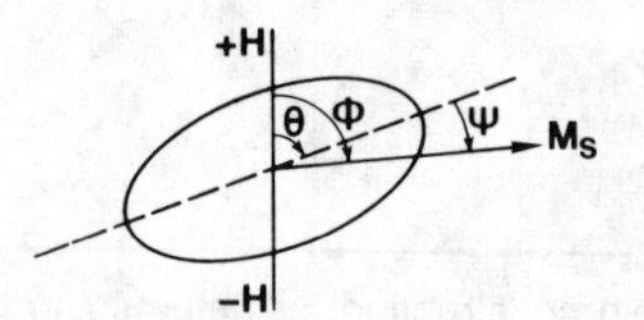

the magnetostatic energy associated with the demagnetization field may be written

$$E_D = -\tfrac{1}{2}\boldsymbol{M}_s \cdot \boldsymbol{H}_D = -\tfrac{1}{2}\boldsymbol{M}_s \cdot (\boldsymbol{N} \cdot \boldsymbol{M}_s)$$
$$= (M_s^2/4)\left[(N_a + N_b) + (N_a - N_b)\cos 2\psi\right]. \tag{2.41}$$

The Zeeman energy is

$$E_z = -M_s H \cos\varphi. \tag{2.42}$$

Introducing the reduced field

$$h = \frac{H}{(N_b - N_a)M_s} \tag{2.43}$$

and minimizing $E_D + E_z$ with respect to φ gives

$$\tfrac{1}{2}\sin 2(\varphi - \theta) + h\sin\varphi = 0 \tag{2.44}$$

and

$$\cos 2(\varphi - \theta) + h\cos\varphi > 0. \tag{2.45}$$

Setting the second equation equal to zero and eliminating the angle φ gives the field at which the magnetization jumps to a new energy minimum for a given particle orientation θ,

$$h_0 = -\frac{(1 - t^2 + t^4)^{1/2}}{(1 + t^2)} \tag{2.46}$$

where $t = \tan^{1/3}\theta$. The average values of the reduced magnetization in the direction of the applied field for a random assembly of particle axes is given by

$$\overline{\cos\varphi} = \frac{\int_0^{\pi/2} 2\pi \cos\varphi \sin\theta \, d\theta}{\int_0^{\pi/2} 2\pi \sin\theta \, d\theta}. \tag{2.47}$$

Expressing φ in terms of θ and evaluating the average numerically, Stoner and Wohlfarth obtained the hysteresis loop shown in Fig. 2.13. The coercivity is given by

$$H_c = 0.479(N_a - N_b)M_s. \tag{2.48}$$

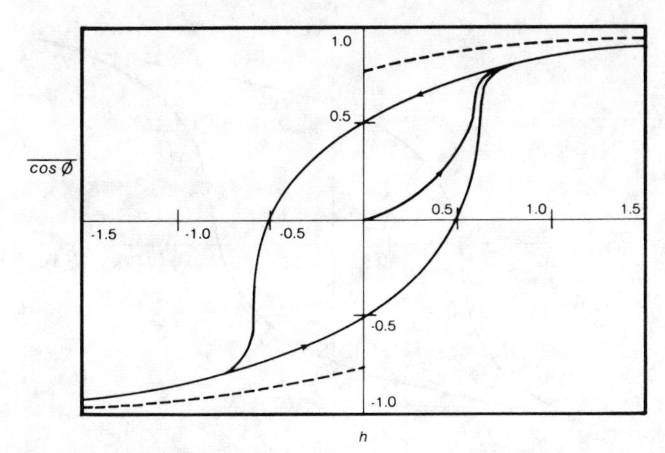

Fig. 2.13: The hysteresis curves of noninteracting prolate ellipsoidal particles oriented at random.

If we decrease h from some value greater than 1 to 0, then $\varphi \rightarrow \theta$, so

$$\overline{\cos\varphi} = \int_0^{\pi/2} \cos\theta \sin\theta \, d\theta = 1/2. \qquad (2.49)$$

Thus, the remanence M_r, according to the Stoner–Wohlfarth theory, is $M_s/2$.

Experimentally, the values of the coercivity tend to be smaller than this result predicts. For example, for $\gamma - Fe_2O_3$, with an axial ratio of 5, $N_b - N_a = 5.23$ and therefore H_c should be 930 Oe. But measured values are closer to 300 Oe. There are a number of reasons for such a discrepancy. First of all, "real" particles are neither ellipsoidal nor do they all have the same axial ratios. But one of the important effects not included in the Stoner–Wohlfarth theory is the role of interactions.

d) Particle Interactions: The magnetic dipole–dipole interaction has the form

$$E_D = \frac{\mu_1 \cdot \mu_2}{r_{12}^3} - 3\frac{(\mu_1 \cdot r_{12})(\mu_2 \cdot r_{12})}{r_{12}^5}. \qquad (2.50)$$

Let us consider two particles having equal volumes, V, and magnetizations separated by a distance r along the z axis:

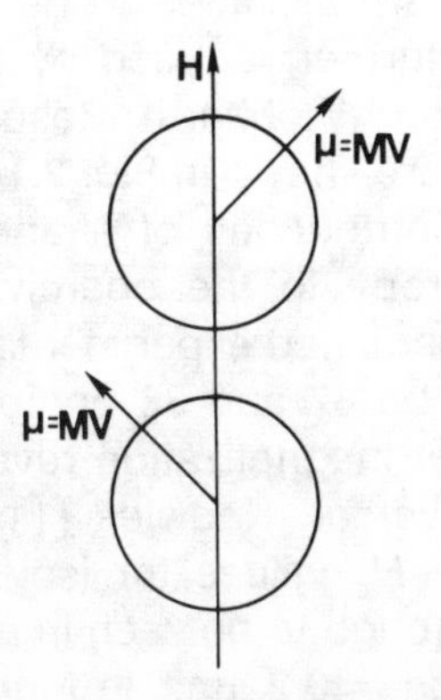

If each particle possesses uniaxial anisotropy along the z direction then the total energy in the presence of an applied field along the z axis is

$$E = \left[-K(\gamma_1^2 + \gamma_2^2) + (M_s^2/r^3)(\alpha_1\alpha_2 + \beta_1\beta_2 - 2\gamma_1\gamma_2) - HM_s(\gamma_1 + \gamma_2)\right] V \qquad (2.51)$$

where α, β, γ are the direction cosines of the magnetizations. The anisotropy constant K contains both shape and crystalline contributions. Minimizing E with respect to the direction cosines subject to the constraint

$$\alpha^2 + \beta^2 + \gamma^2 = 1$$

gives

$$\gamma_1 = \gamma_2 = \pm 1 \qquad \text{for } |H| < H_c$$

and

$$\gamma_1 = \gamma_2 = 1 \qquad \text{for } |H| > H_c$$

and

$$\gamma_1 = \gamma_2 = -1 \qquad \text{for } |H| < -H_c$$

where

$$H_c = \frac{2K}{M_s} + \frac{M_s V}{r^3} = \frac{2K_0}{M_s} + (N_b - N_a)M_s + (V/r^3)M_s. \qquad (2.52)$$

Thus, for this simple example, the dipole–dipole interaction tends to increase the coercivity. V/r^3 is the fraction of the total volume occupied by magnetic material, i.e., the "packing fraction," p. When K is dominated by shape anisotropy this result would suggest that the coercivity should increase linearly with the packing fraction. Experimentally, however, it is generally found that $H_c(p) = H_c(0)\ (1 - p)$. It can be shown that such a relation holds for an assembly of infinitely long, parallel cylinders with arbitrary cross sections in a field parallel to the cylinder axes, and with homogeneous and parallel rotation of the magnetization. The switching of an assembly of interacting particles is very complex. It is generally accepted that one cannot simply calculate the field required to rotate the moment of one particle holding the others fixed. Each environment must be treated simultaneously.

Another simple model that includes interactions was proposed by Preisach in 1935 [19]. He characterized each particle by a rectangular hysteresis loop, and assumed that the coercivities, or switching fields, were described by a distribution. He incorporated interactions by assuming that each loop was shifted by an interaction field which was also described by a distribution function. Thus, each particle could be plotted on the "Preisach plane" whose axes corresponded to the interaction field H_i and the switching field H_c as illustrated below for the demagnetized state.

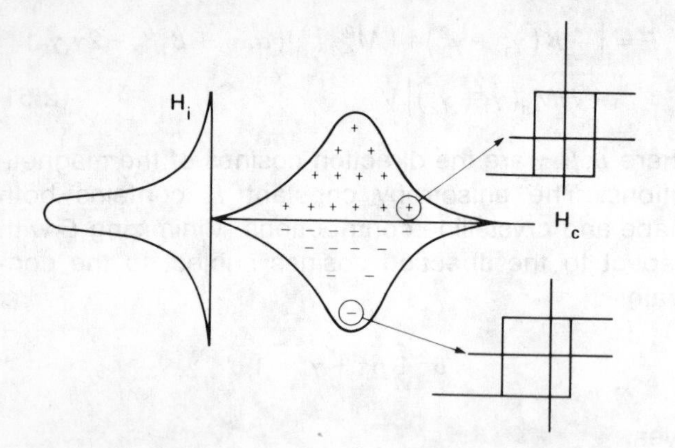

If an external field is now applied, those particles lying within the region bounded by $H = H_i + H_c$, which corresponds to a line at 45° to the axes, will switch. If the field is reduced, then those regions for which $H \leqslant H_i - H_c$ will switch back. This boundary is again a 45° line but perpendicular to the first, thereby defining a triangle with the axes.

The recording process may be thought of as a change in the size and location of such a triangular region. This Preisach description is useful in understanding different magnetization processes, such as the anhysteretic process we shall discuss later. It does, however, represent a one-dimensional approach to what is basically a vectoral problem. Furthermore, it is not clear whether the initial distribution functions for the switching fields or the interaction fields will be stable as the material is cycled around a loop.

e) Incoherent Rotation: So far our discussion has assumed that the magnetization of the particle remains uniform during the reversal process. There are, however, reversal mechanisms involving nonuniform configurations that give lower coercivities. Two of these are "buckling" and "curling," and are illustrated as follows:

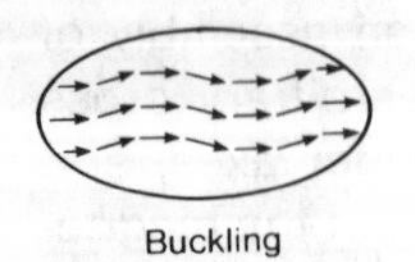
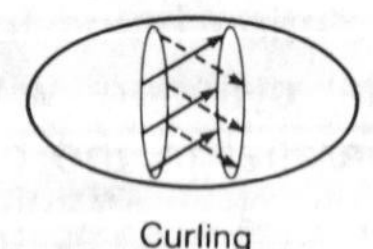
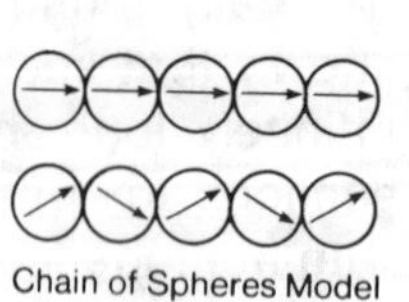

Buckling Curling Chain of Spheres Model

Another model for reversal has been proposed by Jacobs and Bean [9]. They approximate the particle by a chain of spheres which couple only through their dipolar fields. The energy of n spheres whose moments rotate coherently through an angle θ is

$$E_n = \frac{\mu^2}{a^3} nK_n (1 - 3\cos^2\theta) + n\mu H \cos\theta \quad (2.53)$$

where

$$nK_n = \sum_{i=1}^{n}{}' \sum_{j=1}^{n} \frac{1}{|i-j|^3} = \sum_{j=1}^{n} (n-j)/j^3.$$

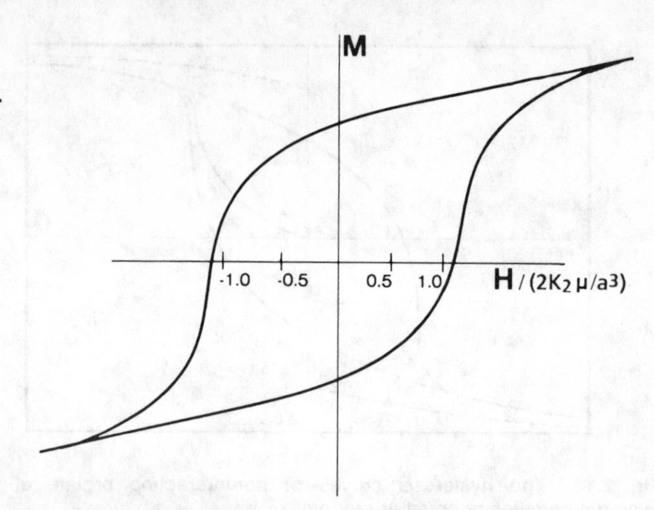

Fig. 2.14: Hysteresis loop of an assembly of randomly oriented two-sphere chains assuming a fanning reversal mechanism.

The field for reversal is then found from $\partial E_n / \partial\theta = 0$ as

$$H_{c,n} = (\mu / a^3) 6K_n. \quad (2.54)$$

For $n = 2$ this just becomes $4\pi M$ where $M = \mu/(4/3)\pi a^3$. However, if the reversal occurs through a "fanning," then the dipolar energy is

$$E_n = (\mu^2/a^3) nL_n (\cos^2\theta - 3\cos^2\theta) + (\mu^2/a^3) nM_n \cdot (1 - 3\cos^2\theta) + n\mu H \cos\theta \quad (2.55)$$

where

$$L_n = \sum_{j=1}^{(n-1)/2 < j \leqslant (n+1)/2} \frac{n - (2j-1)}{n(2j-1)^3}$$

and

$$M_n = \sum_{j=1}^{(n-2)/2 < j < n/2} \frac{(n-2j)}{n(2j)^3}.$$

The field for reversal now becomes

$$H_{c,n} = (\mu / a^3)(6K_n - 4L_n). \quad (2.56)$$

For $n = 2$, fanning reduces H_c by 33 percent.

If these chains are assumed to be oriented at random, then H_c is further reduced by the factor 0.479 found by Stoner and Wohlfarth. Results for chains of only two spheres are shown in Fig. 2.14.

The relative contributions of shape and magnetocrystalline anisotropy to the coercivity of $\gamma - Fe_2O_3$ particles were determined experimentally by Eagle and Mallinson [10] to be 67 and 33 percent, respectively. The mechanism of magnetization reversal appears to be incoherent reversal. Knowles [11] recently measured directly the H_c values for isolated particles of $\gamma - Fe_2O_3$ large enough to be seen in an optical microscope of resolution $= 0.2\ \mu m$. He found a range from 300 to 1100 Oe. The lower values were consistent with a fanning mode and the upper ones with a buckling mode of incoherent rotation.

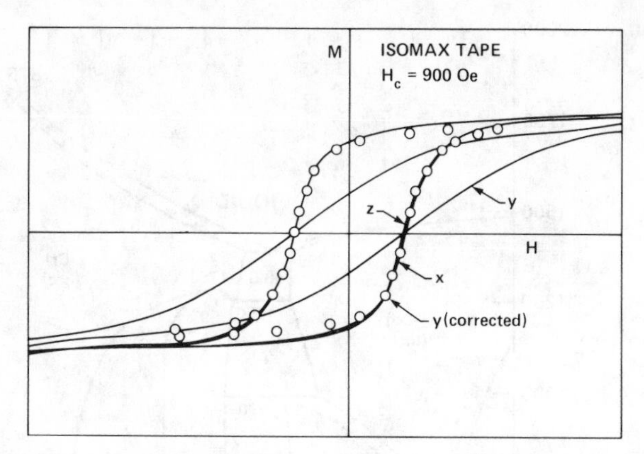

Fig. 2.15: Hysteresis loops in the three directions obtained with a vibrating sample magnetometer. The y curve has been corrected for demagnetizing effects.

f) Remanence: A large remanence leads to a large signal from the medium. Therefore, let us now consider how a large remanence might be obtained. Suppose we have a collection of particles each with m easy axes. When the external field is removed, the magnetization of each particle rotates to the nearest easy axis.

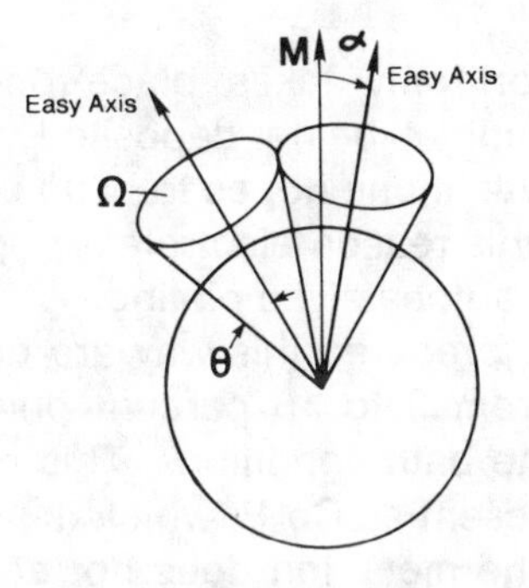

Each easy axis "controls" a solid angle $2\Omega = 4\pi/m$. This is related to the cone angle θ by $\cos\theta = 1 - \Omega/2\pi$, so $\cos\theta = 1 - 1/m$. The remanent magnetization associated with a rotation through α is

$$M_r(\alpha) = M_s \cos\alpha.$$

This must be averaged over all allowed values of α which lie between 0 and θ,

$$\frac{M_r}{M_s} = \frac{\int_1^{\cos\theta} \cos\alpha \, d(\cos\alpha)}{\int_1^{\cos\theta} d(\cos\alpha)} = \frac{1/2(\cos^2\theta - 1)}{\cos\theta - 1}$$

$$= 1 - 1/2m. \tag{2.57}$$

For a uniaxial particle $m = 1$, and we recover the Stoner-Wohlfarth result. If one wished to "design" a medium which was isotropic with a high remanence one would choose particles with many easy axes, such as a cubic material which would have 3([100]) or 4([111]) easy axes. Kodak has exploited this idea in the development of its ISOMAX medium [12]. This consists of "stubby" particles of $\gamma-Fe_2O_3$ with an acicularity ratio of only 2.5. The cubic anisotropy is increased by the addition of a few percent of Co. The result is a magnetically isotropic medium as shown by the hysteresis loops in Fig. 2.15.

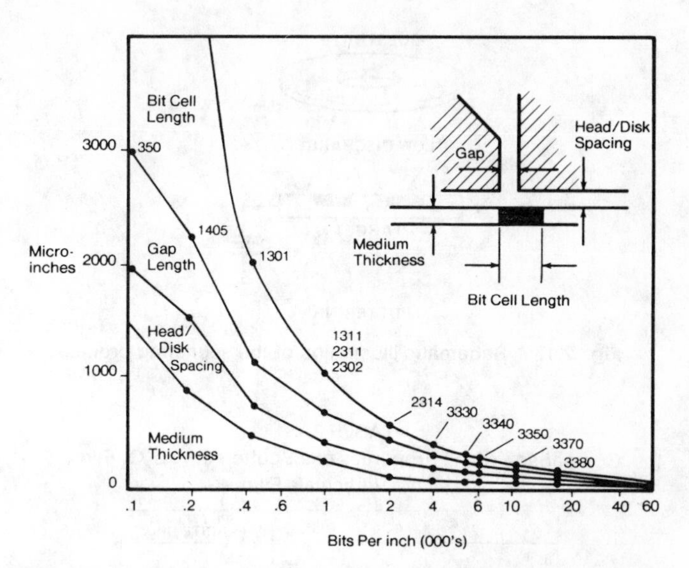

Fig. 2.16: Evolution of characteristic lengths governing longitudinal recording. As in Fig. 1.2, the numbers associated with the points refer to IBM DASD products.

Attempts to enhance the properties of $\gamma-Fe_2O_3$ by adding cobalt actually began over 25 years ago. Although the incorporation of cobalt does increase the coercivity it also increases the temperature dependence of the coercivity. It was subsequently discovered that one could enhance the coercivity of $\gamma-Fe_2O_3$ without a strong temperature dependence by simply coating the particle with cobalt. The mechanism for this enhancement is not yet understood. For a review of particulate media, see Bate [13].

2.2.2 Thin Films

Fig. 2.16 shows that the higher recording densities, at least for horizontal recording, are associated with thinner media. As we shall see in Chapter 5, the readback voltage is proportional to the thickness of the medium. It is also proportional to the magnetization. Therefore, as the medium becomes thinner we would like to increase its magnetization. One way of doing this is to grow $\gamma-Fe_2O_3$ as a film. Fujitsu Laboratories Ltd. in Kawasaki, Japan, has developed a reactive sputtering process to obtain such thin films of ferric oxide [14]. Sputtering is a process in which a target, consisting of the material to be deposited, is placed opposite a substrate in a vacuum chamber (see Fig. 2.17).

The chamber is evacuated to a base pressure on the order of 10^{-6} torr and then filled with argon gas to a pressure of about 10 mtorr. When a potential of several kilovolts is applied between the target and an anode, the argon gas forms a glow discharge. Argon ions are accelerated to the target, where they strike the surface and create an atomic collision cascade in

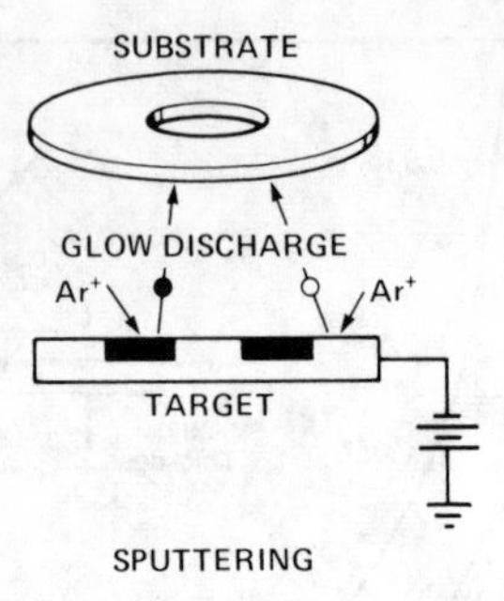

Fig. 2.17: Schematic illustration of the sputtering process.

TABLE 2.1
Comparison of the Properties of a Sputtered $-Fe_2O_3$ Film with a Particulate Film

	COATED DISK	SPUTTERED FERRITE DISK
STRUCTURE	γ-Fe_2O_3 FINE PARTICLE; RESIN; Al SUBSTRATE	γ-Fe_2O_3; ANODIZED Al LAYER; Al SUBSTRATE
RECORDING DENSITY (BPI)	12,000	24,000
FILM THICKNESS (μm)	0.7	0.18
SURFACE ROUGHNESS (μm)	0.12	0.02
COERCIVE FORCE (Oe)	400	700

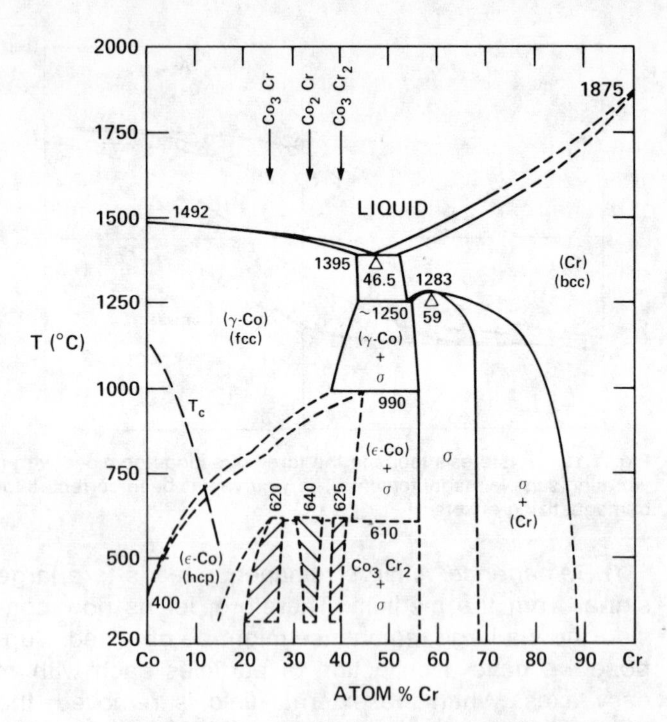

Fig. 2.18: Phase diagram of cobalt–chromium system (from W. G. Mofatt, *The Handbook of Binary Phase Diagrams*, General Electric, 1978).

the bulk. Some atoms reach the surface, travel through the discharge, and are deposited on the substrate, which may or may not be electrically biased. In the Fujitsu process, an iron target is sputtered in an argon-oxygen atmosphere, resulting in a film of Fe_3O_4. This film is then heat-treated in an oxygen atmosphere to convert the Fe_3O_4 into Fe_2O_3. Typical performance parameters for sputtered oxide films are listed in Table 2.1.

Another way of obtaining a large magnetization is to consider pure magnetic metals such as Fe and Co. As we mentioned above, Co has a magnetic moment of 1.6 μ_B/atom. Carrying through the same calculation that we did for $\gamma-Fe_2O_3$, the magnetic moment per gram is 151 EMU/g. The density is 8.9 g/cm^3. Therefore, the magnetization is 1343 EMU/cm^3, three times that of $\gamma-Fe_2O_3$.

There are a variety of techniques for fabricating metal films. Of the plating techniques we generally distinguish two, depending on the sources of the electrons which reduce the metal ions in solution. In *electrodeposition* these electrons are provided by an electric current. In *electroless* deposition they are provided by a chemical reducing agent in the solution.

Electroless plating was discovered by Brenner and Riddell in 1946 when they found that sodium hypophosphite was a reducing agent for Ni^{2+}. The corresponding reaction for plating cobalt is

$$\underset{\text{hypophosphite ion}}{2H_2PO_2^-} + 2H_2O + Co^{2+} \rightleftharpoons Co + H_2 + 4H^+ + \underset{\text{phosphite ion}}{2HPO_3^=}.$$

This reaction only takes place on a "catalytic" surface. The metal being deposited must therefore itself be catalytic if one hopes to build up a film by this process. For this reason electroless deposition is also referred to as autocatalytic plating.

Cobalt films prepared this way are not pure cobalt, but contain from 3 to 15 percent phosphorous, depending on the bath conditions. This phosphorous is likely to be present as Co_2P, which is nonmagnetic.

Generally, the metal ion does not occur by itself in solution, but is "complexed" with some agent such as the citrate radical to control the plating rate. The pH of the solution may also have to be stabilized by the addition of a buffering agent such as NH_4OH. The bath conditions such as the pH, the temperature, etc., govern the microstructure of the resulting film. And the microstructure, in turn, governs the magnetic properties. There is a strong correlation between the size and nature of the microcrystallites making up the film and the coercivity and squareness of the hysteresis loop [15].

Ampex uses electroless deposition in its so-called ALAR process for producing metallic media. In this process a 6 μ layer of nonmagnetic nickel-phosphorous is electrolessly deposited on an aluminum substrate for protection against head impacts. This is then polished to 0.05 μ, peak to valley, before a 0.003–0.06 μ film of cobalt phosphorous is electrolessly plated.

The other technique mentioned above for obtaining metal films is electroplating, wherein an applied potential is used to reduce the cobalt ions. Again, a buffering agent must be added to maintain a constant pH.

One advantage of electroplating is that the current density is an additional parameter for controlling the deposition process. This also permits fast deposition, and the solution can be reused repeatedly. On the other hand, neither electroplating nor electroless deposition are convenient for obtaining alloy films.

The sputtering process described above is particularly well suited for alloys. This has been employed by Iwasaki and Ouchi [16] to prepare films with perpendicular anisotropy. Any film which is perpendicularly magnetized will experience a demagnetization field of $-4\pi M_s$. Thus, even if there is a uniaxial crystalline anisotropy field, $2K/M_s$, favoring the perpendicular direction, unless this exceeds the demagnetization field, $-4\pi M_s$, the magnetization will remain in-plane. The sputtering process tends to induce perpendicular anisotropy through a columnar morphology. To reduce $-4\pi M_s$, Iwasaki and Ouchi alloyed cobalt with chomium. From Fig. 2.2, or the moment relation, $\mu = 10 - Z$, we see that chromium should indeed reduce the moment of cobalt. However, the Co–Cr phase diagram reproduced in Fig. 2.18 shows that there exists an intermetallic compound, Co_3Cr, in the vicinity of 13 percent Cr. Therefore, it has been suggested [17] that it is not the reduced magnetization that is responsible for the perpendicular orientation, but rather that phase separation produces needle-like columns of nearly pure Co isolated by Co_3Cr. It is difficult to resolve such chemical differences on the micron scale. However, there are two pieces of evidence that indirectly support this suggestion. First, the Co-Re system, whose phase diagram does not show intermetallic compounds, also does not exhibit perpendicular anisotropy. And, secondly, it is possible [18] to obtain perpendicularly oriented Co films by electrodeposition without any Cr.

Problems

1) Magnetite, Fe_3O_4, also has the inverse spinel structure. Calculate the moment per molecular unit and the magnetization (the density is 5.197 g/cm^3).

2) Use the functional derivative relation

$$H_x = -\frac{\delta E_x}{\delta M}$$

to calculate an effective exchange field. For a magnetization of the form

$$M_x(x) = -\frac{2M}{\pi}\tan^{-1}(x/a)$$

sketch the exchange field as a function of x/a. For cobalt $AM^2 \sim 10^{-6}$ erg/cm and $M = 1400$ G. What is the maximum exchange field in oersteds with a in units of microns?

3) A typical tensile stress in a deposited metallic film might be 10^9 dyn/cm^2. What would be the order of magnitude of the corresponding anisotropy field in, say, a cobalt film?

4) The "initial susceptibility" is the slope of the magnetization curve just as it leaves the $M_r = 0$ state,

$$\chi = \left.\frac{dM}{dH}\right|_{H=0}.$$

Calculate χ in the Stoner–Wohlfarth model.

5) Consider a thin film with perpendicular anisotropy energy density

$$E_A = K\cos^2\theta$$

where θ is the angle between M and the plane of the film. What is the demagnetization energy density as a function of θ? Prove that for $2K < 4\pi M_s$, the magnetization will lie in the plane.

References

[1] A. R. Williams, V. L. Moruzzi, A. P. Malozemoff, and K. Terakura, *IEEE Trans. Magn.*, vol. MAG-19, 1983.
[2] See, for example, R. M. White, *Quantum Theory of Magnetism*. New York: Springer-Verlag, 1983, p. 119.
[3] C. Kittel, *Rev. Mod. Phys.*, vol. 21, p. 541, 1949.
[4] S. Middelhoek, "Ferromagnetic domains in thin NiFe films," Ph.D. thesis, Amsterdam, 1961.
[5] H. J. Williams, R. M. Bozorth, and W. S. Shockly, *Phys. Rev.*, vol. 75, p. 155, 1949.
[6] J. B. Huber, E. E. Smith, and D. O. Goodenough, *J. Appl. Phys.*, vol. 29, p. 294, 1958.
[7] F. E. Luborsky, *J. Appl. Phys.*, vol. 32, p. 1918, 1961.
[8] E. C. Stoner and E. P. Wohlfarth, *Phil. Trans. Roy. Soc. London*, vol. A-240, p. 599, 1948.
[9] I. S. Jacobs and C. P. Bean, *Phys. Rev.*, vol. 100, p. 1060, 1955; also, I. S. Jacobs and F. E. Luborsky, *J. Appl. Phys.*, vol. 28, p. 467, 1957.
[10] D. F. Eagle and J. C. Mallinson, *J. Appl. Phys.*, vol. 38, p. 995, 1967.
[11] J. E. Knowles, *IEEE Trans. Magn.*, vol. MAG-16, p. 62, 1980.
[12] J. U. Lemke, *J. Appl. Phys.*, vol. 53, p. 2561, 1982.
[13] G. Bate, *J. Appl. Phys.*, vol. 52, p. 2447, 1961.
[14] S. Yoshi *et al.*, *J. Appl. Phys.*, vol. 53, p. 2556, 1982.
[15] See, for example, T. Chen, D. A. Rogowski, and R. M. White, *J. Appl. Phys.*, vol. 49, p. 1816, 1978.
[16] S. Iwasaki and K. Ouchi, *IEEE Trans. Magn.*, vol. MAG-14, p. 849, 1978.
[17] T. Chen, G. B. Charlan, and T. Yamashita, *J. Appl. Phys.*, vol. 54, p. 5103, 1983.
[18] T. Chen and P. Cavallotti, *Appl. Phys. Lett.*, vol. 41, p. 205, 1982.
[19] See, for example, G. Schwantke, *J. Aud. Eng.*, vol. 9, p. 37, 1961.

Chapter 3
The Recording Head

THE transducer that converts electrical signals into magnetization patterns in the media is referred to as the "head." The heads of today are the same as that in Poulsen's telegraphone, namely they are small electromagnets. Writing is accomplished by the fringing magnetic fields from the narrow gap.

3.1 Head Requirements

3.1.1 Hysteresis Loop

In the writing process, the magnetic core in the head serves to generate a concentrated field. As we shall see below this field is limited by the saturation magnetization. Therefore M_s should be large. However, when the head is not energized, we want the field to be zero in order to avoid unwanted writing. This implies a low remanence. Finally, to avoid hysteresis losses the coercivity should also be very small. Such low coercivity materials are said to be magnetically "soft." The hysteresis loop associated with a head material is shown in Fig. 3.1.

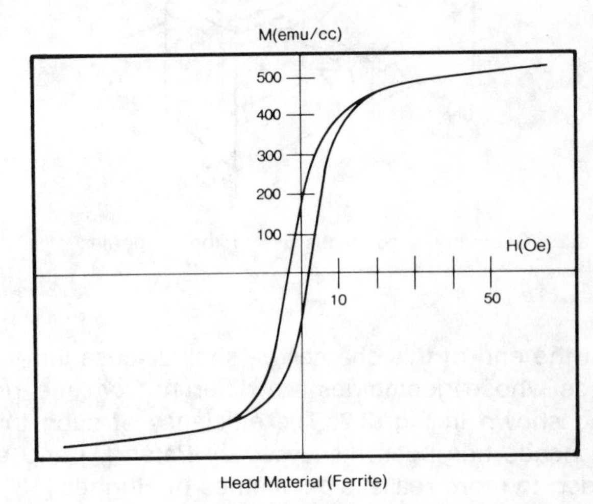

Fig. 3.1: Typical hysteresis loop for a ferrite recording head material.

3.1.2 Head Efficiency

Another important requirement of a head is that it have a high efficiency, i.e., that a large portion of the magnetomotive force—the ampere-turns in the coil—appears across the gap. Let us consider the geometry shown in Fig. 3.2. Since the permeability of the air gap is 1,

$$B_g = H_g. \tag{3.1}$$

Inside the core

$$B_c = \mu H_c. \tag{3.2}$$

We also have Ampere's law,[1]

$$\oint H \cdot dl = (4\pi/c) NI \tag{3.3}$$

Or,

$$H_c l_c + H_g g = (4\pi/c) NI \tag{3.4}$$

where l_c is the pathlength indicated in the figure. As defined above, the efficiency is

$$\eta = H_g g/(4\pi/c) NI. \tag{3.5}$$

This may be expressed in geometric terms by considering the magnetic flux. The flux in the core, Φ_c, is $B_c A_c$. If we assume that the lines of B follow the core, i.e., that there is no leakage of flux, then B will be increased to compensate any taper and $\Phi_g = \Phi_c$. The ratio between the magnetic potential φ and the flux Φ is called the *reluctance*:

$$\mathbb{R}_c = \frac{\varphi_c}{\Phi_c} = \frac{H_c l_c}{\mu H_c A_c} = \frac{l_c}{\mu A_c} \tag{3.6}$$

$$\mathbb{R}_g = \frac{\varphi_g}{\Phi_g} = \frac{H_g g}{H_g A_g} = \frac{g}{A_g}. \tag{3.7}$$

In terms of the reluctances the efficiency becomes

$$\eta = \frac{\mathbb{R}_g \Phi_c}{\mathbb{R}_c \Phi_c + \mathbb{R}_g \Phi_c} = \frac{\mathbb{R}_g}{\mathbb{R}_c + \mathbb{R}_g} = \frac{g/A_g}{g/A_g + l_c/\mu A_c}. \tag{3.8}$$

To the extent that a large gap field implies a large fringing field we therefore want a large permeability and a large ratio A_c/A_g. In Chapter 5 we shall see that the readback signal has a null when the gap length becomes of the order of the recording wavelength λ. If we set $g = \lambda/3$ and require an efficiency of 90 percent then (3.8) tells us that the permeability must exceed

$$\mu \gtrsim \frac{30 l_c}{\lambda}.$$

[1]We have chosen to work in CGS units. This means the current is in units of statamperes, where 1 statampere = (10/c) ampere. In MKS units magnetic field is expressed in amp-turns/m. Since the relation between field and current density in CGS involves the factor $4\pi/c$ we have the relation 1 amp-turn/m = c/10 statamp-turn/m = $c/10^3$ statamp-turn/cm = $(c/10^3)(4\pi/c)$ Oe. Another system of units, the Systeme Internationale (SI) is also frequently used. (See L. H. Bennet, C. H. Pope, and L. J. Swartzendruber, in *AIP Conf. Proc.*, vol. 29, p. xix, 1976.)

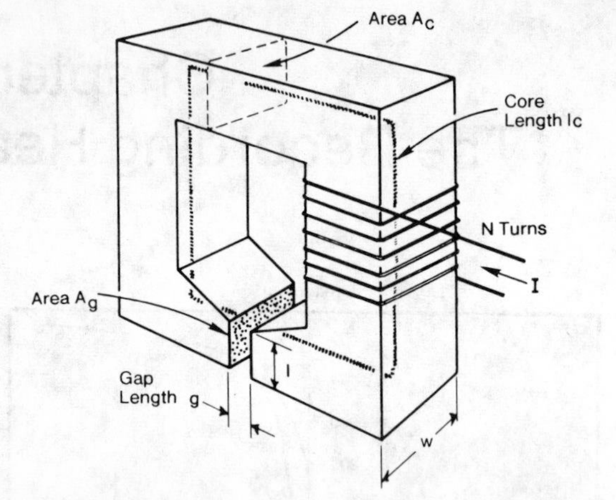

Fig. 3.2: Geometry used in discussing the properties of a ring head.

At the end of this chapter we shall discuss thin-film heads whose geometries are different from the ring head shown in Fig. 3.2. The efficiency of such thin-film heads has been discussed by Paton [1] and extended to more realistic geometries by Hughes [13].

3.1.3 Impedance

Ideally the head is simply an inductance. Therefore its impedance is $Z = i\omega L$. The inductance L is defined as the ratio of the time rate of change of the flux through the coils to that of the current,

$$L(dI/dt) = N(d\Phi/dt). \tag{3.9}$$

The flux threading each turn is

$$\Phi = B_c A_c = \mu H_c A_c. \tag{3.10}$$

According to our argument above this flux is also equal to

$$\Phi = B_g A_g = H_g A_g. \tag{3.11}$$

Combining this result for H_g with Ampere's law above gives H_c in terms of I and leads to

$$L = (4\pi/c)N^2/(\mathbb{R}_c + \mathbb{R}_c) = (4\pi/c)N^2 \eta A_g/g. \tag{3.12}$$

3.1.4 Frequency Response

Ideally we want the head to respond to all frequencies. Fig. 3.3 shows typical permeability spectra. We see that there is a characteristic frequency above which the permeability falls rapidly. There are several mechanisms that can lead to such behavior. We shall describe three of these.

a) Eddy Currents: A time-varying magnetic flux density produces an electric field, and hence, a voltage according to the Maxwell equation.

$$\nabla \times \boldsymbol{E} = -1/c(\partial \boldsymbol{B}/\partial t). \tag{3.13}$$

This field results in a current density by Ohm's law

$$\boldsymbol{j} = \sigma \boldsymbol{E}. \tag{3.14}$$

These "eddy currents" also produce a magnetic field,

$$\nabla \times \boldsymbol{H} = (4\pi/c)\boldsymbol{j}. \tag{3.15}$$

Combining these equations with the constituent relation

$$\boldsymbol{B} = \mu \boldsymbol{H} \tag{3.16}$$

where μ is the permeability, gives a diffusion equation for the field H. As a lowest order approximation we take the static value of the permeability μ_0. The diffusion equation contains a characteristic length, the skin depth,

$$\delta = \frac{c}{\sqrt{2\pi\mu_0\omega\sigma}} \tag{3.17}$$

where ω is the frequency of the time-varying field. This equation must be solved for the appropriate geometry and with given boundary conditions. For the cross-sectional area A_c of the head above we find the following current distribution:

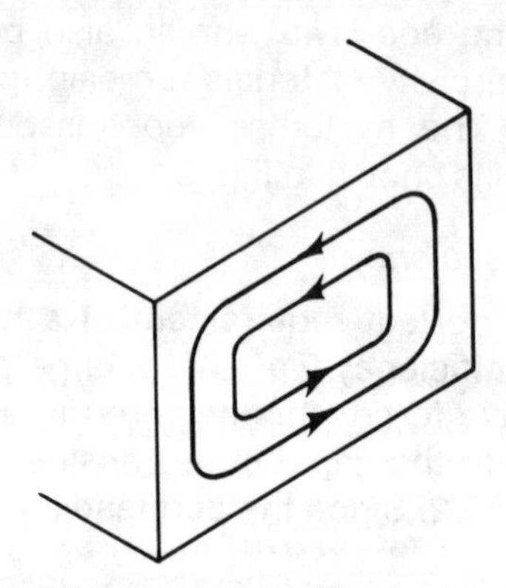

The appearance of eddy currents changes the complex impedance of the head in two ways. First of all, these currents lead to ohmic losses. And, secondly, the flux-carrying capacity of the core is reduced, which reduces the inductance of the head. These changes may be described by a complex permeability. This permeability has a frequency dependence similar to that shown in Fig. 3.3. The rolloff in μ' occurs when the characteristic length δ becomes equal to the characteristic dimension of the conductor, which in our case would be the trackwidth w. Thus,

$$f_{max} = c^2\rho/(2\pi w)^2 \mu_0. \tag{3.18}$$

Thus, one obviously wants a head material with a large resistivity, ρ, or one must laminate the head in order to reduce the characteristic dimension.

b) Spin Resonance: In high resistivity materials the frequency response lies in the gyromagnetic nature of the magnetic moment. That is, a magnetic moment, being proportional to angular momentum, responds to

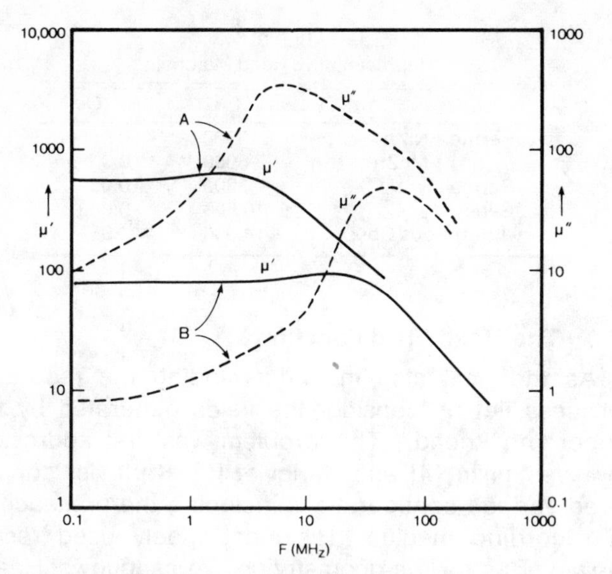

Fig. 3.3: Real (μ') and imaginary (μ'') parts of the complex permeability as a function of frequency for two NiZn ferrites, *A* and *B*, having different compositions.

an applied force just like a spinning top. This is described by the equation of motion

$$d\boldsymbol{M}/dt = \gamma \boldsymbol{M} \times \boldsymbol{H}. \tag{3.19}$$

where γ is the gyromagnetic ratio relating the magnetic moment to the angular momentum, $\gamma\hbar = g\mu_B$. If we wish to rotate a magnetization M_s we apply a field at right angles. Let us assume this applied field has a time dependence $\exp(i\omega t)$. Then

$$\boldsymbol{H} = \boldsymbol{H}_i + \boldsymbol{h}\exp(i\omega t). \tag{3.20}$$

We assume $\boldsymbol{h}$ is much less than the internal field $\boldsymbol{H}_i$. In many cases this internal field is an anisotropy field. Assuming the response of the magnetization is small we may write

$$\boldsymbol{M} = \boldsymbol{M}_s + \boldsymbol{m}\exp(i\omega t) \tag{3.21}$$

where $\boldsymbol{m} \ll \boldsymbol{M}_s$ and $\boldsymbol{m} \times \boldsymbol{M}_s = 0$. Substituting $\boldsymbol{M}$ and $\boldsymbol{H}$ into the equation of motion, the first order equation becomes

$$i\omega \boldsymbol{m} = \gamma(\boldsymbol{M}_s \times \boldsymbol{h} + \boldsymbol{m} \times \boldsymbol{H}_i). \tag{3.22}$$

Cross-multiplying from the right by $\boldsymbol{H}_i$, and taking the scalar product with $\boldsymbol{H}_i$, gives two equations that may be combined to give

$$(i\omega/\gamma)[i\omega \boldsymbol{m} - \gamma(\boldsymbol{M}_s \times \boldsymbol{h})] = \gamma(\boldsymbol{M}_s \cdot \boldsymbol{H}_i)\boldsymbol{h} - \gamma \cdot (\boldsymbol{h} \cdot \boldsymbol{H}_i)\boldsymbol{M}_s - \gamma H_i^2 \boldsymbol{m} \tag{3.23}$$

or

$$\boldsymbol{m} = [i\omega\gamma(\boldsymbol{M}_s \times \boldsymbol{h}) + \gamma^2(\boldsymbol{M}_s \cdot \boldsymbol{H}_i)\boldsymbol{h} - \gamma^2(\boldsymbol{H}_i \cdot \boldsymbol{h})\boldsymbol{M}_s]/(\omega_0^2 - \omega^2) \tag{3.24}$$

where $\omega_0 = \gamma H_i$. Introducing the permeability by $\boldsymbol{B} = \ddot{\mu}\boldsymbol{H}$ we find

$$\ddot{\mu} = \begin{pmatrix} \mu & -i\kappa & 0 \\ i\kappa & \mu & 0 \\ 0 & 0 & 1 \end{pmatrix} \tag{3.25}$$

where

$$\mu = 1 + \omega_0\omega_M/(\omega_0^2 - \omega^2)$$

and

$$\kappa = \omega\omega_M/(\omega_0^2 - \omega^2).$$

Here $\omega_M = 4\pi\gamma M_s$. Equation (3.25) is known as the Polder tensor, and governs the propagation of electromagnetic waves in gyrotropic media. In this lossless approximation the complex permeability has the form:

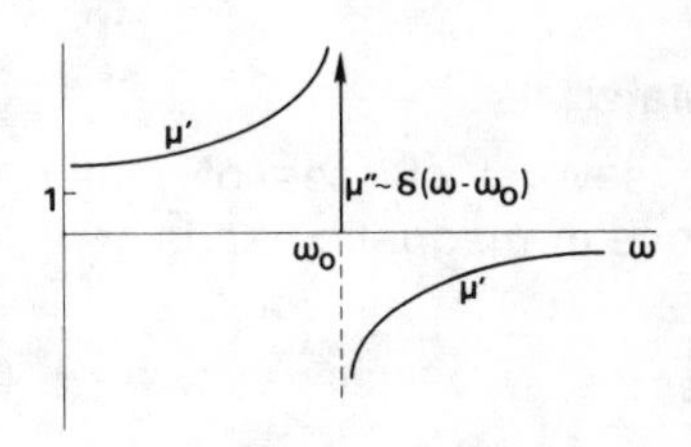

If damping is introduced, the complex permeability develops a behavior more like that shown in Fig. 3.3. The rolloff now occurs for frequencies above the resonance frequency ω_0. In Chapter 2 we pointed out that a small anisotropy favored a large (static) permeability. However, if ω_0 is proportional to the anisotropy field, a small anisotropy means that losses will begin to occur at a lower frequency. This relationship between μ_0 and the maximum in μ'' is evident in Fig. 3.3. The fact that μ'' does not go to zero at low frequencies in the NiZn ferrites is due to the existence of an additional loss mechanism, presumably associated with the hopping of an electron from a divalent to a trivalent iron site and back again.

c) Domain-Wall Resonance: Measurements of the complex permeability of a ferrite by Rado *et al.* [1] showed *two* peaks in μ''. The lower frequency peak was identified as being associated with the resonance of a domain wall.

The point is that when a domain wall moves, the moments precess. This may be thought of as the result of an effective magnetic field. For example, consider a Bloch wall in which the magnetization rotates through an angle Θ along the z axis. The effective field is

$$H_{\text{eff}} = (\omega/\gamma) = (d\Theta/dt)/\gamma = (v/\gamma)\, d\Theta/dz. \tag{3.26}$$

The "field" energy associated with this effective field is

$$E_{\text{eff}} = (1/8\pi)\int H_{\text{eff}}^2\, dz. \tag{3.27}$$

From the solution for the static wall [3]

$$d\Theta/dz = \sqrt{K/A}\,\sin\Theta(z) \tag{3.28}$$

and the fact that $d/dz = (1/v)d/dt$ we obtain

$$E_{\text{eff}} = \left(\frac{1}{2}\frac{1}{8\pi\gamma}\sqrt{K/A}\right)v^2. \tag{3.29}$$

Because of the quadratic dependence on velocity we may think of this as a kinetic energy with an effective mass given by the quantity in parenthesis. If the wall is confined to a local energy minimum in space with a restoring force that is approximately linear in displacement, then the wall will behave like a harmonic oscillator. When damping is introduced we again obtain a permeability similar to that already discussed.

3.2 Head Materials

There are several classes of materials that have evolved for use in magnetic heads.

3.2.1 Ferrites

MnZn and NiZn ferrites having the spinel structure are the most widely used materials for recording heads. They are produced by cold or hot pressing powders obtained by ballmilling or precipitation from solution. NiZn is more sensitive to strain, and when used in contact recording develops a nonmagnetic layer at the head–medium interface.

3.2.2 Al–Fe and Al–Fe–Si Alloys

Known as Alfenol and Sendust these alloys have magnetic and electrical properties similar to the permalloys but they are mechanically hard. Consequently, they are often used as pole tips.

3.2.3 Permalloys

The general term permalloys refers to alloys of Fe and Ni. As we mentioned in Chapter 2, the particular composition of 78 percent Ni has very small anisotropy and magnetostriction. This leads to a low coercivity and high permeability. Their low resistivity, however, requires a laminated structure in a conventional ring head in order to minimize eddy current losses. They are also mechanically soft which limits their use to noncontact applications.

3.2.4 Amorphous Co–Zr

Due to the absence of grain boundaries, which tend to pin magnetic domain walls, amorphous magnetic films have very low coercivities. This makes them candidates for head materials. It has been found that sputtered films of $Co_{90}Zr_{10}$ also have very high permeability and high saturation magnetization. In Table 3.1 we compare the various head materials.

TABLE 3.1
Representative Head Materials

	$4\pi M_s$ G	H_c (Oe)
Ferrite NiZn	3300	0.5
Ferrite MnZn	5000	0.05
Permalloy	8300	0.02
Sendust	10 500	0.002
Amorphous Co-Zr	14 000	0.05

3.3 The Head Field Function

As the next step in understanding the recording process, let us consider the fields generated by the recording "head." This problem was first addressed by Westmijze [4] and Karlqvist [5]. Karlqvist considered various configurations including the presence of a recording medium. His most widely used result, however, is for the geometry shown as follows. Heads in which the pole-tip lengths are large compared with the gap are typically referred to as "Karlqvist" heads. Today's Winchester heads are a good example.

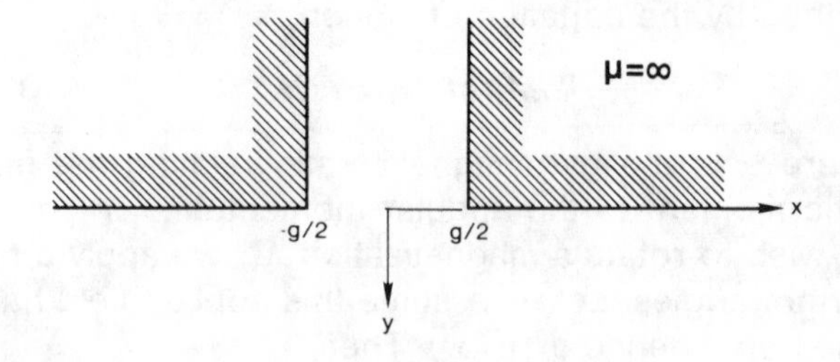

3.3.1 The Karlqvist Field

Since $\nabla \times \boldsymbol{H} = 0$ we may describe the field by a scalar potential,

$$\boldsymbol{H} = -\nabla\varphi$$

and the problem becomes a problem in potential theory. The source of the fields lies in the current energizing the head. Deep in the gap the field H will be horizontal, i.e., $\boldsymbol{H} = H_g\boldsymbol{x}$. Integrating the field around the path indicated and using Amplere's law,

$$\oint \boldsymbol{H}\cdot d\boldsymbol{l} = (4\pi/c)\oiint \boldsymbol{j}\cdot\boldsymbol{n}\,da = (4\pi/c)NI. \tag{3.30}$$

But since the normal component of B must be continuous at the gap interfaces and since $\boldsymbol{H} = \boldsymbol{B}/\mu$, the fact that $\mu = \infty$ means that the field inside the head is zero. Thus the "deep-gap" field H_g is given by

$$H_g = (4\pi/c)NI/g. \tag{3.31}$$

This argument also implies $H_g = 4\pi M$ where M is the magnetization induced in the head. If the current becomes too large, M will saturate and the head will no longer respond. In practice, the corners of the gap saturate first. Thus, for MnZn ferrite with $4\pi M_s = 5000$ G the practical limit on H_g is found experimentally to be about 2500 Oe. One way of overcoming this limita-

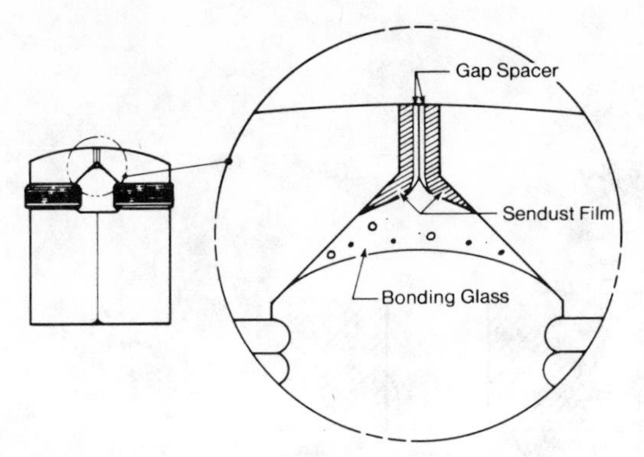

Fig. 3.4: Cross section of a "MIG" tape drive recording head showing the Sendust pole tips [6].

tion is to cover the pole tips with Sendust which has a $4\pi M_s = 9500$ G.

Fig. 3.4 shows such a "metal-in-gap" or MIG composite head. Such an elaborate solution emphasizes the fact that present head materials do not possess enough $4\pi M_s$ to write on high coercivity media and thereby achieve high recording densities.

Returning to the Karlqvist head, the potential in the gap varies linearly across the gap:

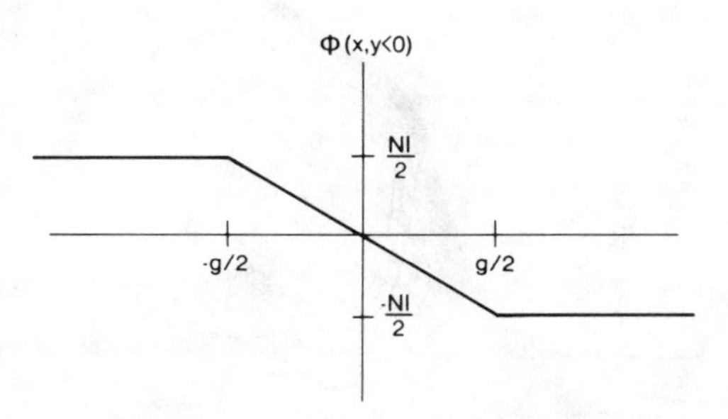

Karlqvist assumed that this also described the potential along the $y=0$ plane as well. Thus, we have a potential problem in which the potential itself is specified along the boundary. This is called a Dirichelet boundary condition. The equation for the potential is obtained by substituting $\boldsymbol{H} = -\nabla\varphi$ into $\nabla\cdot\boldsymbol{H} = -4\pi\nabla\cdot\boldsymbol{M}$ giving

$$\nabla^2\varphi = -4\pi\rho_m \tag{3.32}$$

where ρ_m is an effective magnetic charge

$$\rho_m = -\nabla\cdot\boldsymbol{M}. \tag{3.33}$$

This is Poisson's equation and it may be converted into an integral equation by using Green's theorem:

$$\iiint(\varphi\nabla^2\psi - \psi\nabla^2\varphi)\,d^3\boldsymbol{r} = \oiint\left(\varphi\frac{\partial\psi}{\partial n} - \psi\frac{\partial\varphi}{\partial n}\right)da. \tag{3.34}$$

Writing

$$\psi = G(\boldsymbol{r},\boldsymbol{r}')$$

where

$$\nabla^2 G(\boldsymbol{r},\boldsymbol{r}') = -4\pi\delta(\boldsymbol{r}-\boldsymbol{r}') \tag{3.35}$$

then

$$\varphi(\boldsymbol{r}) = \iiint\rho(\boldsymbol{r}')G(\boldsymbol{r},\boldsymbol{r}')\,d^3\boldsymbol{r}' + \frac{1}{4\pi}\oiint\left[G(\boldsymbol{r},\boldsymbol{r}')\frac{\partial\varphi}{\partial n'} - \varphi(\boldsymbol{r}')\frac{\partial G}{\partial n'}\right]da'. \tag{3.36}$$

$G(\boldsymbol{r},\boldsymbol{r}')$ is called the Green's function for this potential problem. For Dirichelet boundary conditions we know $\varphi(\boldsymbol{r}')$. Since there is some freedom in the definition of $G(\boldsymbol{r},\boldsymbol{r}')$, we require $G(\boldsymbol{r},\boldsymbol{r}')=0$ for $\boldsymbol{r}'$ on the boundary. Equation (3.36) for $\varphi(\boldsymbol{r}')$ then reduces to

$$\varphi(\boldsymbol{r}) = \iiint\rho(\boldsymbol{r}')G(\boldsymbol{r},\boldsymbol{r}')\,d^3\boldsymbol{r}' - \frac{1}{4\pi}\oiint\varphi(\boldsymbol{r}')\frac{\partial G}{\partial n}\cdot da'. \tag{3.37}$$

In the particular case we are considering there is no magnetic media for $y>0$ so the first term on the right vanishes. The assumed potential along the boundary $y=0$ has the form

$$\varphi(x,0) = \begin{cases} H_g g/2 & x < -g/2 \\ -H_g x & -g/2 < x < g/2 \\ -H_g g/2 & g/2 < x \end{cases}. \tag{3.38}$$

The problem now remains to find the Green's function. From its definition, the Green's function is the potential associated with a unit point charge at $\boldsymbol{r}'$. In our problem everything extends to infinity in the z direction. Thus, we must consider a *line* of charge at $\boldsymbol{r}$. The potential at r associated with such a line varies as $\log|\boldsymbol{r}-\boldsymbol{r}'|$. When $\boldsymbol{r}$ lies on the boundary this must vanish if it is to be a Dirichelet Green's function. This is accomplished by adding an image line beyond the boundary. From the geometry we can see that

$$G(\boldsymbol{r},\boldsymbol{r}') = \log\frac{(x-x')^2+(y-y')^2}{(x-x')^2+(y-y')^2}. \tag{3.39}$$

The normal derivative on the boundary is

$$\left.\frac{\partial G}{\partial y'}\right|_{y'=0} = -\frac{4y}{(x-x')^2+y^2}. \tag{3.40}$$

Thus

$$\varphi(x',0) = \frac{1}{4\pi}\int_{-\infty}^{\infty}\varphi(x',0)\frac{4y}{(x-x')^2+y^2}\,dx'. \tag{3.41}$$

Substituting for $\varphi(x',0)$ and carrying out this integral

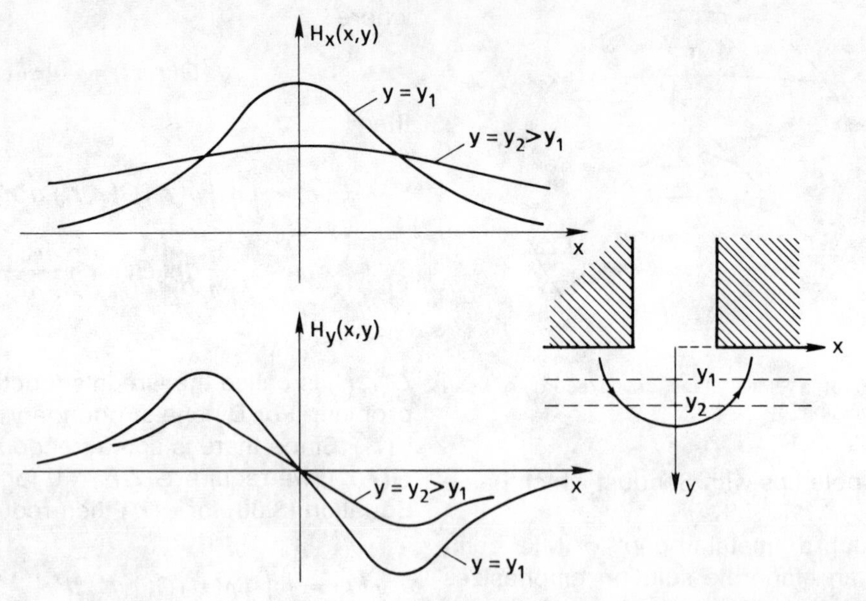

Fig. 3.5: Vertical and horizontal fields associated with a Karlqvist head.

gives

$$\varphi(x, y) = -\frac{H_g}{\pi}\left[(x+g/2)\tan^{-1}\left(\frac{x+g/2}{y}\right) - (x-g/2)\tan^{-1}\left(\frac{x-g/2}{y}\right) - \frac{y}{2}\log\frac{(x+g/2)^2+y^2}{(x-g/2)^2+y^2}\right]. \quad (3.42)$$

The Karlqvist fields are then given by

$$H_x(x, y) = \frac{H_g}{\pi}\left[\tan^{-1}\left(\frac{x+g/2}{y}\right) - \tan^{-1}\left(\frac{x-g/2}{y}\right)\right] \quad (3.43)$$

$$H_y(x, y) = \frac{-H_g}{2\pi}\log\frac{(x+g/2)^2+y^2}{(x-g/2)^2+y^2}. \quad (3.44)$$

These are plotted in Fig. 3.5.

The relatively simple analytic expressions (3.43) and (3.44) were Karlqvist's reason for taking a linear variation of the magnetic potential across the gap. He verified the accuracy of this solution by taking the next (quadratic) term in the expansion of the potential and found the fields (3.43) and (3.44) to be accurate to 0.5 percent for $y > g/2$.

Although we have used the Green's function approach to obtain the head fields, there are a variety of techniques available for solving such potential problems. Both Westmijze and Karlqvist used the Schwartz–Christoffel *conformal mapping* technique. This exploits the two-dimensional nature of the problem and also assumes that the head has an infinite permeability. Elabd [7] appears to have been the first to apply this technique to a head of finite length. Lindholm [8] has extended this to a three-dimensional head of finite width by using superposition. Other methods, such as the *finite element method* [9] or the *boundary element method* [10] require numerical computation but produce intuitively appealing results. Fig. 3.6, for example, is a typical result for the horizontal field component off to the side of the gap.

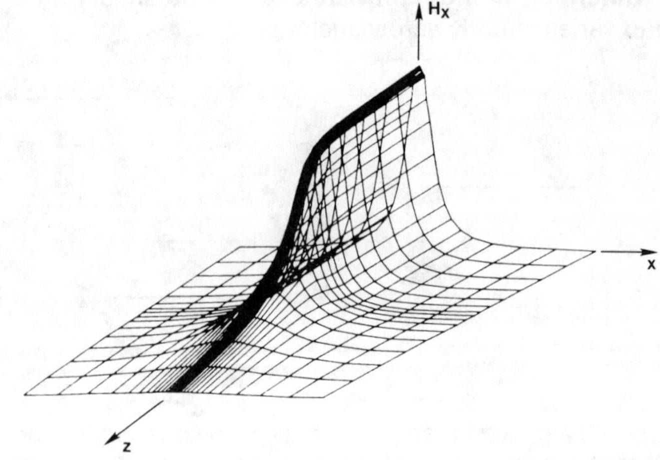

Fig. 3.6: Calculation of the head field H_x using the finite element method.

It is sometimes convenient to work in the small gap limit, e.g., $g \to 0$. Expanding the arctangent,

$$\tan^{-1}\left(\frac{x+g/2}{y}\right) = \tan^{-1}\left(\frac{x}{y}\right) + \frac{yg/2}{x^2+y^2} + \cdots \quad (3.45)$$

and similarly for the logarithm. Now recalling that

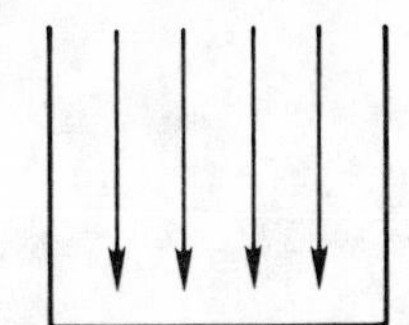

Fig. 3.7: Pole-tip head.

$H_g = (4\pi/c)(NI/g)$, we obtain

$$H_x(x,y) = \frac{4NI}{c}\frac{y}{x^2+y^2}$$

$$H_y(x,y) = \frac{-4NI}{c}\frac{x}{x^2+y^2} \qquad (3.46)$$

which is just the field around a wire in the z direction! This should not be surprising, for the Karlqvist fields could also have been obtained from an equivalent current distribution according to

$$\boldsymbol{j} = c\nabla \times \boldsymbol{M}. \qquad (3.47)$$

3.3.2 Pole-Tip Heads

The horizontal configuration assumed in the Karlqvist problem above was motivated by the fact that most magnetic films magnetize more easily in-plane than perpendicular to the plane simply due to demagnetization effects. It is possible, however, to prepare films with their easy axis perpendicular to the plane of the film. In this case, one might consider writing on such films with a "pole-tip" head as illustrated in Fig. 3.7.

As Fig. 3.8 illustrates, the fringing fields from the Karlqvist head could just as well have been obtained from a magnetization in the opposite direction in the gap. Thus, the difference between the pole-tip head and the Karlqvist head is simply a 90° rotation of the source of the fields! Mallinson [11] has pointed out that *in two dimensions* if the magnetization is rotated through an angle θ, then the corresponding field remains constant and *counter*rotates through an angle θ. Let us prove this for a 90° rotation.

The magnetic charge density associated with a magnetization

$$\rho_1(\boldsymbol{r}) = -\nabla\cdot \boldsymbol{M}_1(\boldsymbol{r}) = -\left(\frac{\partial M_{1x}}{\partial x} + \frac{\partial M_{1y}}{\partial y}\right). \qquad (3.48)$$

In two dimensions, the field associated with this charge distribution is

$$\boldsymbol{H}_1(\boldsymbol{r}) = \int \frac{\rho_1(\boldsymbol{r}')(\boldsymbol{r}-\boldsymbol{r}')}{|\boldsymbol{r}-\boldsymbol{r}'|^2}\, d^2r' \qquad (3.49)$$

where $\boldsymbol{r} = x\hat{\boldsymbol{x}} + y\hat{\boldsymbol{y}}$.

Let us now rotate the magnetization through $-90°$ so

$$M_{2x} = M_{1y} \text{ and } M_{2y} = -M_{1x}. \qquad (3.50)$$

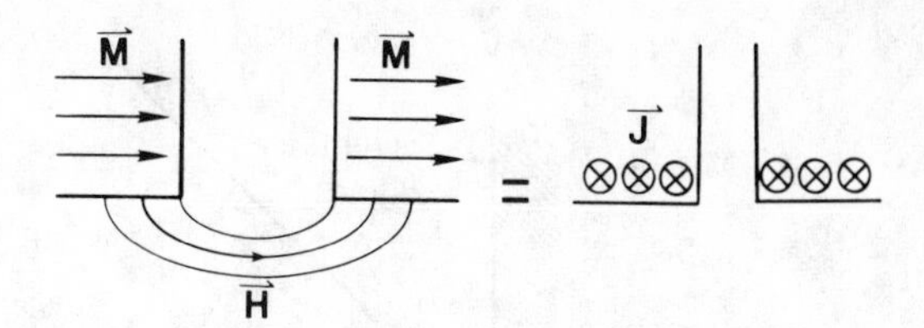

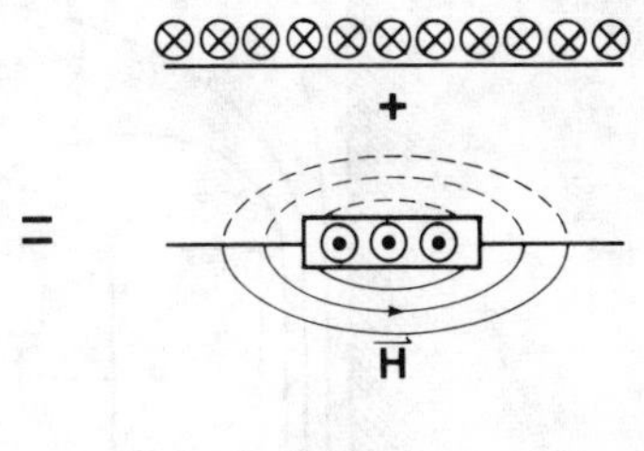

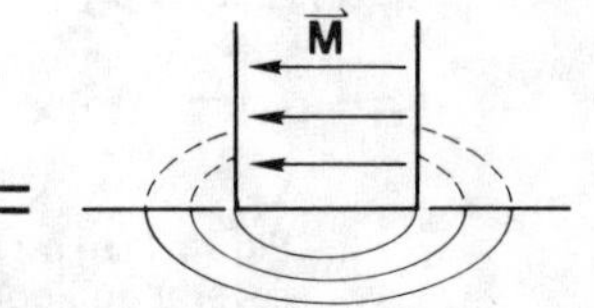

Fig. 3.8: Illustration of the equivalence of currents or a "magnetic gap" in producing the fringing magnetic fields.

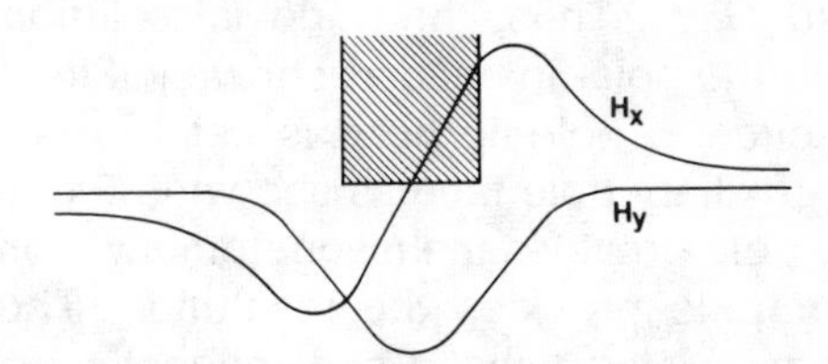

Fig. 3.9: Head field functions for a pole-tip head.

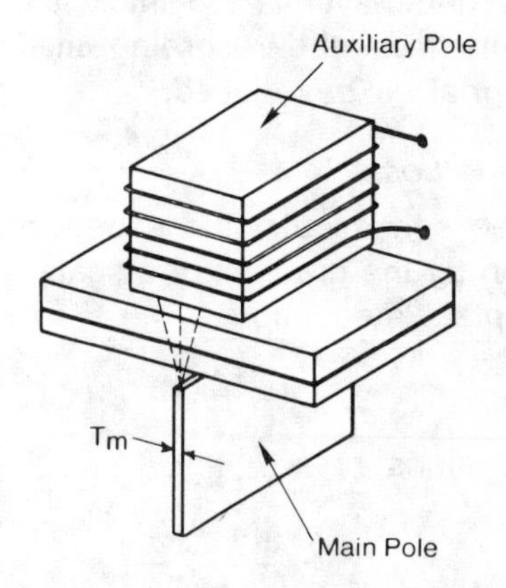

Fig. 3.10: Pole tips for recording on vertical media.

The current density associated with this new magnetization distribution is

$$\boldsymbol{j}_2 = c\nabla \times \boldsymbol{M}_2 = c\left(\frac{\partial M_{2y}}{\partial x} - \frac{\partial M_{2x}}{\partial y}\right)\hat{\boldsymbol{z}} = c\rho_1(\boldsymbol{r})\hat{\boldsymbol{z}}. \qquad (3.51)$$

The field associated with this current density is

$$\boldsymbol{H}_2(\boldsymbol{r}) = B(\boldsymbol{r}) = \int \frac{\boldsymbol{j}_2 \times (\boldsymbol{r}-\boldsymbol{r}')}{|\boldsymbol{r}-\boldsymbol{r}'|^2}\, d^2\boldsymbol{r}'. \qquad (3.52)$$

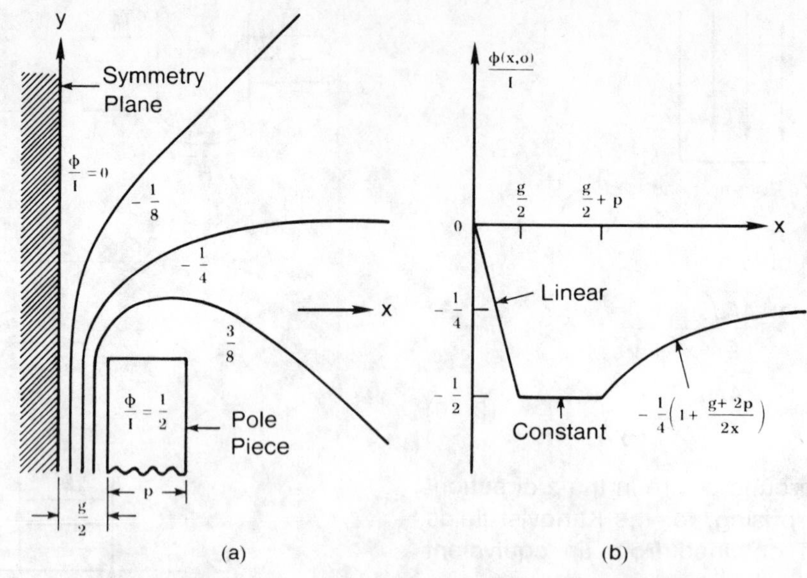

Fig. 3.11: (a) Cross section of finite pole-tip length head with several equipotential lines schematically shown. (b) Assumed magnetic scalar potential at $y=0$ is shown.

From the form of $\boldsymbol{j}_2(\boldsymbol{r})$, we see that $\boldsymbol{H}_2(\boldsymbol{r})$ is at right angles to $\boldsymbol{H}_1(\boldsymbol{r})$. Thus, the Karlqvist solution may be applied to the pole-tip head as shown in Fig. 3.9.

In practice, a pole-tip head is used in conjunction with an auxiliary pole face as shown in Fig. 3.10. The auxiliary pole provides the magnetomotive force, while the main pole provides the resolution. The vertical medium may also have a soft magnetic underlayer, such as permalloy, to provide a return flux path. However, if this underlayer is too thick it will shield the auxiliary pole from the recording medium and the readback signal will be reduced.

3.3.3 Spacing Loss

In the free space outside the head, the magnetic field is given by the gradient of a scalar potential that satisfies Laplace's equation,

$$\nabla^2\varphi = 0. \tag{3.53}$$

In two dimensions

$$\nabla^2\varphi = \frac{\partial^2\varphi}{\partial x^2} + \frac{\partial^2\varphi}{\partial y^2}. \tag{3.54}$$

The Fourier transform $f(k)$ of a function $F(x)$ is defined by

$$f(k) = \int_{-\infty}^{\infty} e^{-ikx}F(x)\,dx. \tag{3.55}$$

Applying this to Laplace's equation,

$$-k^2\varphi(k,y) + \frac{\partial^2\varphi(k,y)}{\partial y^2} = 0. \tag{3.56}$$

Therefore

$$\varphi(k,y) = \varphi(k,0)\,e^{-ky} \tag{3.57}$$

and

$$H_{x,y}(k,y) = H_{x,y}(k,0)\,e^{-ky} \tag{3.58}$$

where $H_{x,y}(k,0)$ is the field along the surface of the head. This reveals the general "spacing loss behavior," of

$$20\log_{10}(e^{-ky}) = 20\log_{10}(e^{-2\pi y/\lambda}) = -54.6(y/\lambda)\,\text{dB}. \tag{3.59}$$

We could also directly Fourier transform the Karlqvist solution (3.43),

$$H_x(k,y) = H_g g\frac{\sin(kg/2)}{kg/2}e^{-ky}. \tag{3.60}$$

This not only gives the spacing loss, but also gives the "gap loss" factor, which we shall discuss later.

3.3.4 Finite Pole Tips

From this expression for the Fourier transform of the Karlqvist field we are led to an inconsistency that reveals an unphysical aspect of the Karlqvist solution. At long wavelengths, i.e., $k=0$, $H_x(0,y) \Rightarrow H_g g$. But $H_x(0,y)$ is also given by

$$H_x(0,y) = \int_{-\infty}^{\infty} H_x(x,y)\,dx. \tag{3.61}$$

This integral may be written

$$\int_{-\infty}^{\infty} H_x\,dx = \oint \boldsymbol{H}\cdot d\boldsymbol{s} - \int_{\frown} \boldsymbol{H}\cdot d\boldsymbol{s}. \tag{3.62}$$

The closed integral is zero since the path of integration may be closed at infinity in such a way that it is not threaded by a current loop. The semicircular integral at infinity is also zero because the fields from

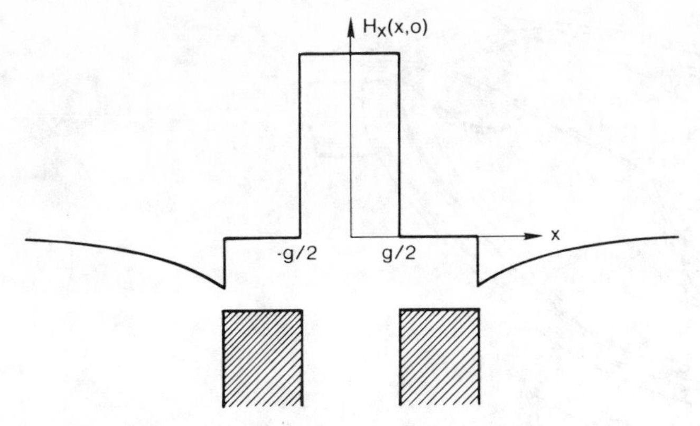

Fig. 3.12: Horizontal field component along the pole faces of a head with finite pole tips.

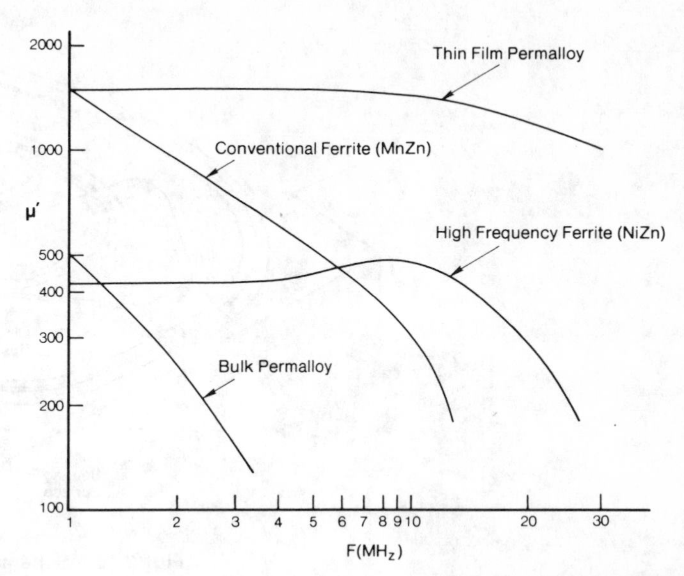

Fig. 3.14: Real part of the permeability for several head materials as a function of frequency.

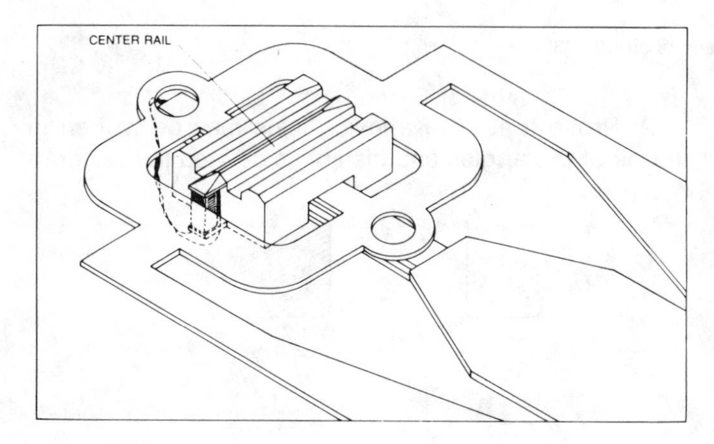

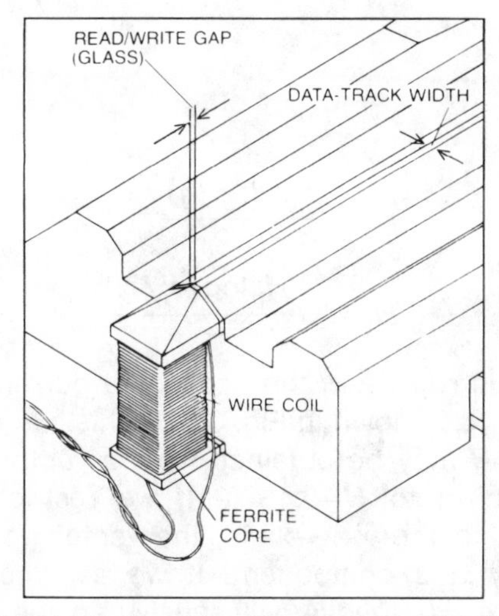

Fig. 3.13: Winchester head showing the tapered air-bearing slider with a ring head mounted on the back.

the current fall off as $1/r^2$ in two dimensions. The fact that we then obtain two different values for $H_x(0, y)$ lies in the model of infinite pole tips and the assumption that the potential is constant out to infinity.

To resolve this inconsistency we must consider a head having pole tips of finite length p. The magnetic potential drop across the gap is NI. If we take the symmetry plane between the pole tips to be at zero potential, then the pole tips are at $+NI/2$. The other equipotential surfaces have the form shown in Fig. 3.11. Potter [12] noted that at infinity the potential (on this side) is $-NI/4$. He therefore suggested an analytic form for the potential along the plane $y=0$ which has the Karlqvist form in the gap and along the pole tip, but falls off as $1/x$ to $-NI/4$ beyond this pole tip. Using this potential with the Green's function (3.39) gives the horizontal field component shown in Fig. 3.12. Notice the negative values at the back of the head! These negative regions insure that the integral over $H_x(x, 0)$ is zero.

In tape and flexible disk drives the heads are nominally in contact with the recording medium. This presents a wear problem which is minimized by operating at a relatively low head-to-medium velocity. In the case of disks this means a longer time for the disk to make half a revolution (the latency time) and a lower data rate. In higher performance drives the head is maintained out of contact with the disk surface by a cushion of air, which is "self-generated" by a properly designed air bearing slider. Fig. 3.13 shows the slider, as viewed from the disk side, of a Winchester head.

3.4 Thin-Film Heads

If we look back at Fig. 2.16 in Chapter 2 we see that higher linear recording densities have been accompanied by smaller gaps. For many years heads have been hand-wound and assembled under a microscope. As the gap lengths become smaller and smaller, however, it becomes increasingly difficult to consistently produce heads to specification. The thin-film techniques developed by the electronics industry offer excellent control over critical head dimensions. Furthermore, the use of thin films reduces eddy current losses and enables the use of high permeability head media. Fig. 3.14, for example, shows the advantage in frequency response that thin-film permalloy has over

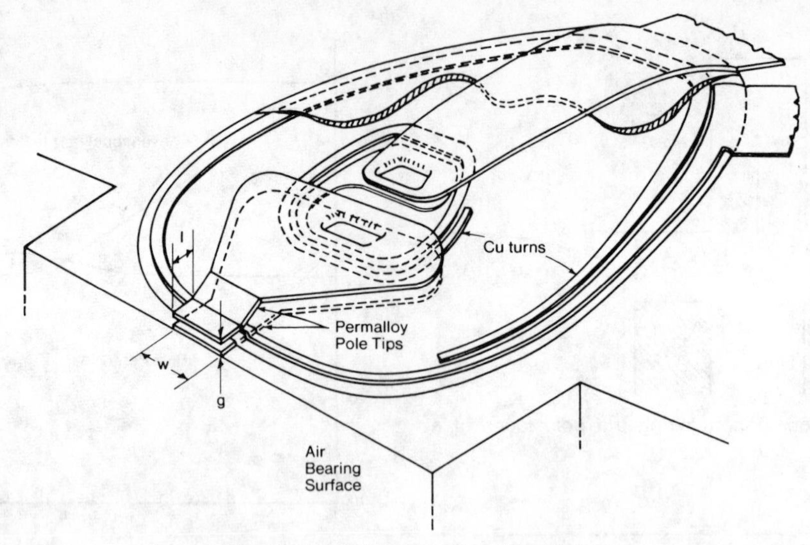

Fig. 3.15: Schematic of an IBM 3370.

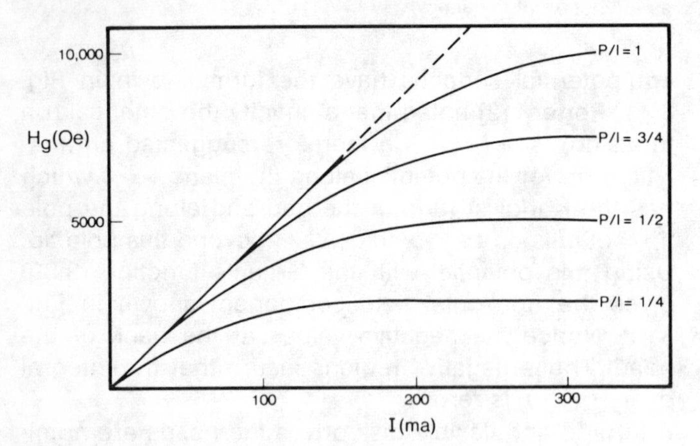

Fig. 3.16: H_g versus I for a single turn thin-film head with $g = 0.25$ μ and various values for p/l.

ferrite heads. Also, the finite pole tips tend to enhance the resolution as the wings of the head field function are pulled in to produce the effect shown in Fig. 3.12.

As a result, in 1979 IBM introduced thin-film heads in its 3370 disk drive. Fig. 3.15 shows a diagram of a thin-film head [14].

The pole thickness p is typically a few microns. The gap *depth* l must be made small in order to maximize H_g. Fig. 3.16 shows how H_g depends upon the ratio p/l. H_g saturates when the gap flux, $H_g lw$, where w is the track width, equals the pole layers saturation magnetization flux $4\pi M_s pw$, i.e.,

$$H_g^{\max} \equiv 4\pi M_s(p/l).$$

Problems

1) Show that the product of a large static permeability and the frequency at which the imaginary part of the permeability has a maximum is a constant which depends only on material parameters. Evaluate this product for MnZn.

2) Prove that the Karlqvist fields may be written in terms of the angles and distances defined in the figure

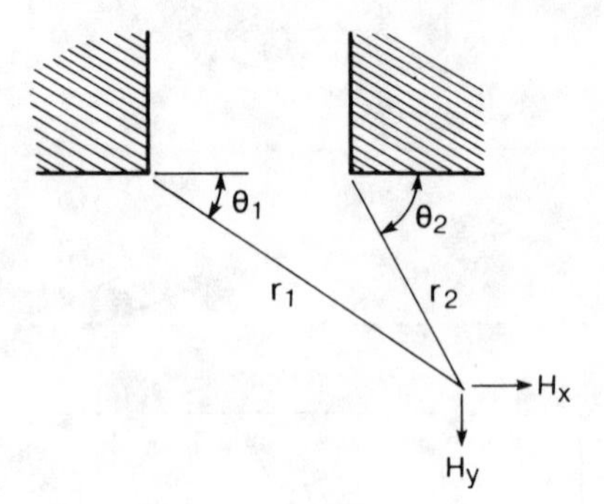

as

$$H_x = \frac{H_g(\theta_1 - \theta_2)}{\pi}$$

$$H_y = \frac{-H_g \log(r_1/r_2)}{\pi}.$$

3) Let us represent the head by a current sheet of density $\boldsymbol{K}_0$ as shown in Fig. 3.8. In this exterior region $\boldsymbol{B} = \boldsymbol{H}$ so $\boldsymbol{H}$ may be obtained from a vector potential. $\boldsymbol{A}$, according to $\boldsymbol{H} = \nabla \times \boldsymbol{A}$. If we restrict the field variations to the x–y plane, the vector potential will have only a z component. If we assume a gauge where $\nabla \cdot \boldsymbol{A} = 0$, then A_z will satisfy Laplace's equation in two dimensions. Taking our discussion of spacing loss into account, write $A_z(x, y)$ as a Fourier cosine expansion. Integrate

$$\nabla \times \boldsymbol{H} = (4\pi/c)\boldsymbol{K}_0 \delta(y) \qquad (-g/2 \leqslant x \leqslant g/2)$$

to obtain the boundary condition on $\partial \boldsymbol{A}/\partial y|_{y=0}$. Use this boundary condition to evaluate the coefficients in the cosine expansion and thereby derive the Karlqvist head fields in terms of $\boldsymbol{K}_0$.

4) Discuss what you might expect to happen to the Karlqvist field if the gap were filled with a material

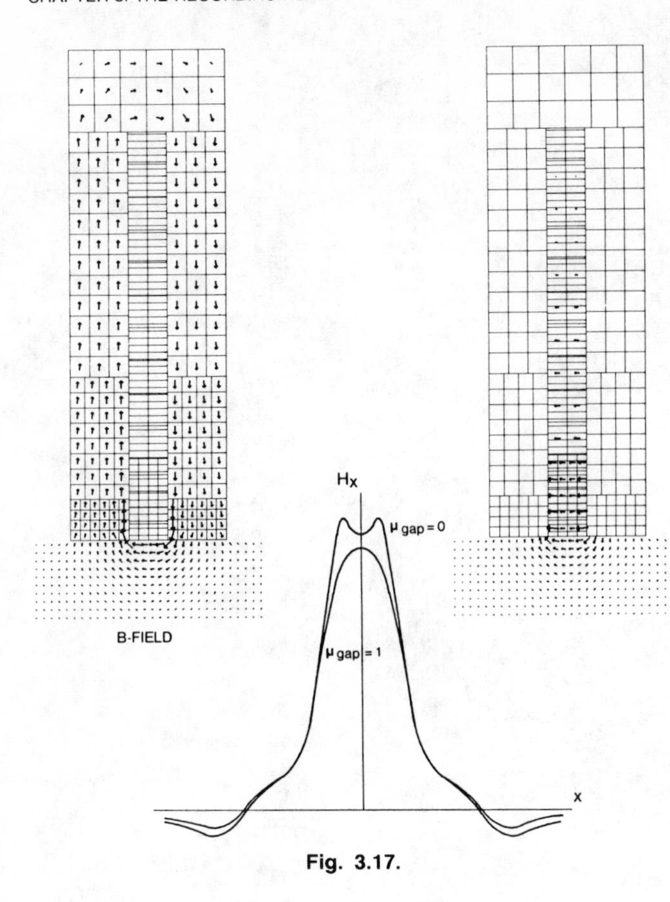

Fig. 3.17.

whose permeability was zero! How might you model this situation to calculate the field function? Hint: Fig. 3.17 shows the result of a numerical computation for such an "eddy current" head. (Kindly provided by Dr. D. Bloomburg of Xerox.)

References

[1] A. Paton, *J. Appl. Phys.*, vol. 42, p. 5868, 1971.

[2] G. T. Rado, R. W. Wright, and W. H. Emerson, *Phys. Rev.*, vol. 80, p. 273, 1950.

[3] See, for example, R. M. White and T. H. Geballe, *Long Range Order in Solids.* New York: Academic, 1979, p. 398.

[4] W. K. Westmijze, *Philips Res. Rep.*, vol. 8, p. 161, 1953.

[5] O. Karlqvist, *Trans. Roy. Inst. Technol. Stockholm*, p. 86, 1954.

[6] After F. J. Jeffers, R. J. McClure, W. W. French, and N. J. Griffith, *IEEE Trans. Magn.*, vol. MAG-18, p. 1146, 1982.

[7] I. Elabd, *IEEE Trans. Audio*, vol. AU-11, p. 21, 1963.

[8] D. Lindholm, *IEEE Trans. Magn.*, vol. MAG-13, p. 1460, 1977.

[9] P. Silvester and M. V. K. Chari, IEEE Trans. Power App. Syst., vol. PAS-89, p. 1642, 1970.

[10] M. H. Lean and A. Wexler, *IEEE Trans. Magn.*, vol. MAG-18, p. 331, 1982.

[11] J. C. Mallinson, *IEEE Trans. Magn.*, vol. MAG-17, p. 2453, 1981.

[12] R. I. Potter, *IEEE Trans. Magn.*, vol. MAG-11, p. 80, 1975.

[13] G. Hughes, *J. Appl. Phys.*, vol. 54, p. 4168, 1983.

[14] R. E. Jones and W. Nystrom, U.S. Patent 4 190 872, Feb. 26, 1980.

Chapter 4
The Writing Process

LET us now consider how the medium responds to the writing fields from the head. This is a difficult problem, for the magnetization of the medium also acts as a source for an additional field, the demagnetization field, which must be incorporated in the writing process.

4.1 Concept of a Transition Length

Let us begin by considering the effect of a head field pulse on a nearby stationary medium. Let us assume that this medium only responds to the longitudinal component of the head field. The contours of the field lines for H_x being constant are circles with centers located along the *y* axis.

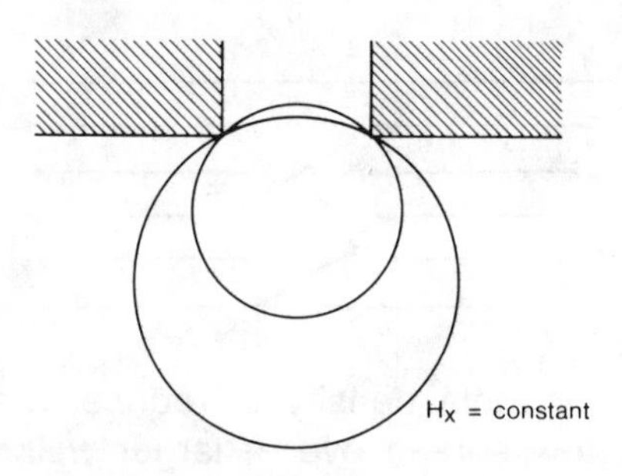

This is easily seen from the narrow gap expression, (3.46), for H_x, but is true for the Karlqvist expression, (3.43), as well. Let us suppose that the medium is initially magnetized in the "positive" direction, i.e., it sits in the remanent state *A* on the hysteresis loop below. As the head field builds up in the "negative" direction, the medium at any given point is taken down along its hysteresis loop. Suppose at the peak of the pulse the contour where $H_x = H_c$ has penetrated through the medium:

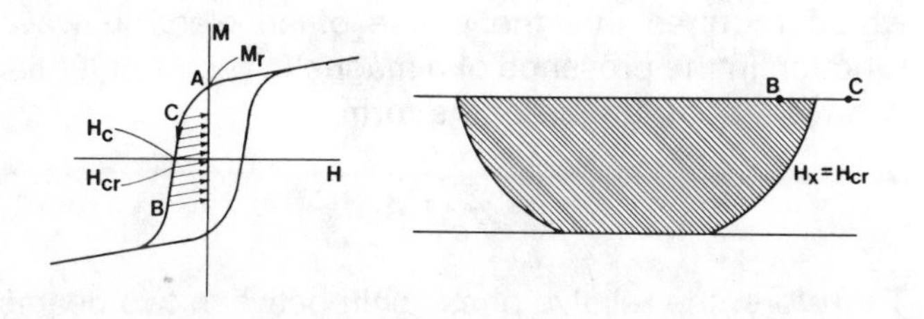

Now when the head field decreases to zero the magnetization will return to the $H=0$ axis along a *minor* loop as indicated by the arrows in the figure above. We generally approximate these minor loop segments by straight lines parallel to the slope of the major loop at $H=0$. Thus, the contour in the medium that had reached the coercivity point will actually remagnetize slightly. The contour that ends up in a completely demagnetized state had to have been driven slightly beyond H_c to a field value we refer to as the *remanent coercivity* H_{cr}. Because of the spatial variation of the field, different points of the medium will be at different points on the hysteresis loop. Thus, at the upper surface of the medium, for example, we will find a point *C* whose magnetization is reduced from the remanent value, but is still in the "positive" direction. Point *B*, on the other hand, has been reversed but not completely. Thus, there is a length over which the transition occurs. We can estimate this length by writing

$$dM/dx = (dM/dH)(dH/dx). \qquad (4.1)$$

Assuming the slope of the hysteresis loop and the head field gradient are constant, integrating this relation gives a transition length $2a$ where

$$a = M_r/(dM/dH)(dH/dx). \qquad (4.2)$$

This shows the importance of a square loop and a high field gradient in writing sharp transitions.

As the medium moves past the gap, notice that the "writing" is occurring at the trailing edge of the pole face.

Since the magnetization varies spatially through the transition region, a demagnetization field will be generated. Thus, the final field at points *B* and *C*, for example, will *not* be zero as implied by the minor-loop segments in the hysteresis loop above. Only at the center of the transition will the field be zero.

4.2 The Demagnetization Field

The demagnetization field is obtained from the Maxwell equation

$$\nabla \cdot \boldsymbol{H} = -4\pi \nabla \cdot \boldsymbol{M}. \qquad (4.3)$$

In the absence of currents $\nabla \times \boldsymbol{H} = 0$ and the field is then derivable from a scalar potential, $\boldsymbol{H} = -\nabla\varphi$. Combining this with the Maxwell equation above gives the Poisson equation

$$\nabla^2\varphi = -4\pi \nabla \cdot \boldsymbol{M} \qquad (4.4)$$

which has the solution

$$\varphi(\boldsymbol{r}) = -\int \frac{\nabla' \cdot \boldsymbol{M}(\boldsymbol{r}')}{|\boldsymbol{r} - \boldsymbol{r}'|} d^3\boldsymbol{r}'. \qquad (4.5)$$

Therefore, the demagnetization field becomes

$$\boldsymbol{H}_d(\boldsymbol{r}) = -\int \frac{\nabla' \cdot \boldsymbol{M}(\boldsymbol{r})(\boldsymbol{r}-\boldsymbol{r}')}{|\boldsymbol{r}-\boldsymbol{r}'|^3} d^3\boldsymbol{r}'. \quad (4.6)$$

We have already mentioned in Chapter 2 that the demagnetizing field associated with a uniformly magnetized ellipsoidal sample is $-4\pi NM$ inside the sample. For a sphere, the demagnetizing factor N is 1/3. This may be obtained from the general solution, (4.6). In spherical coordinates,

$$\nabla \cdot \boldsymbol{M} = \frac{1}{r^2}\frac{\partial}{\partial r}(r^2 M_r) + \frac{1}{r\sin\theta}\frac{\partial}{\partial\theta}(\sin\theta M_\theta) + \frac{1}{r\sin\theta}\frac{\partial M_\varphi}{\partial\varphi}. \quad (4.7)$$

For a uniformly magnetized sphere of radius a

$$M_r = M\cos\Theta\,\theta(a-r)$$
$$M_\Theta = M\sin\Theta\,\theta(a-r) \quad (4.8)$$
$$M_\varphi = 0$$

where $\theta(a-r)$ is the theta function,

$$\theta(a-r) = \begin{cases} 1 & r<a \\ 0 & r>a. \end{cases} \quad (4.9)$$

Therefore, $\nabla\cdot\boldsymbol{M} = -M\cos\Theta\,\delta(r-a)$. Combining this with the spherical harmonic expansion,

$$\frac{1}{|\boldsymbol{r}-\boldsymbol{r}'|} = 4\pi\sum_{l=0}^{\infty}\sum_{m=-l}^{l}\frac{1}{2l+1}\frac{r_<^l}{r_>^{l+1}} \cdot Y_l^{m*}(\theta',\varphi')Y_l^m(\theta,\varphi). \quad (4.10)$$

Only the $l=1$, $m=0$ term survives the integral, (4.5), for $\varphi(r)$. The result is

$$\varphi(r) = \frac{4\pi M}{3}a^2\left(\frac{r_<}{r_>^2}\right)\cos\theta \quad (4.11)$$

where $r_<(r_>)$ is the smaller (larger) of r and a. This potential gives the constant demagnetizing field inside the sphere and the familiar dipole field outside.

4.2.1 The Arctangent Transition

Suppose we had an infinitely sharp transition in the horizontal magnetization:

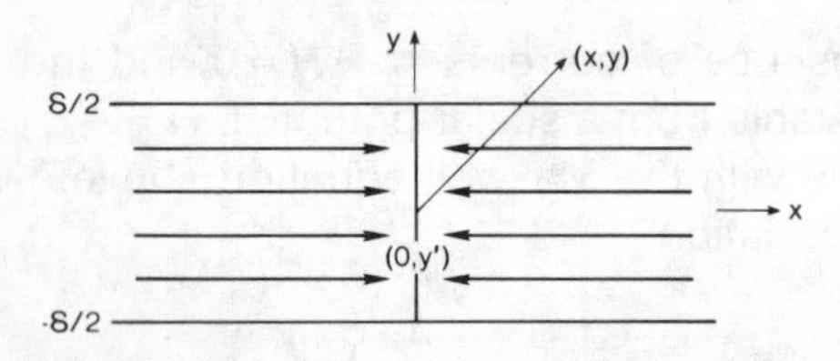

Then

$$\nabla\cdot\boldsymbol{M} = \nabla\cdot\{M[\Theta(-x)] - \Theta(x)]\}\,\hat{\boldsymbol{x}} = -2M\delta(x) \quad (4.12)$$

and the demagnetization field is

$$H_d(x,y)_x = -2M\int_{-w/2}^{w/2}\int_{-\delta/2}^{\delta/2}\frac{x\,dy'\,dz'}{\left[x^2+(y-y')^2+z'^2\right]^{3/2}}. \quad (4.13)$$

If the trackwidth w is assumed to be the longest dimension in the problem, then we take $w\to\infty$ and the problem becomes two-dimensional and the integral simplifies to

$$H_d(x,y)_x = -4M\int_{-\delta/2}^{\delta/2}\frac{x\,dy'}{x^2+(y-y')^2}. \quad (4.14)$$

In the midplane of the medium, where $y=0$,

$$H_d(x,0)_x = 8M\tan^{-1}(\delta/2x) \quad (4.15)$$

which has its maximum value, $4\pi M$, at the discontinuity (see Fig. 4.2).

The energy associated with this discontinuity can be lowered by having the magnetization adopt a different configuration. One such configuration is a zig-zag wall:

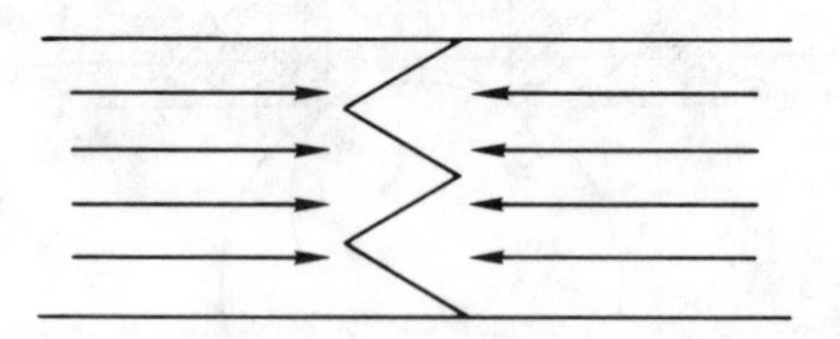

The magnetic pole density is reduced because the poles are now spread over a larger wall area. However, there will be an increase in the wall energy that will determine the detailed shape of the wall. Evidence for such a configuration has been seen in 300–500 Å Co films by electron holography. Electron holography is an exciting new technique developed by Hitachi [1] for imaging this magnetic state of thin films. This technique is based on the wave nature of a coherent electron beam. If such a beam is split into two paths and subsequently recombined, the intensity may exhibit interference effects if the two beams enclose magnetic flux. This arises from gauge invariance, which requires that the phase of an electron wave function in the presence of a magnetic vector potential $\boldsymbol{A}$ have a contribution of the form

$$\Delta\varphi = \frac{e}{\hbar c}\int\boldsymbol{A}\cdot d\boldsymbol{s}. \quad (4.16)$$

Therefore, the relative phase shift between two beams is

$$\frac{e}{\hbar c}\oint\boldsymbol{A}\cdot d\boldsymbol{s} = \frac{e}{\hbar c}\iint\nabla\times\boldsymbol{A}\cdot d\boldsymbol{S} = \frac{e}{\hbar c}\iint\boldsymbol{B}\cdot d\boldsymbol{S} = \pi(\Phi/\Phi_0) \quad (4.17)$$

where Φ_0 is the flux quantum $hc/2e$. In the Hitachi implementation, the electron source is a field emission

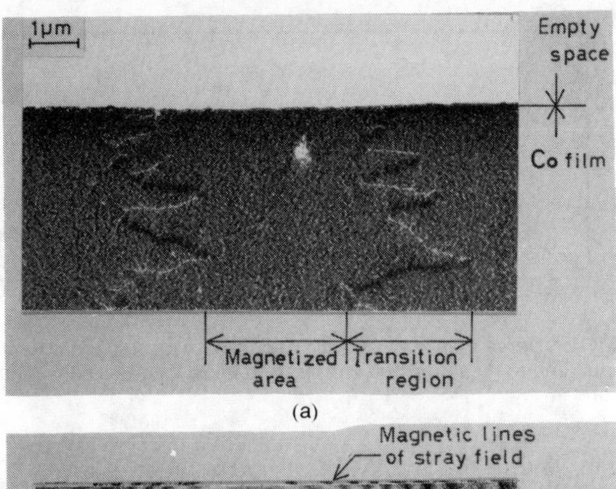

(a)

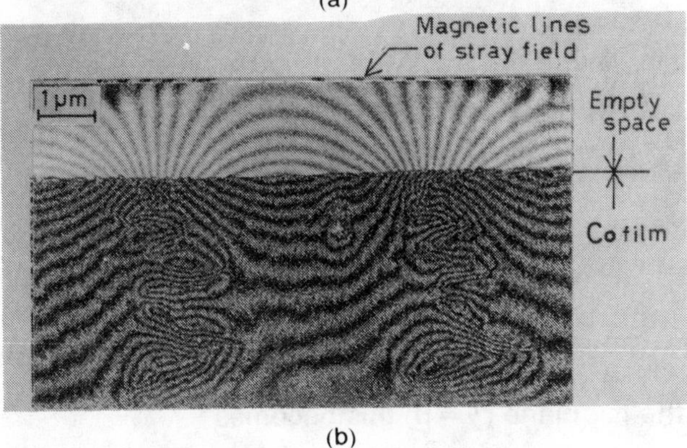

(b)

Fig. 4.1: Recorded magnetization pattern on a Co film (film thickness = 450 Å; coercivity = 340 Oe. The bit length is 5 μm. (a) Lorentz micrograph. (b) interference micrograph. (Courtesy of Hitachi.)

tip. Part of the beam passes through the magnetic film while the other part is unaffected. The two are overlapped and the interference pattern is then magnified and recorded on film as a hologram. The hologram must then be optically reconstructed to obtain the interference micrograph as shown in Fig. 4.1.

Rather than deal with a complex wall configuration, it is common to assume a simple analytic description. The most popular is to assume that the *amplitude* of the magnetization has the arctangent form,

$$M_x(x) = -\frac{2M}{\pi}\tan^{-1}(x/a) \tag{4.18}$$

where the transition length a is taken as an adjustable parameter. Substituting this form for $M_x(x)$ into the equation for $\boldsymbol{H}_d(\boldsymbol{r})$ and carrying out the integrals we obtain

$$H_d(x,y)_x = 4M\left[\tan^{-1}\frac{(\delta/2+y)x}{x^2+a^2+|\delta/2+y|a} + \tan^{-1}\frac{(\delta/2-y)x}{x^2+a^2+|\delta/2-y|a}\right]$$

$$H_d(x,y)_y = 2M\ln\frac{x^2+(|\delta/2+y|+a)^2}{x^2+(|\delta/2-y|+a)^2}. \tag{4.19}$$

For calculational convenience, H_{dx} may also be written

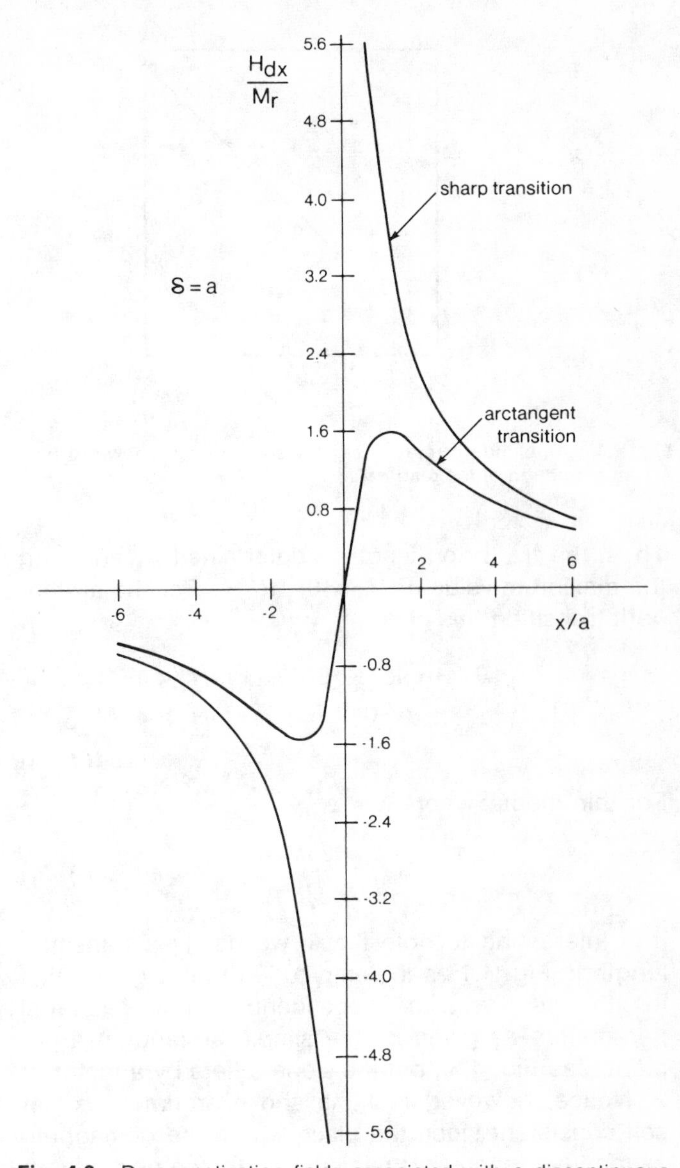

Fig. 4.2: Demagnetization fields associated with a discontinuous transition and an arctangent transition.

$$H_d(x,y)_x = -4M\left[\tan^{-1}\frac{a+\delta/2+y}{x} + \tan^{-1}\frac{a+\delta/2-y}{x} - 2\tan^{-1}a/x\right]. \tag{4.20}$$

The value of $H_{dx}(x,0)$ is plotted in Fig. 4.2. The maximum value of $H_d(x,0)$ occurs at

$$x/a = \sqrt{1+\delta/2a}. \tag{4.21}$$

Notice that there is also a *vertical* demagnetization field, $H_d(x,y)_y$, which is largest at the surface of the medium. This is generally neglected because it is assumed that the medium only responds longitudinally.

It is now argued that this demagnetization field cannot exceed the coercivity of the medium. Otherwise the magnetization will readjust itself to lower this field.

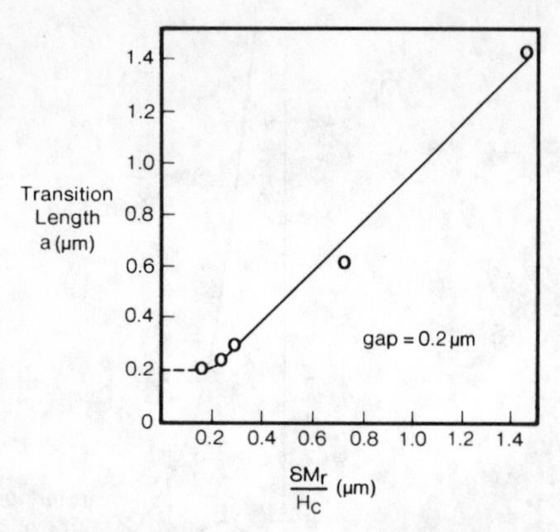

Fig. 4.3: Amplitude of the zig-zag transition region shown in Fig. 4.1 as a function of the ratio $\delta M_r/H_c$.

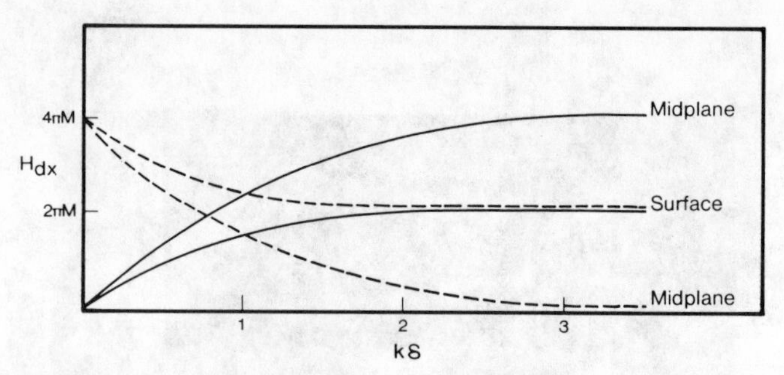

Fig. 4.4: Demagnetization fields associated with horizontal (solid curves) and vertical (dashed curves) sinusoidal magnetization distribution.

Thus, the transition length is determined by equating the maximum value of $H_d(x,0)_x$ to H_c. For the arctangent transition this gives

$$a_{\min} = \begin{cases} (\delta/4)(\csc H_c/8M - 1) & H_c < 4\pi M \\ 0 & H_c > 4\pi M \end{cases}. \tag{4.22}$$

For thin media, where $\delta \ll a$,

$$a_{\min} = \frac{2\delta M}{H_c}. \tag{4.23}$$

It is interesting to note that if we define a transition length in Fig. 4.1 as the amplitude of the zig-zag, this length has the *same* dependence on the material parameters as given by the simple arctangent argument (see Fig. 4.3), but the slope differs by a factor of 2. Notice, however, that the above argument is not self-consistent since the place where the demagnetization field reaches its minimum is not the point where the magnetization is zero. Since $a_{\min}$ represents the closest approach of any two magnetic reversals, high densities require a small value of $a_{\min}$. This tells us we want thin media with a high coercivity and a small remanent magnetization. Later we shall see that the readback signal is proportional to $M_r\delta$. Therefore, a higher coercivity is a more appropriate route to higher densities.

4.2.2 Wavelength Dependence

Another magnetization pattern for which the demagnetization field can be readily evaluated is that of a sinusoidal variation,

$$M_x(x) = M \sin kx \tag{4.24}$$

where $k = 2\pi/\lambda$. The demagnetization field is

$$H_d(x,y,z=0)_x = -kM \iiint \frac{\cos kx'(x-x')\,dx'\,dy'\,dz'}{\left[(x-x')^2 + (y-y')^2 + z'^2\right]^{3/2}}. \tag{4.25}$$

Again, taking the trackwidth to be very large gives

$$\begin{aligned} H_d(x,y)_x &= 2kM \int_{-\infty}^{\infty} \cos kx' \left[\tan^{-1} \frac{(y-\delta/2)}{x-x'} - \tan^{-1} \frac{(y+\delta/2)}{x-x'} \right] dx' \\ &= -2\pi M \sin kx \left[2 - e^{-k(y+\delta/2)} - e^{k(y-\delta/2)} \right]. \end{aligned} \tag{4.26}$$

In the midplane ($y = 0$) this becomes

$$H_d(x,0)_x = -4\pi M \sin kx \left[1 - e^{-k\delta/2} \right] \tag{4.27}$$

while at either surface ($y = \pm\delta/2$),

$$H_d(x, \pm\delta/2)_x = -2\pi M \sin kx \left[1 - e^{-k\delta} \right]. \tag{4.28}$$

This behavior is plotted in Fig. 4.4.

It is also interesting to carry out this calculation for a vertical magnetization,

$$M_y(x) = M \sin kx \left[-\Theta(y-\delta/2) + \Theta(y+\delta/2) \right]. \tag{4.29}$$

Then

$$\begin{aligned} H_d(x,y,z=0)_y &= M \int_{-\infty}^{\infty} \sin kx' \left[\frac{-(y-\delta/2)}{(x-x')^2 + (y-\delta/2)^2} + \frac{y+\delta/2}{(x-x')^2 + (y+\delta/2)^2} \right] dx' \\ &= 2\pi M \sin kx \left[e^{k(y-\delta/2)} + e^{k(y+\delta/2)} \right]. \end{aligned} \tag{4.30}$$

In the midplane,

$$H_d(x,0)_y = -4\pi M \sin kx e^{-k\delta/2}. \tag{4.31}$$

While at the surfaces

$$H_d(x, \pm\delta/2)_y = -2\pi M \sin kx \left[1 + e^{-k\delta} \right]. \tag{4.32}$$

These are plotted as the dashed curves in Fig. 4.4. Notice that the demagnetization field in the vertical case goes to zero at short wavelengths, i.e., high recording densities. This is the origin of the interest in vertical recording. Notice, however, that the demagnetization fields *at the surface* are the same in the high density limit for the two configurations. Thus, any conclusions as to the relative merits of these two configurations must be based on a more detailed analysis that incorporates the head fields as well.

4.3 Williams–Comstock Model

If one is willing to assume a functional form for the magnetization then one can obtain an analytic form for the transition length that incorporates both the head field and the demagnetization field. Such an approach is very useful in understanding the role of various parameters and in comparing different recording configurations. Let us consider the model introduced by M. L. Williams and R. L. Comstock [2] in 1971.

4.3.1 Slope Criterion

The Williams–Comstock model makes use of the fields we have already discussed. In the figure below we sketch the Karlqvist field and the demagnetization field associated with an arctangent transition.

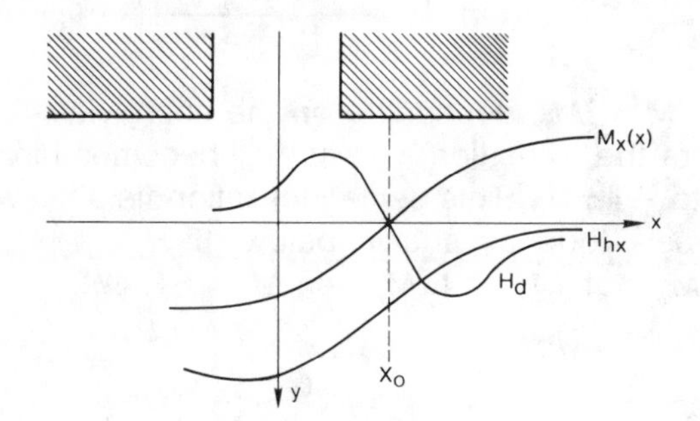

Notice that the demagnetization field is zero at the center of the transition. The transition length is obtained by assuming an arctangent transition with a parameter *a* which is determined through a self-consistency condition based on the following derivative relation:

$$\left.\frac{dM}{dx}\right|_{x_0} = \frac{dM}{dH}\left(\frac{dH_h}{dx} + \frac{dH_d}{dx}\right). \quad (4.33)$$

The factor dM/dH brings in the role of the hysteresis loop. As the head approaches the point x_0, the magnetization at that point moves away from its remanent value as illustrated on the hysteresis loop. If this point is to have zero remanence after the field is removed then it must overshoot as discussed above to some point *I* and return to zero along a minor loop path characterized by a slope χ. The slope of the major loop at the point *I* is

$$\left.\frac{dM}{dH}\right|_{I} = \frac{M_r}{H_c(1-S^*)} \quad (4.34)$$

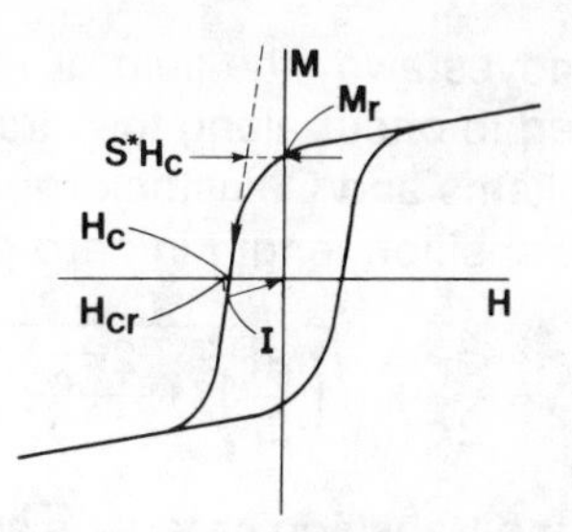

Fig. 4.5: Definition of parameters characterizing hysteresis loop.

where S^* measures the squareness of the loop as defined in Fig. 4.5. And the value of the field at *I* is

$$H_{cr} = \frac{-H_c}{1-\chi(1-S^*)H_c/M} = -\frac{H_c}{r}. \quad (4.35)$$

We have already seen that to obtain a sharp transition the head field gradient should be at maximum. Let us assume that the current into the head has been adjusted such that the maximum gradient also occurs at the transition, i.e., that $H_h = H_I$. We shall investigate what happens when this assumption is relaxed in the next chapter. The derivative of the head field function

$$H_{hx} = \frac{H_g}{\pi}\left[\tan^{-1}\left(\frac{x+g/2}{y}\right) - \tan^{-1}\left(\frac{x-g/2}{y}\right)\right] \quad (4.36)$$

is

$$\frac{dH_{hx}}{dx} = \frac{H_g}{\pi y}\left[\frac{1}{1+\left(\frac{x+g/2}{y}\right)^2} - \frac{1}{\left(\frac{x-g/2}{y}\right)^2}\right]. \quad (4.37)$$

Maximizing this derivative with respect to *x* gives the distance x_0 where this maximum occurs. x_0/g depends only on y/g. However, since we require that $H_{hx} = -H_I$ at this value of *y*, H_g must be adjusted at each value of *y*. This suggests that we write the maximum of the gradient as

$$\left.\frac{dH_{hx}}{dx}\right|_{max} = \frac{H_I Q}{y} \quad (4.38)$$

where *Q* is a function of the head-medium spacing *y*. The gradient of the demagnetizing field, for small medium thickness, is

$$\frac{dH_d}{dx} = -\frac{4M_r\delta}{a^2}. \quad (4.39)$$

Combining all these derivatives leads to the transition length at *I*,

$$\frac{a_1}{r} = \frac{(1-S^*)y}{\pi Q} + \sqrt{\left[\frac{(1-S^*)y}{\pi Q}\right]^2 + \frac{2M_r\delta}{H_c}\frac{2y}{Qr}}. \quad (4.40)$$

As the head moves away the field at l returns to zero. This is assumed to occur along the straight line shown in Fig. 4.5. Williams and Comstock repeat the analysis to obtain the transition length in zero field:

$$a_2 = \frac{a_1}{2r} + \sqrt{\left(\frac{a_1}{2r}\right)^2 + \frac{2\pi\chi\delta a_1}{r}}. \quad (4.41)$$

Since $a_2 > a_1$ the transition broadens as it moves away from the head field. Typical numbers for a metallic disk are

$$a_0 = 2M\delta/H_c = 0.094\ \mu m$$
$$a_1 = 0.293\ \mu m$$
$$a_2 = 0.321\ \mu m.$$

This analytical result shows the importance of having a sharp head field gradient (large Q), a very square loop ($S^* \to 1$), a small demagnetization field ($4\pi M_r \ll H_c$), and a thin medium in obtaining high densities.

4.3.2 Perpendicular versus In-Plane Recording

Middleton and Wright [3] have recently used the Williams–Comstock model to compare the transition lengths for perpendicular and in-plane recording. This constitutes a nice example of the usefulness of this model.

a) In-Plane Case: For the head field, Middleton and Wright take the small gap limit of the Karlqvist head field,

$$H_{hx} = \frac{gH_g}{\pi}\frac{y}{x^2+y^2}. \quad (4.42)$$

The gradient is

$$\frac{dH_{dx}}{dx} = -\frac{gH_g}{\pi}\frac{2xy}{(x^2+y^2)^2} \quad (4.43)$$

which has its maximum at $x = y\sqrt{3}$. If H_g is again adjusted such that H_{hx} has the value $-H_l \approx -H_c$ at the point where the gradient has its maximum, then

$$\left.\frac{dH_{dx}}{dx}\right|_{max} = \frac{\sqrt{3}}{2}\frac{H_c}{y}. \quad (4.44)$$

Recalling the Williams–Comstock expression, (4.33), of dM/dx, if the hysteresis loop is very square so that dM/dH is very large, then if dM/dx is to have a reasonable value we must have

$$\frac{dH_{hx}}{dx} = -\frac{dH_{dx}}{dx}. \quad (4.45)$$

Using the demagnetization field obtained we find

$$\left.\frac{dH_{dx}}{dx}\right|_{\substack{x=0\\y=0}} = -4M_r\frac{\delta}{a_l(a_l+\delta/2)} \quad (4.46)$$

where a_l denotes the "longitudinal" transition length. Equating the field gradients then gives

$$a_l = -\frac{\delta}{4} + \sqrt{\frac{\delta^2}{16} + \frac{8M_r(d+\delta/2)\delta}{\sqrt{3}\,H_c}}. \quad (4.47)$$

In the limit of a thin medium this is consistent with the Williams–Comstock result, (4.40).

b) Perpendicular Case: In the case of the perpendicular media we can write with either a ring head or a pole-tip head. Using the head field results derived in Chapter 3 we find that the gradients are

$$\left.\frac{dH_{hy}^{ring}}{dx}\right|_{max} = \frac{H_c}{2\sqrt{3y}} \quad (4.48)$$

$$\left.\frac{dH_{hy}^{pole}}{dx}\right|_{max} = \frac{\sqrt{3}\,H_c}{2y}. \quad (4.49)$$

Notice that the pole-tip head has a larger field gradient at the point where the transition occurs.

The demagnetization field associated with a vertical arctangent transition is

$$H_{dy} = -4M_0\left[\tan^{-1}\left(\frac{x_0}{a_v+|y_0-\delta/2|}\right) + \tan^{-1}\left(\frac{x_0}{a_v+|y_0+\delta/2|}\right)\right] \quad (4.50)$$

where $M_0 \leqslant M_r$, depending on the coercivity. That is, far from the transition, i.e., as $|x_0|$ becomes large, H_d goes to $-4\pi M_0$. This demagnetization field skews the hysteresis loop as shown below. If $H_c > 4\pi M_r$ then $M_0 = M_r$, but if $H_c < 4\pi M_r$ then $M_0 = H_c/4\pi$.

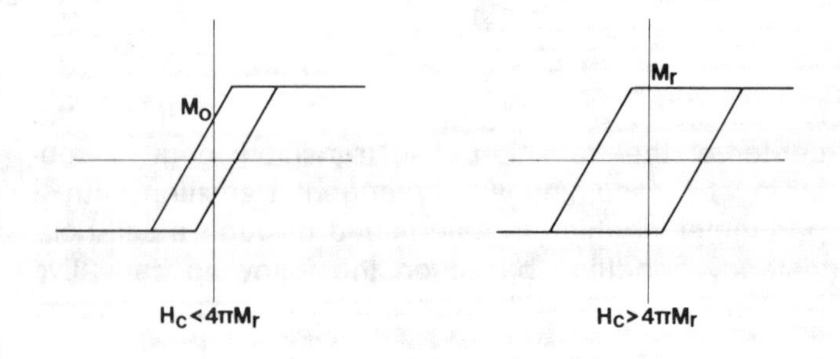

The demagnetization field gradient is

$$\frac{dH_{dy}}{dx} = -8M_0\frac{1}{a_v+\delta/2}. \quad (4.51)$$

Again, equating the field gradients gives the transition lengths

$$a_v^{ring} = \frac{-\delta}{2} + 16\sqrt{3}\,\frac{M_0}{H_c}(d+\delta/2)$$
$$= \frac{1}{2}\left(\frac{16\sqrt{3}\,M_0}{H_c} - 1\right)\delta + \frac{16\sqrt{3}\,M_0 d}{H_c} \quad (4.52)$$

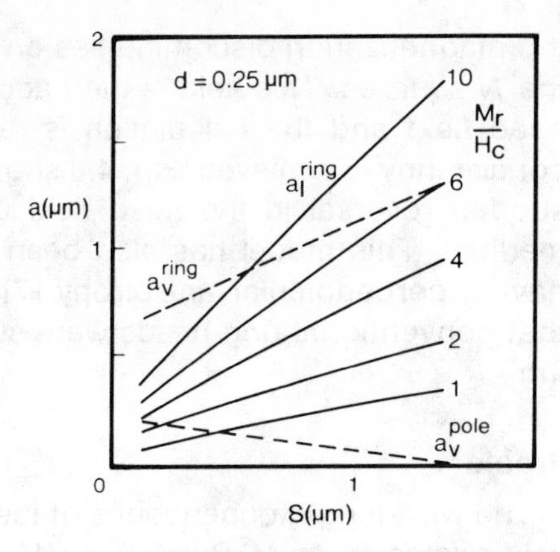

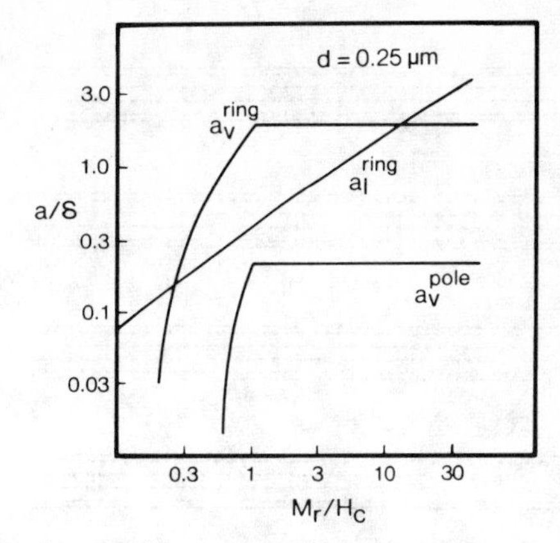

Fig. 4.6: Transition lengths as functions of the medium thickness and M_r/H_c.

$$a_v^{\text{pole}} = \frac{-\delta}{2} + \frac{16}{\sqrt{3}} \frac{M_0}{H_c}(d+\delta/2)$$

$$= \frac{1}{2}\left(\frac{16}{\sqrt{3}} \frac{M_0}{H_c} - 1\right)\delta + \frac{16 M_0 d}{\sqrt{3}\, H_c}. \tag{4.53}$$

These are compared with the longitudinal result in Fig. 4.6 as functions of the medium thickness δ and the ratio M_r/H_c. Notice that in the case of writing with a pole head on a perpendicular medium, the transition length gets *shorter* as the medium gets thicker! This arises from the fact that $4\pi M_0$ is always less than H_c, and therefore $(4/\sqrt{3}\,\pi)(4\pi M_0/H_c) < 1$. From this analysis we see that it is not the demagnetization fields themselves which are important, but rather their *gradients*. Notice that its larger gradient makes the pole-tip head preferable for writing on a vertical medium.

Whether perpendicular recording really possesses inherent advantages is still a subject of great debate. Theoretical approaches invariably involve assumptions open to criticism. The Middleton–Wright analysis, for example, completely ignores the columnar morphology of typical perpendicular recording media. Experiments [4] in which a sputtered Co–Cr medium was progressively ion-milled away and the magnetic state studied at each step by the Bitter technique show that a ring head writes only in a thin layer at the surface, while a pole head with an underlayer writes much more deeply.

4.4 Self-Consistent Calculations

Characterizing the magnetization during the writing process by a transition length parameter is a great oversimplification. Efforts have been made to improve upon this, for example, by allowing the transition length to vary with depth into the medium. However, to really describe the writing process the microscopic nature of the medium and its response to the *total* field must be considered as the head passes over.

One of the first attempts to treat the head field and the demagnetization field simultaneously was made by Iwasaki and Suzuki [5]. They characterized the medium

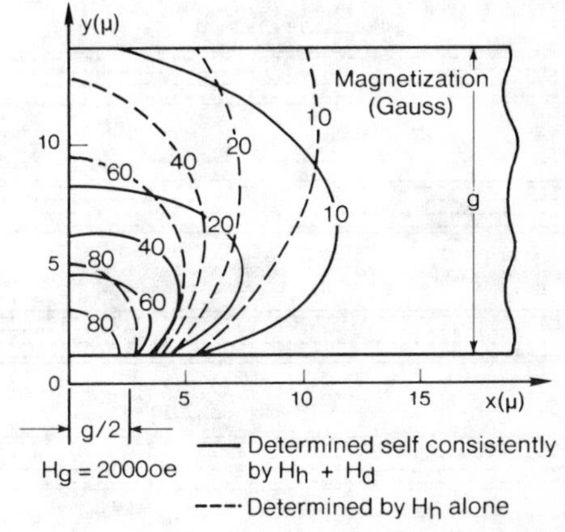

Fig. 4.7: Longitudinal magnetization distribution when head field is applied to pre-erased medium.

by a nonlinear susceptibility

$$M = \chi(H)H. \tag{4.54}$$

The Karlqvist head field H_h was used to calculate M. This magnetization was then used to calculate the demagnetization field H_d, and $H_h + H_d$ was then used to recalculate M. This iteration process was carried to convergence. Fig. 4.7 shows the results of such a calculation. These results show that near the gap the demagnetization field acts against the head field, while far from the gap H_d adds to H_h, enlarging the written region.

A more *ab initio* approach has been developed by I. Ortenburger (now Beardsley) and R. Potter [6]. They partition the medium cross section into N squares. Each square contains an assembly of Stoner–Wohlfarth particles. The orientation of the axes of these particles is described by a distribution function $g(\theta,\varphi)$, which is chosen to produce the desired hysteresis loop. Initially, we might consider the Karlqvist field from the head. For a particle with an orientation θ,φ, the asteroid construction described in Chapter 2 is then used to determine the direction of the particle magnetization, $m(\theta,\varphi;H)$. The magnetiza-

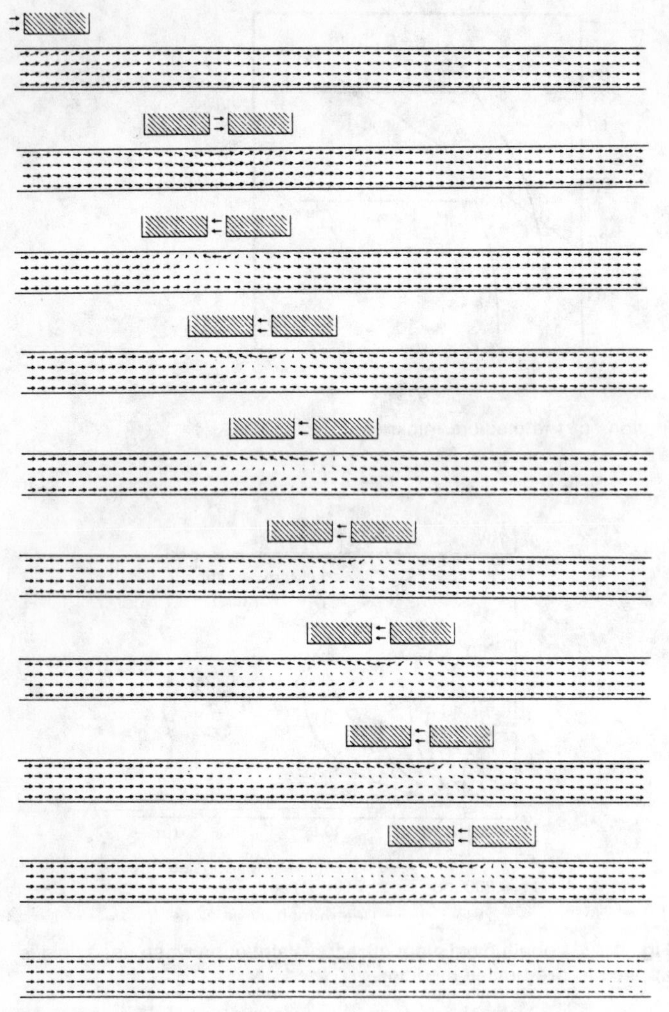

Fig. 4.8: Computer simulation of an isolated transition being written.

tion of the cell is then given by

$$M_n = M_s/4\pi \int_0^{\pi} d\theta \sin\theta \int_0^{2\pi} d\varphi g(\Theta,\varphi) m(\Theta,\varphi;H).$$

This is carried out for each cell. The demagnetization field at the center of each cell is then calculated from the magnetization discontinuities on the four edges of the N squares. This field is then added to the original head field and the calculation is repeated until self-consistency is achieved. Fig. 4.8 shows the results of a sudden reversal in the head field upon a horizontal medium. This model has also been applied to media having perpendicular anisotropy [7] where it showed that conventional ring heads were suitable for recording.

Problem

The wavelength dependence of the demagnetization field relates H_d to M. But H_d and M are also related to the hysteresis loop. As the demagnetization field increases (in the negative direction), the energy product $-MH_d$ goes through a maximum prior to reversal. The point at which this maximum occurs is approximately given by

$$\left|\frac{M}{H}\right| = \frac{M_r}{H_c}.$$

Using this as a criterion, estimate the maximum recording density that a medium with a given M_r and H_c could sustain.

References

[1] K. Yoshida, T. Okuwaki, N. Osakabe, H. Tanabe, Y. Horiuchi, T. Matsuda, K. Shinagawa, A. Tonomara, and H. Fujiwara, *IEEE Trans. Magn.*, vol. MAG-18, p. 1600, 1983.

[2] M. L. Williams and R. L. Comstock, in *AIP Conf. Proc.*, no. 5, 1971, p. 738.

[3] B. K. Middleton and C. D. Wright, in *IERE Conf. Proc.*, vol. 54, 1982, p. 181.

[4] K. Kobayashi, J. Toda, T. Yamamoto, *Fujitsu Sci. Tech. J.*, vol. 19, p. 99, 1983.

[5] S. Iwasaki and T. Suzuki, *IEEE Trans. Magn.*, vol. MAG-4, p. 269, 1968.

[6] I. B. Ortenburger and R. I. Potter, *J. Appl. Phys.*, vol. 50, p. 2395, 1979.

[7] R. I. Potter and I. Beardsley, *IEEE Trans. Magn.*, vol. MAG-16, p. 957, 1980.

Chapter 5
The Readback Voltage

THE presence of a magnetization distribution in the medium generates "demagnetization" fields which extend beyond the medium. If these field lines can be made to pass through a coil which is moving relative to the medium, a voltage will be induced in the coil. One of the first investigations of this readback voltage was that of R. L. Wallace [1].

5.1 The Wallace Solution

Wallace assumed that the magnetization had the longitudinal sinusoidal form we have already discussed in Chapter 4:

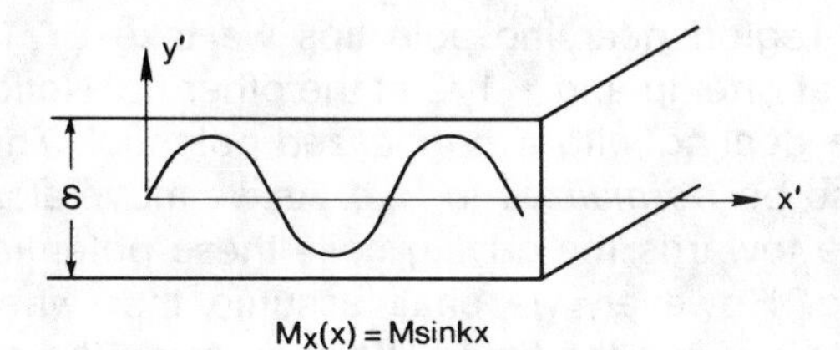

In the present situation, however, we are interested in the magnetic fields *above* the medium. The same expression (4.6) developed for $H_d(x, y)$ applies. In particular, the fields at a point (x, y) above the medium are

$$H_x(x,y) = -2\pi M \sin(kx) e^{-ky}[e^{k\delta/2} - e^{-k\delta/2}]$$
$$H_y(x,y) = -2\pi M \cos(kx) e^{-ky}[e^{k\delta/2} - e^{-k\delta/2}]. \tag{5.1}$$

Notice the exponential dependence on y which we already pointed out was a general feature of Poisson's equation in two dimensions.

Wallace assumed that the reproduce head consisted of a semiinfinite block of high permeability material with a flat face spaced a distance d above the recording medium. This block was assumed to be infinitely thick so that it "collected" all the field lines. (See Fig. 5.1). These lines then thread through a coil wound around the block. The value of the induction B inside this block is easily obtained by the method of images and is found to be $2\mu/(\mu+1)$ times the value of H in the absence of the block,

$$B_x(x,y) = -(2\mu/(\mu+1))2\pi M \sin(kx) \cdot e^{-ky}(e^{k\delta/2} - e^{-k\delta/2}). \tag{5.2}$$

The flux per unit width is then

$$\Phi_x = \int_{d+(\delta/2)}^{\infty} B_x\, dy = -[2\mu/(\mu+1)]2\pi\delta M \sin kx[(1-e^{-k\delta})/k\delta]e^{-kd}. \tag{5.3}$$

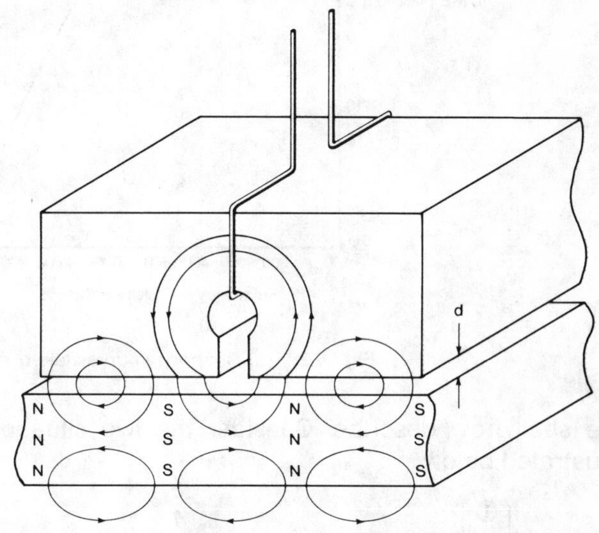

Fig. 5.1: Schematic illustration of "flux capture" by a high permeability head.

If $x = vt$ and the width of the head is w

$$d\Phi/dt = -[\mu/(\mu+1)]4\pi wvM(1-e^{-k\delta})e^{-kd}\cos\omega t. \tag{5.4}$$

For a realistic head with N turns and an efficiency of η, the voltage across the coil is

$$V(t) = N\eta\, d\Phi_x/dt. \tag{5.5}$$

This result reveals a number of features that are characteristic of magnetic recording. First of all there is the *spacing loss:*

$$20\log_{10} e^{-kd} = -54.6(d/\lambda)\,\text{dB} \tag{5.6}$$

which we noted before. Wallace confirmed this behavior experimentally by inserting spacers between a ring head and a rotating magnetic disk during readback. The writing was done in contact. His data are shown in Fig. 5.2. Another feature to note is the *thickness dependence* of the output voltage:

$$[1-e^{-k\delta}] \approx k\delta = 2\pi\delta/\lambda = \delta\omega/v. \tag{5.7}$$

This corresponds to an increase with frequency of

$$20\log_{10} 2 = 6\ \text{dB/octave}.$$

5.2 Reciprocity Theorem

The Wallace solution is a special case of a more general and powerful approach to computing the voltage associated with a moving magnetic medium which

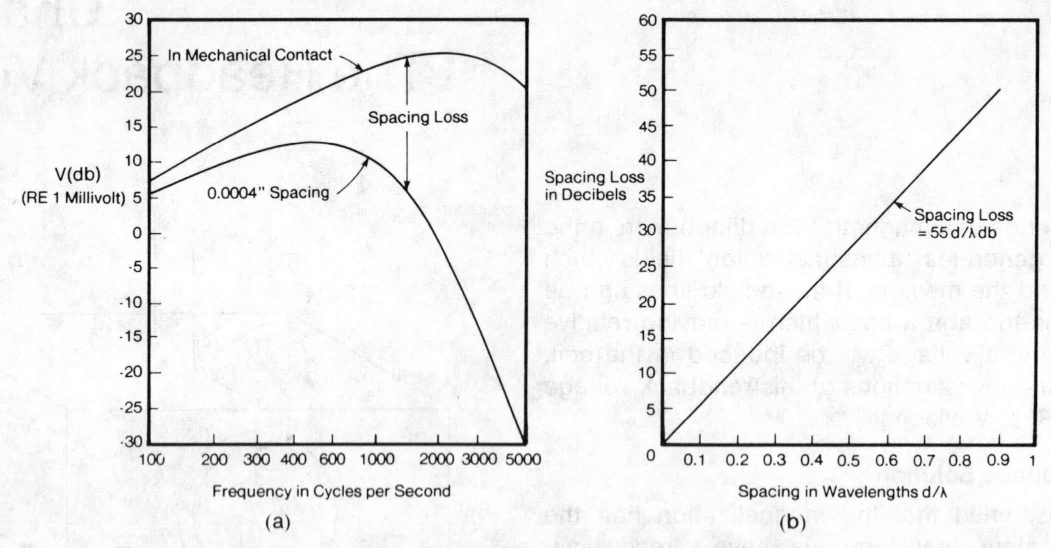

Fig. 5.2: (a) Response curves taken at 21 in/s. (b) Spacing loss as a function of d/λ.

we shall now describe. Consider the two situations illustrated below:

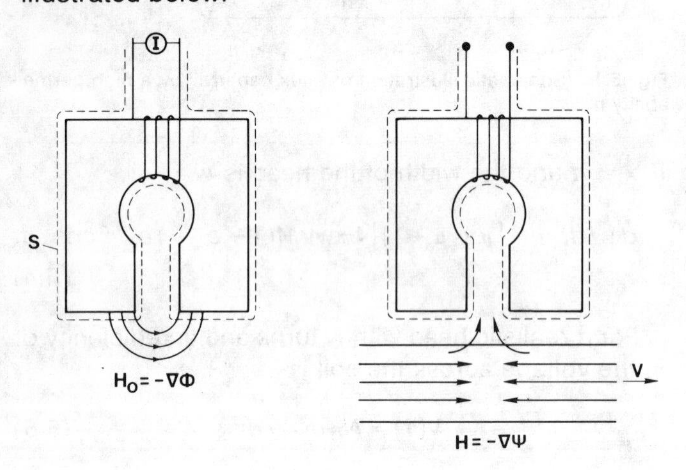

Case (a) represents the head field problem we solved in Chapter 3, the Karlqvist field. In case (b) the high permeability head draws in the fringe fields associated with the recorded medium and introduces a flux in the head. The surface S defines the space exterior to the magnetic heads in both cases. Green's theorem, which we met in Chapter 3, states:

$$\oiint (\varphi(\boldsymbol{r})\,\partial\psi/\partial n(\boldsymbol{r},t)-\psi(\boldsymbol{r},t)\,\partial\varphi(\boldsymbol{r})/\partial n)\,da$$

$$=\iiint[\varphi(\boldsymbol{r})\nabla^2\psi(\boldsymbol{r},t)-\psi(\boldsymbol{r},t)\nabla^2\varphi(\boldsymbol{r})]\,d^3\boldsymbol{r}. \tag{5.8}$$

This is true for any functions φ and ψ and is what enables us to relate the solutions of the two cases illustrated above. The functions we have chosen above have the features that $\psi(\boldsymbol{r},t)=0$ on S while $\varphi(\boldsymbol{r})$ is nonzero due to the magnetomotive force as we discussed in deriving the Karlqvist field. Therefore, the left side of (5.8) becomes

$$\oiint \varphi(\boldsymbol{r})\,\partial\psi(\boldsymbol{r}t)/\partial n\,da=\oiint \varphi(\boldsymbol{r})\boldsymbol{H}(\boldsymbol{r},t)\cdot\hat{\boldsymbol{n}}\,da. \tag{5.9}$$

In the region near the pole tips we take $\varphi(\boldsymbol{r})$ to be $+1/2$ at one tip and $-1/2$ at the other tip. Notice that we are dealing with a *normalized* potential. Thus, $\boldsymbol{H}_0$ will also be *normalized* to $H_g g$. As we move along the surface towards the driving coils these potentials will change. However, we shall assume that when this starts to happen the fields $H(r,t)$ crossing the surface at this point are zero. Then, since $\boldsymbol{H}(\boldsymbol{r},t)=\boldsymbol{B}(\boldsymbol{r},t)$ on this external surface, the integral (5.9) just represents the flux entering the head, Φ.

Now consider the right-hand side of (5.8). Since the region in which $\boldsymbol{H}_0(\boldsymbol{r})$ is defined contains no magnetization,

$$\nabla^2\varphi(\boldsymbol{r})=0 \tag{5.10}$$

while

$$\nabla^2\psi=-4\pi\nabla\cdot\boldsymbol{M}(\boldsymbol{r},t) \tag{5.11}$$

where $\boldsymbol{M}(r,t)$ is the magnetization of the medium.

Integrating this nonzero term by parts,

$$4\pi\iiint\varphi(\boldsymbol{r})\cdot\nabla\cdot\boldsymbol{M}(\boldsymbol{r},t)\,d^3\boldsymbol{r}=4\pi\iiint\boldsymbol{H}_0\cdot\boldsymbol{M}(\boldsymbol{r},t)\,d^3\boldsymbol{r} \tag{5.12}$$

which finally gives the result

$$\Phi=4\pi\iiint\boldsymbol{H}_0(\boldsymbol{r})\cdot\boldsymbol{M}(\boldsymbol{r},t)\,d^3\boldsymbol{r}. \tag{5.13}$$

Thus, the flux is a convolution of the normalized head field function with the magnetization in the medium!

Wallace's solution follows immediately. We write the magnetization in the form

$$\boldsymbol{M}(\boldsymbol{r},t)=M\sin(kx-\omega t)\,\hat{\boldsymbol{x}} \qquad d<y<d+\delta. \tag{5.14}$$

We also recall that the normalized Karlqvist head field function in the limit of a small gap is [see (3.46)]

$$H_0(x,y)_x = (1/\pi)y/(x^2+y^2). \tag{5.15}$$

Since

$$\int_{-\infty}^{\infty} dx\,[\sin(kx-\omega t)]/(x^2+y^2) = (\pi/y)\sin\omega t e^{-ky} \tag{5.16}$$

we then have

$$\Phi = 2wM\lambda e^{-kd}[1-e^{-k\delta}]\sin\omega t. \tag{5.17}$$

Taking the time derivative, and recalling that we are assuming a high permeability head, gives (5.4).

For future purposes it is convenient to express the reciprocity theorem in terms of Fourier components. Writing $vt = x'$ we have

$$\Phi(x) = 4\pi\int H(x')M(x-x')\,dx'. \tag{5.18}$$

The Fourier transform is

$$\Phi(k) = \int_{-\infty}^{\infty}\Phi(x)e^{ikx}dx$$

$$= 4\pi\int\int H(x')M(x-x')e^{ikx}dx\,dx'. \tag{5.19}$$

Inserting $e^{ikx'}e^{-ikx'}$ on the right gives

$$\Phi(k) = 4\pi H(k)M(k). \tag{5.20}$$

$H(k)$ is sometimes referred to as the spectral response of the head.

5.3 Gap Loss

As we continue to decrease the length of the magnetized segments, the associated flux lines will begin to close in the head gap and not "flow" through the head circuit:

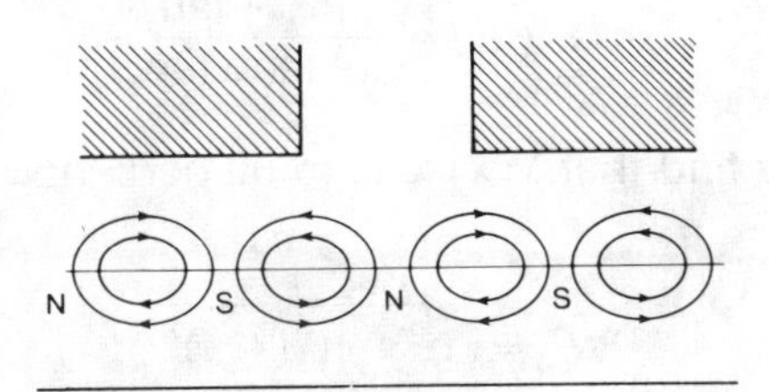

This leads to a loss of signal. To estimate this effect let us consider a sheet of sinusoidal magnetization in contact with the head:

$$M_x(x,y) = M\cos(kx\cdot\omega t)\,\delta(y). \tag{5.21}$$

Using the Karlqvist field in the reciprocity theorem this

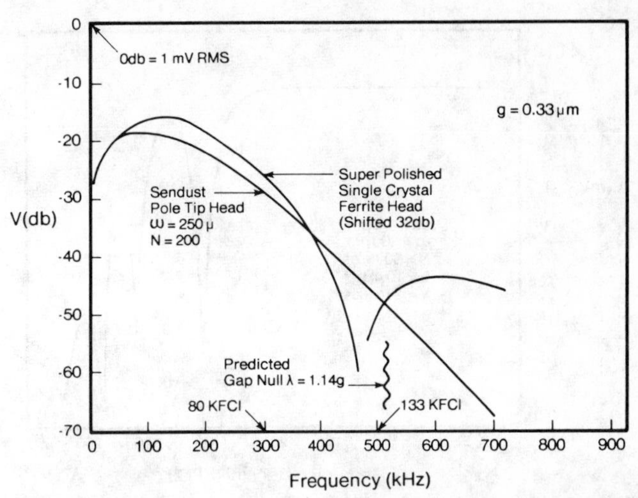

Fig. 5.3: Response curve of two heads showing gap null behavior. The ferrite head curve has been shifted up to compensate for differences in track width, turns, and core efficiency. The data were taken on Isomax tape (H_c = 760 Oe, B_r = 1600 G) at 7.5 in/s.

gives

$$\Phi = w\int_0^{\infty}dy\int_{-\infty}^{\infty}dx\frac{1}{\pi g}\cdot\left[\tan^{-1}\left(\frac{x+g/2}{y}\right)-\tan^{-1}\left(\frac{x-g/2}{y}\right)\right]\cdot M\cos(kx-\omega t)\delta(y)$$

$$= w\int_{-g/2}^{g/2}\frac{M}{g}\cos(kx-\omega t)\,dx$$

$$= Mw\frac{\sin kg/2}{kg/2}\cos\omega t. \tag{5.22}$$

We already obtained this result when we Fourier transformed the Karlqvist head function. The factor $\sin(kg/2)/(kg/2)$ is illustrated in the following figure:

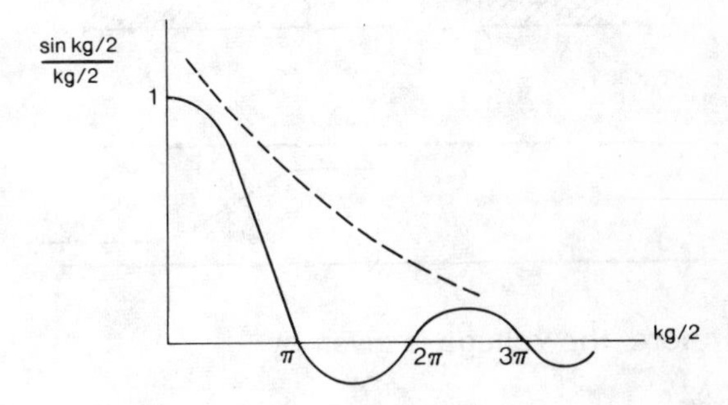

More accurate treatment of the head field function using the conforming mapping techniques mentioned in Chapter 3 shows that the first gap null actually occurs at $kg/2 = 0.88\,\pi$. These "gap resonances" are a familiar feature of spectral response curves which plot the rms signal as a function of frequency. The gap loss envelope decreases as $1/k \sim 1/\omega$. Therefore, the signal decreases 6 dB/octave.

Fig. 5.3 shows this gap null behavior. The Sendust pole-tip head, although having the same nominal gap, does not show a gap null. This is due to the presence

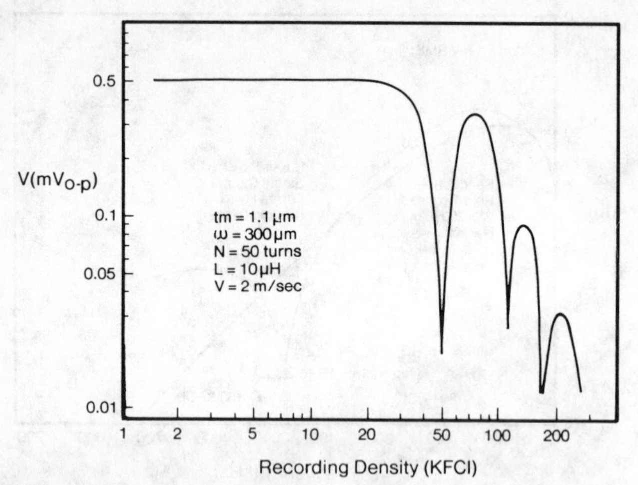

Fig. 5.4: Response curve for a pole-tip head on a vertically oriented flexible disk.

of scratches in the pole faces which give a gap length variation that washes out the gap null.

In Chapter 4 we showed that a pole-tip head is preferable for writing on vertical media because of its large field gradient. However, in readback these heads exhibit gap nulls associated with the width t_m of the main pole tip. Fig. 5.4 shows such gap nulls.

5.4 Pulse Recording

5.4.1 Pulse Width

Let us now consider the readback voltage arising from an arctangent transition. The magnetization $M(r,t)$ entering the reciprocity theorem may be written $M(x-x',y)$ where $x'=vt$.

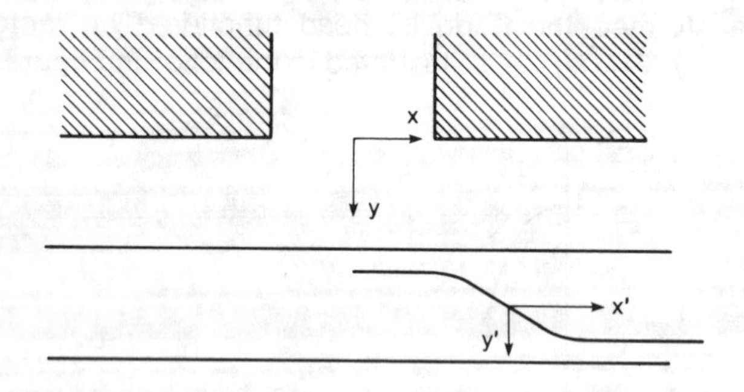

Therefore, the voltage is given by

$$V(x') = 4\pi N\eta vw \int\int \left[\partial M(x-x',y)/\partial x'\right] H(x,y)\, dx\, dy. \tag{5.23}$$

For an arctangent transition

$$M(x-x',y) = -(2M_x/\pi)\tan^{-1}\left[(x-x')/a\right]. \tag{5.24}$$

Taking the derivative and carrying out the integration gives the result

$$V_x(x') = 8N\eta vwM_x\left[f(x')+f(-x')\right] \tag{5.25}$$

where

$$f(x') = \left(\frac{d+a+x}{g}\right)\tan^{-1}\left(\frac{x'+g/2}{d+a+x}\right) - \left(\frac{d+a}{g}\right) \cdot \tan^{-1}\left(\frac{x'+g/2}{d+a}\right) + \frac{1}{2g}(x'+g/2) \cdot \log\frac{(x'+g/2)^2+(d+a+x)^2}{(x'+g/2)^2+(d+a)^2}. \tag{5.26}$$

Since the head field function also has a vertical component, any vertical component of the magnetization will also give rise to a voltage

$$V_y(x') = 8N\eta wvM_y\left[h(x')-h(-x')\right] \tag{5.27}$$

where

$$h(x') = \left(\frac{d+a+\delta}{2g}\right)\log\left[(x'+g/2)^2+(d+a+\delta)^2\right] - \left(\frac{d+a}{2g}\right)\log\left[(x'+g/2)^2+(d+a)^2\right] - \left(\frac{x'+g/2}{g}\right)\left[\tan^{-1}\left(\frac{x'+g/2}{d+a+\delta}\right) - \tan^{-1}\left(\frac{x'+g/2}{d+a}\right)\right]. \tag{5.28}$$

These voltages are sketched in Fig. 5.5. Note that the spacing d and the transition length a always appear additively in this model and therefore their roles cannot be separately distinguished.

An important measure of this readback is the pulse width, PW_{50}. For simplicity let us consider the thin medium limit. If δ is small compared with d or a, then

$$f(x) \to \frac{\delta}{g}\tan^{-1}\left(\frac{x+g/2}{d+a}\right). \tag{5.29}$$

Then using the identify

$$\tan(\alpha+\beta) = \frac{\tan\alpha+\tan\beta}{1-\tan\alpha\tan\beta} \tag{5.30}$$

we readily find that $V(x)$ falls to 50 percent of its peak value at

$$PW_{50} = \sqrt{g^2+4(d+a)^2}.$$

When the medium thickness is not small this should be replaced by

$$PW_{50} = \sqrt{g^2+4(d+a)(d+a+\delta)}. \tag{5.31}$$

This shows the relation between the gap width, the medium thickness, the flying height, and the transition length in determining the pulse width.

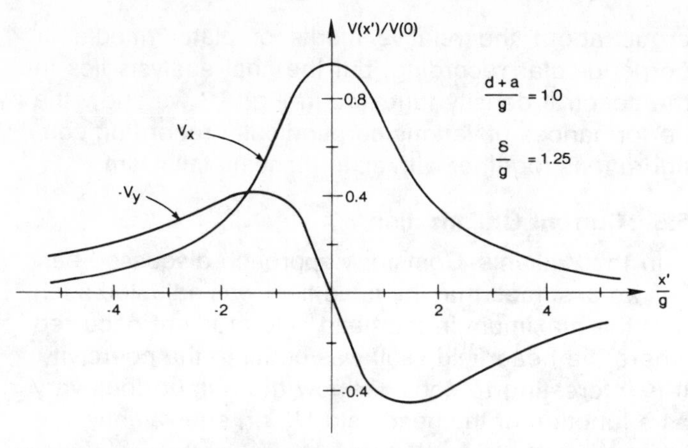

Fig. 5.5: Output voltages arising from horizontal and vertical components of the magnetization.

The role of these characteristic lengths may also be seen by considering the Fourier version of the reciprocity theorem,

$$\Phi(k) = 4\pi H(k) M(k). \tag{5.32}$$

The Fourier transform of the head field function was given in (3.60)

$$H_x(k,y) = \frac{H_g g \sin kg/2}{kg/2} e^{-ky}. \tag{5.33}$$

The Fourier transform of an arctangent magnetization

$$M(x) = (2M/\pi)\tan^{-1}(x/a) \tag{5.34}$$

is not straightforward since the integral of $|\tan^{-1} x/a|$ over the interval $(-\infty,\infty)$ is not convergent. If we write

$$\tan^{-1}(x/a) = \pi/2 - \tan^{-1}(a/x) \tag{5.35}$$

then

$$M(k) = i\pi e^{-|k|a}/k + \pi^2\delta(k). \tag{5.36}$$

Since the voltage is proportional to the *rate of change* of flux this introduces a factor of k which eliminates the $1/k$ divergence in the first term and eliminates the delta function completely.

Since M does not depend upon y (in this approximation), the integral over y converts e^{-ky} into $e^{-kd}(1 - e^{-k\delta})/k$. Thus, $\Phi(k)$ is the product of

writing process loss	$e^{-\lvert k\rvert a}/k$
thickness loss	$(1-e^{-k\delta})/k.$
spacing loss	e^{-kd}
gap loss	$(\sin kg/2)/(kg/2).$

Notice that even if the recorded transition is very sharp, i.e., a very small transition length, the output pulse still has a finite width. In the sketches which follow we illustrate how the readback signal varies with the thickness, δ.

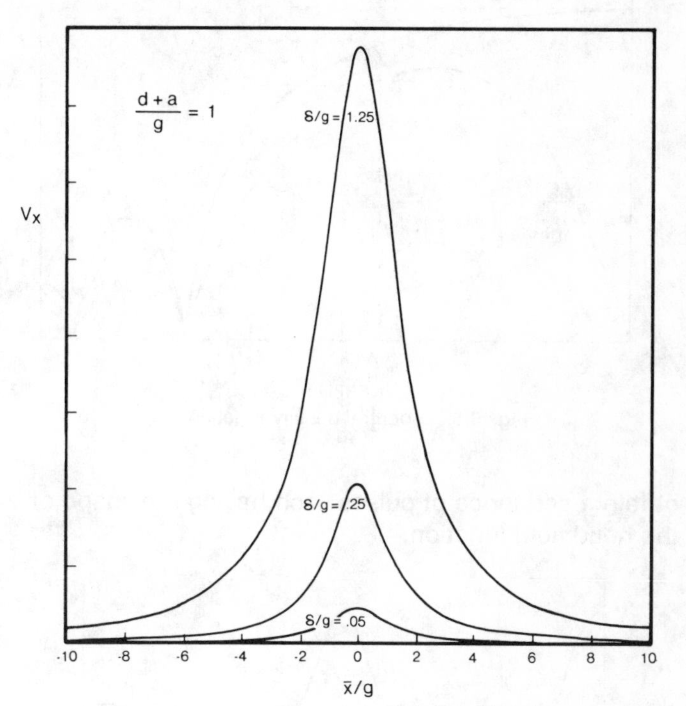

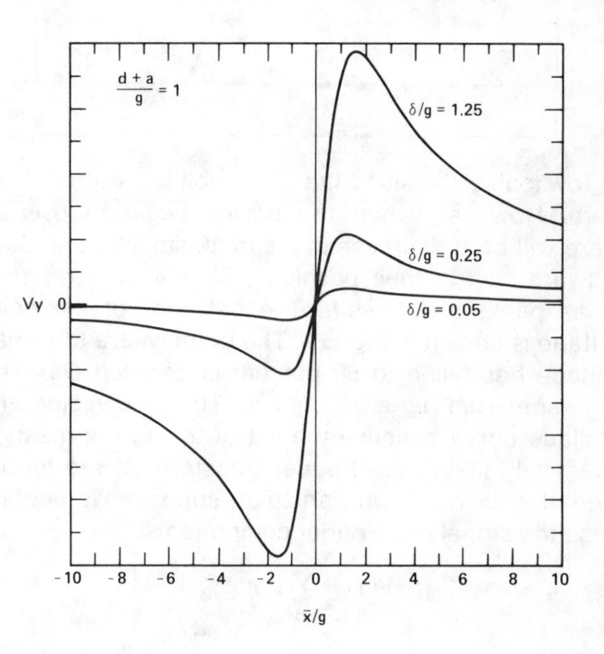

5.4.2 *Intersymbol Interference*

Since the readback process is linear, the output from a sequence of transitions will be a superposition of their individual amplitudes. Two sequential transitions will always have opposite signs. Let us consider the output from an assumed square wave magnetization distribution. This corresponds to a sequence of infinitely sharp transitions. When these are differentiated to obtain the voltage they give a sequence of alternately positive and negative delta functions. When these are convolved with the head field function we

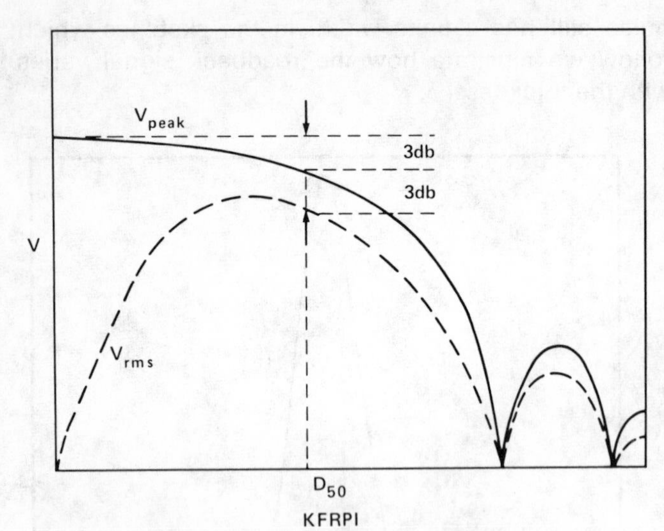

Fig. 5.6: Spectral density function.

obtain a sequence of pulses each having the shape of the head field function:

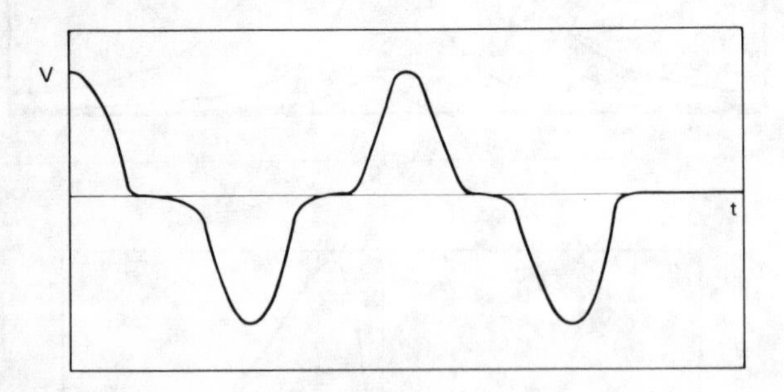

At low pulse densities the *peak* voltage will be constant. However, when the pulses begin to overlap there will be a decrease in the peak amplitude and an outward shift in their positions. This latter is referred to as peak, or bit, shift. The behavior of the peak voltage is shown in Fig. 5.6. The point where this peak voltage has fallen to 50 percent is denoted D_{50}. The rms spectrum is also shown. This resembles the Wallace curve because the act of taking the mean is effectively taking the Fourier transform at the fundamental wavelength. In particular, suppose we decompose the signal into Fourier components,

$$V(t)=\sum_n V_n e^{in\omega t}. \tag{5.37}$$

Then

$$\overline{V(t)^2}=\sum_n V_n \frac{1}{T}\int_0^T V(t)\, e^{in\omega t}\, dt \tag{5.38}$$

where $T=2\pi/\omega$. Since the largest of the coefficients will be that associated with the fundamental, i.e., $n=1$, then

$$\overline{V(t)^2}\simeq V_1^2. \tag{5.39}$$

The "proof of the pudding" in playback amplitude is in the spectral density function, as it were. One can argue about the relative merits of plated media or perpendicular recording, but the final analysis lies in the spectral density function. In Fig. 5.7 we show the performances of various perpendicular recording configurations together with an in-plane metallic film.

5.5 Current Optimization

In the Williams–Comstock approach discussed earlier we assumed that the head field was adjusted such that the maximum in the head field gradient occurred where the head field itself was equal to the coercivity. It is interesting to consider how the output does vary as a function of the head field H_g, or, equivalently, the input current.

The longitudinal head field function is

$$H_{hx}=\frac{H_g}{\pi}\left[\tan^{-1}\left(\frac{x+g/2}{y}\right)-\tan^{-1}\left(\frac{x-g/2}{y}\right)\right]. \tag{5.40}$$

The center of the transition occurs when $H_h=-H_l=-H_{cr}$ as discussed in the Williams–Comstock model. For a reasonably square loop $H_{cr}\simeq H_c$. The gradient of the head field function is

$$\frac{dH_{hx}}{dx}=\frac{H_g}{\pi y}\left[\frac{y^2}{(x+g/2)^2+y^2}-\frac{y^2}{(x-g/2)^2+y^2}\right]. \tag{5.41}$$

If $H_h=-H_c$ then one of the coordinates may be eliminated. If we eliminate x then

$$\frac{dH_{hx}}{dx}=\frac{H_g}{\pi y}\sin^2\left(\frac{\pi H_c}{H_g}\right). \tag{5.42}$$

H_g must be at least as large as H_c, and probably more like $2H_c$. As H_g/H_c increases beyond 2 the head field gradient goes through a maximum near 2.7. The center of the transition also moves further from the gap edge as illustrated in Fig. 5.8.

If we again take an arctangent transition,

$$M=(2M_r/\pi)\tan(x'/a) \tag{5.43}$$

then at the center of the transition

$$dM/dx'|_{x'=0}=2M_r/\pi a. \tag{5.44}$$

The demagnetization field is given by (4.19). The gradient of this field at the top surface of the medium ($y=\delta/2$) and at the center of the transition is

$$\left.\frac{dH_d}{dx'}\right|_{\substack{x'=0\\ y=\delta/2}}=\frac{4M_r\delta}{a(a+\delta)}. \tag{5.45}$$

Using the Williams–Comstock squareness parameter S^*,

$$\frac{dM}{dH}=\frac{M_r}{H_c(1-S^*)}. \tag{5.46}$$

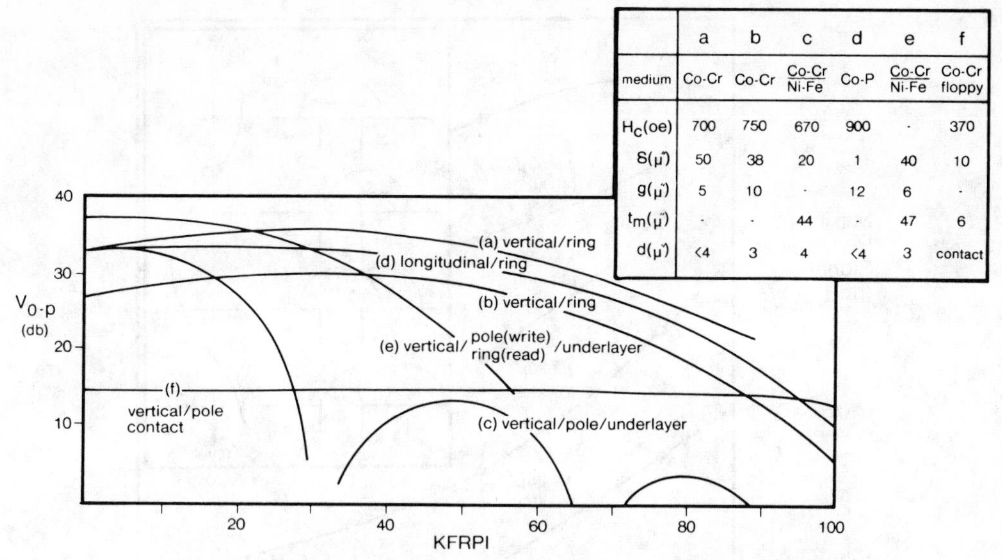

Fig. 5.7: Spectral density curves for the media indicated in the insert. All the data have been referenced to 0 dB = 1 nv/turn/mil/in/s. Curve (a) is from Hitachi (*J. Appl. Phys.*, vol. 53, p. 2588, 1982); curves (b), (c), and (e) are from Fujitsu (*Proc. 1982 Sendai Conf. Perp. Magn. Rec.*, p. 177, 1982); curve (d) is from Ampex (*IEEE Trans. Magn.*, vol. MAG-19, p. 1605, 1983); curve (f) is from Iwasaki's group (*IEEE Trans. Magn.*, vol. MAG-20, p. 105, 1984).

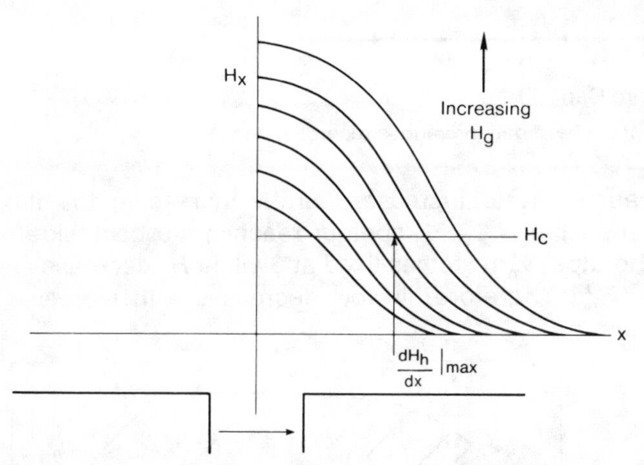

Fig. 5.8: Longitudinal field intensity at some point above the gap as a function of the deep-gap field.

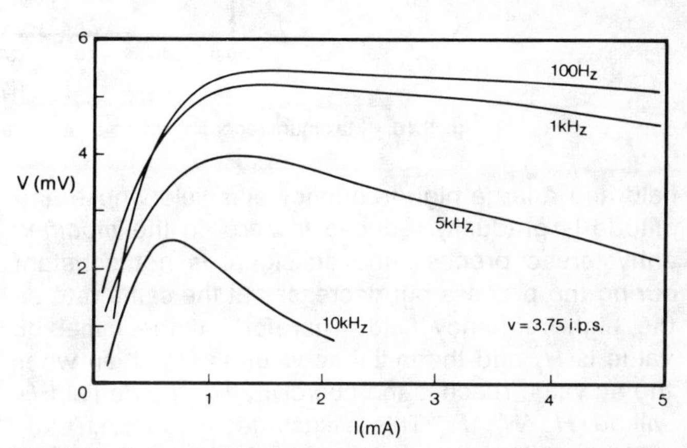

Fig. 5.9: Output pulse amplitude as a function of record current for different record frequencies. Data [2] were taken on a tape system with $M_r = 90$ G, $H_c = 250$ Oe, $\delta = 10\ \mu$m.

Substituting these expressions into the slope criterion,

$$\frac{dM}{dx} = \frac{dM}{dH}\left(\frac{dH_h}{dx} + \frac{dH_d}{dx}\right) \tag{5.47}$$

leads to the expression for the transition length

$$a_1 = \frac{2d(1-S^*)}{\dfrac{H_g}{H_c}\sin^2\left(\dfrac{\pi H_c}{H_g}\right)}. \tag{5.48}$$

The reciprocal dependence on the head field gradient means that the transition length will have a minimum when this gradient is a maximum. A sharper transition means a larger output voltage. Fig. 5.9 shows this output maximum as a function of the input current [2].

Another mechanism that can lead to a maximum in the output signal as a function of the recording current is the so-called "snowshoe" effect. This corresponds to an overwriting of the just-written medium. This is illustrated in the insert of Fig. 5.10. As this figure also shows, the maximum signal can be increased by going to a smaller gap.

5.6 AC Bias Recording [3]

When one is recording analog information, such as an audio signal, it is important to have a medium that responds linearly to the record signal. This is accomplished in tape recording by a process known as anhysteresis.

The anhysteretic magnetization is the magnetization developed by the simultaneous application of a dc

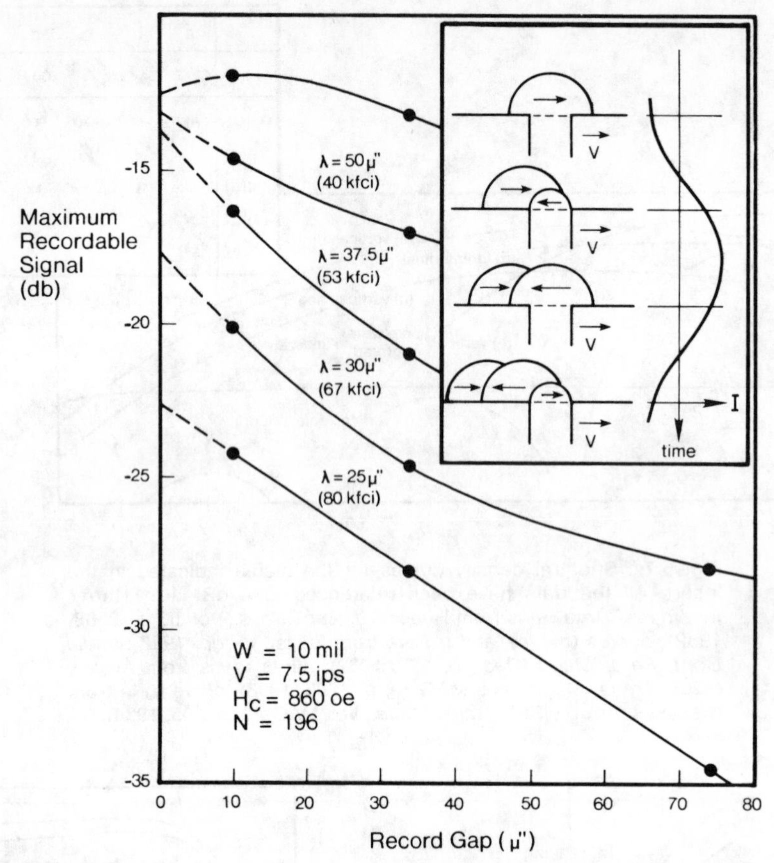

Fig. 5.10: Maximum recording signal as a function of the head gap for various recording densities.

field and a large high-frequency bias field whose amplitude is gradually reduced to zero. In the *modified* anhysteretic process, the dc signal is not constant during the process but decreases at the same rate as the high-frequency field. Therefore, if the initial dc value is H_s^0 and the initial ac value is H_{ac}^0, then when the ac value reaches the coercivity H_c, the dc field H_s will be $(H_c/H_{ac}^0)H_s^0$. This is illustrated in Fig. 5.11(a).

The initial ac field must exceed the coercivity. The resulting anhysteretic magnetization is single-valued and remains linear up to about 40 percent of the saturation value. If the value of the signal field is H_s, then

$$M = \chi H_s \tag{5.49}$$

where χ is the anhysteretic susceptibility. Let y_m be the depth into the tape where $H_{ac} = H_c$. At y_m, H_x has the value H_s. The flux per unit track width in the tape is then given by

$$\Phi = 4\pi H_s y_m \chi. \tag{5.50}$$

From the Karlqvist field expression

$$y_m = (g/2)\cot[(\pi H_c)/(2H_{ac}^g)] \tag{5.51}$$

where H_{ac}^g is the deep gap bias field. Suppose we increase the *bias* current. Then H_{ac}^g increases, which causes y_m to increase, thereby increasing the flux. When the $H_{ac} = H_c$ contour reaches the back side of the tape, y_m becomes fixed at δ while H_s decreases as $1/H_{ac}^0$. Therefore, Φ now decreases with increasing bias:

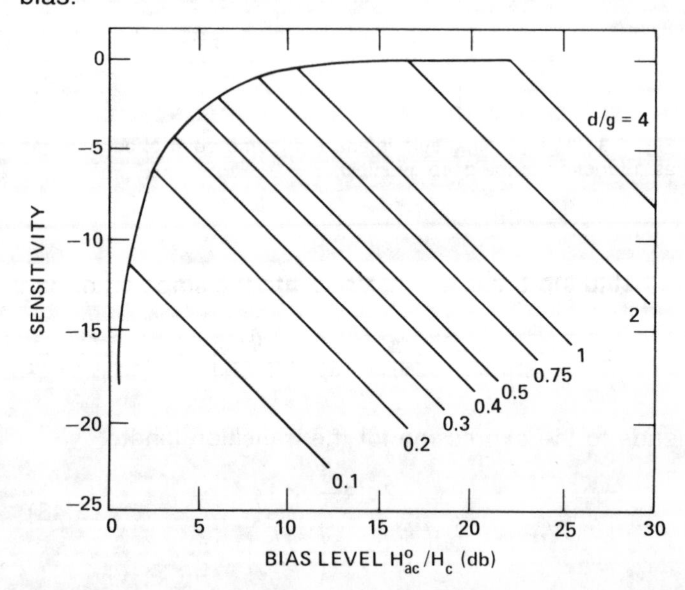

Experimentally, the overbias behavior is much stronger than that shown above. Bertram has shown that better agreement is obtained if one includes the vertical component of the recording field.

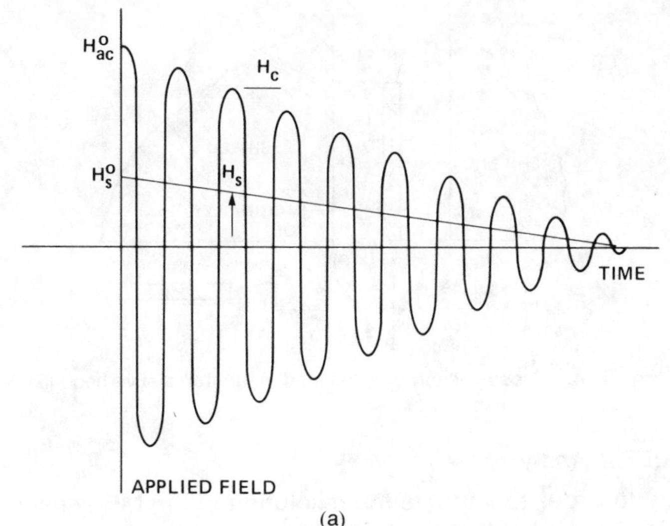

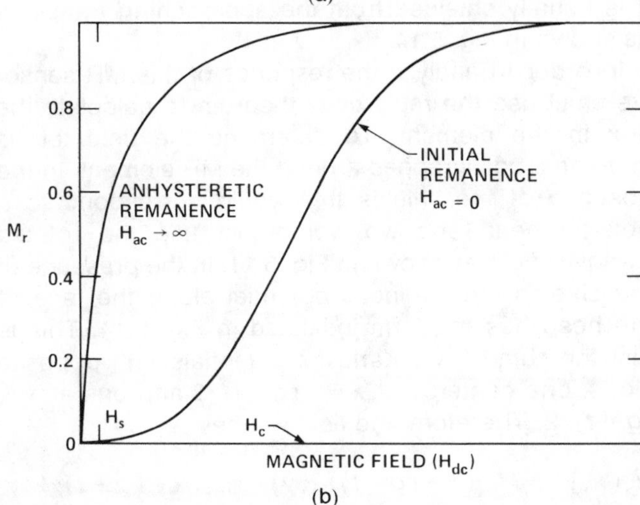

Fig. 5.11: (a) Time dependence of the total anhysterizing field, $H_{dc} + H_{ac}$. (b) Anhysteretic remanent magnetization and initial remanent magnetization as a function of dc field.

5.7 Magnetoresistive Readback

The ring and single-pole heads we have been discussing all rely on Faraday's law: the output voltage is proportional to rate of change of magnetic flux linking the head. As the recording media and track widths become smaller, and as we move to smaller diameter disks, these induction signals become weaker. Although we might consider increasing the relative velocity between head and medium, this is not practical at high bit densities. A more promising approach for detection lies in the magnetoresistive effect.

Magnetoresistive sensors were first suggested for magnetic recording readout by Hunt [4]. Shortly thereafter they were employed in magnetic bubble sensing.

5.7.1 Magnetoresistivity

Generally, when we speak of magnetoresistance we mean the increase in the resistance of a metal or semiconductor when placed in a magnetic field. In the "standard" geometry, the field is transverse to the current and $\Delta\rho/\rho \sim H^2$.

In a magnetic material this magnetoresistance is *anisotropic*, depending upon the direction of M. The origin of this anisotropy lies in spin-dependent scattering [5].

To see how the resistance depends upon an applied field, let us consider a film of permalloy which has been deposited in a field so as to give it an easy axis:

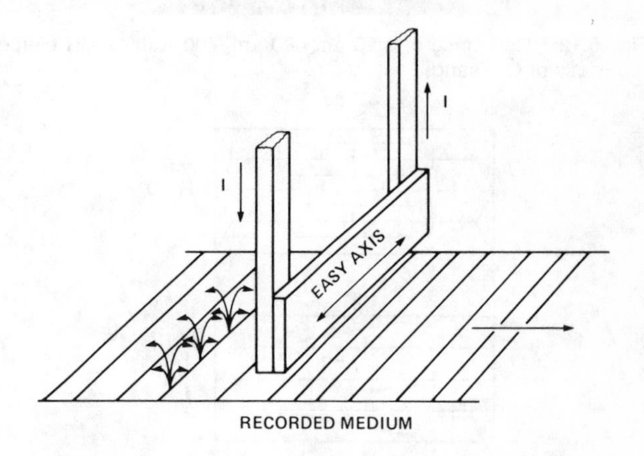

The anisotropy energy density may then be written

$$E_A = K(\sin^2\theta) \tag{5.52}$$

where θ is the angle the magnetization makes with this easy axis. If an external field H_0 is applied perpendicular to this axis then the Zeeman energy is

$$E_z = -M_s H_0 \sin\theta. \tag{5.53}$$

Minimizing the total energy,

$$\sin\theta = H_0/(2K/M_s).$$

Suppose current now flows along the easy axis. This may be resolved into components parallel and perpendicular to the magnetization:

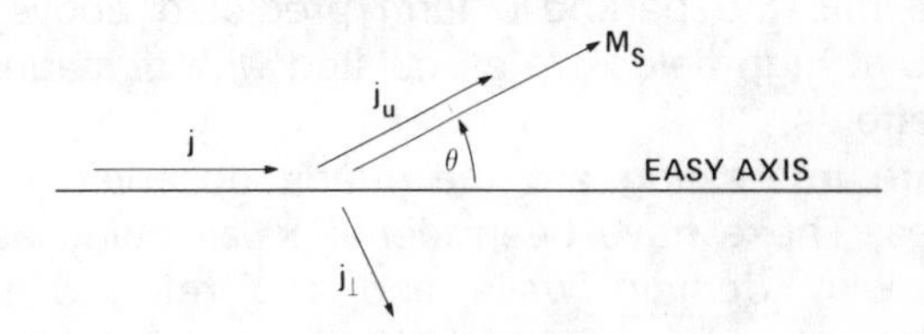

Since the resistivities associated with these directions are different, the electric field components are

$$\mathscr{E}_{\parallel} = \rho_{\parallel} j_{\parallel}$$

$$\mathscr{E}_{\perp} = \rho_{\perp} j_{\perp}. \tag{5.54}$$

Therefore, the electric field that is measured is

$$\mathscr{E}_z = \mathscr{E}_{\parallel}\cos\theta + \mathscr{E}_{\perp}\sin\theta$$

$$= \rho_{\parallel} j\cos^2\theta + \rho_{\perp} j\sin^2\theta \tag{5.55}$$

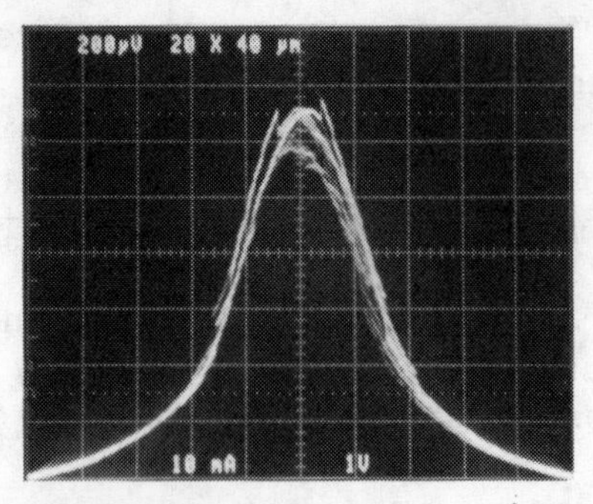

Fig. 5.12: Response of a 20 μm$\times$80 μm, 400 Å thick MR element (courtesy of C. Tsang).

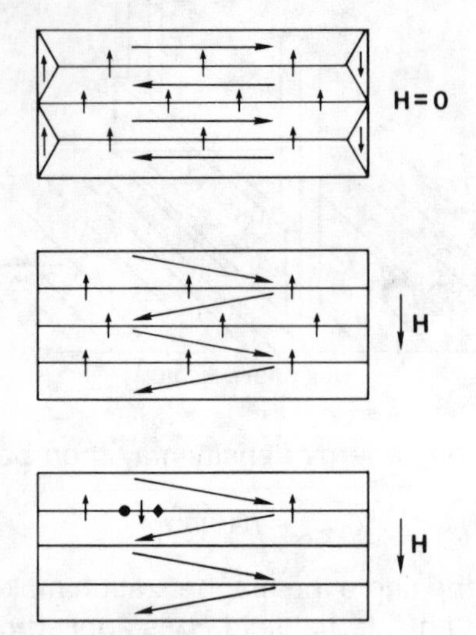

Fig. 5.13: Sequence of domain patterns as a small MR element is magnetized perpendicular to its easy axis.

and the resistivity becomes

$$\rho = \rho_0 + \Delta\rho_{max} - \Delta\rho_{max}\left(\frac{H_0}{2K/M_s}\right)^2 \qquad (5.56)$$

where $\rho_0 = \rho_{\parallel}$ and $\Delta\rho_{max} = \rho_{\perp} - \rho_{\parallel}$.

Fig. 5.12 shows the actual resistance of a permalloy film as a function of field. In the low field region the curve has the parabolic form predicted above. The wings at high fields are associated with demagnetization effects.

More interesting are the jumps observed on the curves. These have been identified as being associated with domain walls and are referred to as Barkhausen jumps. Figure 5.13 shows the magnetization at zero field *after having been magnetized in the "up" direction.*

There are domains separated by Néel walls which have an "up" sense. As the field is increased in the "down" direction these "N^+" walls eventually transform into segments of "N^-" walls bounded by a Block point and a crosstie. These wall singularities then propagate away from one another converting the entire wall into N^-. The noise associated with these wall processes is eliminated by biasing the MR sensor with a longitudinal field sufficient to remove the domain structure [6].

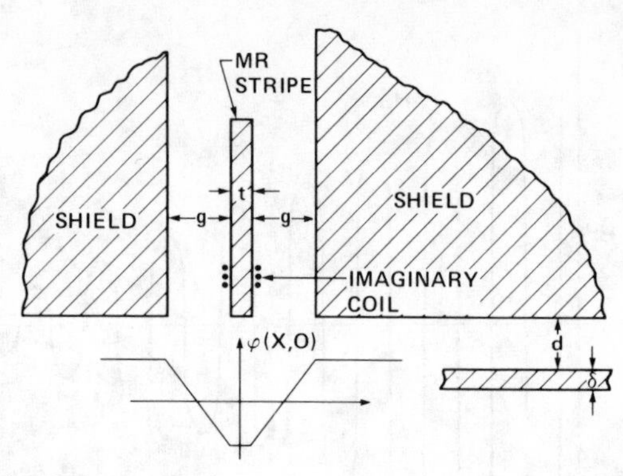

Fig. 5.14: Cross section of a shielded magnetoresistive head [7].

5.7.2 Analysis

In order to improve the resolution of the MR sensor, it is usually shielded from the approaching transition as shown in Fig. 5.14.

In order to analyze the response of this MR sensor we shall use the reciprocity theorem to calculate the flux in the element. To determine the field let us imagine a coil wrapped around the MR element. In the absence of the shields this would correspond to a pole-tip head and we would just use the rotated Karlqvist field as shown in Fig. 5.11. In the presence of the shields the magnetic potential along the face of the head has the form indicated in Fig. 5.14. This is just the sum of two Karlqvist potentials with opposite signs, one centered at $x = -(g+t)/2$ and one at $x = (g+t)/2$. Therefore, the field will be

$$H_x(x,y) = H_{hx}(x+(g+t)/2, y) - H_{hx}(x-(g+t)/2, y). \qquad (5.57)$$

When this result is used in the reciprocity relation the voltage will be the difference of two Wallace-like expressions. Since the voltage is the time rate of change of the flux, we can integrate the voltage to obtain the flux,

$$\Phi(\bar{x}) = (1/v)\int_{-\infty}^{\bar{x}} \left[V(x'+(g+t)/2) - V(x'-(g+t)/2)\right] dx' \qquad (5.58)$$

where $x = vt$.

In practice, an MR sensor is also biased with a transverse field H_0 such that $\theta = \pi/4$ where the response is linear. Thus

$$[H_0/(2K/M_s)]^2 \to [H_0/(2K/M_s)]^2 + (2H_0\Delta H)/(2K/M_s)^2 = 1/2 + \Delta H/H_0. \qquad (5.59)$$

Since H_0 induces the transverse magnetization $M_y = M_s/\sqrt{2}$, and since $M_y \gg H_0$,

$$\Delta H/H_0 = \left[\sqrt{2}\,\Phi(\bar{x})\right]/(4\pi M_s tw). \qquad (5.60)$$

Therefore

$$\rho = \rho_0 + (1/2)\Delta\rho_{max} - \sqrt{2}\,\Delta\rho_{max}(\Phi(\bar{x})/(4\pi M_s tw). \qquad (5.61)$$

Here $\Phi(\bar{x})$ is the flux *entering* the sensor and ρ is the resistivity along this face. As one moves up the sensor element the flux leaks out of the element into the shields. If the height of the element is of the order of this characteristic "decay" length then the average flux is $\Phi/2$. The voltage across the element is then

$$V = -[j\Delta\rho_{max}\Phi(\bar{x})]/[\sqrt{2}\,4\pi M_s t]. \qquad (5.62)$$

Typical numbers for permalloy give a peak-to-peak amplitude of the order of 90 μV/μm track width. This is about ten times that of an inductive head and is the reason that such sensors are being developed for recording.

Problems

1) A current distribution $\boldsymbol{J}(x)$ exists in a medium of unit permeability adjacent to a semiinfinite slab of material having permeability μ and filling the half-space, $z > 0$.

a) Show that for $z < 0$ the magnetic induction can be calculated by replacing the medium of permeability μ by an image current distribution $\boldsymbol{J}^*$ with components:

$$\left(\frac{\mu-1}{\mu+1}\right)J_x(x,y,-z), \left(\frac{\mu-1}{\mu+1}\right)J_y(x,y,-z),$$
$$-\left(\frac{\mu-1}{\mu+1}\right)J_z(x,y,-z).$$

b) Show that for $z > 0$ the magnetic induction appears to be due to a current distribution

$$\left(\frac{2\mu}{\mu+1}\right)\boldsymbol{J}$$

in a medium of unit permeability.

2) Use the reciprocity theorem to calculate the voltage from an infinitely sharp vertical transition using a ring head.

3) Within the context of the reciprocity theorem, discuss the implications of a head field function such as that shown in Fig. 3.13.

4) Calculate the voltage from an unshielded MR head.

References

[1] R. L. Wallace, *Bell Syst. Tech. J.*, vol. 30, p. 1145, 1957.
[2] B. K. Middleton and P. L. Wisely, in *IEEE Conf. Proc.*, no. 35, 1976, p. 35.
[3] H. N. Bertram, *IEEE Trans. Magn.*, vol. MAG-10, p. 1039, 1974.
[4] R. P. Hunt, *IEEE Trans. Magn.*, vol. MAG-7, p. 150, 1971.
[5] T. R. McGuire and R. I. Potter, *IEEE Trans. Magn.*, vol. MAG-11, p. 1051, 1975.
[6] C. Tsang, *J. Appl. Phys.*, vol. 55, p. 2226, 1984.
[7] After R. Potter, *IEEE Trans. Magn.*, vol. MAG-10, p. 502, 1974.

Chapter 6
The Air Bearing

6.1 Introduction

OUR previous discussions have emphasized the importance of the head-to-medium spacing. In tape recorders, the tape runs in contact with the head, which is contoured to enhance this effect. Tape velocities range from 20 in/s for cartridges, to 200 in/s for large tape drives.

The standard flexible disk drive also employs contact recording. The head is in the shape of a button, and pushes the medium onto a pressure pad on the back side. Again, typical linear velocities are in the range of 100 in/s.

The time it takes a disk to make half a revolution is the "latency," which is an important component of the access time. To achieve shorter access time, one would like to rotate the disk at a higher rotational speed. However, at higher speeds, wear, both of heads and media, becomes more severe. The answer is to operate the head slightly out of contact with the disk. This is not as easy as it appears, for spinning disks have a certain amount of runout, or wobble. The head must follow this runout, or the resulting variations in spacing will produce large variations in the signal. The solution, which was, in fact, employed in the first disk drive, the RAMAC, is to use an air bearing.

In an air bearing, a pressure between the head and the medium maintains a constant separation against a load applied to the head. In the RAMAC, air was forced into the interface region. As Fig. 1.2 shows, hydrodynamic, or self-acting, air bearings were soon adopted. In such bearings, the air flow associated with the relative velocity between the two surfaces is sufficient to produce the bearing action. A good deal of research on air bearings was carried out at IBM and culminated in the excellent book by Gross [1]. In the Winchester-type drives, the head loading is small enough that they start and stop in contact with the disk. Previous designs required that the heads be retracted from the disk until the operating velocity was established.

We often speak of heads as "flying" above the disk. However, the physics of the air bearing is very different from that governing the performance of a 747. Most papers in air bearings today begin with Reynolds equation. In particular, most of the efforts today involve improved solutions to more complex geometries. In keeping with our emphasis on the physics, we shall look into the derivation of the Reynolds equation in order to understand its limitations and discuss a few simple examples.

6.2 Definitions

Let us begin by introducing a few definitions and concepts. Imagine a differential volume element of the material. When this material is deformed, this volume element is distorted:

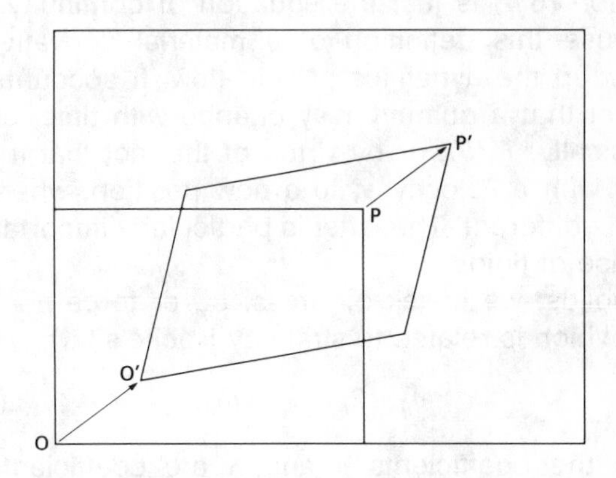

The point P moves to P', the *difference* being described by the functions (ξ, η, ζ). *Strain* is defined by the derivatives

$$\epsilon_{xx} = \frac{\delta \xi}{\delta x} \tag{6.1}$$

$$\epsilon_{xy} = \epsilon_{yx} = 1/2\left(\frac{\partial \xi}{\partial y} + \frac{\partial \eta}{\partial x}\right) \text{ etc.}$$

Notice that the strain is a tensor.

The fractional change in volume of this differential unit is called the dilation Θ, and can be shown to be equal to the trace of the strain tensor

$$\Theta = \frac{\Delta V' - \Delta V}{\Delta V} = \epsilon_{xx} + \epsilon_{yy} + \epsilon_{zz}. \tag{6.2}$$

An incompressible fluid is one for which $\Theta = 0$. Notice that since the velocity of the point P is

$$v_x = \frac{\partial \xi}{\partial t} \text{ etc.}$$

then $\nabla \cdot \boldsymbol{v} = \partial \Theta / \partial t$. Thus, $\nabla \cdot \boldsymbol{v} = 0$ for an incompressible fluid. If the fluid is also homogeneous, then the density will be constant in time and space.

Even if a fluid is compressible, mass must still be conserved. Thus, the mass moving across a surface S in time Δt,

$$\int \rho \boldsymbol{v} \cdot d\boldsymbol{S} \Delta t$$

which, for a differential element, is $\nabla \cdot (\rho \boldsymbol{v}) \Delta V \Delta t$, and must equal the change of mass inside,

$$-\frac{\partial \rho \Delta V}{\partial t}\Delta t.$$

Therefore,

$$\frac{\partial \rho}{\partial t}+\nabla\cdot(\rho \boldsymbol{v})=0. \tag{6.3}$$

Expanding the divergence enables us to write

$$\frac{d\rho}{dt}=-\rho\nabla\cdot\boldsymbol{v} \tag{6.4}$$

where the total, or "material" derivative is defined by

$$\frac{d}{dt}=\frac{\partial}{\partial t}+\boldsymbol{v}\cdot\nabla. \tag{6.5}$$

Equation (6.4) is just the equation of continuity. We shall use this definition of a material derivative in setting up the dynamics of fluid flow. It accounts for the fact that a quantity may change with time, either intrinsically, $\partial/\partial t$, or by virtue of the fact that it may move, with a velocity $\boldsymbol{v}$, to a new position, where its value is different. The latter is particularly important in the case of fluids.

In solids, we speak of stress, σ_{ij}, or force per unit area, which is related to strain by Hooke's law:

$$\sigma_{ij}=2\mu\epsilon_{ij}+\lambda\Theta\delta_{ij} \tag{6.6}$$

where the coefficients μ and λ are coefficients of elasticity (or Lamé's moduli). For a fluid, one talks of the *pressure* p_{ij}, which is a negative stress, but has the same dimensions of the force per unit area.

If we imagine a fluid in which the velocity, say v_x, increases in the z direction,

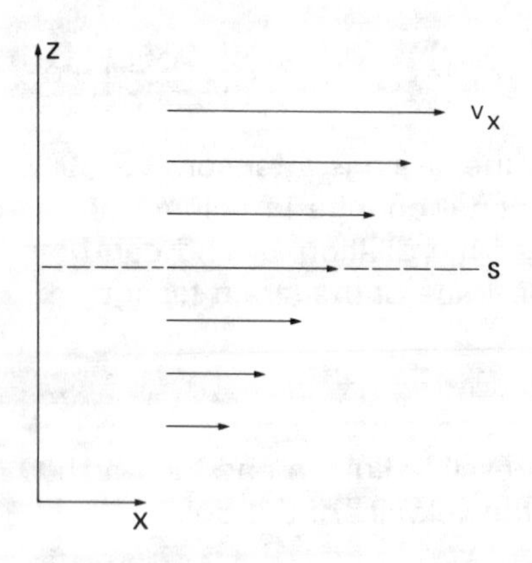

there will be a shearing force on any surface S. That is, the medium above S will feel a drag from the medium below it, and vice versa. Newton suggested that this *viscous* pressure be proportional to the rate of change of the velocity, i.e.,

$$p_{zx}=-\mu\frac{\partial v_x}{\partial z}.$$

From our definition of strain, this may also be written

$$p_{ij}=-2\mu\dot{\epsilon}_{ij}$$

where the dot denotes $\partial/\partial t$. From this stress–strain relation for a solid, (6.6), this may be generalized to a compressible fluid, according to

$$p_{ij}=-2\mu\dot{\epsilon}_{ij}-\lambda\dot{\Theta}\delta_{ij} \tag{6.7}$$

where μ and λ are now coefficients of viscosity. The coefficient μ is sometimes referred to as the shear or laminar viscosity. The bulk viscosity λ is generally set equal to $-2\mu/3$ on the basis of a gas-kinetic argument, which is only valid for monatomic gases. In what follows, we shall adopt this approximation, although we are not aware of any justification of its application to the diatomic constituents of air. As long as $\nabla\cdot\boldsymbol{v}$ is small this is not serious. In addition to this viscous pressure, there will be the usual scalar pressure p associated with the external force when the fluid is at rest.

6.3 The Navier–Stokes Equation

We now have the definitions and concepts to write the equation of motion for a fluid. Let us begin by considering a static situation. Under the influence of an external force (per unit volume), the equilibrium condition for an elastic solid in terms of its stresses is

$$\boldsymbol{F}=\mathrm{Div}\,\sigma$$

where the use of a capital "D" in Div stands for "vector" divergence, which converts a tensor to a vector according to

$$F_x=\frac{\partial \sigma_{xx}}{\partial x}-\frac{\partial \sigma_{yx}}{\partial y}-\frac{\partial \sigma_{zx}}{\partial z}.$$

The analogous relation for a fluid is

$$\boldsymbol{F}=\mathrm{Div}\,p.$$

Using Newton's assumption, (6.7), the x component, for example, becomes

$$F_x=\frac{\partial p_{xx}}{\partial x}+\frac{\partial p_{yx}}{\partial y}+\frac{\partial p_{zx}}{\partial z}+\frac{\partial p}{\partial x}$$
$$=\mu\nabla^2 v_x-(\mu+\lambda)\frac{\partial}{\partial x}(\nabla\cdot\boldsymbol{v})+\frac{\partial p}{\partial x}.$$

To include dynamical effects, the inertial force term, $-\rho\,d\boldsymbol{v}/dt$, must be added to $\boldsymbol{F}$. Here, d/dt is the material derivative we introduced above. The result is the Navier–Stokes equation,

$$\rho\frac{\partial \boldsymbol{v}}{\partial t}+\rho(\boldsymbol{v}\cdot\nabla)\boldsymbol{v}-\mu\nabla^2\boldsymbol{v}-(\mu+\lambda)\nabla(\nabla\cdot\boldsymbol{v})+\nabla\rho=\boldsymbol{F}. \tag{6.8}$$

6.4 Lubricating Film

When applied to the case of a self-acting bearing, the Navier–Stokes equation simplifies a great deal. Let us consider the following slider bearing geometry:

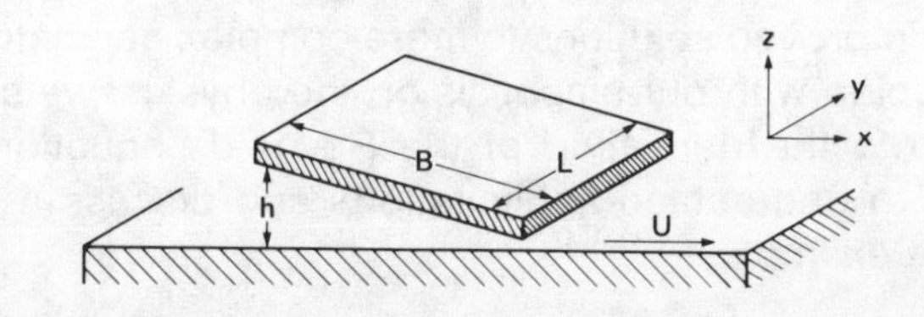

The lower surface is moving with a velocity U. If the mean free path λ of the molecules of the fluid is short compared with the length h, then the fluid next to the moving surface will be in equilibrium with this surface, and will be dragged along with the same velocity. This ratio, λ/h, is called the Knudsen number K. When $0.1<K<10$, the fluid exhibits "slip," i.e., the fluid velocity at the moving surface begins to slip below that of the surface. For air molecules at room temperature and atmospheric pressure, $\lambda \sim 2.5\ \mu$in. Thus, the air bearings in present-day Winchester drives are already entering the slip regime. As we move away from this surface, the velocity decreases, eventually becoming zero at the surface of the slider. Our objective is to determine the velocity profiles and the resulting pressure.

To simplify the Navier–Stokes equation, we shall use a scaling argument. The coordinates scale as the characteristic dimensions:

$$x=0(B), \qquad y=0(L), \qquad z=0(h).$$

The horizontal velocities scale as U, i.e.,

$$v_x=0(U), \qquad v_y=0(U).$$

The characteristic time is B/U. Therefore, if the continuity equation (6.4) is to remain invariant under these scale changes, we must have

$$\frac{\partial v_z}{\partial z}=0\left(\frac{U}{B}\right), \qquad v_z=0\left(\frac{hU}{B}\right).$$

These scaling relations are now used in the Navier–Stokes equation. The x-component equation, with the order of magnitude of each term indicated below that term, is

$$\begin{array}{l}
\dfrac{\partial v_x}{\partial t}+\left(v_x\dfrac{\partial}{\partial x}+v_y\dfrac{\partial}{\partial y}+v_z\dfrac{\partial}{\partial z}\right)v_x \\
0\left(\dfrac{U^2}{B}\right) \quad 0\left(\dfrac{U^2}{B}\right) \quad 0\left(\dfrac{U^2}{B}\right) \quad 0\left(\dfrac{U^2}{B}\right) \\
-\dfrac{\mu}{\rho}\left(\dfrac{\partial^2 v_x}{\partial x^2}+\dfrac{\partial^2 v_x}{\partial y^2}+\dfrac{\partial^2 v_x}{\partial z^2}\right) \\
\dfrac{\mu}{\rho}0\left(\dfrac{U}{B^2}\right) \quad \dfrac{\mu}{\rho}0\left(\dfrac{U}{B^2}\right) \quad \dfrac{\mu}{\rho}0\left(\dfrac{U}{h^2}\right) \\
-\dfrac{\mu}{3\rho}\dfrac{\partial}{\partial x}\left(\dfrac{\partial v_x}{\partial x}+\dfrac{\partial v_y}{\partial y}+\dfrac{\partial v_z}{\partial z}\right)+\dfrac{1}{\rho}\dfrac{\partial p}{\partial x}=0 \\
\dfrac{\mu}{\rho}0\left(\dfrac{U}{B^2}\right) \quad \dfrac{\mu}{\rho}0\left(\dfrac{U}{B^2}\right) \quad \dfrac{\mu}{\rho}0\left(\dfrac{U}{B^2}\right).
\end{array}$$

The ratio of the inertial term, $v_x\partial v_x/\partial x$, to the viscous term $(\mu/\rho)\partial^2 v_x/\partial z^2$ is

$$\frac{\text{inertial}}{\text{viscous}}=0\left(\frac{\rho BU}{\mu}\frac{h^2}{B^2}\right).$$

The combination $\rho BU/\mu$ is called the Reynolds number R, and Rh^2/B^2 the modified Reynolds number R^*. For a practical bearing, with breadth and length of order unity, we have $U\sim 10^3$, $h\sim 10^{-4}$, and $\mu/\rho\sim 10^{-2}$. Therefore, $R^*\sim 10^{-3}$ and the pressure gradient is governed largely by the viscous term, $0(10^7)$.

In the z-component equation, both the inertial and incompressible viscous terms are reduced by h/B, while the compressible viscous terms are increased by B/h. However, the result is still an order of magnitude less than that for horizontal terms. The Navier–Stokes equation therefore reduces to

$$\begin{aligned}
\frac{\partial p}{\partial x}&=\mu\frac{\partial^2 v_x}{\partial z^2}\\
\frac{\partial p}{\partial y}&=\mu\frac{\partial^2 v_y}{\partial z^2}\\
\frac{\partial p}{\partial z}&=0.
\end{aligned} \tag{6.9}$$

From this latter relation, the pressure gradient in the z direction may be neglected, and $p=p(x,y)$.

It is important to note that the scaling argument used in obtaining the equations of (6.9) breaks down when the amplitude of the surface roughness becomes comparable with the clearance h. This occurs for flying heights less than 10 μin.

In the opposite limit, when $R^*\to\infty$, the pressure is governed by the inertial terms, and we obtain Bernoulli's equation, which provides the lift for the 747 mentioned above.

The equations of (6.9) may be integrated with the boundary conditions $v_x(x,y,0)=U$, $v_x(x,y,h)=0$, $v_y(x,y,0)=0$, $v_y(x,y,h)=0$, $v_z(x,y,0)=0$, and $v_z(x,y,h)=0$, to give

$$\begin{aligned}
v_x&=\left[U-\frac{h(\partial p/\partial x)z}{2\mu}\right](1-z/h)\\
v_y&=\left[h\frac{(\partial p/\partial y)z}{2\mu}\right](1-z/h).
\end{aligned} \tag{6.10}$$

The next step is to integrate the continuity equation (6.4) across the thickness of the bearing:

$$\int d(\rho v_z)=-\int_0^h\left[\frac{\partial}{\partial x}(\rho v_x)+\frac{\partial}{\partial y}(\rho v_y)+\frac{\partial \rho}{\partial t}\right]dz.$$

From the boundary conditions on v_z, the left side vanishes. Interchanging the integration and differentiation, and using (6.10), gives

$$\frac{\partial}{\partial x}\left(\frac{\rho h^3}{\mu}\frac{\partial p}{\partial x}\right)+\frac{\partial}{\partial y}\left(\frac{\rho h^3}{\mu}\frac{\partial p}{\partial y}\right)=6\frac{\partial}{\partial x}(\rho Uh)+12\frac{\partial}{\partial t}(\rho h).$$

When a fluid is compressed, work is done, which could result in a change in the temperature of the fluid. However, the thermal conductivity of the gas film is sufficient to "short circuit" the temperature profile generated by the adiabatic compression giving an isothermal situation. This enables us to relate the density to the pressure, in particular, p/ρ is a constant. This gives the *isothermal* Reynolds equation

$$\frac{\partial}{\partial x}\left(\frac{h^3 p}{\mu}\frac{\partial p}{\partial x}\right)+\frac{\partial}{\partial y}\left(\frac{h^3 p}{\mu}\frac{\partial p}{\partial y}\right) = 6U\frac{\partial}{\partial x}(ph)+12\frac{\partial}{\partial t}(ph). \tag{6.11}$$

For incompressible fluids, the density is constant, and we obtain the *incompressible* Reynolds equation

$$\frac{\partial}{\partial x}\left(\frac{h^3}{\mu}\frac{\partial p}{\partial x}\right)+\frac{\partial}{\partial y}\left(\frac{h^3}{\mu}\frac{\partial p}{\partial y}\right)=6U\frac{\partial h}{\partial x}+12\frac{\partial h}{\partial t}. \tag{6.12}$$

Let us recall the approximations that went into these equations:

- the fluid is Newtonian, i.e., it satisfies (6.7);
- we neglected inertial contributions;
- we assumed no-slip boundary conditions;
- we assumed isothermal (or incompressible) flow.

In spite of these approximations, the Reynolds equation is still very complex:

- the powers of h and p mean that the forces are a nonlinear function of the geometry;
- this nonlinearity also means that the response function of the film is a complicated function of frequency;
- for a compressible fluid, it takes the pressure a finite time to respond to a change in geometry. Typically, this time is of the order of the time it takes the surface to move a distance $2B$.

A great deal of effort has gone into developing numerical methods for solving the Reynolds equation, including effects such as slip and surface roughness. We shall not go into these techniques, but refer to the reader to Gross' book [1].

6.5 Plane Slider Bearing

As the simplest example of an air bearing, let us consider a plane slider of infinite length. Let the height of the film at the entrance be h_1, and that at the exit, h_2. It is convenient to normalize the Reynolds equation by introducing the normalized variables

$$X=x/B$$
$$H=h/h_2$$
$$P=p/p_a.$$

where a refers to ambient.

Since the slider is infinitely long, there is no variation in the y direction. For simplicity, let us also consider an incompressible fluid, so that the density is constant. The equation in steady state is then

$$\frac{\partial}{\partial X}\left(H^3\frac{\partial P}{\partial X}\right)=\Lambda\frac{\partial H}{\partial X}$$

where $\Lambda=(6\mu_a UB)/h_2^2 p_a)$ is called the bearing number. When Λ becomes very small the isothermal (6.11)

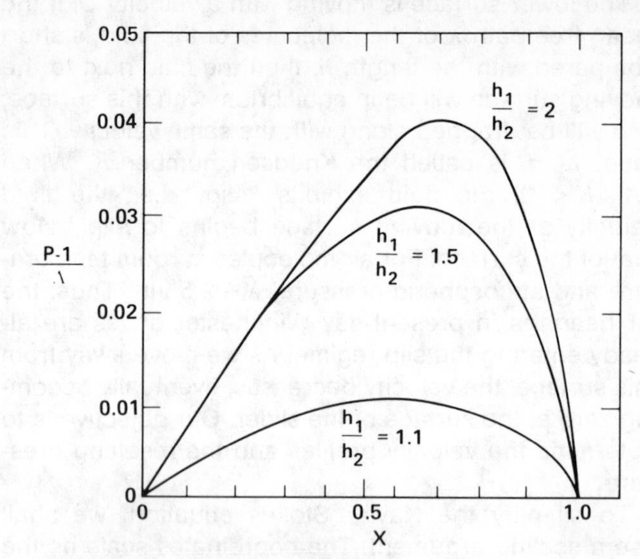

Fig. 6.1: Pressure profiles for a plane slider bearing.

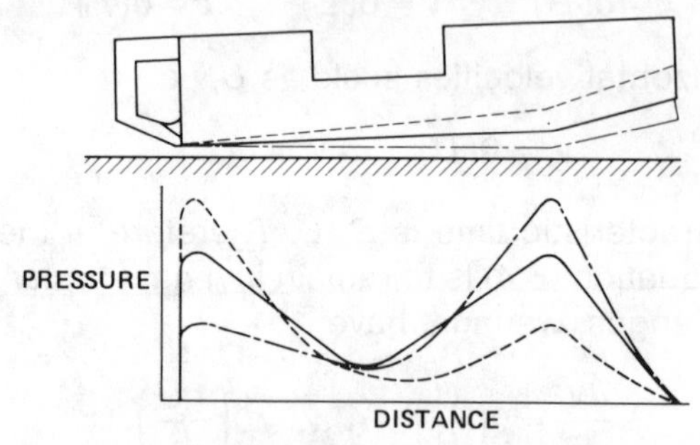

Fig. 6.2: Longitudinal pressure profiles for a Winchester slider.

takes the form of the incompressible (6.12). Therefore, at small Λ the compressible film behaves like an incompressible film.

Integration of this equation gives

$$\frac{P-1}{\Lambda}=\frac{kX(1-X)}{(2+k)(1+k-kX)^2}$$

where $k=(h_1-h_2)/h_2$. Pressure profiles are shown in Fig. 6.1 for several ratios of h_1/h_2.

6.6 Winchester Air Bearing

The geometry of the Winchester slider was shown in Fig. 3.13. This slider has the opposite dimensions from that just considered. In particular, it has a very small length, the width of the rail. This leads to a much different pressure profile, as shown in Fig. 6.2. The pressure decreases at the center, because the air can flow laterally out from under the bearing, into the "bleed slots" separating the rails. Thus, the profile has the double-peaked shape down.

This design has the consequence that if the leading edge of the slider pitches up or down in response to an irregularity on the surface, the pressure profile readjusts itself in such a way as to force the slider back into equilibrium. Furthermore, the pivot point for this motion is about an axis that nearly passes through the gap itself. Therefore, the head-to-medium spacing

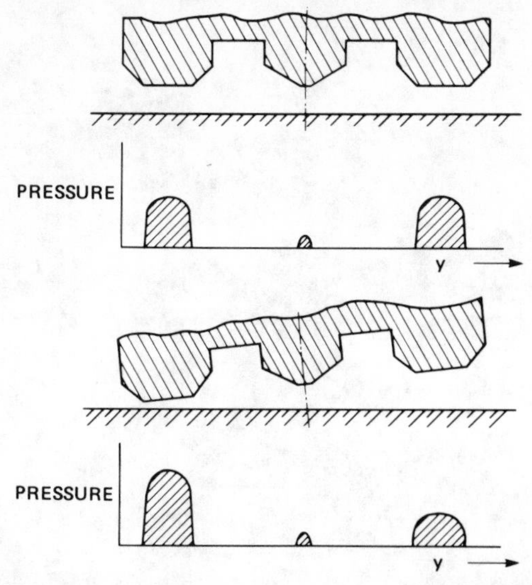

Fig. 6.3: Transverse pressure profiles for a Winchester slider.

is not affected to lowest order by this pitching. Fig. 6.3 shows how the three-rail design also stabilizes the head to rolling motion. The center rail serves simply to protect the magnetic gap. It does not contribute to lift.

Problem

1) For the infinitely long slider bearing considered in the notes, calculate a) the bearing load, and b) the position of the center of pressure.

2) What is the role of the inlet ramps on the Winchester bearing? (Hint: What happens to the plane slider bearing when $h_1 < h_2$, i.e., the bearing noses down for some reason?)

References

[1] W. A. Gross *et al.*, *Fluid Film Lubrication.* New York: Wiley Interscience, 1980.

Chapter 7
The Recording Channel

OUR emphasis in these notes has been on the physics of magnetic recording which lies primarily in the heads and media. A storage system, however, has many of the characteristics of a communication system: the data to be stored are generally encoded before they are fed into the head; the heads, the media, and the electronics all contribute noise: upon readback the signal may be processed in various ways to remove linear distortion, reject noise, etc., before it is decoded. As the density of disk drives increases it also becomes essential to provide error detection and correction. This electronic chain, at the heart of which are the heads and media, is referred to as the recording channel. In this chapter we shall briefly consider those channel aspects that relate to the transition density.

7.1 Codes

In digital recording the data are represented by a sequence of symbols, usually the binary "0" and "1." In digital recording the magnetic medium is completely saturated parallel or antiparallel to the track direction.

The simplest scheme that one might imagine for storing binary digits (bits) is to identify a "1" with one polarization, a "0" with the other. Following, we show the corresponding current waveform associated with a particular sequence of bits:

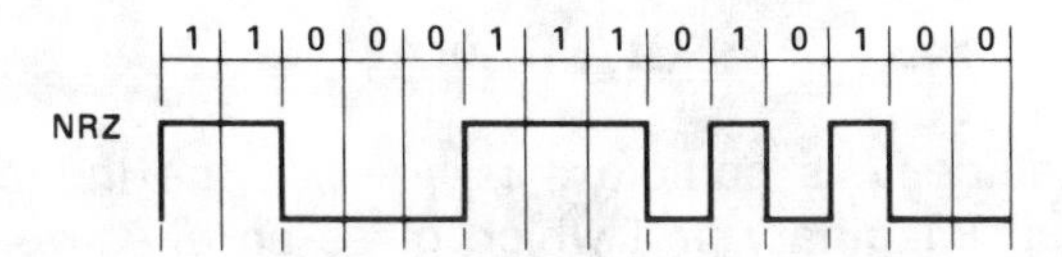

This particular scheme is called the non-return-to-zero, or NRZ, *code*. In readback however, the head only detects *transitions*. Therefore, if a flux transition were missing, all successive bits would be exactly opposite to what they should be! This suggests that we use the transition itself to represent the data. Thus, in the NRZI code a "1" is given by a transition and a "0" by no transition. This has the following current waveform:

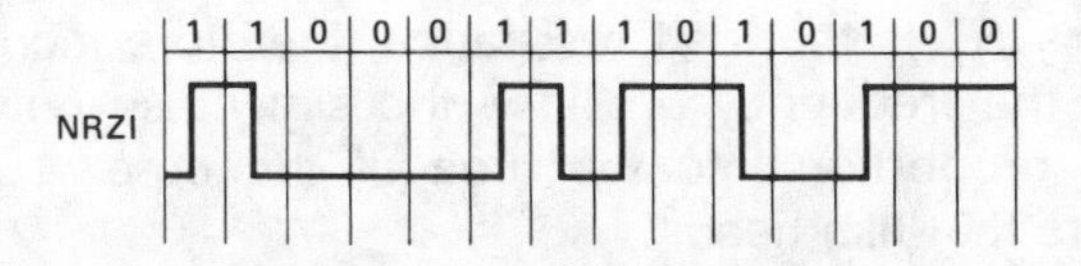

The voltage output from the head would have the following form:

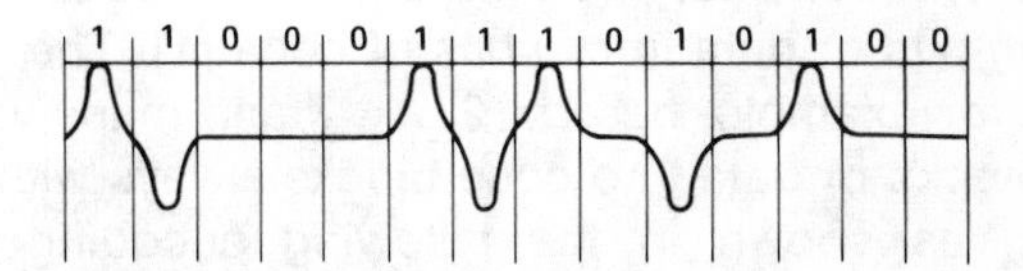

Since it is the location of the transition that carries the information, we differentiate the output and look for the zero crossings. With the NRZI code, however, a long string of "0's" will produce extra zero crossings as is obvious from the output above. The appearance of such strings also makes clocking difficult. To avoid these difficulties we insert extra transitions if the number of consecutive "0's" exceeds some value denoted by k. Such codes are called *run-length-limited* (RLL) codes. One of the first RLL codes to be used in disk codes drives was the FM or Manchester code. In the FM code, the "1" or "0" is always followed by an extra "1." This corresponds to $k=1$. The Manchester code is a variation on this. These are illustrated as follows:

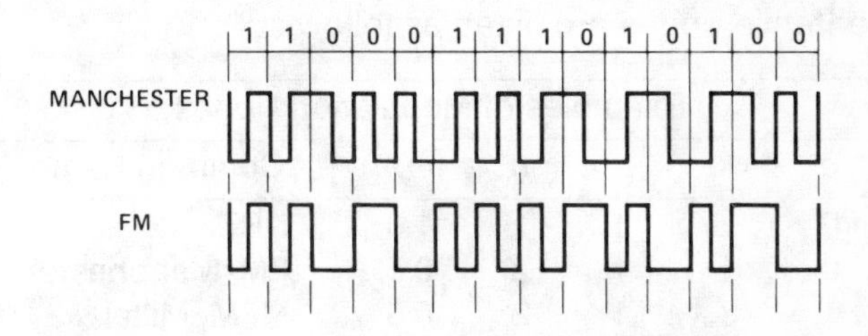

Notice that this corresponds to the mapping of one data bit into two coded bits. Notice also that in the coded sequence we have "1's", i.e., transitions, appearing without any "0's" separating them. This is not good from the point of view of intersymbol interference. This suggests using a code that also has a minimum number d of "0's" appearing between "1's."

The modified FM (MFM) code has $d=1$ and $k=3$. The encoding prescription is given as follows:

MFM Encoding Table

Data Sequence	Code Sequence
0	X0
1	01

where X stands for the complement of the preceding bit in the code sequence.

The MFM waveform has the form

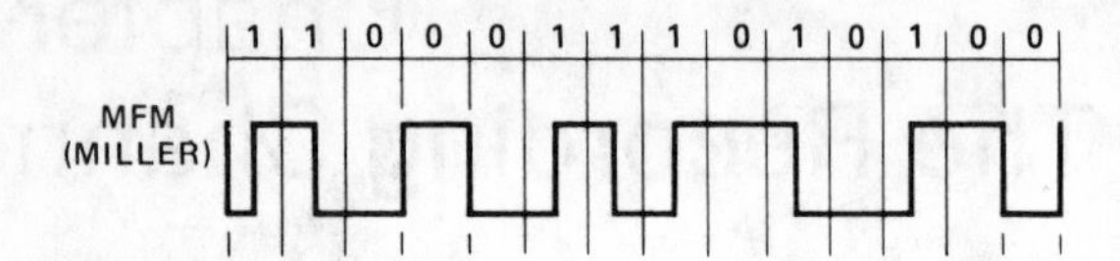

The reduction in pulse crowding compared to the FM waveform is very obvious.

One can consider even more complex codes which map m bits of data into n bits of code [1]. The "2–7" code, for example, has $d = 2$, $k = 7$ and maps various size blocks of data into code blocks always with $m/n = 1/2$, as shown in the following encoding table. (Strictly speaking, this code should be denoted as 2, 7, 1, 2 for d, k, m, n.)

2,7 Encoding Table

Data Sequence	Code Sequence
11	1000
10	0100
000	000100
010	100100
011	001000
0010	00100100
0011	00001000

The 2,7 code is in wide use today. The 1,7 code with $m = 2$, $n = 3$ is being used by Hewlett-Packard in its new disk drive products. The parameters of representative codes are listed as follows:

Examples of Modulation Codes

d	k	m	n	m/n	Common Name
0	∞	1	1	1.0	NRZ
0	1	1	Z	0.5	FM, Manchester
1	3	1	2	0.5	MFM, Miller
		2	4		
2	7	3	6	0.5	2,7
		4	8		
		2	3		
1	7	4	6	0.67	1,7

The ratio m/n is the *code rate*. Of particular interest is the relation of the BPI to the actual number of flux changes, FCI. Suppose the data rate is $1/T$. This means a data bit every T s. Since we are mapping m data bits into n code bits, the code "window," $T_w = (m/n)T$. If transitions are separated by at least d "0's" then the minimum time between transitions is $(m/n)T(d+1)$. Therefore, the ratio of BPI to FCI is

$$\text{BPI/FCI} = [(1/T)/v)]/[1/(m/n)T(d+1)/v] = (m/n)(d+1). \tag{7.1}$$

For the MFM code this ratio is 1, while for the 2,7 code it is 1.5. One might think that this could be increased simply by increasing d. However, the value of m/n is also a function of d. In code design the code rate, m/n, is generally maximized subject to the d, k constraints.

The code window may be expressed in terms of the number of flux transitions according to

$$T_w = 1/[v(d+1)\text{FCI}]. \tag{7.2}$$

Thus, as the FCI increases, the code window decreases. Since the code window is also the detection window for transitions, as this becomes smaller we eventually encounter unacceptable errors due to the inevitable presence of noise.

7.2 Noise

In a magnetic recording system erroneous signals can arise from many sources: there may be defects in the media; the head may pick up signals from adjacent tracks or from incompletely overwritten data. Mallinson classifies such sources as "interference." He reserves for "noise" those fluctuations having their origin in nondeterministic, or random, processes.

7.2.1 Head Noise

Johnson thermal noise is the best example. And, indeed, at high frequencies it is the Johnson noise associated with the head that becomes the dominant noise source. The impedance of the head is $i\omega L$. In Chapter 3 we derived the expression for the inductance

$$L = (4\pi/c)\left(N^2/(\mathbb{R}_g + \mathbb{R}_c)\right). \tag{7.3}$$

At high frequencies, or high recording densities, we are dealing with very small gaps. Therefore, the core reluctance dominates the inductance. Since

$$\mathbb{R}_c = l_c/\mu A_c \tag{7.4}$$

and since μ is complex, $\mu = \mu' + i\mu''$, the inductance has an imaginary part which gives an effective resistance

$$R = (4\pi/c)\left(N^2 A_c \mu'' \omega\right)/l_c. \tag{7.5}$$

The Johnson thermal noise associated with this resistance has an rms voltage

$$V_{\text{rms}} = \sqrt{8\pi k_B T R\, \Delta\omega} \tag{7.6}$$

where $\Delta\omega$ is the frequency "slot" over which this noise is measured. The "slot" noise power, which is V_{rms}^2/R_L where R_L is the load resistance, therefore increases with the frequency ω. Since the slot noise power is also proportional to the area of the core A_c, this favors thin-film heads.

Another source of noise lies in the particulate nature of the medium.

7.2.2 Particulate Noise

The figure below emphasizes the fact that each particle in a particulate medium contributes flux which threads the reproduce head.

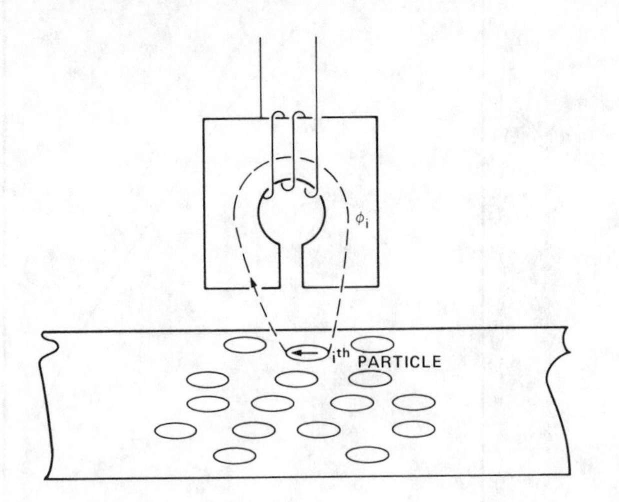

The reciprocity theorem tells us that the flux in the head is

$$\Phi_i(x) = \int\int\int \boldsymbol{H}(\boldsymbol{r}) \cdot \overline{\boldsymbol{M}}_i(\boldsymbol{r}, \boldsymbol{t})\, d^3\boldsymbol{r}. \qquad (7.7)$$

Assuming a longitudinally oriented medium, $\overline{\boldsymbol{M}}_i(\boldsymbol{r}, \boldsymbol{t}) = M_i(x - x', y, z)$ where $x' = vt$. Changing variables, this may be written

$$\Phi_i(x) = \int\int\int H(r' + x) M(r')\, d^3r'$$

$$= M_i A_i \alpha_i \int_{x_i-(l/2)}^{x_i+(l+2)} H(x' + x, y_i)\, dx' \qquad (7.8)$$

where M_i, A_i, and α_i are the magnetization, cross-sectional area, and polarity (± 1) of the ith particle, respectively. Each particle is also assumed to be of length l. Now suppose the medium is in an erased state. We shall take this to mean that the polarities α_i are random. Thus, the average flux will be zero. We therefore consider the noise power spectral density, $NP\Phi(k)$. To calculate this we first Fourier transform $\Phi_i(x)$ which gives

$$\Phi_i Z(k) = M \alpha_i \Omega_i H(k, y_i)[(\sin kl/2)/(kl/2)]\, e^{ikx_i} \qquad (7.9)$$

where Ω_i is the particle volume and $H(k, y_i)$ has the typical spacing loss behavior e^{-ky_i}. Thus

$$NP\Phi(k) = \lim (1/L) \left\langle \sum_i |\Phi_i(k)|^2 \right\rangle. \qquad (7.10)$$

Substituting for $\Phi_i(k)$ and writing $\Sigma_i \rightarrow n \int\int\int d^3r$ where n is the particle density which is related to the packing fraction p by $n = p/\Omega$, thus

$$NP\Phi(k) = pM^2 w\delta\Omega[(1 - e^{-2k\delta})/2k\delta] \cdot e^{-2kd}[(\sin kl/2)/kl/2)^2. \qquad (7.11)$$

Recall that the *signal* power is

$$SP\Phi(k) = [pMw\delta[(1 - e^{-k\delta})/k\delta]\, e^{-kd}]^2. \qquad (7.12)$$

Therefore, the "slot" signal-to-noise ratio, for $k\delta \gg 1$, i.e., at high spectral frequencies

$$SNR \sim pw/\Omega k \Delta k \qquad (7.13)$$

and the wide-band SNR, i.e., $\Delta k \rightarrow k$ is

$$SNR \sim (pw\lambda^2)/\Omega = nw\lambda^2. \qquad (7.14)$$

This very simple result is the consequence of the fact that the head only senses a volume of material of the order of $w\lambda^2$. Since the signal voltage depends upon the number of particles n while the noise goes as $\sqrt{n}$, this gives the SNR result in (7.14).

The areal density goes as $1/w\lambda$. Therefore, if we wish to increase the areal density, without paying an undue penalty in SNR, it is best to do this by decreasing w, *not* λ!

7.2.3 Modulation Noise

In the presence of a signal there is an additional noise contribution which is generally believed to be associated with inhomogeneities whose correlations produce a spectrum of "sidebands" around the signal frequency:

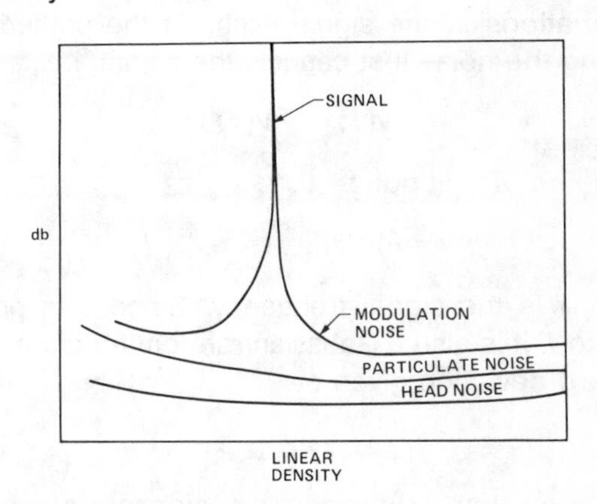

7.3 Error Rates [2]

In digital systems noise is reflected in the error rate, which is one of the most critical parameters of the system. To see how the error rate relates to the signal-to-noise ratio let us consider the region of a zero crossing within a detection window of width $2T_w$:

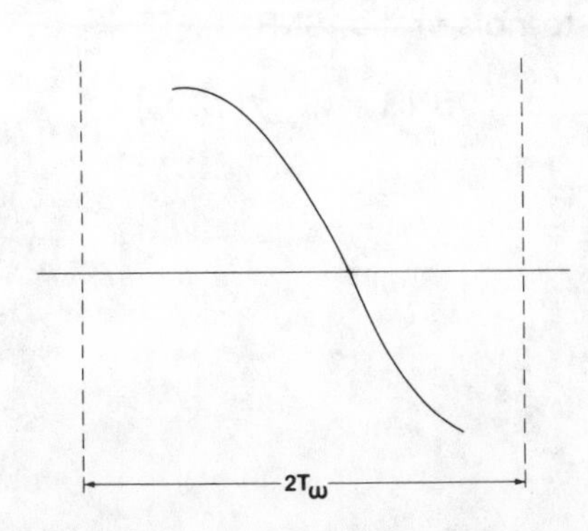

Now add noise to this signal:

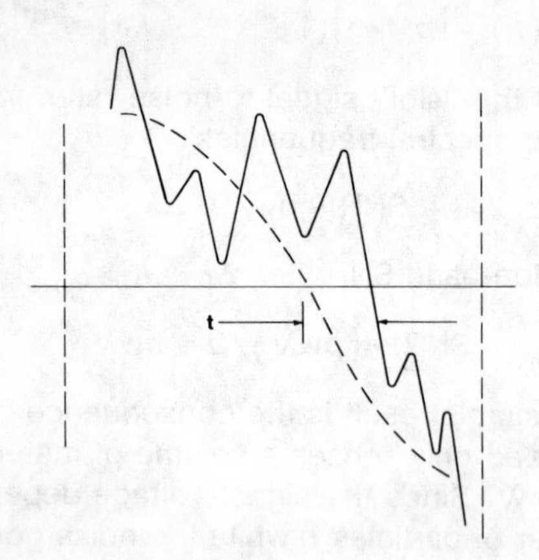

We see that the zero crossing has been shifted by an amount t. We assume that t has a Gaussian distribution about zero,

$$P(t) = 1/\left(\sqrt{2\pi}\,\sigma_t\right)e^{-t^2/(2\sigma_t^2)} \tag{7.15}$$

where σ_t is the standard deviation, i.e.,

$$\sqrt{\overline{t^2}} = \int_{-\infty}^{\infty} t^2 P(t)\,dt = \sigma_t.$$

This variation in the zero crossing may be related to the variations in the signal itself. At the shifted zero crossing the noise just cancels the signal, i.e.,

$$v(t) = -V(t).$$

Expanding $V(t)$ about $t = 0$,

$$v = \partial V/\partial t|_{t=0}\, t \approx \omega V_{\max} t \tag{7.16}$$

where ω is the signal frequency. Since v is proportional to t, it is also a Gaussian random function with a standard deviation given by

$$\sigma_v = \omega V_{\max} \sigma_t. \tag{7.17}$$

If t is larger than T_w, the zero crossing falls outside the detection window and results in an error. Therefore, the number of errors is

$$1 - \int_{-T_w}^{T_w} P(t)\,dt \equiv 1 - \operatorname{erf}(T_w/\sqrt{2}\sigma_t) \tag{7.18}$$

which defines the *error function*, erf(x). If we define the signal-to-noise ratio, SNR, as

$$\mathrm{SNR} = V_{\max}/\left(\sqrt{2}\,\sigma_v\right) \tag{7.19}$$

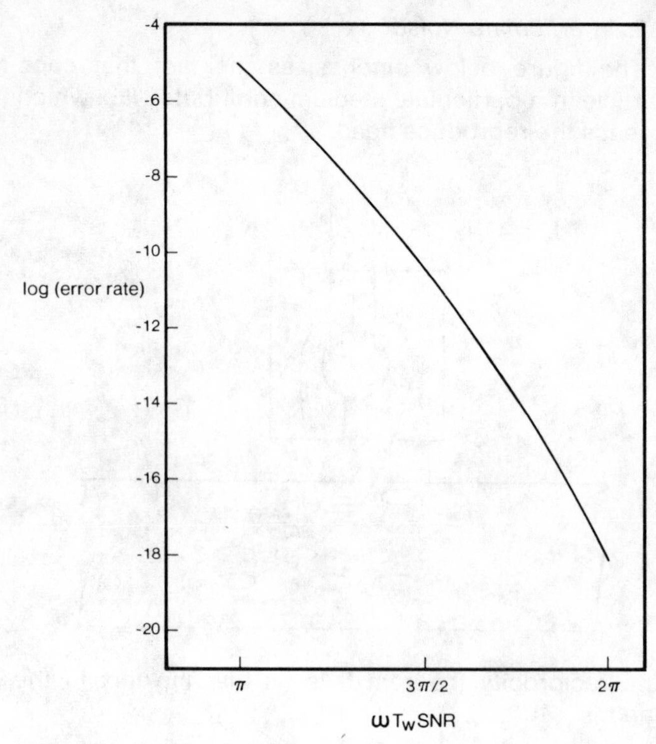

Fig. 7.1: Relation between error rate and signal-to-noise ratio.

then

$$\text{error rate} = 1 - \operatorname{erf}(\omega T_w \mathrm{SNR}). \tag{7.20}$$

The error function is a very strong function of its argument. For large values of this argument the error rate may be written as

$$\text{error rate} = e^{-x^2}/\sqrt{\pi}\,x$$

where $x = \omega T_w \mathrm{SNR}$. This function is plotted in Fig. 7.1. Computer applications generally require an error rate less than 10^{-12}.

Problem

1) Since the NRZI code has the same minimum distance between transitions but has twice the window, would a modified NRZI (with an added k constant) be preferable to MFM?

References

[1] T. Horiquchi and K. Morita, *IEEE Trans. Magn.*, vol. MAG-12, p. 740, 1976.

[2] See also, G. F. Hughes and R. K. Schmidt, *IEEE Trans. Magn.*, vol. MAG-12, p. 752, 1976.

Reprint Papers

Part I
General

Disk-Storage Technology

On the surface of a spinning disk the data for a computer are stored as minute magnetic regions. In devices being developed the data will be "written" and "read" by a laser

by Robert M. White

The speed with which computers process data has increased a hundredfold in recent years. Such progress has been supported by progress in other technologies. In particular the technology of data storage has also advanced dramatically. The information in a book the size of a large unabridged dictionary can now be stored in the form of minute magnetic patterns on a rapidly rotating disk. Even that capacity will be surpassed. In the near future an optical system in which information is stored and retrieved by laser will make possible the storage on a disk of the contents of a library of several thousand books.

In many ways the magnetic and optical systems are alike. Both require disks. Both require a head to "write" and "read" the data onto and off of the surface of the disk. In the one case the head is an electromagnetic device that resembles in principle the head of a conventional tape recorder. In the other case the head is a laser and its associated optics. Furthermore, both magnetic and optical systems require mechanisms that properly position the head in relation to specific tracks of data at particular radii on the disk, and both require an electronic system that mediates between the disk memory and a computer. The electronic system encodes the data in a way that makes them appropriate for storage and decodes them for their return to the computer. In many designs it adds signals to the data entering the memory that can be used later to locate any errors made in storage.

It is likely in addition that magnetic and optical disk memories will be part of the same information system. Perhaps, for example, a large optical disk memory will serve as an archive. It could store a year's accumulation of documents and replace filing cabinets throughout a large organization. Meanwhile smaller magnetic disk memories distributed throughout the organization would serve individual users. The disks in these smaller systems could have the advantage of being removable. The disadvantage would be a greater aptness to "crash," or lose data, because access to the disks makes them vulnerable to dirt. Even a dust particle the size of a bacterium can disrupt the writing or reading of the data. Nonremovable disks could offer better performance (and lower cost), but a power failure would make their data unavailable.

The storage of documents is not the only use for the disk technology. An audio signal such as the electric current generated by a microphone can be sampled many thousands of times per second and the intensity of each sample can be represented as a number. In this way a signal that is continuous in time, an analogue signal, is converted into a discontinuous, or digital, signal. A picture can likewise be converted into a map of digital data by sampling the picture in two spatial dimensions. One application already being examined is the storage on disks and in digital form of the thousands of X-ray images that accumulate in a hospital.

The fundamental advantage of the conversion of signals from analogue into digital is that data in digital form can be processed, transmitted and stored almost without error, even if there is no recourse to error-locating signals. The reason is as follows. Digital information is customarily represented by a sequence of digits in which each digit is a bit: either a 0 or a 1. Small areas of the medium in which such binary information is stored (for example the iron oxide coating of a magnetic disk) must be put in one state or another to represent the data. For magnetic storage the two states are the extremes at which the dipole magnetic field in the medium is saturated, or at a maximum, in one orientation or in the opposite orientation. The two states can be distinct to such an extent that the stored information is virtually never corrupted by va-

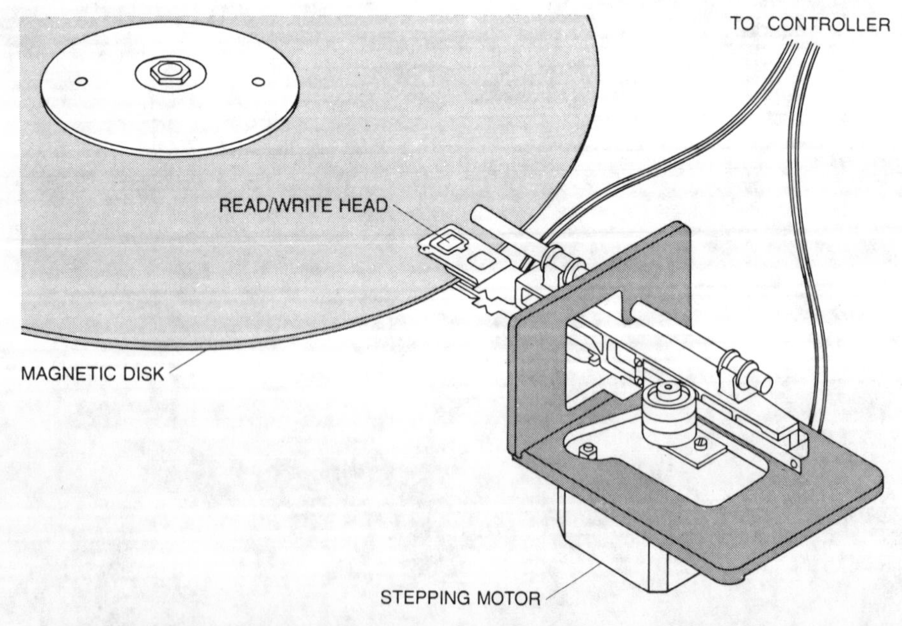

MAGNETIC-DISK DRIVE is the secondary memory in a modern computer system. In the system's primary memory (part of the central computer itself) the retrieval of data is faster. On the other hand, the secondary memory costs less for its storage capacity. The configuration here is typical. A motor positions a read/write head over tracks of data on a spinning disk.

Reprinted with permission from *Sci. Amer.*, vol. 243, pp. 138–148, Aug. 1980.

garies such as a stray electromagnetic field from the writing and reading head or from the neighboring stored information. In a high-performance magnetic-disk memory the errors caused by dirt and other mechanical problems corrupt approximately one bit in 10 billion bits. Coding for the detection of errors reduces the error rate to one bit in 10 trillion bits.

The Danish engineer Valdemar Poulsen exhibited the first magnetic recorder at the Paris Exposition of 1900. It came 23 years after Thomas Edison had built the phonograph. In Poulsen's device a steel piano wire followed a spiral groove around the surface of a drum. An electromagnet made contact with the wire and was free to slide along a rod positioned parallel to the axis of the drum. The drum's rotation pulled the electromagnet down the rod. When current from a microphone passed through the electromagnet, a segment of the wire was magnetized in proportion to the current. Although Poulsen's invention created a sensation, the recorded signal was weak. It was not until the invention of the vacuum-tube amplifier in the 1920's that magnetic recording began its steady evolution. The piano wire evolved into plastic tape with a coating of magnetic material. In another configuration a rotating drum was coated with a magnetic medium on which signals could be recorded on numerous circular tracks. Each track had its own electromagnet. Such devices became memories for the first modern computers. Indeed, they were primary memories: devices for the fastest storage and retrieval of data.

Today technologies based on semiconductor elements provide the primary memory of a computer. The secondary memory is larger and slower but stores information at a lower cost per bit. The devices serving this function are now based on magnetic disks. The tertiary memory, which is still larger, slower and cheaper, is based on magnetic tape at present. This hierarchy of memories enables the users of computers to optimize performance and cost. In fact, the flow of information can be managed by certain computer systems so that data anywhere in the hierarchy is in the primary level of storage when it is needed. It is as if all the data were on that level. The principle is known as virtual memory.

Magnetic writing—the recording of data in a magnetic medium—is based on the same principle today that applied in Poulsen's device. If a current flows in a coil of wire, it produces a magnetic field. The field is largely confined in a ring-shaped core of magnetic material around which the wire is wound. A narrow slot is cut in the magnetic material. It is the field in the vicinity of the slot that magnetizes the magnetic me-

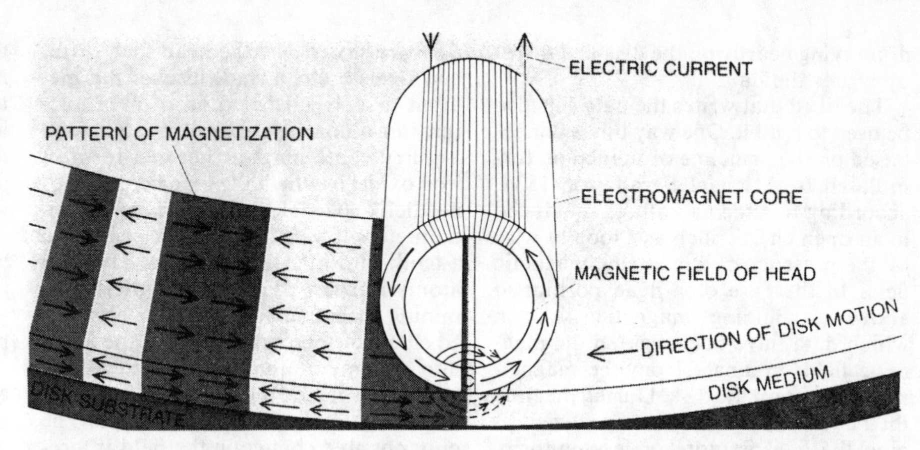

WRITING OF DATA into storage in the magnetic medium on a disk requires a head that resembles in principle the head of a tape recorder. Electric current supplied to the head flows through a coil around a core of magnetic material. The core throws a magnetic field into the disk, thereby magnetizing the medium. When current is reversed, the magnetization reverses.

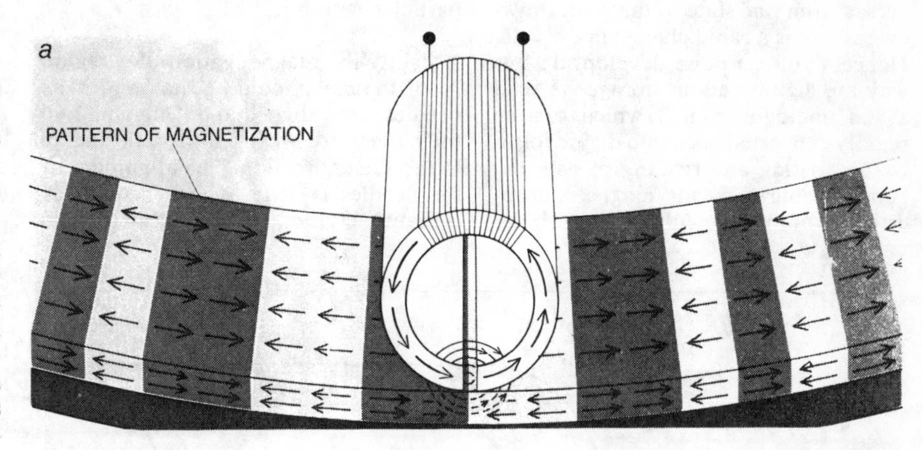

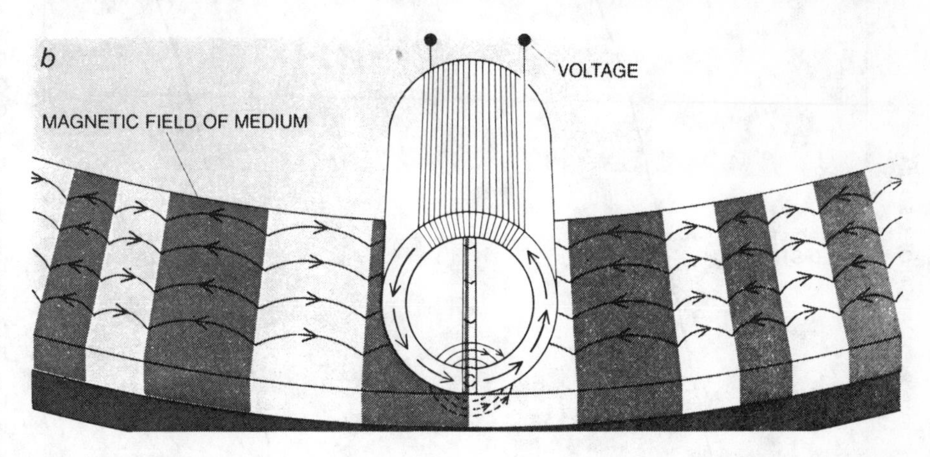

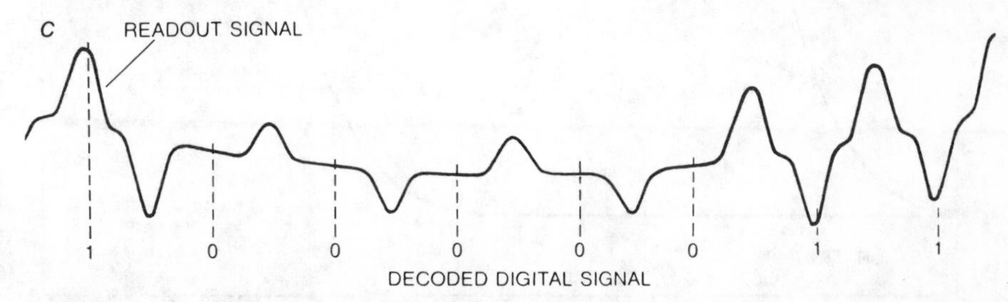

READING OF THE DATA employs the head that wrote them. The data in storage (*a*) are digital; each datum, or bit, is a 0 or a 1. Here a 1 is stored as a place on the disk where the magnetization reverses and a 0 as the absence of a reversal. A reversal also lies between each stored bit. As the disk spins, the magnetic fields of the stored data (*b*) pass successively under the head. The changing fields induce in the head a voltage signal (*c*), which is converted back into a digital form. The first seven bits of the decoded signal are 1000001, which represent the letter *A*. An eighth, or check, bit is 1 because the preceding seven bits add up to an even number.

dium lying nearby on the disk and thereby writes the data.

The head that writes the data can also be used to read it. One way this is done is based on the principle of induction, formulated by Michael Faraday in 1831, according to which a voltage is induced in an open circuit such as a loop of wire by the presence of a changing magnetic field. In the case of a head positioned above a spinning magnetic disk on which data have been written the magnetic fields emanate from the magnetized regions on the disk. During the time the head is over a single magnetized region the field is more or less uniform. Hence no voltage develops across the coil that is part of the head. When a region passes under the head in which the magnetization of the medium reverses from one state to the other, however, there is a rapid change in the field. Hence a voltage pulse develops. In this way the digital data in storage are read as an analogue signal, which can be readily converted back into digital form.

The writing and reading of data depend of course on the magnetic properties of both the medium in which the data are stored and the head that writes and reads them. I shall discuss the medium first. In one method of manufacture an aluminum disk is coated with a slurry containing the gamma form of iron oxide, in which the oxide consists of needlelike particles approximately a micrometer (10^{-4} centimeter) in length and a tenth of a micrometer wide. The iron atoms in each particle have their own minute magnetic fields, but the elongated shape of the particle forces the fields into alignment along the particle's long axis. Each needle is therefore a bar magnet and has a dipole magnetic field. The only possible change in the field is a reversal of the north and south poles at the ends of the needle. The overall magnetization in any given region of the disk is the sum of the fields of the needlelike particles within it.

Plainly the magnetization of a region of the disk would be maximal if its needles were aligned and if they all had their north (or their south) pole facing in the same direction. The alignment of the needles is achieved when the disk is manufactured, by rotating the disk in the presence of a magnetic field before the slurry has dried. The needles come to lie in the plane of the disk and more or less perpendicular to a radius of the disk. In an operating disk memory they are more or less aligned with the direction of motion of the disk.

The alignment of the poles is achieved when data are written. Specifically, it is achieved when the head applies a magnetic field to the medium. The magnetic particles are sufficiently far apart so that their own fields do not interact appreciably with one another. As the strength of the applied field increases, however, some of the magnetic particles whose dipoles are opposite to the direction of the applied magnetic field reverse their dipole field. Ultimately the applied field becomes strong enough to polarize all the particles. Two complications must be noted. First, the field of the head falls off rapidly with distance from the head. Second, the medium is moving. It therefore passes out of the region in which the field is strong enough to polarize the medium. It is the trailing edge of the field that governs the final orientation of the magnetization.

When the field of the head is removed, the region of polarized medium remains. That is why the data can be stored. The magnetization can be driven back to zero, and beyond it to saturation in the opposite sense, by reversing the flow of current through the coil in the head and thereby applying to the magnetic medium a reversed magnetic field. Since the magnetization persists in the medium, the reversal of the magnetic field does not immediately reverse the course of events by which the medium was magnetized in the first place. The term hysteresis is applied to this phenomenon. In general hysteresis signifies the delay of an event because of something that has happened previously, in this case the earlier magnetization. As the magnetic field is cycled the degree of magnetization in the medium follows a pattern called a hysteresis loop.

For a magnetic medium it is desirable that the remanent magnetism (the magnetism that persists when the magnetic field is absent) be large. It also is desirable that a moderately large field strength be necessary to demagnetize the medium. Both of these requirements help to ensure the permanence of the stored data. In addition it is desirable that the reversal of the magnetization of the medium be accomplished over a small range of applied field strength. This helps to ensure that the states of the medium that are used for data storage will be well defined. All these criteria are summarized by the requirement that the hysteresis loop for the magnetic medium be large and nearly square.

The two possible saturations of the magnetization of the medium are the two states that serve for the storage of

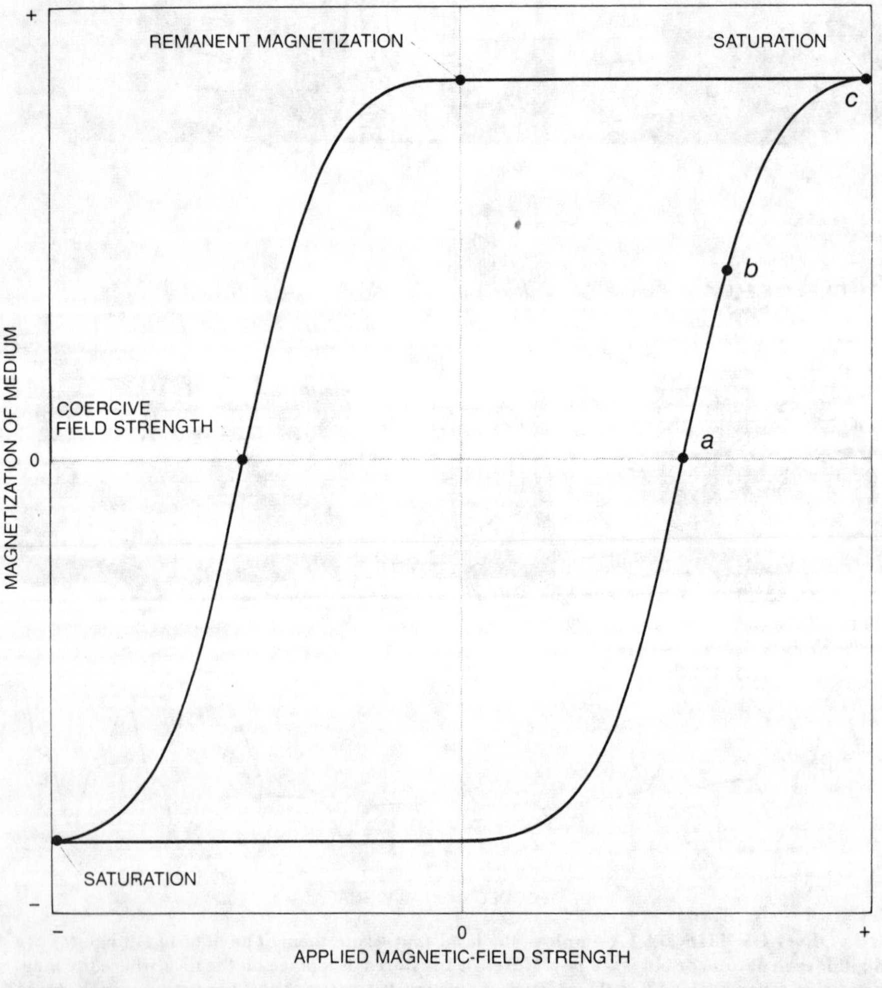

HYSTERESIS LOOP plots the magnetization of a material to which a magnetic field is applied. At first the material's magnetization increases from zero (*a*) to a maximum, or saturation (*c*). If the strength of the field is then decreased, the magnetization persists. A reversal of the field drives the material to saturation with magnetism oriented in the opposite direction.

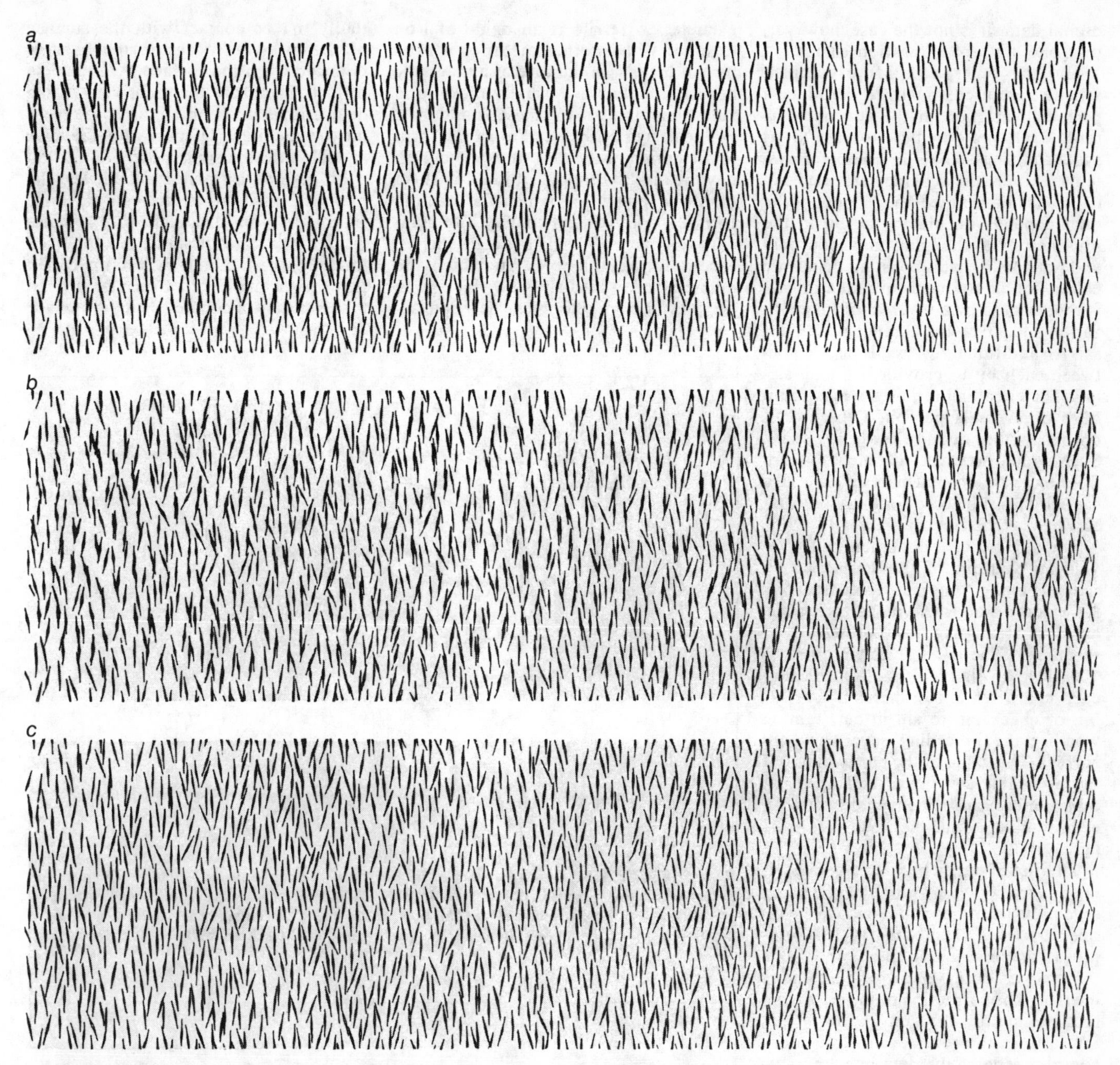

THREE STAGES IN MAGNETIZING the medium on a disk are shown schematically; they correspond to the points labeled *a*, *b* and *c* in the hysteresis loop on the opposite page. The medium here consists of needlelike particles of iron oxide. Each particle has its own magnetic field. In *a* the net magnetization is zero. In *c* the magnetization is saturated in a region that measures eight by 40 micrometers.

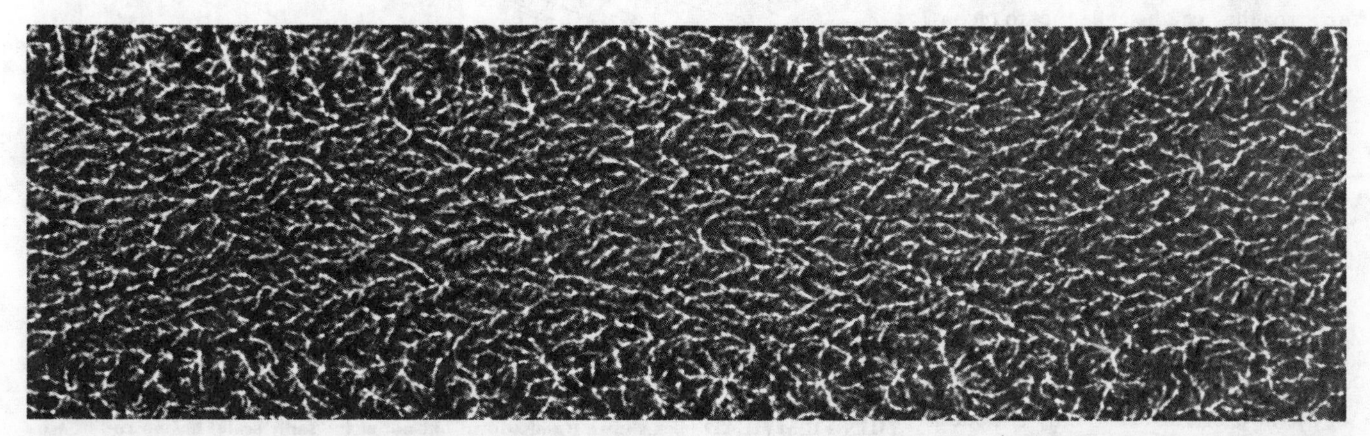

ACTUAL PATTERN OF MAGNETIZATION is suggested by this Lorentz electron micrograph. To produce the image electrons were beamed through a magnetized cobalt-rhenium film; their trajectories were bent by the magnetization. Featherlike ripples are evident in the image because the magnetization (hence the bending) is not uniform. Branches of each feather are approximately half a micrometer long.

digital data. It is not the case, however, that one state of saturation signifies a 1 and the other a 0. More complex schemes are used, in part to introduce a periodic variation in the readout current. Such periodicities are useful as synchronizing signals in computer systems. In part the schemes are used because, as I have noted, the readout of the data depends on placing a coil in the presence of a changing magnetic field and not a field that is constant. In a code called double frequency modulation a 1 is represented by a reversal of magnetization and a 0 by the absence of a reversal. An additional reversal is inserted between each bit to provide a timing signal. The encoding requires a maximum of two reversals per bit. Other codes do better: they require fewer reversals. Such codes, however, are more susceptible to error, so that part of the extra capacity must be used for extra bits that serve for error correction.

The storage of a document will provide an example of the simplest such scheme. In the American Standard Code for Information Interchange each character in a language is represented by seven bits. The letter *A*, for instance, is 1000001. An eighth bit is often added to each character in storage as a "parity" bit, or check bit, to aid in determining whether the preceding seven bits are correct. The value of the parity bit is 0 if the preceding seven bits add up to an odd number and 1 if they add up to an even number. Thus the letter *A* in storage is 10000011.

This use of parity bits can aid in locating an error only to within the preceding seven bits. A more complex code developed by R. W. Hamming in 1950 employs a greater number of bits and yields the precise address of a single bit that is in error. The correction of the error then requires simply the conversion of a 1 into a 0, or the reverse. In a still more complex scheme the data bits are treated as the coefficients of a polynomial. The polynomial is manipulated algebraically to yield a smaller set of bits. These bits are put in storage. In the event of an error they can be called up to reconstruct the data. The last scheme is particularly well suited to the correction of bursts of errors, which is generally the way errors develop on a magnetic disk.

For the magnetic material that is the core of the electromagnet in the head the requirements are distinctive. Here a small flow of current through the coil around the core should yield a large magnetization, and when the flow of current stops, the magnetization should return as nearly as possible to zero. Moreover, a reversal of the direction of the flow of current to only a modest value should yield a reversal of the magnetization. In many heads the core is a ceramic consisting of spherical ferrite particles. A ferrite is an oxide of iron together with another metal or a mixture of metals. In the case of a magnetic head the metals are usually nickel and zinc; sometimes manganese is added.

The design of the head must conform to the design of the disk. In one technology the disk is "floppy": it is a thin sheet of Mylar plastic on which the gamma form of iron oxide is coated. The standard diameter of the disk is eight inches, except for minifloppies, in which the diameter of each disk is $5\frac{1}{4}$ inches. In floppies and minifloppies the head actually makes contact with the surface of the disk. On occasion the head is bounced from the surface by a particle of dirt. Therefore the error rate of the device is relatively high, as is the wear on the medium. Of necessity the disk in such a device spins slowly.

In high-performance memories the magnetic medium is the coating on a rigid aluminum disk eight or 14 inches in diameter, and the head is kept from touching the medium by what is called the air-bearing effect. Consider a head that is nearly in contact with the surface

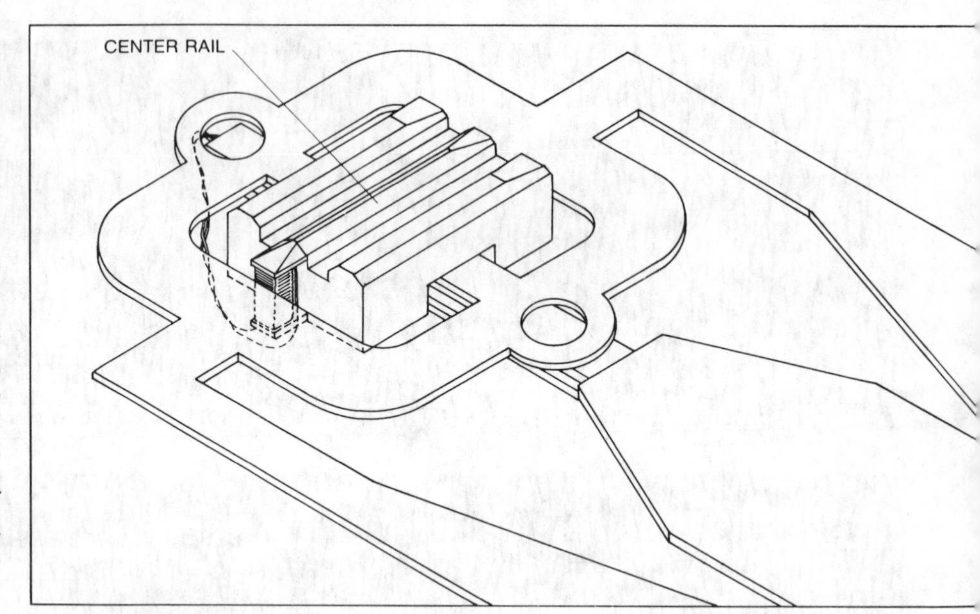

WINCHESTER HEAD was introduced by the International Business Machines Corporation in 1973. It is shown upside down; actually the rail-like surfaces of the head confront the disk at a distance of half a micrometer. The flow of air under the outside rails generates an aerodynamic force that supports the head; the trailing end of the center rail holds the electromag-

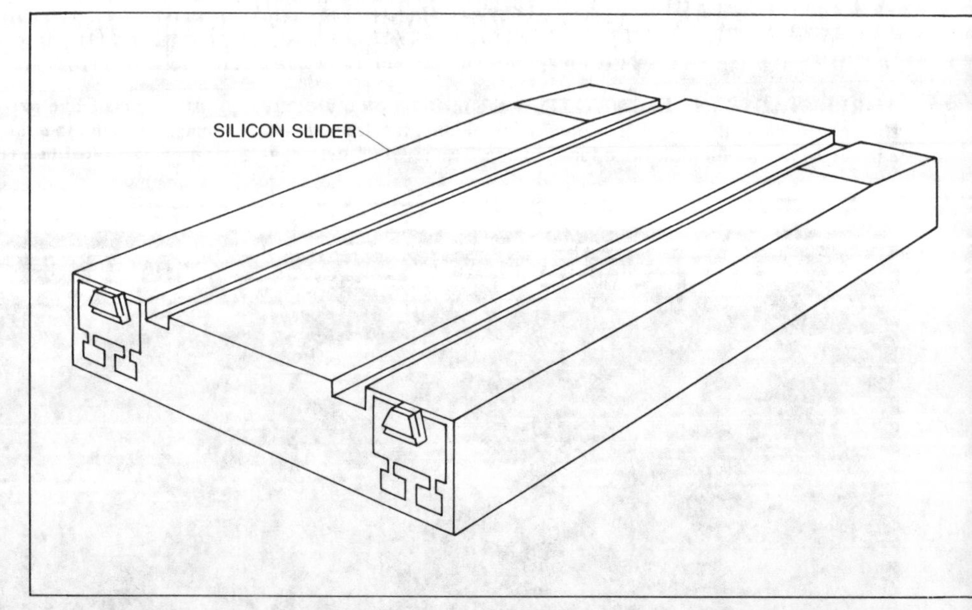

THIN-FILM HEAD (also shown upside down but not at the same scale) has no coil of wire in its electromagnet. Instead it employs a spiral film of electrical conductor. The core of the electromagnet is Permalloy, a mixture of nickel and iron. (The core in a Winchester head is ferrite: an oxide of iron in combination with other metals.) The electromagnet again is at the

of a 14-inch disk spinning at 3,000 revolutions per minute. The velocity of the head with respect to a data track in the medium on the disk is approximately 100 miles per hour. If the length of the head along the direction of relative motion is two orders of magnitude longer than the separation between the head and the medium, the flow of air between the head and the medium provides support for a head weighing up to several grams.

In 1973 the International Business Machines Corporation introduced the Model 3340 disk memory. The technology of the system has since been adopted by many manufacturers. It is now known generically as Winchester technology, that being the code name under which the device was developed at IBM. A Winchester disk memory has one or more rigid disks, either eight or 14 inches in diameter. Each disk is coated on both sides with a magnetic medium, so that two surfaces per disk are available for the storage of data. Each Winchester head has three rails, or raised surfaces. The trailing end of the middle rail holds a magnetic core with wire coiled around for writing and reading the data. The two outer rails govern the flow of air. The force that results is sufficient to support a weight of 10 grams at a height of half a micrometer above the disk. The disks and the head assemblies in such a memory are sealed in a small "clean room": a chamber approximately the size of a hatbox, in which the air is continuously recirculated and filtered to exclude any dust particles larger than .3 micrometer in diameter.

The quantity of data that can be

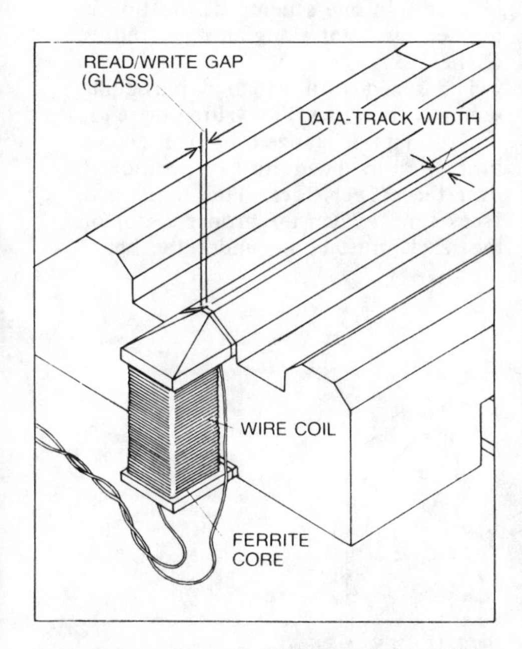

net that writes and reads the data. At the right the electromagnet is seen in closer view. The width of the beveled part of the center rail corresponds to the width of a track of data.

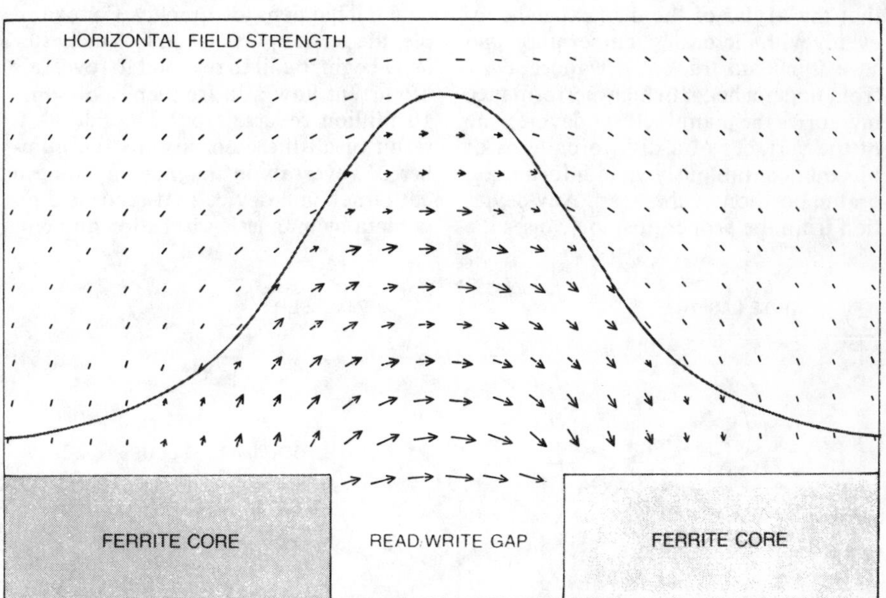

FRINGE FIELD OF WINCHESTER HEAD is the magnetic field that lies outside a gap in the core of the electromagnet. It is the field that writes the data. The arrows show the orientation of the field; their lengths show the intensity. The curve shows the intensity too; it plots the horizontal field strength at a distance from the head equal to half the width of the gap.

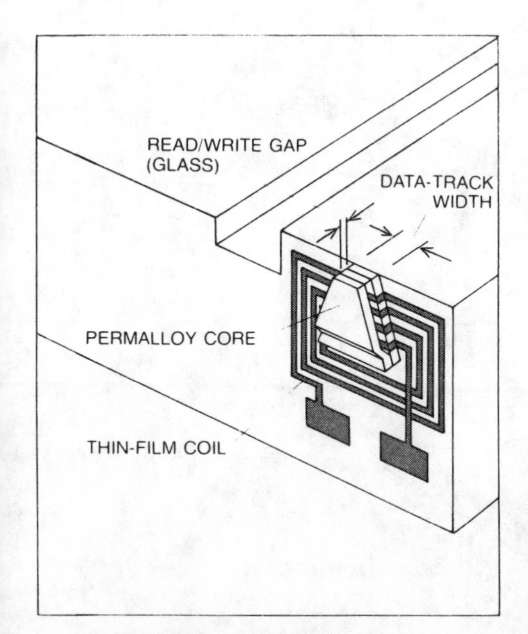

trailing end of a structure designed to generate a supporting aerodynamic force when a disk is spinning under it. A memory with thin-film heads was introduced this year by IBM.

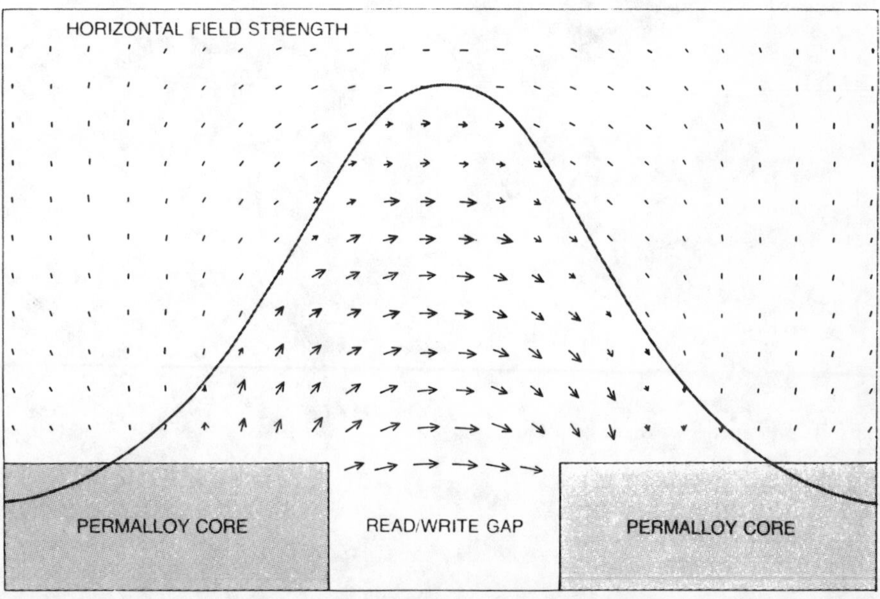

FRINGE FIELD OF THIN-FILM HEAD decreases steeply at each side of the electromagnet's gap. The steepness of the decrease allows the writing (and subsequent reading) of data at a greater packing density on a disk. Specifically, a Winchester head can record about 10,000 reversals of magnetization per inch along a data track. A thin-film head can record 15,000.

stored on a disk depends on how much of its surface area is magnetized for the storage of a bit. In the first place the width of a magnetized region, or equivalently the width of a data track, is affected by limitations on both the head and the disk. For example, the width of the center rail of a Winchester head is limited by a problem of durability: if the rail is too thin, it is fragile. This limit on the width is some 20 micrometers, which corresponds to a track density on the disk of about 1,000 per inch of the radius.

On a floppy disk the track density is only 48 tracks per inch. The reason is that the Mylar of the disk expands unevenly with increasing temperature, and so a thin data track can wander away from under a head. In high-performance memories the manufacturer devotes one of the surfaces of a disk to patterns of bits that continuously yield information on the position of the head. Any deviation from the proper position causes the generation of a signal in the head that actuates a motor for repositioning. A new strategy is to embed such patterns of bits within the stored data itself.

The number of bits that can be written along a track also is affected by limitations on both the head and the disk. For one thing a reversal in the magnetization of the medium cannot be infinitely sharp, because then it would correspond to the confrontation of north and north (or south and south) magnetic poles, and like poles repel each other. Then too a reversal in the direction of the flow of current through the coil in the head ceases to reverse the head's magnetic field if it happens too quickly. For example, the ferrite particles in a Winchester head begin to fail to respond to reversals of current flow at a frequency of some 10 million reversals per second. As a result of all these constraints the number of reversals in magnetism along a data track in a device that records digital data by magnetic saturation and employs a Winchester head is about 10,000 per inch. The quantity of data stored ranges from about 20 million bits for one surface of a floppy disk to billions of bits for high-performance rigid disks.

The rate at which bits are written or read along a track is called the data rate. It ranges from hundreds of thousands of bits per second for floppy-disk systems to 10 million bits per second for rigid-disk systems. The main reason for the difference is the fact that floppy disks must rotate at lower speeds. The bits are written onto what are called sectors of a track. A sector is typically hundreds of bits long. In one scheme its location is marked by a slot along an inner radius of the disk.

The speed with which a particular sector is found for the writing or reading of data is gauged by the access time. First the head must be positioned over the proper track. This requires a "seek time." Then the proper sector of the track must come under the head.

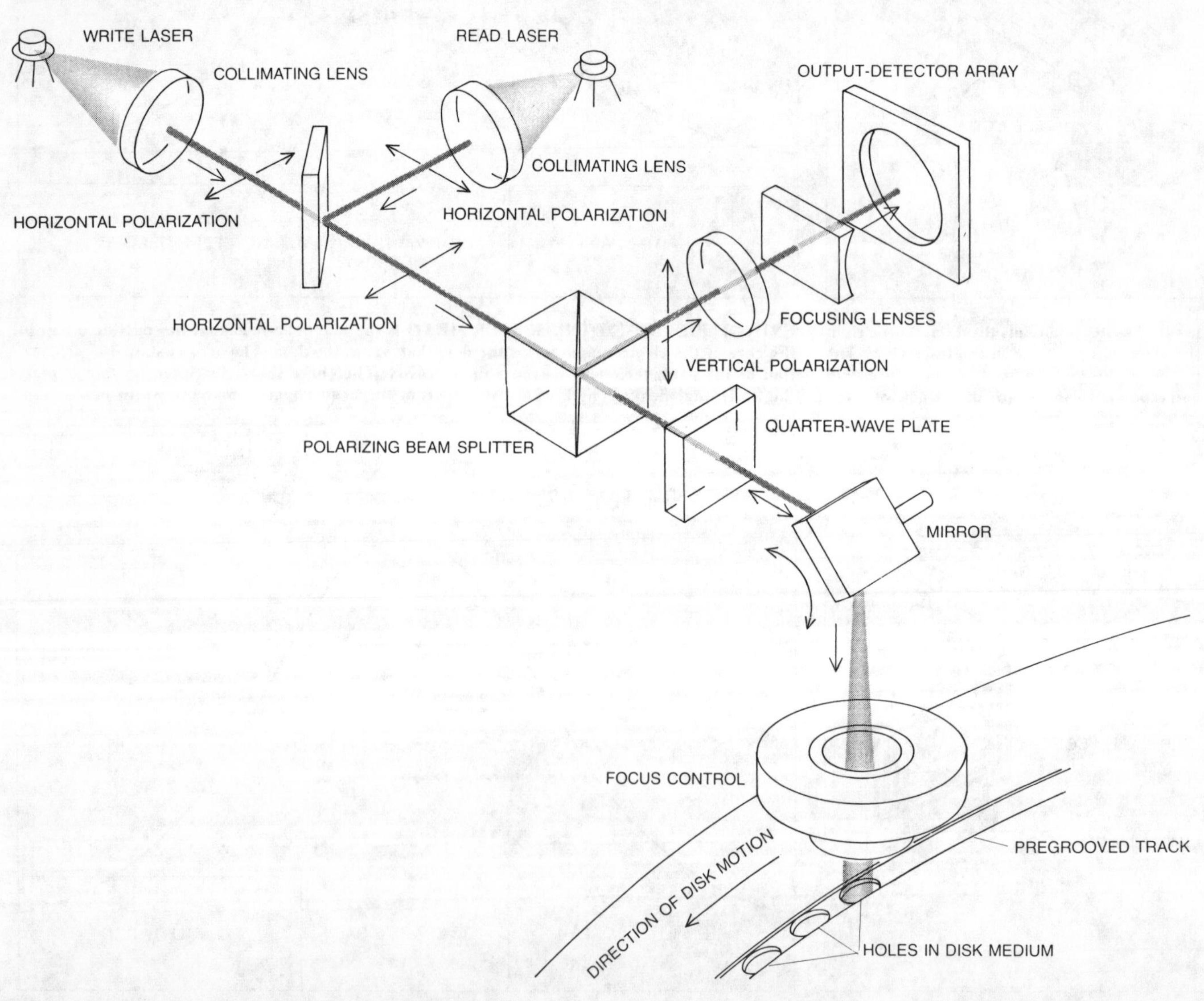

SIMPLEST OPTICAL DISK MEMORY employs laser light to write data by burning holes in the medium on a spinning disk. The laser for writing the data is at the far left. A laser for reading the data is at the top. Its light reflects from the disk in places where a hole has not been burned, then makes a second passage through the optical train of the device, arriving back at a beam splitter with a vertical polarization. This reflects the beam to a detector array. The reflected light also yields feedback signals for control of tracking and focus.

This requires a "latency time." For disk memories the access time is typically between 10 and 100 milliseconds.

Plainly each advance in disk memory must be a constellation of changes made throughout the technology. Nevertheless, I shall describe certain possible changes as if they could be made individually.

Over the past two decades improvements in disk mediums have taken the form of increases in the density of the magnetic particles and in the homogeneity of their size, but the material has remained the gamma form of iron oxide. More recently an effort has been made to increase the coercivity of the medium (the resistance of its magnetism to erasure) by modifying the chemistry of the particles themselves. It turns out that if particles of the gamma form of iron oxide are added to a solution containing cobalt ions and then are heated, the particles acquire a thin surface layer of cobalt, which increases the coercivity for reasons that are not yet completely understood. The particles are being employed in magnetic tape but have not yet appeared in disks, whose high rate of spin means that the magnetic medium must be remarkably durable.

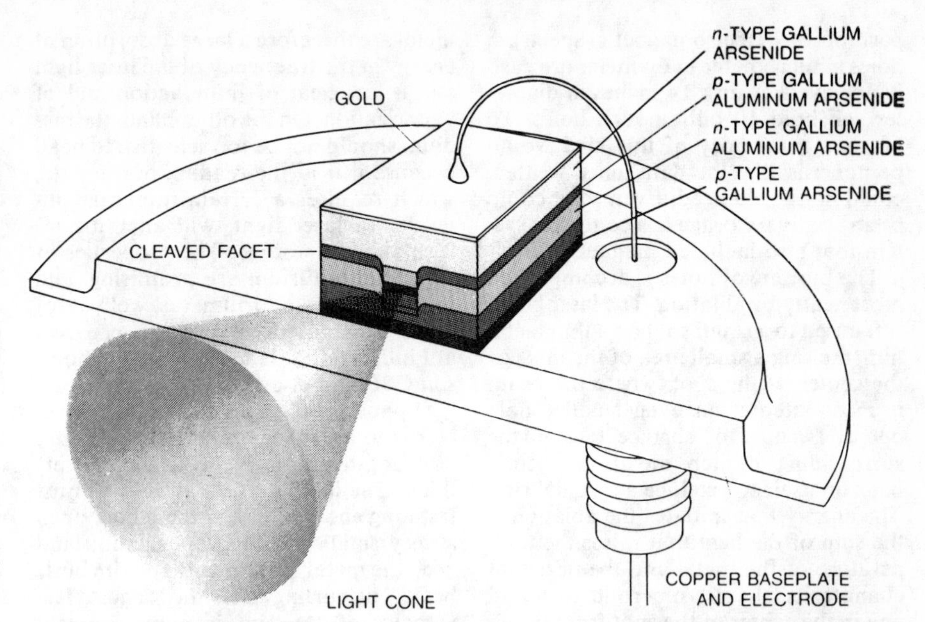

SEMICONDUCTOR LASER the size of a pinhead will probably be used in future optical disk technology. The device consists of layers of gallium arsenide or gallium aluminum arsenide, each with impurities that give the material either an excess of electrons (*n*-type semiconductor) or an excess of "holes," which are states of energy that electrons can occupy (*p*-type semiconductor). When a voltage is applied, electrons flow downward into a narrow layer of *p*-type gallium arsenide at the center of the laser. There they combine with holes and give up energy in the form of photons: quanta of light. Flanking the light-emitting region are semiconductor layers whose electrical and optical properties keep electrons and photons from moving sideways.

The magnetization of the medium (and thus the strength of the readout signal) can be increased by using a material that attains a greater magnetization than iron oxide. Pure metallic cobalt, for example, has a magnetization much greater than that of the gamma form of iron oxide, and many research groups have made disks in which it is the medium. In one method of fabrication ions of an inert gas, typically argon, are accelerated into a target consisting of cobalt. The bombarding ions cause the ejection of cobalt atoms, which deposit themselves as a film on a metal disk. By controlling various aspects of the process it is possible to control the microscopic structure of the film and therefore its magnetic properties.

What has kept such disks from being manufactured commercially is economic inertia. The production of disks based on iron oxide requires the entirely different technology where the medium is applied to the disk as a slurry. In addition, however, thin metal films are intolerant of defects in the metal disk that is their substrate. They also appear to be less durable than the iron oxide disks. On the other hand, the greater magnetization of cobalt means that the medium can be thinner. As a result the magnetic field of the head varies less through the medium, and so the magnetized regions acquire greater definition. In particular the transition length between reversals in the magnetization is shorter. The need for greater storage densities for data will make metallic disks more attractive.

In an effort to improve the head the manufacturers of disk memories are beginning to exploit thin-film technology, in which layers of materials are deposited by a plating method. IBM, for example, has recently introduced a thin-film head in its Model 3370 disk memory. Here the inductive circuit is not a coil of wire; it is a thin-film conductor deposited as a spiral that makes eight turns on the face of a silicon substrate that becomes the rail of the head. The magnetic core of the head is Permalloy: a mixture of nickel and iron. The head can respond to reversals of current as rapidly as 100 million times per second. Moreover, the pattern of the magnetic field thrown off by the head makes it possible to record 15,000 magnetic reversals per inch along a data track. (If the magnetic medium were a metal film, 25,000 reversals per inch might be attained.) It happens that the electrical resistance of Permalloy increases notably in the presence of a magnetic field. The degree of the increase depends on the strength of the magnetic field and not on its rate of change. Hence in future thin-film heads the readout could be independent of the rate of spin of the disk.

S. Iwasaki of Tohoku University has suggested that data might be stored perpendicular to the plane of the disk. Here the end of the core of a head, and therefore a pole of a dipole magnetic field, confronts the disk and throws field lines deep into the magnetic medium. As a result the data are stored, so to speak, on end. More than 100,000 magnetic reversals per inch might be possible, but the implementation awaits the development of both the head and the medium for it. By the technique described above the magnetic dipoles of iron oxide can be oriented in the plane of a disk. The problem is to find a way in which such dipoles can be oriented perpendicular to the plane of a disk.

Even with all the advances I have described it is inescapable that in magnetic memories the energy responsible for the readout signal is contained in the medium itself. To put it another way, the strength of the readout signal depends on the strength of the medium's remanent magnetization. Digital memories employing beamed energy would have the advantage that for the readout of data the beam itself provides all the required energy and the stored data act as "gates." A further advantage is that the head giving rise to the beam would not have to be nearly in contact with the medium. As a result the medium could be protectively enclosed, so that the disk would be removable without the penalty of an increased error rate. Data could then be safely stored off line (outside the memory) for merely the cost of the memory's disks.

The simplest possibility for a beam-addressed memory is a device in which the data are written when a laser beam burns holes in the coating of a disk. If the coating is a metallic film on a transparent substrate, each hole will transmit light when the laser is later used at lower power for readout. Elsewhere the silvery metal will reflect the readout beam. For a laser whose energy lies in the red part of the electromagnetic spectrum holes a micrometer in diameter are easily achieved. A disk 14 inches in diameter can hold 10 billion such holes. To be sure, a memory of this kind would be nonerasable: the data, once written, could not be changed. Even so, it could be an archival storage in a computer system that also included magnetic disks.

The burning of holes is accomplished most neatly by ablation. The laser beam is focused to a small spot on a thin metal film, melting a small area of the film. At the center of the spot, where the beam is most intense, an even smaller hole opens. Because of surface tension the surrounding molten metal then curls back on itself to produce a toroidal rim. The energy that initiates the ablation is the sum of the heat that raises the temperature of the metal and the heat that changes its phase from solid to liquid and in the center of the spot from liquid to gas. The desirable properties of the metal are therefore a large absorption of energy at the frequency of the laser light and a low heat of liquefaction and of vaporization. On the other hand, the medium should not be too sensitive to heat, because if it is, the reading of the data, which requires a certain minimum intensity of laser light, will alter the information in storage. The properties of the metal tellurium are promising, and so studies of tellurium (as well as of many other mediums) are in progress at Philips, RCA, Hitachi, Xerox, Thomson CSF and elsewhere.

Suppose a hole a micrometer in diameter is to be written on a tellurium film 300 angstrom units (.03 micrometer) thick. The laser pulse that does it must last long enough to heat the metal, open a very small hole by vaporization and keep the metal molten until the rim curls back. The curling takes the longest. The velocity of the rim is approximately 1,000 centimeters per second, and so a hole a micrometer in diameter opens in some 50 billionths of a second. The laser power needed is 20 milliwatts. During the time the laser is on, the disk advances by approximately a micrometer, giving the hole an elliptical shape. A scheme for encoding data might capitalize on such elongation. The holes, for example, might overlap to produce slots. In Magnavision, a device manufactured by the Magnavox Company that "reads" prerecorded disks but cannot make recordings, variations in the length of holes modulate a laser beam to yield a video signal.

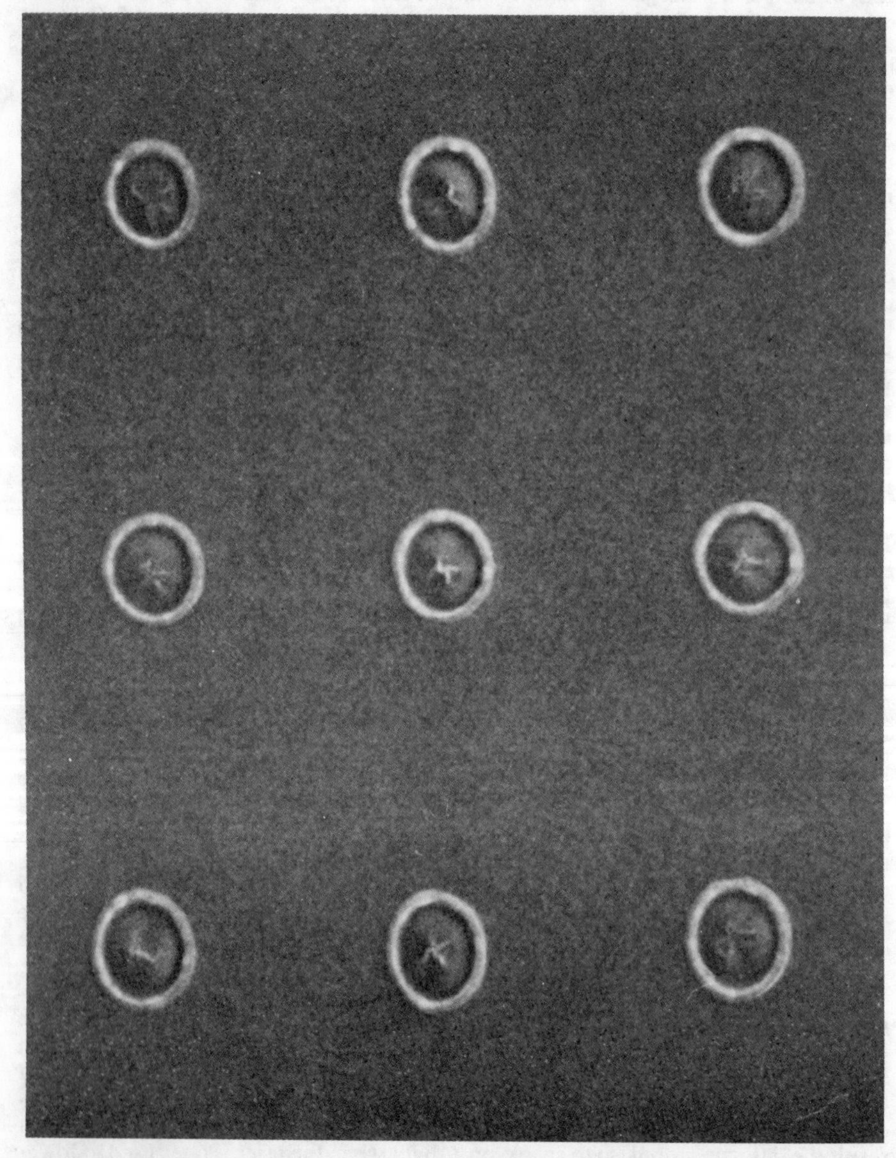

HOLES BURNED BY A LASER in a film of the metal tellurium are shown in an electron micrograph. To make each hole the laser melted a circle of metal and vaporized the center of the circle. The metal then curled back on itself. The distance between holes is five micrometers.

Devices such as Magnavision rely on gas lasers, of which even the smallest are rather large and require an apparatus resembling a shutter to modulate the beam. The rapid development of the much smaller semiconductor lasers makes it likely that in data-storage systems they will be used instead. The simplest semiconductor laser consists of two layers of semiconducting material, each layer "doped" in a different way with a small amount of impurity so that the energies of the electrons in one layer are higher than those of the electrons in the other. When the layers are in contact, the electrons soon stop flowing, because their initial motions establish an electrostatic field that prevents any further movement. If a voltage is applied, however, the field is overcome. The electrons with the higher values of energy move into the other layer, where they give up their energy in the form of light.

A cascade of events then gives rise to the laser beam. Specifically, some of the light that has been generated propagates to the front and the back of the semiconductor, where it is partially reflected by mirrorlike surfaces. As a result of these reflections the light wave grows in amplitude, because each photon, or quantum of light, interacts with electrons in the semiconductor and gives rise to the emission of additional photons.

The only semiconductors now known to be suitable for lasers are gallium arsenide and its related alloys, whose laser light is red or infrared. That is somewhat unfortunate for a disk memory, because tellurium is only weakly absorbing of energy at those wavelengths. On the other hand, the semiconductor lasers are becoming more efficient. The light-producing layers are being sandwiched between additional layers of semiconductor that reflect the moving electrons and thereby confine them to the active regions of the device. The same layers confine the photons. By means of such designs Hitachi has obtained 80 milliwatts of continuous laser output from an experimental semiconductor laser the size of a pinhead. The properties of the medium on the disk thus may cease to be limitations.

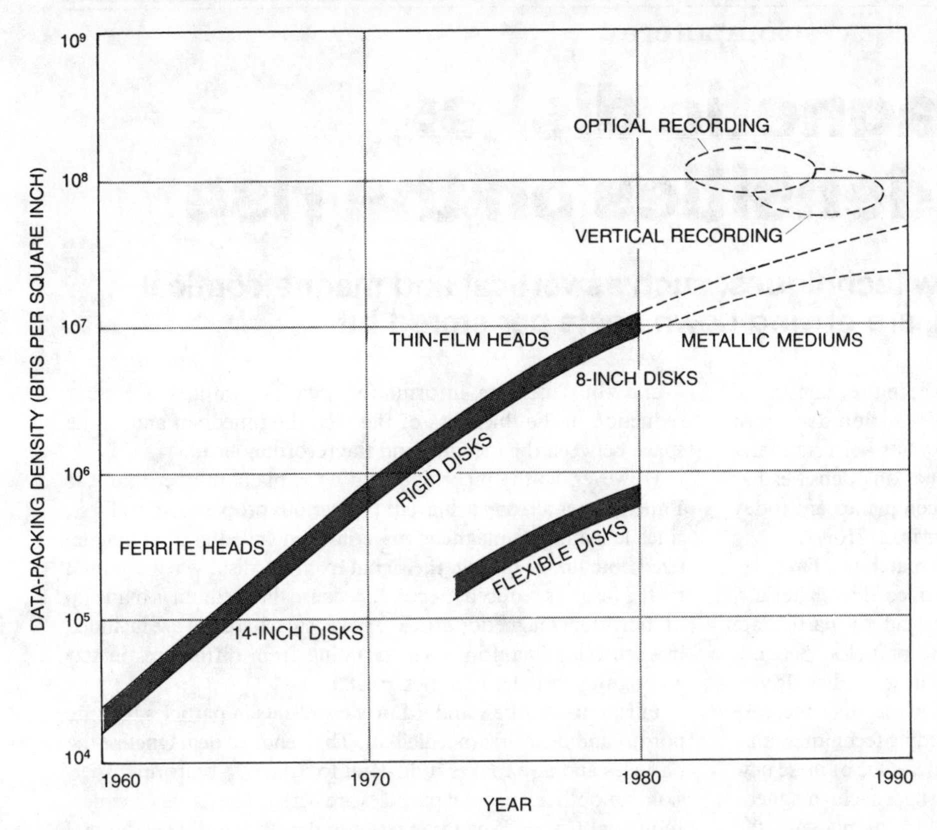

ADVANCES IN DISK TECHNOLOGY are reflected by advances in the packing density of data, expressed in this chart as bits per square inch on the surface of a disk. In rigid disks the magnetic medium is coated on an aluminum substrate; in "floppy," or flexible, disks it is coated on Mylar plastic. Two improvements are foreseen for the magnetic technology: the use of metal film instead of iron oxide as the medium on rigid disks and the recording of data in regions of magnetization oriented vertically, or perpendicular to the plane of the disk, instead of horizontally, the current practice. Optical disk technology might attain the greatest storage density.

Alan E. Bell and Robert A. Bartolini of RCA have proposed another possibility: a disk where the tellurium is coated on a transparent substrate so that the thickness of the tellurium and the substrate is a quarter of the wavelength of the light. Under the substrate is an aluminum reflector. The light reflected from the aluminum would be half a wavelength out of phase with the light reflected from the tellurium and so the destructive interference between the two would be complete. An equivalent description is that the tellurium is black as seen from the viewpoint of the laser. Hence all the incident energy is absorbed; it contributes to the heating of the medium.

It is still hoped that an optical disk memory can be devised in which the data can be erased and rewritten. One attempt to develop such a memory is based on the interaction of a magnetic medium with light. For the writing of data the medium is heated above its Curie temperature: the temperature at which its magnetism disappears. As the medium cools, a magnetic field is applied, so that when the magnetism reappears, its orientation is established preferentially. For the reading of the data light is either reflected from the magnetic medium or transmitted through it. The magnetism slightly changes the polarization characteristics of the light. To maximize the effect the magnetization must be perpendicular to the plane of the disk. That is the orientation also required for the vertical recording proposed by Iwasaki.

In the late 1960's IBM experimented with an erasable optical disk memory employing europium oxide as the medium. Unfortunately the Curie temperature of this substance is less than room temperature, and so the memory must be refrigerated. Later the properties of manganese bismuth were investigated. Then it was discovered that the atoms in manganese bismuth rearrange themselves at the Curie temperature, so that the repeated writing of data degrades the material.

Optical disk technology is therefore just emerging. It promises to increase the capacity of disks a hundredfold and to reduce the cost of data storage correspondingly. This is not to say that optical disks will promptly replace their magnetic precursors. Thin-film heads and metallic mediums will improve the performance of magnetic disks, and so can vertical recording. Many issues remain to be resolved that will affect the integration of the optical technology into computer systems.

Computers

Magnetic disks: storage densities on the rise

New materials and new techniques, such as vertical and magnetooptical recording, are driving down costs per stored bit

For 30 years, magnetic-disk memories have been the mainstay of computer information storage, and the information density of the disks has steadily increased to 2 million bits per centimeter squared. But the demand for ever greater packing densities has not slackened. As a result, a number of companies are today pushing for higher disk densities along four main fronts.

Two involve the development of better materials. First, researchers are improving existing ferric oxide recording materials, either by using continuous thin films instead of particulate coatings or by adding cobalt to the oxide particles. Second, metallic coatings are replacing ferric oxides in some disk drives.

But though improvements in materials alone may increase recording densities fivefold, changes in recording techniques may raise densities as much as twenty or thirty times. One of these new techniques is vertical recording. With this approach, magnetization is perpendicular rather than parallel to the plane of the recording media. The great advantage is that it reduces self-demagnetization, which blurs the magnetic cell.

Another effort, being pressed by Philips GmbH in Hamburg, West Germany; the Xerox Corp. in Palo Alto, Calif.; and a number of Japanese companies, is toward perfecting magnetooptical disks. These have densities comparable to those of optical disks, but the disks are also fully erasable.

With so many attractive options for the future, the obvious question is: Which is superior? At present there is no clear answer. Many factors are involved. Enhanced oxides and metallic media are already appearing in Winchester drives because they are compatible with existing systems. In combination with thin-film heads, these disks will permit much higher densities.

But the application of vertical recording will require new heads or special electronics to handle the asymmetric readback pulses from conventional ring heads. And as for the magnetooptic option, it requires tracking and focusing servo systems to exploit its potential. In time all these media will likely appear in the marketplace, just as there are many video-disk technologies.

Making oxides better

Until recently the basic recording medium for magnetic disks has been ferric oxide particles dispersed in a binder. The particulate dispersion is spin-coated onto the underlying aluminum disk [see "Disk basics," p. 36.]. The magnetic head, a C-shaped electromagnet with a small gap between the poles, writes information on the disk by altering the magnetization of the particles. This is done through a fringing field, which falls off by 50 percent at a distance comparable to the gap length. Therefore any reduction in this gap to reduce the spacing between magnetic domains —and thus increase information density—implies a similar reduction in the thickness of the recording medium and in the space between the medium and the recording head.

However, as the recording coating is made thinner than the 1 micrometer already achieved, two serious problems arise. First, since less and less magnetic material is included in each magnetized domain, or bit cell, the signal from the dish, when it is read by the head, is reduced. Second, because the medium is made up of individual magnetic particles, fewer such particles are included in each cell. Therefore noise resulting from differences in size among the particles becomes greater.

Furthermore, the standard process results in particles that are porous and dendritic (needlelike). This tends to demagnetize the particles and also makes it difficult to achieve a uniform dispersion. Smooth, ellipsoidal particles are better. The Sakai Chemical Industrial Co. in Japan has recently developed a hydrothermal process for producing gamma-Fe_2O_3 (gamma referring to the crystalline structure) particles that appear very ellipsoidal with smooth surfaces.

One way of getting bigger signals is to replace the particulate dispersion with a continuous thin film of ferric oxide. Since these films consist purely of magnetic material, rather than magnetic particles in a nonmagnetic base, they have more magnetic material in a given volume, thus increasing the readback signal. In addition, the individual grains of oxide that make up the film are much smaller than the particles in the conventional medium, so noise due to granularity is reduced.

Sputtering is a key process

Fujitsu Laboratories Ltd. in Kawasaki, Japan, has developed a reactive sputtering process to obtain such thin films of ferric oxide [Fig. 1]. Sputtering is a process in which a target, consisting of the material to be deposited, is placed opposite a substrate in a vacuum chamber. The chamber is evacuated to a base pressure on the order of 10^{-6} torr and then filled with argon gas to a pressure of about 10 millitorr. When a potential of several kilovolts is applied between the target and an anode, the argon gas forms a glow discharge. Argon ions are accelerated to the target, where they strike the surface and create an atomic collision cascade in the bulk. Some atoms reach the surface, travel through the discharge, and are deposited on the substrate, which may or may not be electrically biased.

In the Fujitsu process, an iron target is sputtered in an argon-oxygen atmosphere, resulting in a film of Fe_3O_4 . This film is then heat-treated in an oxygen atmosphere to convert the Fe_3O_4 into Fe_2O_3 . Typical performance parameters for sputtered oxide films are listed in the table. Oxide media prepared in this manner are expected to appear in subsequent generations of Fujitsu's Eagle disk drives.

This process of sputtering is central not only to continuous

Robert M. White Xerox Corp.

Reprinted from *IEEE Spectrum*, vol. 20, pp. 32-38, Aug. 1983.

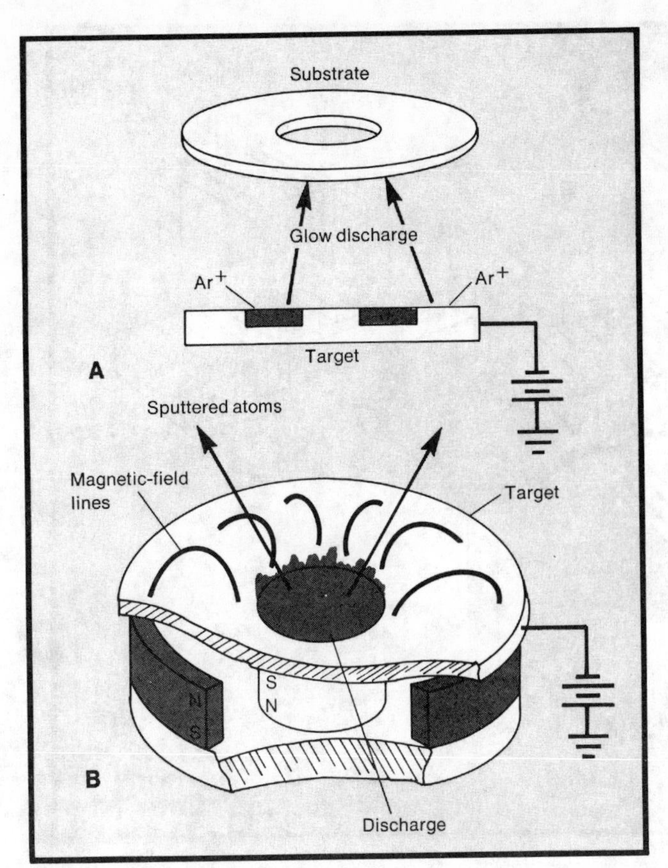

[1] Sputtering is a key technique in the production of several new magnetic-disk media, such as thin-film oxides, metallic media, and vertical recording media. In the standard sputtering process (A), argon ions in an evacuated chamber are accelerated to a target cathode where they knock out (sputter) atoms from the target and deposit them on the substrate. To speed the very slow sputtering process, a magnetic field is used to confine the argon plasma (B), creating a higher-density plasma with more argon ions. This magnetron approach also prevents electrons emitted from the cathode from reaching and heating the substrate.

thin-film oxides, but also to several of the other new disk technologies. Unfortunately sputtering is a rather slow process and requires expensive equipment. The problem of either improving the sputtering process or of developing effective substitutes is key in the development of these new disk materials.

One approach to speeding the sputtering is to use a magnetic field to confine the argon plasma. With this "planar magnetron" geometry, a higher-density plasma can be created. Furthermore, the electrons emitted from the cathode are prevented from reaching the substrate where they would produce unwanted heating, but the presence of a magnetic target material complicates the design of such a magnetron source.

Fujitsu has been able to demonstrate recording densities of 2.5 megabits per centimeter squared with oxide film disks. With special, narrow-gap heads (0.15-micrometer gaps), the 0.12-micrometer-thick film allowed even higher densities—up to 6 Mb/cm^2. Signal-to-noise ratios of 36 decibels—higher than conventional disk ratios of 25 dB—have been achieved. The Fujitsu researchers believe that densities as high as 10 Mb/cm^2, more than five times current maximum densities, can be reached.

Cobalt for more coercivity

Fujitsu has added cobalt to its thin films to increase the coercivity—that is, the value of the magnetic field at which the medium reverses its magnetization. Coercivity should be high enough so the medium will not demagnetize spontaneously, but no higher than the field produced by the writing head. Coercivities in the range of 300 to 700 oersteds are typically required.

In general, however, the higher coercivities ensure that there will be a sharp demarcation between magnetic cells of different orientation, increasing the sharpness of the pulse that detects the cell boundary and thus the signal-to-noise ratio.

Since high coercivity is connected with highly anisotropic magnetic properties—that is, with properties that vary greatly with direction—and since cobalt has such properties, it is a natural additive to increase coercivity. Cobalt has, in fact, been used to increase the coercivity of magnetic tapes—in particular, video tapes—by coating the outside of oxide particles. Recently such cobalt-coated particulates have been applied to disks. The Verbatim Corp. of Sunnyvale, Calif., for example, makes such high-coercivity disks for use in the Apple office computer system Lisa, and Dysan Corp. in Santa Clara, Calif., is providing disks to various manufacturers for evaluation.

Metallic media explored

An inherent drawback of oxide media is that the oxygen dilutes the magnetization. Since the output signal is proportional to the magnetization, as researchers go to thinner and thinner coatings in the quest for higher linear densities, the signal-to-noise ratio eventually becomes a limiting factor. This has prompted much work on pure magnetic metals, such as iron, cobalt, and nickel. Their magnetizations are an order of magnitude larger than those of oxides.

But the anisotropy associated with iron and nickel, which determines the coercivity, is in general too small for recording purposes. Therefore the focus has been on cobalt films. The Ampex Corp. of Redwood City, Calif., has just begun limited marketing of cobalt film metallic disks at costs matching those of Winchester disks. Though the technology for metallic disks has existed for several years, IBM's resistance to the disks has until recently inhibited their commercialization. IBM's main objection was that the disks were liable to corrode, although in Winchester

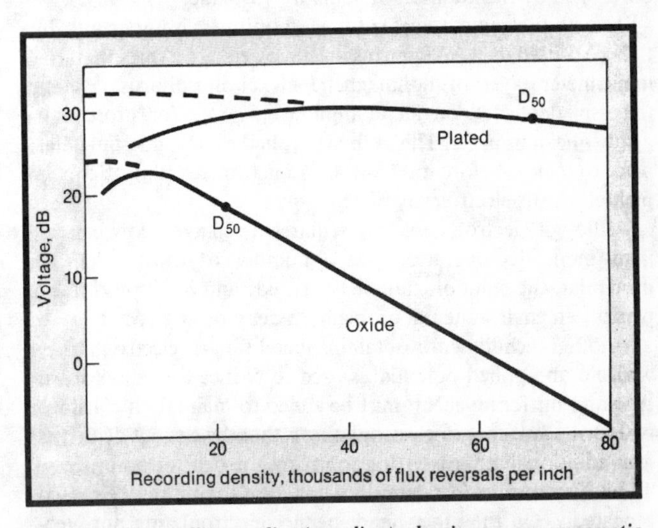

[2] Plated metallic media (top line), having more magnetic material per unit volume than oxide media (bottom line), show a more gradual decrease in signal-to-noise ratio as recording density increases and the size of magnetic domains decreases. The signal-to-noise ratio of the oxides declines by 6 decibels (D_{50}, which denotes half the signal strength) at only about 25 000 flux reversals per inch, while this decline is not obtained until 70 000 flux reversals per inch for plated media.

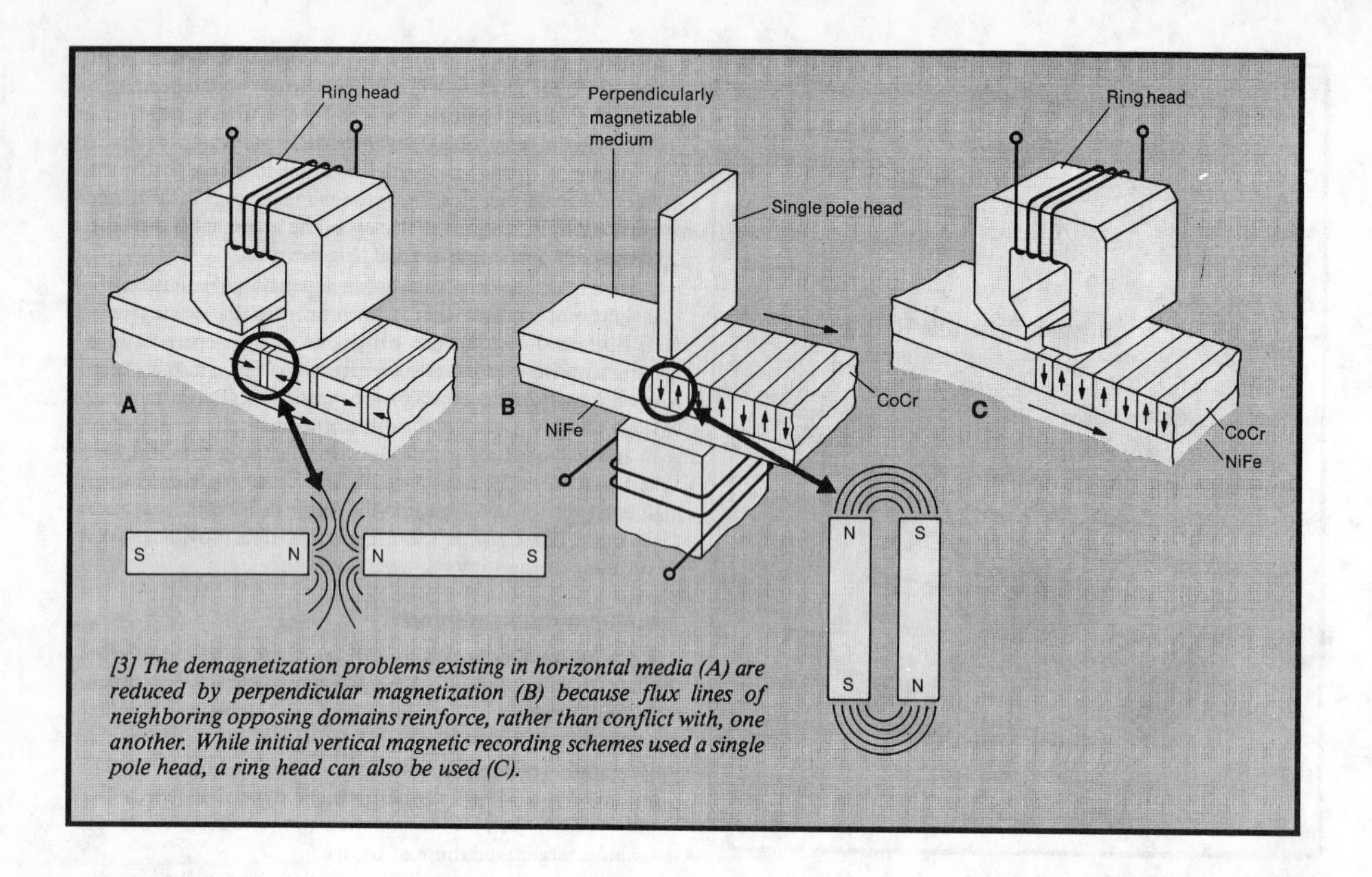

[3] The demagnetization problems existing in horizontal media (A) are reduced by perpendicular magnetization (B) because flux lines of neighboring opposing domains reinforce, rather than conflict with, one another. While initial vertical magnetic recording schemes used a single pole head, a ring head can also be used (C).

drives, where the heads and media are enclosed in a controlled environment, this is not a great problem.

A number of processes are used to produce metallic film disks. Ampex uses electroless deposition—in which a solution containing cobalt ions is prepared from cobalt chloride, for example. A reducing agent, such as sodium hypophosphite, is added. The microstructure of the resulting film depends upon the conditions of the bath—the pH and the temperature. The pH is generally controlled by the addition of sodium hydroxide.

As with the gamma-Fe_2O_3 films, smoothness is important. In the so-called ALAR process developed by Ampex, a 6-micrometer layer of nonmagnetic nickel-phosphorus is electrolessly deposited on an aluminum substrate for protection against head impacts. This is then polished to 0.05 micrometers, peak to valley, before a 0.03-to-0.06-micrometer film of cobalt phosphorus is electrolessly plated.

Although electroless plating requires a relatively low capital investment, its operation has a number of drawbacks. In particular, the solution cannot be reused, and its disposal may present an environmental problem, thereby raising costs.

Another technique for obtaining metal films is electroplating, wherein an applied potential is used to reduce the cobalt ions. Again, a buffering agent must be added to maintain a constant pH. One advantage of electroplating is that the current density is an additional parameter for controlling the deposition process. This also permits fast deposition, and the solution can be reused repeatedly. On the other hand, neither electroplating nor electroless deposition are convenient for obtaining alloy films.

The sputtering process previously described is particularly well suited for alloys. Being a vacuum-deposition process, it is also inherently "cleaner" than the plating processes.

Evaporation can also be used to obtain metal films. IBM has been exploring the use of electron-beam heating to evaporate films of $(Fe_{0.5}Co_{0.5})_{0.85}Cr_{0.15}$. By evaporating the substrate at an oblique angle, the easy axis of magnetization can be oriented circumferentially. The 15 percent chromium inhibits corrosion.

Because metallic media contain more magnetic materials per unit volume than do oxide media, just as thin-film oxides contain more than particulate oxides, they can support a higher recording density without diminishing the signal-to-noise ratio [Fig. 2]. Experiments have confirmed that metallic disks can achieve densities at least five times higher than those in conventional particulate oxide disks, or about 10 million b/cm^2, comparable to what can be achieved with thin-film oxides.

Vertical recording

In looking into changes in recording techniques to increase disk densities, researchers have focused on vertical recording and its advantage of reduced self-demagnetization. In horizontal recording, when a transition is made from one magnetic orientation to another, the north or south poles of the neighboring cells are adjacent. Just as the poles of two magnets laid end to end in this manner weaken each other, so the cells tend to demagnetize each other, and the tendency gets worse with smaller cells. In vertical recording, however [Fig. 3], the opposing poles of the two cells lie next to each other, so the magnetic fields reinforce each other. This produces an extremely sharp cell boundary and thus the possibility of smaller cells.

But there are difficulties in creating media with vertical magnetic orientation. Since the media are far wider than they are tall, shape anisotropy makes it much easier for magnetic moments to lie in the plane of the media—a lower energy situation than the perpendicular orientation. This tendency must be overcome for vertical media.

One route to at least partially achieving vertical alignments is to develop isotropic media—media that have no intrinsic orientation and that therefore allow themselves to be vertically oriented by an external field. The Spin Physics Division of Eastman

Kodak in San Diego, Calif., has successfully developed such an isotropic medium based on particulate oxides [Fig. 4]. It is currently being used in tapes that Spin Physics is marketing for instrumentation recorders. Several disk-drive manufacturers are now assessing its potential for application to disks.

To achieve isotropy, the iron oxide particles are doped with cobalt in such a way that the cobalt's natural crystalline anisotropy competes with the shape anisotropy of the particles. The net result is a uniform magnetization.

When an isotropic medium is subjected to the fringing field of a recording head, both horizontal and vertical components are recorded [Fig. 5]. Since there is a phase shift between the two components, reinforcement can be achieved if the gap is made very small. James Lemke of Spin Physics has shown that very high recording densities can be obtained with extremely small (0.25-micrometer) gap heads.

The Spin Physics medium has demonstrated a signal-to-noise ratio of 48 dB at a linear density of 15 000 flux reversals per centimeter. Since this is far above the 25 dB generally required for digital recording, it can be calculated that tracks only 2-micrometers wide would still give good performance. This implies a maximum density of 75 million b/cm^2—almost eight times better than the density promised by any other medium so far.

One possible drawback of the Spin Physics medium, however, is that particles doped with cobalt (as opposed to coated with it) have coercivities that are temperature-dependent, about 10 oersteds per degree Celsius.

Cobalt-chromium alloy tried

An alternative, and more widely pursued, approach is to develop a medium that will support only vertical magnetization. The first successful effort to do this occurred in the mid-1970s when S. Iwasaki and his colleagues at Tohuku University in Japan showed that when a cobalt-chromium alloy is sputtered onto a substrate, a vertical anisotropy can be induced.

In the case of pure cobalt, the anisotropy induced in the sputtering process is not enough to overcome the demagnetizing field. Dr. Iwasaki, however, recognized that by alloying the cobalt with chromium, which is easily done in sputtering, the magnetization, and hence the demagnetizing field, would be reduced. Indeed, films containing more than 13 percent atomic chromium have a perpendicular magnetization. Fortunately they also have coercivities in a range that makes them practical.

Dr. Iwasaki proposed writing on such media with a single pole head resembling a bar magnet. He also suggested a permalloy underlayer beneath the Co-Cr film to provide a return path for the magnetic flux. As these features all represented major departures from the existing technology, vertical recording remained simply a curiosity for a number of years. However, in 1980 Robert Potter and Irene Beardsley of IBM Research Laboratories in San Jose, Calif., showed, by computer modeling, that one could reap some of the higher-density benefits from a vertical medium with a standard Winchester head. This finding accelerated interest in this technology, and it is now being actively developed. Dr. Potter left IBM to help found the Lanx Corp. in San Jose, Calif., which plans to offer sputtered vertical media for rigid disk drives. Another company, Vertimag Systems in Minneapolis, Minn., is developing a flexible disk drive on vertical media. But the most ambitious efforts in developing this technology are being conducted in Japan.

Almost every well-known Japanese electronics company is investigating vertical recording. Some are focusing on consumer applications, others on data storage. Toshiba has already announced a flexible disk drive based on the Co-Cr medium that it hopes to have in production in two years. It reports a linear density of 20 000 flux reversals per centimeter. On a 3½-inch disk with a track density of 57 tracks per centimeter, this translates into 3 megabytes per side.

The Anelva Corp. in Toyko is developing a continuous roll-feed sputtering process and expects to produce flexible disks with vertical magnetization. However, at present this medium is not durable enough to be used in contact recording.

Electroplating instead of sputtering

Alternatives to sputtering are also being explored. Tu Chen of Xerox and Pietro Cavallotti of the Milan Polytechnical Institute in Italy have developed an electroplating process for obtaining a vertically oriented medium. Cobalt films were electroplated from solutions of cobalt sulfate. It was found that at a pH of 6.5 and a current density of between 10 and 20 milliamperes per centimeter

Jim Lemke

[4] Scanning electron micrographs (magnified 20 000 times) of the surface of Spin Physics' isotropic magnetic medium (top) shows an absence of the graininess evident in the conventional floppy disk such as one from IBM (bottom). Symmetric particles disperse more uniformly than the needlelike particles in conventional medium. The small grain size, combined with the medium's ability to use both vertical and horizontal magnetization, increases the recording densities.

squared, the cobalt was deposited in the form of needlelike crystals, with both the axis of the particle and the hexagonal crystal axis normal to the film.

As a result, both the crystalline and the shape anisotropy favor vertical orientation of the magnetization. The particles are isolated by nonmagnetic material that is segregated in the grain boundaries during the cellular growth. At higher current densities the film growth is dendritic, much like ice crystals, with a mixture of orientations. Magnetically this oriented cobalt is a candidate for vertical recording, but its recording properties have not yet been determined.

Another alternative is based on barium ferrite particles. These particles are single-domain hexagonal platelets less than 0.2 micrometer in diameter. The crystalline anisotropy of these particles results in the magnetization being perpendicular to the plane of the platelet. Thus, when such platelets are coated onto a substrate, they tend to lie flat and a vertically oriented medium is obtained. This approach has the advantage of being compatible with conventional coating techniques and is corrosion-resistant. Toshiba is establishing a tape-manufacturing capability based on these particles.

There is debate as to just how advantageous vertical recording will turn out to be. Some resolution curves show linear densities in excess of 50 000 b/cm. However, the readback voltage for normalized track width tends to be small with chromium-cobalt alloys. LANX's first disks, for example, designed for use with a Winchester head, will have a conservative linear density of 8000 to 10 000 flux reversals per centimeter.

Vertical media need not be extremely thin to have high recording density, and this may be a great advantage in manufacturing, if the problems of sputtering and its alternatives can be overcome.

Magnetooptical recording: a radical approach

The most radical departure from conventional magnetic disks now under investigation is magnetooptical recording. In this ap-

Disk basics

Magnetic recording occurs when a current passes through an electromagnet, producing a magnetic field that orients the direction of magnetization along a track in a nearby magnetic medium—a disk or tape. When such an oriented magnetic pattern passes under the electromagnet, it produces a voltage in the coil, allowing the record to be read. The combination of the electromagnet and the structures on which it is mounted along the radius of the disk is called the head. In flexible—or floppy—disk drives, the heads make contact with the medium. To minimize wear, such disks typically turn at 300 revolutions per minute. In higher-performance drives, where much faster access time is desired, rigid disks that turn at 3600 rpm are used and the heads "fly" above the disk on an airflow cushion (A). The most popular rigid disk drives today are the Winchesters, which are derived from IBM's 1973 storage module drive.

For digital applications, where it is desirable to have the field of the medium fully parallel or fully antiparallel to the magnetizing direction, the medium must be strongly anisotropic—that is, it must respond much more readily to magnetic fields along a certain axis than along any other axis. Such anisotropy can be caused either by the crystalline structure of the magnetic material, which creates preferred magnetization directions along a crystalline axis, or by the shape of magnetic particles. An ellipsoidal particle, for example, orients magnetic moments along the long axis of the ellipse.

In the material commonly used for disks, gamma-Fe_2O_3 (gamma referring to the crystalline structure), shape anisotropy is used, since the natural form has the highly symmetric cubic crystal structure. To grow needlelike particles, crystals of oxohydroxide alpha-(FeO)OH are first grown, because they have a naturally elongated shape. These are then chemically converted into gamma-Fe_2O_3.

To construct a recording medium with these particles, they are first mixed with an epoxy. A phenolic is then added to crosslink the epoxy into a solid. "Paint additives" control the flow of the coating. After filtering, the mixture is coated onto Mylar sheets or spin-coated onto aluminum disks. Because of the needlelike shape of the particles, they tend to lie in the plane of the medium.

The familiar floppy disks are cut from the coated Mylar. The particles here are irregular and unoriented in the plane of the disk. In the case of rigid disks, they are spun in a magnetic field during the drying process to orient the particles tangentially. The disks are then cured, polished, and cleaned.

In 1979 IBM replaced the hand-wound, ferrite-core inductive heads with a thin-film version, consisting of a permalloy yoke threaded by a copper spiral winding (B). The head is fabricated so that when it is not energized, the magnetization of the permalloy lies parallel to the track width and the plane of the disk. When a write current flows through the coils, the magnetization rotates 90° to generate a magnetic field across the gap. In readback, however, the fields from the medium merely tip the magnetization in the domains up or down. This permits a much higher frequency response than what the conventional Winchester head gives. —R.M.W.

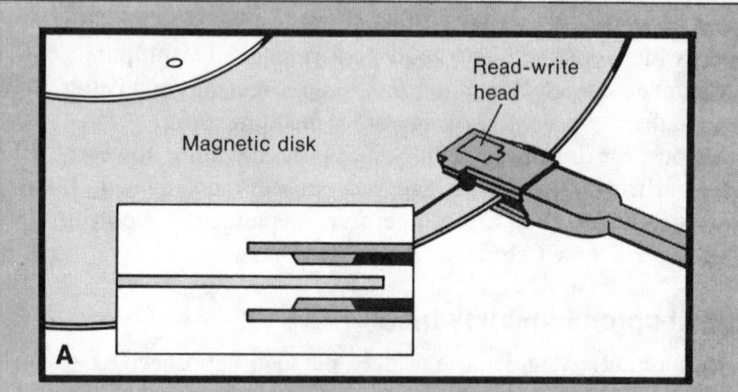

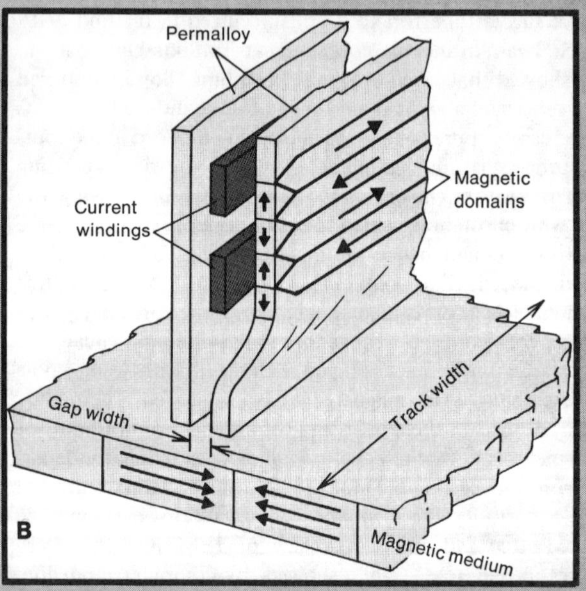

An electromagnet mounted on a sliding device (the "head" in A) is used to both record on and read from a magnetic disk (the inset shows the profile of the head and disk, and the spacing between them). The thin-film head (B) generates a magnetic field across its gap to record data when it is energized by a current through its coils, and generates a current when reading the fields in the magnetic disk.

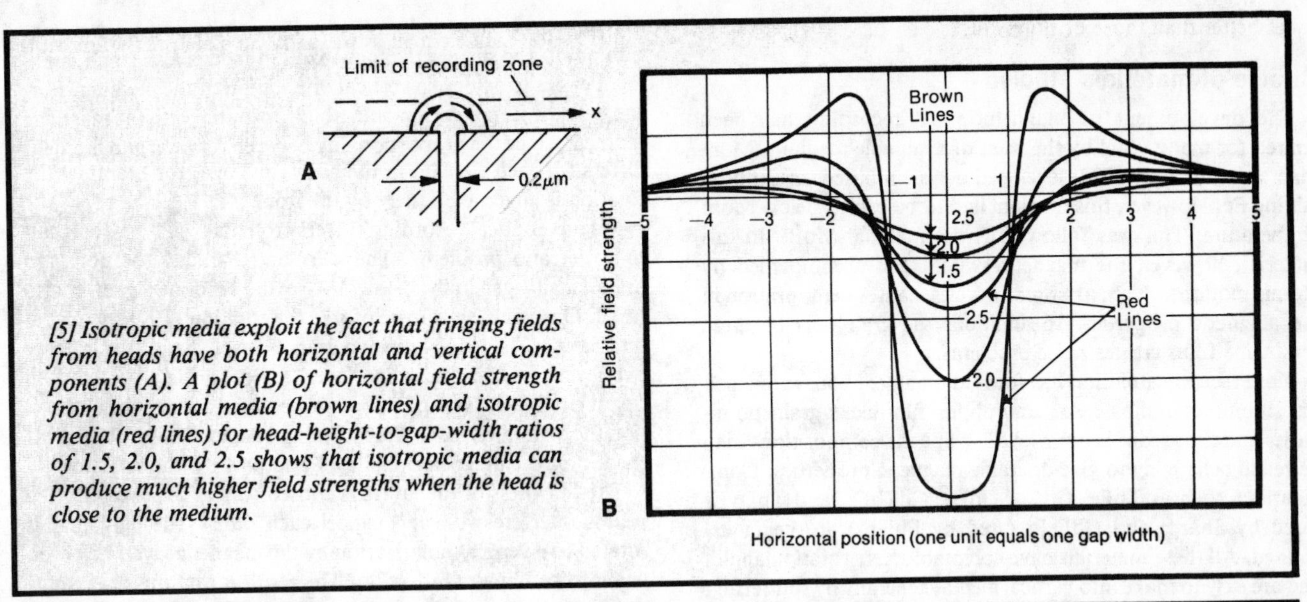

[5] Isotropic media exploit the fact that fringing fields from heads have both horizontal and vertical components (A). A plot (B) of horizontal field strength from horizontal media (brown lines) and isotropic media (red lines) for head-height-to-gap-width ratios of 1.5, 2.0, and 2.5 shows that isotropic media can produce much higher field strengths when the head is close to the medium.

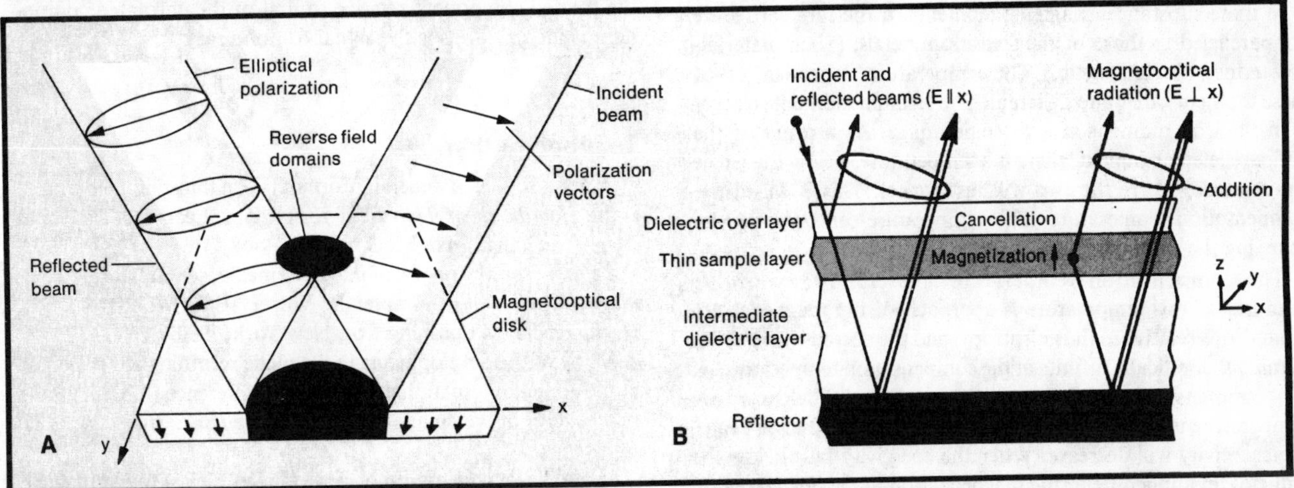

[6] In magnetooptical recording, a writing laser beam heats a small region so that it can be magnetized. Reading (A) uses another weaker linearly polarized beam. Because of the Kerr effect, a beam reflected off a vertically magnetized material will be elliptically polarized with a small component in the y-direction. The orientation of the component—plus y or minus y—will indicate the magnetic field orientation in the media. Since the signal is weak, some magnetooptical disk designs use a four-layer structure (B), which cancels out the portion of the incident beam reflected from the surfaces above and below the magneto-optical layer while allowing the signal originating within the layer to be reinforced, thus increasing the signal-to-noise ratio.

proach, recording is done by raising the temperature of the recording medium by way of a focused laser beam, thus reducing its coercivity. A magnetic field is simultaneously imposed, the beam passes on, the temperature drops, and the magnetization is then resistant to change, except by the same process. Reading is done by either the Kerr or the Faraday effect.

In the Kerr effect [Fig. 6], a linearly polarized beam reflected off a vertically magnetized surface will have its polarization partly rotated. This creates an elliptical polarization, with the axis of the ellipse depending on the direction of magnetization. The perpendicular components of the polarization can then be separated and measured to detect the direction of magnetization. The Faraday effect is similar, but the polarized beam passes through the material rather than being reflected from its surface.

Vertically oriented media must be used in magnetooptical techniques to achieve the maximum effect. However, the magnetooptical approach has advantages over simple vertical recording. Its reading technique does not depend on the volume of magnetized material, so potentially better signal-to-noise ratios can be obtained for equivalent aerial densities. The size of the magnetized cell is limited only by the wavelength of light used and, with ultraviolet lasers, could yield densities as high as several hundred Mb/cm². In addition the laser can be quite distant (several millimeters from the disk), eliminating the problems of small head-to-disk clearances.

A number of companies are working to develop magneto-optical disks. NV Philips Gloeilampenfabrieken of Eindhoven, the Netherlands, and the Sharp Corp. in Japan have already announced prototypes of small disk drives. The Philips system stores 10 megabytes on a disk 5 centimeters in diameter—between 10 and 100 times more than a conventional 13-cm floppy disk. The packing density of 4 Mb/cm² is twice that achieved with rigid-disk systems. Philips has said that its product, to be available in about two years, is aimed at such applications as personal computers and text-processing units.

The Philips disk consists of a 500-angstrom-thick magnetic film deposited over a transparent, 3-mm-thick plastic substrate. A 3-milliwatt laser records 2-by-5-micrometer spots in 5-micrometer-wide grooves and reads the disk by the Faraday effect at a power of 1 mW. Access times are about 50 ms—nearly 10 times better than those of floppy disks.

Choice of materials studied

The development of magnetooptical recording has been limited for many years by the want of a suitable medium. More than a decade ago, IBM developed a system based on Eu-containing Fe. However, this medium had to be cooled below room temperature. This was followed by the material MnBi. In this material, however, the magnetic switching is accompanied by atomic motion, which, after many cycles, causes deterioration in the magnetic properties. In addition, the graininess of these crystalline films creates noise problems.

More recently attention has focused on amorphous rare-earth transition-metal alloys. An amorphous film lacks grain boundaries and thus reduces noise levels, while the compositions may be readily changing to give desirable magnetic properties. Combinations such as TbFe, GdCo, GdFe, GdTbFe, and TbDyFe (used by Sharp) and GdFeBi (used by Philips) are being examined. All these materials have acceptable Kerr rotation angles and are easy to make into vertical media, generally by sputtering.

In these substances magnetic moments of the rare earths are antiparelleled to those of the transition metals. (Such materials are termed ferrimagnetic.) The temperature dependencies of these moments are quite different. The terbium moment is larger than the iron moment at low temperatures. As a result of the antiparelleled alignment, there is a temperature, below the Curie temperature, where the two will just cancel. This is called the compensation temperature. This temperature can be varied by changing the composition.

The compensation temperature is useful for recording because, as this temperature is approached, the magnetic moment drops relative to the anisotropy, and the coercivity rises, becoming theoretically infinite at the compensation temperature. If the compensation temperature is set slightly below room temperature and the medium is heated by a focused laser beam, the coercivity will decrease. When the coercivity falls below the value of an applied field, the magnetization in the heated region will align itself with the field. Notice that this writing process does not require heating to the Curie point. In fact, cobalt may be added to the material to raise its Curie temperature.

This writing scheme is advantageous not only in reducing energy requirements. More important, the Kerr effect decreases as the Curie temperature is approached. With low temperatures, the signal may be reduced because of the heating during readout. But with materials with high Curie points and low compensation temperatures, writing and reading can be accomplished without approaching the Curie point. In addition, when the overall level of magnetization is small, as it is in the vicinity of the compensation point, the surface demagnetization field is also small, making it easier to achieve a vertical orientation of the medium. Fortunately the Kerr rotation does not drop in step with the magnetic moment, since the proportionality between rotation and moment differs for each element in the alloy. In TbFe, for example, the rotation is governed largely by the iron.

Magnetooptical recording suffers from its own type of signal-to-noise ratio problems. The Kerr and Faraday rotations are quite small—generally below 1°—so the readout signal is also small. One solution is to reinforce the rotation effect by optical interference. A four-layer coating is used here: a dielectric overlayer, the thin recording layer, a second dielectric, and an opaque reflector. When the beam is polarized along the X direction (horizontally), as in the incoming beam, part is reflected off the top of the magnetooptic layer and part off the reflector.

The thickness of the layers is chosen so that the two reflected beams interfere with and cancel each other. The light that is rotated to a vertical polarization by the magnetic layer, however, is reinforced by the reflection. The result is that the strength of the rotated component relative to that of the unrotated component is increased, as is the signal-to-noise ratio.

Parameters of coated and sputtered oxide disks compared

	Coated	Sputtered ferrite
Recording density (BPI)	12 000	24 000
Film thickness (μm)	0.7	0.18
Surface roughness (μm)	0.12	0.02
Coercive force (Oe)	400	700

To probe further

A good source of general information on thin-film preparation is the *Handbook of Thin Film Technology,* Leon I. Maissel and Reinhard Glang, Eds., McGraw-Hill Book Co., New York, 1970. For a more specific review of magnetic media, see G. Bate in *Ferromagnetic Materials,* Chapter 7, Vol. 2, E.P. Wohlfarth, Ed., North-Holland Publishing Co., New York, 1980.

Many of the developments in magnetic recording are presented at the annual International Magnetics Conference (Intermag), the proceedings of which are published in the *IEEE Transactions on Magnetics.*

Iwasaki's demonstration of vertical recording is found in *IEEE Transactions on Magnetics MAG-14,* 849 (1978); and Lemke's introduction to isotropic media is given in *IEEE Transactions on Magnetics MAG-15,* 1561 (1979). Another general source of articles on magnetic media and heads is the proceedings of the annual Conference of Magnetism and Magnetic Materials, which are published in the March 1982 issue of the *Journal of Applied Physics.*

Systems aspects of magnetooptical recording are found in the optical recording literature. A particularly good source is proceedings of conferences sponsored by the Society of Photo-Optical Instrumentation Engineers (SPIE). For example, *Optical Disk Technology,* the proceedings of the SPIE 329 (1982) contains a number of very good articles.

A particularly useful collection of papers dealing with detailed aspects of media preparation is the proceedings of the Symposium on Magnetic Media Manufacturing Methods held in Honolulu, Hawaii, May 25-27, 1983 published by Magnetic Media Information Services (Chicago).

◆

The author would like to thank his Xerox colleagues, in particular Dr. Tu Chen, for many enlightening discussions.

Part II
Media

On the Coercivity of γFe_2O_3 Particles

D. F. EAGLE AND J. C. MALLINSON
Ampex Corporation, Redwood City, California

Over the temperature range −180° to 300°C, measurements have been made of the coercivity of 5:1 acicular γFe_2O_3 particles. Samples of two different lengths, 2000 and 6000 Å, were included in the study. The coercive force is linearly dependent upon two terms, shape (αM) and magnetocrystalline ($\beta K/M$). These may be separated by reducing identical particles to Fe_3O_4 and measuring their coercive force at the isotropy point −143°C, yielding the shape factor α. It is found that the shape term accounts for approximately 67% of the observed room-temperature coercive force. The shape term is almost independent of particle diameter and fits well with the "chain-of-spheres" reversal model. The remaining 33% of the coercivity has the same temperature dependence as the coercive force of cubic γFe_2O_3 particles, thus confirming its magnetocrystalline origin.

The room-temperature coercivity of cobalt-doped cubic particles was measured as a function of particle diameter. Over the diameter range 300–1000 Å the data again fits the linear addition of shape and magnetocrystalline terms. The shape term is diameter-dependent and fits the micromagnetic "curling" reversal model which has been found previously only in metallic whiskers.

INTRODUCTION

CONSIDER a fine particle held uniformly magnetized by a large field parallel to an easy axis. As this field is reduced to zero and increased in the opposite direction, there are two critical fields of interest. At some field, called the nucleation field, H_n, the magnetization starts to reverse. Dependent upon the subsequent stability of this state, the magnetization may either reverse completely, so that $H_n = H_c$, or, in the event of partial reversal, a further change in field may be required to reach the coercive force H_c. The equality of the nucleating and coercive fields has been rigorously proven only for infinitely long cylinders of any radius and for uniform rotation in small particles of any shape.[1] The critical radius for uniform rotation, R_0, is found to be close to the value $A^{1/2}M^{-1}$, where A is the exchange constant.

The critical fields are due to the linear addition of two terms: the local effect of crystal anisotropy K, and the effect of magnetostatic fields, arising from particle morphology and inter particle interaction, termed shape anisotropy. Thus,

$$H_c = \alpha M + \beta(K/M).$$

The calculated shape factors, α, fall into two classes. Those associated with reversal modes in which the magnetization remains locally uniform lead to α, dependent only on particle shape. Examples are the "rotation-in-unison"[2] and the "chain-of-spheres"[3] models. On the other hand, the nonuniform reversal models, termed "curling" and "buckling,"[1] lead to size-dependent shape factors of the form $\alpha = kR_0^2R^{-2} - N_{11}$, where N_{11} is the major-axis demagnetizing factor and k is a dimensionless factor of order unity.

For uniform reversal modes, the anisotropy field factor β is of the order unity. In the case of nonuniform reversals, the theoretical situation is more complicated: the nucleation field due to anisotropy is of order unity regardless of particle size; the coercive field factor due to anisotropy is some decreasing function of particle size. The precise form of this decreasing factor has not yet been calculated.[4]

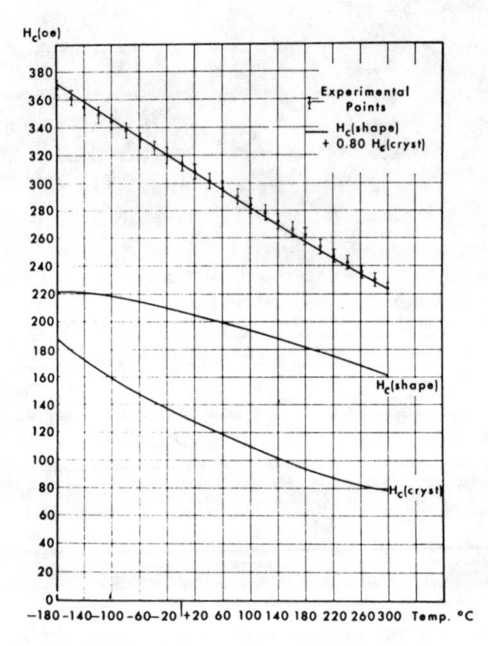

FIG. 1. Coercive force vs temperature for pure γFe_2O_3 acicular and equant particles.

ACICULAR PARTICLES

The coercive force and saturation moment of two commercial acicular γFe_2O_3 powders have been measured over the temperature range −180°C to +300°C.

[1] E. H. Frei, S. Shtrikman, and D. Treves, Phys. Rev. 106, 446 (1957).
[2] E. C. Stoner, E. P. Wohlfarth, Phil. Trans. Roy. Soc. 240, 599 (1948).
[3] I. S. Jacobs and C. P. Bean, Phys. Rev. 100, 1060 (1955).
[4] A. Aharoni, Rev. Mod. Phys. 34, 227 (1962).

Reprinted with permission from *J. Appl. Phys.*, vol. 38, no. 3, pp. 995–997, Mar. 1, 1967.

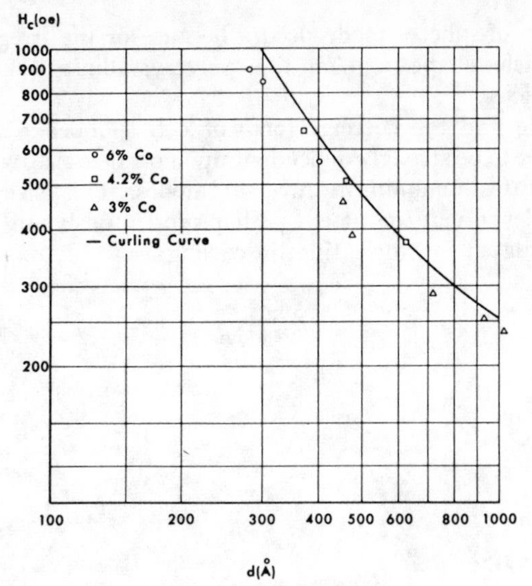

FIG. 2. Coercive force vs size for cobalt-doped γFe_2O_3 equant particles.

Both types showed, by electron microscopy, length-to-width ratios of approximately 5:1. Chemical purities were better than 99%; and the saturation moments at room temperature were between 340 and 350 G (versus the theoretical figure of 370 G). The mean particle lengths were approximately 2000 and 6000 Å, respectively. The coercive force data is similar to that previously reported.[5]

The temperature variation and magnitude of coercive force were found to be quite similar, suggesting a locally uniform reversal mode. In order to separate the shape and anisotropy contributions to the coercive force, one of the samples (6000 Å length) was chemically reduced to Fe_3O_4 for which $K=0$ at $-143°C$.[6] At this temperature the Fe_3O_4 coercive force is determined solely by shape. This shape factor will be exactly correct for γFe_2O_3 if the reversal is locally uniform. If the reversal is nonuniform the shape factor will be slightly different from that applicable to γFe_2O_3 due to the small difference in the critical radius $R_0=A^{1/2}M^{-1}$.

Assuming a locally uniform reversal, we write the shape contribution to the coercive force of γFe_2O_3 as

$$H_c=(H_c/M_s)_0 M_s$$

(where subscript 0 refers to Fe_3O_4 at $-143°C$) as plotted in Fig. 1.

The residual fraction of the coercive force, which varies from 30–40%, is presumed to be of magnetocrystalline origin. This presumption was confirmed as follows. The magnitude and temperature coefficient of the coercive force of 700 Å diameter equant particles of γFe_2O_3, for which $\alpha=0$, was measured (see Fig. 1). By curve fitting it was found that, over the whole temperature range, the residual fraction was equal to 0.80 ± 0.04 of the equant particle coercive force. Thus we have

$$H_c=0.56M_s+0.80H_c(\text{equant})\pm5 \text{ Oe}.$$

The shape factor 0.56 may be compared with the theoretical factors applicable in the absence of particle interactions. For unoriented samples switching by coherent rotation this is $0.479\times2\pi$ and for the chain-of-spheres asymmetric fanning mode 1.08. We may crudely allow for interactions by applying the $(1-p)$ packing factor correction, yielding, for the 40% volume packing measured in these experiments, a corrected chain-of-spheres shape factor of $0.6\times1.08=0.65$, in fair agreement with the data.

EQUANT PARTICLES

Data have been taken at room temperature on the coercive force of several unoriented samples of cobalt-doped γFe_2O_3 equant particles, prepared by P. Hwang of the authors' company. All such samples showed an essentially cubic shape and had saturation magnetizations within 90% that of pure γFe_2O_3. The principal variables studied were the cobalt content and the particle size.

It was found that the cobalt content had no effect upon the coercive force, which depended only upon the particle size (see Fig. 2). This implies that either K/M is not a prime factor controlling the coercivity or that K/M is independent of cobalt doping in the 3–6% range used in this work. The latter was found, by measurements of the ferromagnetic resonance line width,[7] to be the case. These measurements showed that whereas cobalt dopings below 3% had, as expected, anisotropy constants proportional to the cobalt level, cobalt doping in the range 3–6% always produced K values of about $+10^6$ ergs/cc. This behavior was not investigated further.

The coercive force for uniform rotation $(0.64K/M)$ is therefore expected to be approximately 1750 Oe. That the measured coercive force is strongly dependent upon size indicates a nonuniform reversal mode. While the authors are aware of the distinction between nucleating and coercive fields, we should like to point out that the nucleating field calculated for curling in spheres fits the data quite well. The curling equation used was

$$H_n=[2\pi(1.39)(d_0/d)^2-\tfrac{4}{3}\pi]M+0.64\,K/M.$$

The only adjustable parameter in the curve fitting is the critical diameter $d_0=2A^{1/2}M^{-1}$. A value of $d_0=150$ Å was chosen, which assumes a value of 10^{-7} erg/cm for

[5] A. H. Morrish and L. A. K. Watt, J. Appl. Phys. **29**, 1029 (1958).

[6] L. R. Bickford, *et al.*, Proc. Inst. Electr. Eng. **B104**, 238 (1957).

[7] E. Schlomann, Proc. Conf. Magnetism, Magnetic Materials, AIEE Spectrum Publication **T-91**, 600 (1956).

the exchange constant A. However, this constant is not accurately known (the values quoted in the literature for pure iron span a factor of ten).

CONCLUSIONS

Acicular particles switch in a mode similar to the asymmetric fanning chain-of-spheres model with some magnetocrystalline anisotropy added. Presumably the chain-of-spheres mode occurs because of the irregular particle shape and/or the polycrystallinity of the particles.

We find the coercive force of cobalt-doped γFe_2O_3 cubes to be strongly dependent upon particle size, which indicates nonuniform reversal modes. This size dependence may be fitted, perhaps fortuitously, to that calculated for nucleating by curling.

MAGNETIC PROPERTIES OF INDIVIDUAL ACICULAR PARTICLES

J.E. Knowles

Abstract - A review is made of the work on the measurement and interpretation of the properties of individual particles, of the type incorporated into recording tape. It was shown that the remanent coercive force $H_r(0)$ of particles of any of the materials measured covered a range of several hundred Oersted. A high proportion of the particles were multiple, that is they comprised several particles lying side by side: it is known that most of the particles in recording tape are of this type. An investigation was made of the nature of the reversal process, and of the origin of the range in $H_r(0)$, for particles of γ-Fe_2O_3, CrO_2 and metal powder (iron). The variation of $H_r(0)$ with angle between the field and particle axis was determined for γ-Fe_2O_3 and CrO_2. The results of the above work were used in an attempt to synthesize the remanent loop of a tape from the knowledge of the properties of its constituent particles.

1. INTRODUCTION

This paper reviews the work [1,2,3] on the measurement and interpretation of the properties of individual particles, of the type used in the manufacture of recording tape. An immediate objective of the investigation was to attain a better understanding of the magnetisation reversal process in such particles, and this aim was to some extent realized. A long term objective was to be able to predict the magnetic properties of the recording tape from a knowledge of the properties of its individual particles, but attainment of this aim is certainly some way off.

There are many papers in the literature on the theoretical properties of idealized particles, both in isolation and packed together in a compact, and on the actual properties of compacts of acicular particles: see e.g. [4]. However, without a knowledge of the properties of isolated real particles, attempting to reconcile these two viewpoints is an unrewarding task. This was of course recognised a long time ago, and in 1955-6 Morrish and Yu [5] made the first measurement of the coercive force of a single particle of γ-Fe_2O_3. This experiment, which was of exceptional difficulty, showed that the coercive force was 800 Oe, which even allowing for the effect of packing is much greater than that of a recording tape, which was then probably around 250 Oe. Although this result was very valuable, the fact that it referred to only one particle was a limitation which the present work overcame.

In what follows, the experimental method is first briefly described, and results are then presented for the remanent coercive force of large numbers of particles of each of the commonly used materials, e.g. γ-Fe_2O_3, CrO_2 and metal powder (iron). Experiments to determine whether the particles are single or not are also discussed, together with the factors which contribute to the range in coercive force found in particles of a given sample. Observations to find the variation of the coercive force with angle between the particle axis and the applied field are also described. Finally, as a first step in attempting to realize the long term aim mentioned above, an attempt was made to use the results obtained on a sample of γ-Fe_2O_3 particles to synthesize the remanent loop of the corresponding tape.

It is a common practice to use cgs units when discussing recording materials, and this is followed here. Values of field are also given in MKS units.

2. EXPERIMENTAL METHOD

The method of measurement [1,2] is now briefly described.

A very dilute suspension was made of the particles in a transparent viscous lacquer, and the particles observed with the highest power of the optical microscope, the image being displayed on a TV monitor. The dilution was such that in general only one particle was visible on the TV screen. A brief search was made to locate a suitable particle for measurement, i.e. one that was apparently single and remote from its neighbours.

The particles were aligned with a continuously applied field of a few Oersted, and an opposing pulse field applied. If the magnitude of this was sufficiently large to reverse the magnetization of the selected particle, then under the influence of the small field the particle subsequently rotated through 180°, and so by a process of successive trials and interpolation the remanent coercive force was quickly determined: since the packing density p was very nearly zero this was designated $H_r(0)$. After each 'trial' pulse, the particle was put into the remanent state $M_r(\infty)$ by applying a large 'set' pulse, of the same sense as the 'trial' pulse. Since the pulse direction was opposed to that of the small constant field, the particle then rotated, and so was then conveniently ready for the next trial. By measuring the rate of rotation of the particle upon reversing the small field, the reduced remanence $M_r(H)/M_r(\infty)$ could also be obtained. It was noted that if the magnitude of the pulse field happened to be nearly the same as $H_r(0)$, then the particle rotated very slowly and came to rest with its axis making an angle of approximately 90° to the constant field, as predicted by Tjaden: see [6].

The advantages of the above procedure were:

1) The measurement procedure was easy, so that a large number of particles were measured. This was an important advantage, for it was found that there was a wide variation in properties between the particles of a particular sample.
2) The reverse field was automatically applied along the magnetic axis of the sample.
3) The particle under observation could be readily photographed.

This last point leads to a consideration of the limitations of the technique:

4) Whilst optical microscopy was adequate to indicate whether the object under examination was nominally a single particle, it was quite inadequate to reveal any correlation between the appearance of a particle and its properties.
5) The properties of a particle could be measured only in a remanent state, so it was not possible to measure e.g. $M_r(\infty)/M_s$.
6) The observer had to decide whether to measure a particular particle or not, which meant deciding

Manuscript received March 23, 1981.

Philips Research Laboratories,
Redhill, Surrey RH1 5HA, England.

Reprinted from *IEEE Trans. Magn.*, vol. MAG-17, pp. 3008-3013, Nov. 1981.

whether it was single or an agglomerate. Some assistance was found by watching the particle as it rotated, when an erstwhile 'single' particle was sometimes seen to be double. A similar effect was occasionally noted when a large pulse field was applied.

3. RESULTS AND INTERPRETATION

γ-Fe_2O_3

Measurements of remanent coercive force $H_r(0)$ were made on from 25 to 100 particles of γ-Fe_2O_3 from several sources, and the results for each sample plotted as a histogram as shown in Fig. 1. Similar histograms were obtained for all the materials examined (CrO_2 etc.), being approximately Gaussian in shape, and having reduced standard deviations in the range $0.15 < \sigma/H_r(0)(\text{mean}) < 0.25$ but truncated at $\sim 0.6H_r(0)(\text{mean})$. The mean of $H_r(0)$ is not of much significance, for H_r of an assembly of aligned, widely separated particles is given by the median value.

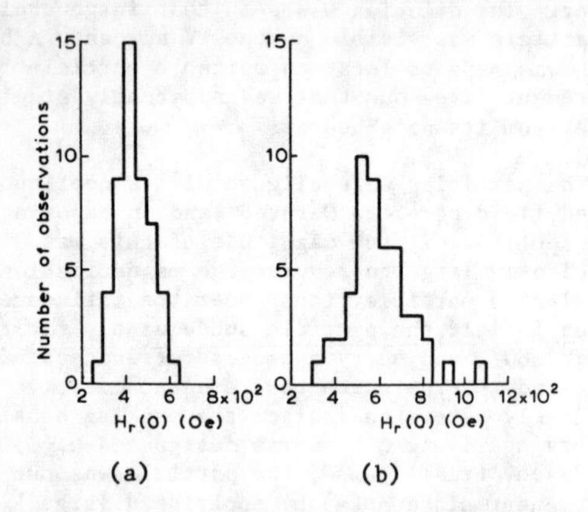

Fig. 1. Histograms of H_r for 50 particles of γ-Fe_2O_3 from (a) sample A; (b) sample C.

The histogram of Fig. 1a (sample A) was obtained for particles suspended in a dilute lacquer which was specially prepared by a tape factory, and further diluted for use in the experiment. The histogram was one of the narrowest found, and it was at first believed that this was due to the particles having been subjected to a very long milling procedure in order to produce a very good dispersion for use in this experiment. Trials using a different grade of oxide indeed showed that a prolonged milling appreciably narrowed the histogram. However, a batch of the same oxide as sample A was subsequently dispersed in the laboratory. This suspension gave a histogram very similar to that of Fig. 1a, indicating that this particular grade of oxide had an intrinsically narrow range of $H_r(0)$.

For sample A, $H_r(0)(\text{max})$ was much less than the value of $H_r(0)$ found by Morrish and Yu [5], which was disquieting. Accordingly, a fresh tape of 'studio' quality was obtained, and the particles removed and dispersed in the laboratory. As shown in Fig. 1b, the histogram then obtained (sample C) had a much greater range than that of Fig. 1a. The value of $H_r(0)(\text{max})$ was 1070 Oe (85 kA/m), which is considerably more than the 800 Oe (64 kA/m) observed by Morrish and Yu: there is then no discrepancy between their result and those of the present work.

Besides the histograms, remanent loops were also obtained for a few particles from samples A and C. Figure 2a shows the loop of an inferior particle of sample A, and Fig. 2b that for the particle of sample C having the largest value of $H_r(0)$. That of sample A has a loop of poor rectangularity, a low value of $H_r(0)$, and the magnetization can adopt states intermediate between $\pm M_r(\infty)$. Obviously this does not accord with the theoretical loop of the conventional particle, which is perfectly rectangular. The loop of the particle of sample C is much closer to this ideal. As for a particle of CrO_2 measured previously [1], $H_r(0)$ was a little irreproducible, and so to obtain the loop shown in Fig. 2 no 'set' pulse was used. It is not known whether the magnetization could be set to values lying on the steep side of the loop.

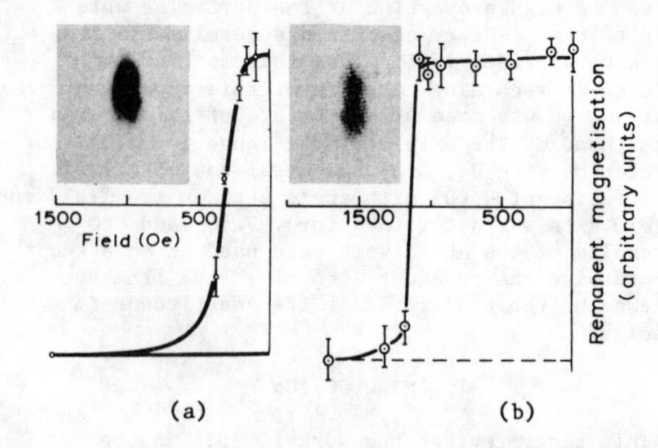

Fig. 2. Remanent loops and photographs of particles of γ-Fe_2O_3. (a) an 'inferior' particle, 0.9 μm long, of sample A; (b) the 'best' particle, 0.6 μm long, of sample C.

The existence of particles having very inferior properties raises questions as to whether these particles are single or not, and whether they are truly representative of those found in the corresponding tape. To determine the first question, the particles in the sample tube were first carefully demagnetized, the tube placed under the microscope, and the remanent magnetization $M_D(ac)$ of a particle determined, in arbitrary units. Without removing the sample tube, a saturating pulse field was applied, and the remanence found of the same particle. Now an ideal single particle cannot be demagnetized, so for such a particle the ratio $M_D(ac)/M_r(\infty)$ is unity. On the other hand, two identical particles lying side by side can in effect be demagnetized by causing their respective magnetizations to lie anti-parallel, and this ratio is then zero. Obviously, any intermediate value is possible. Figure 3 shows experimental points for 28 particles of a sample D, which were contained in a dilute lacquer specially prepared by a tape factory.

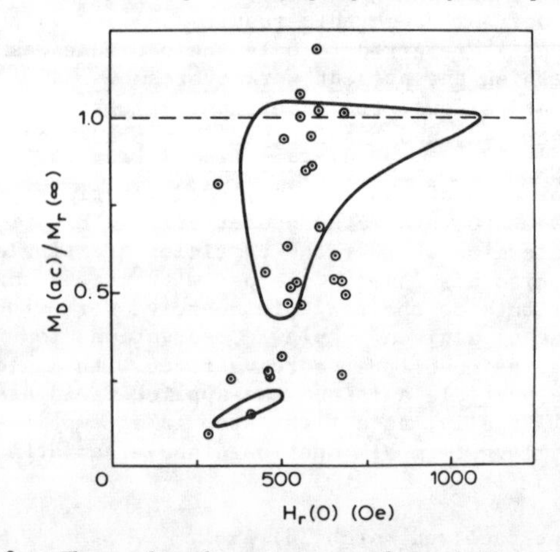

Fig. 3. The reduced remanence after demagnetization as a function of $H_r(0)$, for 28 particles of γ-Fe_2O_3 (sample D). Data points (not shown) for 50 particles of sample C lay within the two delineated areas.

The TV images of these particles of sample D were not very well defined, so the experimental error associated with each point is quite large, being typically ±0.16: thus some of the points lie above the line $M_D(ac)/M_r(\infty) = 1$. It is apparent that a high proportion of the particles can be partially demagnetized, so it seems probable that, as just suggested, these consist of two or more particles lying side by side. Such particles were postulated by Waring [7] who called them 'bundles', but the term 'multiple particles' was used by the present author [2], and is retained here. It is not known whether such particles are agglomerates held together by magnetic forces, or whether they are sintered together. The latter seems likely, for the application of a large magnetic field is expected to disrupt them, but this was seldom observed.

Also drawn in Fig. 3 are two regions enclosing 50 data points (not shown) for particles of sample C [2]. The distribution of the data points is different from that of sample D, but the conclusions are the same as for that material. The data point referring to the particle of Fig. 2b occurs at the extreme right of the enclosed region.

It is then clear that the observed particles were often multiple. They do however appear to be representative of those in the tape, for TEM photomicrographs of thin sections of recording tapes [8] show that a high proportion of the particles in the tape coating are multiple. Figure 4 shows a small region of one of these micrographs, of a tape made from particles of sample A. Unlike a normal TEM photograph, many of the particles do not lie in the plane of the photograph, and furthermore many of such particles have probably been cut short by the microtome.

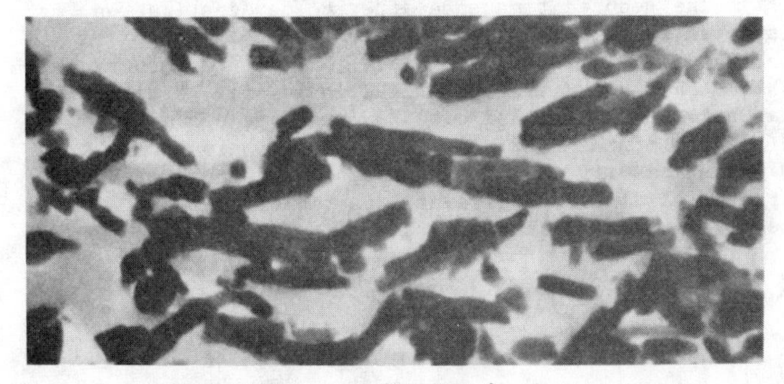

Fig. 4. A TEM photograph (50,000x) of a cross-section of tape of particles of sample A [8].

Reversal Mechanisms in γ-Fe_2O_3. It has been shown that for γ-Fe_2O_3, $H_r(0)$ lies between 250 and 1100 Oe (20 and 88 kA/m), and the origins of this range of values is now discussed. The procedure used is to estimate the values of the nucleation field (which is here equated with $H_r(0)$) predicted by each of the known mechanisms of fanning, buckling and curling: coherent rotation always predicts much larger values than those observed. Table 1 lists data and calculated results for three tape materials.

Table 1.

Material	M_s (emu)	$2K_1/M_s$ (Oe)	A 10^{-7}	$A^{\frac{1}{2}}/M_s$ 10^{-7}	d (μm)	S	$H_r(0)$, (Oe): theory		
							fann.	curl.	buck.
γ-Fe_2O_3	375	-247	0.8	7.5	0.05 0.09	3.3 6.0	(300) (380)	240 80	1360 920
CrO_2	483	950	1.74	8.64	0.03 0.05	1.7 2.9	(400) (480)	1100 400	2710 1930
Iron	1714	560	-	-	-	-	(900) (1750)		-

The values of M_s and $2K_1/M_s$ are taken from the review by Bate [9]. The saturation magnetization is taken to be the intrinsic value, rather than that measured on compacts of particles. The difference may be attributable to the effect of non-magnetic surface layers which lower the measured value of M_s, but the nucleation field is presumably determined by the magnitude of M_s in the interior of the particle. The value of the exchange length $A^{\frac{1}{2}}/M_s$ is due to Eagle and Mallinson [10]. The reduced radius S is given by [4]:

$$S = (d/2)/(A^{\frac{1}{2}}/M_s) \quad (1)$$

where d is the diameter of a single particle. Since considerable variations occur within a given sample, two values of d are given, which were obtained from TEM photomicrographs. The calculated values for $H_r(0)$ listed in Table 1 were obtained from the equations given by e.g. Kneller [4], but the contribution arising from the magneto-crystalline anistotropy field is not included in these figures.

Inspection of Table 1 shows that for γ-Fe_2O_3, the lower range of values found for $H_r(0)$ is consistent with the magnetization reversing by non-symmetric fanning. The bracketed pairs of values refer to particles having length/diameter ratios of from 3/1 to very large. The calculated values of $H_r(0)$ for reversal by buckling are consistent with the observed result of 1070 Oe (85 kA/m) for $H_r(0)(max)$. This raises a difficulty, for as shown in the Table, it is estimated that the curling process is associated with much lower values of $H_r(0)$, and so is energetically favourable. It is believed that this anomaly arises because the morphology of γ-Fe_2O_3 particles is so poor that curling cannot occur. The geometry of the curling process is rigorously defined: the magnetization everywhere remains parallel to the surface of the cylindrical particle so that there is no increase in magneto-static energy. A typical particle of γ-Fe_2O_3 is not cylindrical and contains many pores, so it is impossible to satisfy this condition. The buckling mode on the other hand is associated with a periodic fluctuation of the magnetization about the axis of the particle, resulting in an increase in magneto-static energy. It is expected that the buckling process can still occur in a non-cylindrical particle, although no doubt the nucleation field is somewhat different from the calculated value. It is concluded that for a particle to reverse by buckling, it must be fairly free from pores and irregularities of shape. It is undesirable for the shape to be too close to perfection, for it will then reverse by the curling process, with a consequent large decrease in $H_r(0)$. If, as is often the case, the morphology is very poor, then the particle will reverse by fanning. So far, the effect of the magneto-crystalline anisotropy has not been mentioned. In the theory of reversal by buckling and curling, only the case of uniaxial anisotropy with the easy axis aligned along the particle axis has been considered [4], and for fanning the effect of anisotropy is neglected altogether. Experiments on epitaxial crystals of γ-Fe_2O_3 [11] have shown that the anisotropy is cubic, with $K_1 = -4.64 \times 10^4$ erg/cm^3, so the easy axis is [111]. For particles of this orientation, whether single crystal or textured poly-crystal, then the uniaxial theory probably gives quite a fair estimate for $H_r(0)$. The effective anisotropy field is then $-4K_1/3M_s$ = 165 Oe (13 kA/m), and this figure is to be added to those calculated for buckling and curling, and probably fanning as well. For the much commoner case [9] of a polycrystalline particle having a [110] texture, it was shown [3] that the effective anisotropy field is $-\frac{1}{2}K_1/M_s$ or 62 Oe (5kA/m). This figure is rather small compared with typical experimental results for $H_r(0)$, as shown in Fig. 1.

There is then no difficulty in accounting for the range of values of $H_r(0)$ found in γ-Fe_2O_3.

γ-Fe_2O_3 with adsorbed cobalt.

Measurements were made [3] on particles of γ-Fe_2O_3 before and after cobalt was adsorbed: the details of the treatment are not known, but no doubt followed the general procedure described by Imaoka et al [12]. These particles were received dry and dispersed in the laboratory.

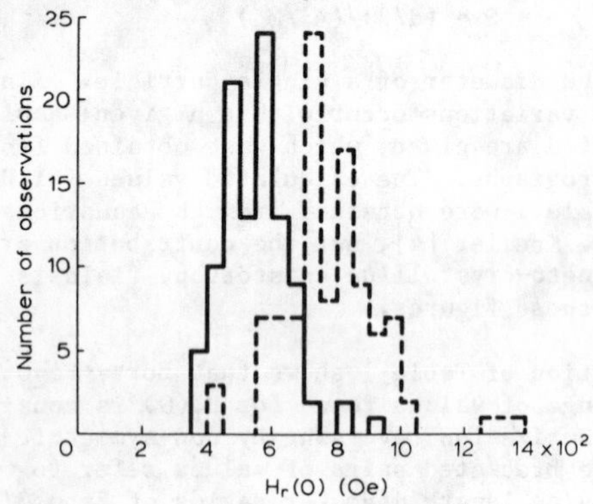

Fig. 5. Histograms of $H_r(0)$ for 100 particles of samples of γ-Fe_2O_3 before (left) and after (right) adsorption of cobalt.

The corresponding histograms are shown in Fig. 5 above. The interesting feature of these is that after treatment the left hand side of the histogram is displaced by only 50-200 Oe (4-16 kA/m), whilst the right-hand side is displaced by 300-450 Oe (24-36 kA/m). This is explained by making use of the conclusion which which has just been reached; that particles of largest $H_r(0)$ reverse by buckling. Now for this mode of reversal, $H_r(0)$ increases rapidly as the particle diameter decreases, which implies that those untreated particles which have a large value of $H_r(0)$ are of relatively small diameter. Such particles have a high surface/volume ratio, and so upon adsorption of cobalt give the largest increase in $H_r(0)$.

The effect of the treatment with cobalt was to increase H_c for the compacted powder by 52%, but the median value of $H_r(0)$ for the particles increased by only 37%. However, these figures are expected to be the same only if the well-known relationship [4] holds:

$$H_c(p) = H_c(0)(1 - p) \qquad (2)$$

Presumably a similar expression applies to $H_r(p)$ and $H_r(0)$. In section 4, a different expression (3) is given for the relationship between $H_r(0)$ and $H_r(p)$. This shows that the increase in H_c for the powder is indeed expected to be larger than the increase in the H_r for the particles.

$(\gamma$-$Fe_2O_3)_x(Fe_3O_4)_{1-x}$

Flanders and Kaganowicz [13] have measured the properties of particles, initially of γ-Fe_2O_3, which had been partially reduced. As mentioned above, the untreated powder was of the same grade as that of Sample A. As for γ-Fe_2O_3 with adsorbed cobalt, the effect of the treatment was most marked upon those particles of large $H_r(0)$, but in addition there was a general broadening of the distribution, accompanied by a reduction in the amplitude of the peak. It has been shown [14] that the anomalous properties of such powders can be explained if it is assumed that the particles consist of an inner core of γ-Fe_2O_3, together with an outer layer of Fe_3O_4: the outer layer is in tension and the core is in compression. The differences in character between the two histograms are consistent with this model.

CrO_2

Measurements of $H_r(0)$ were again made on particles suspended in a very dilute lacquer specially made by the tape manufacturer, and Fig. 6 shows a histogram of $H_r(0)$ obtained on 100 of these particles. $H_r(0)(\max)$ was 1260 Oe (100 kA/m), which was only 90 Oe (7 kA/m) larger than that found for γ-Fe_2O_3. Also shown in the Figure is the differentiated remanence curve of the corresponding tape, which resembles the histogram but is displaced from it by 160 Oe (13 kA/m): it is of course a common practice to use such a curve as an empirical measure of the range of $H_r(0)$.

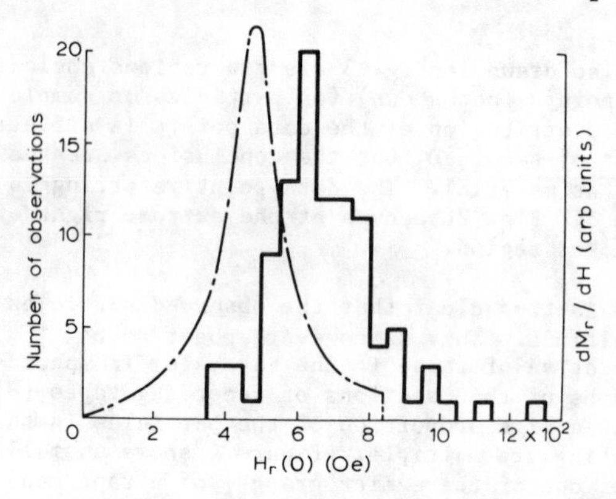

Fig. 6. A histogram of $H_r(0)$ for 100 particles of CrO_2. The differentiated remanence curve of the corresponding tape is also shown.

The reduced remanence after ac demagnetization is shown in Fig. 7. It is apparent that all the observed particles were multiple. TEM photomicrographs showed that the diameters of the individual particles lay between 0.03 µm and 0.05 µm, whilst the overall diameters of the multiple particles lay between 0.06 µm and 0.15 µm. It is possible that particles of small diameter, e.g. single particles, were not resolved by the optical microscope.

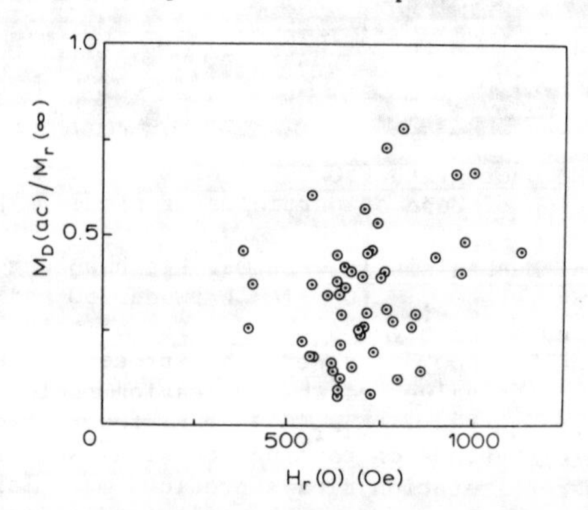

Fig. 7. The reduced remanence after demagnetization as a function of $H_r(0)$, for 50 particles of CrO_2.

These figures are incorporated into Table 1. The exchange constant A was derived from measurements of the anisotropy field [15] and wall energy [16]. Unlike γ-Fe_2O_3, curling now predicts values $H_r(0)$ within the observed range, whilst those predicted by buckling are much greater than this. As the morphology of particles of CrO_2 is far superior to those of γ-Fe_2O_3, there would seem to be every reason for them to reverse by curling, although the values of $H_r(0)$ calculated for non-symmetric fanning also lie around the lower limit of the histogram of Fig.6.

The situation is complicated by the particles being multiple, and by their possessing a large uniaxial anisotropy field of 950 Oe. It has been shown that 67% of one sample of particles had a [001] orientation [17], so that in these particles the 'easy' axis and the particle axis coincided [16]. In a typical multiple particle comprising three or four single particles, the probability that they do not all have a [001] orientation is then 70-80%. As always, the coercive force of the multiple particle is largely determined by that of the 'softest' constituent particle, which in this instance is the particle where the crystallographic axis makes the largest angle with the axis of the particle. A similar argument applies to the effect of the distribution in the diameters of the particles.

Variation of $H_r(0)$ with angle. The measurement technique outlined in Section 2 was modified in order to determine the variation of $H_r(0)$ with angle ψ between the axis of the particle and the applied field. Results for a particle of γ-Fe_2O_3 were reported earlier [1]: $H_r(0)$ for this particle was 570 Oe (45 kA/m), so it is probable that it reversed by fanning, and indeed the experimental curve of $H_r(0)(\psi)$ resembled that derived from the theoretical hysteresis loops of Jacobs and Bean [18].

Similar results were obtained for a particle of CrO_2, and these are shown in Fig. 8. It should be noted that this curve is not a curve of coercivity as a function of angle, which have quite a different shape [4]. For an 'ideal' single particle, which of course does not show minor loops, the method of measurement described in Section 2 determines the field at which the hysteresis loop closes: see e.g. Fig.3 of [18]. Below the critical angle ψ_0 which marks the transition between non-coherent and coherent rotation, this field is equal to the nucleation field. However, the particle of CrO_2 considered here was almost certainly multiple, and so probably could support at least some minor loops. In the limiting case of a particle able to support a very large number of minor loops, the measurement gives the value of $H_r(0)$, and this parameter has been chosen to designate the axis of Fig. 8.

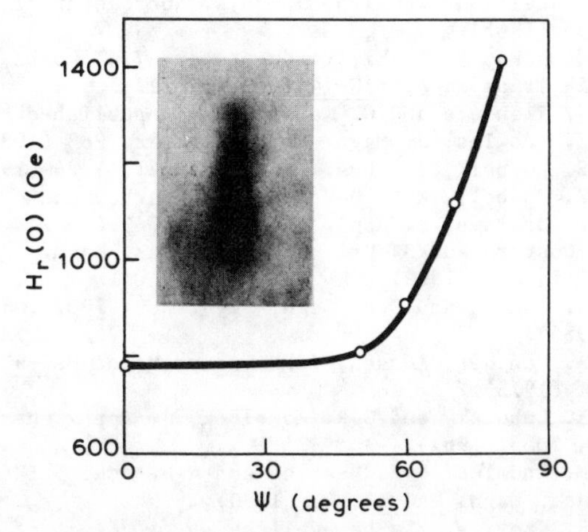

Fig. 8. The variation $H_r(0)$ with angle ψ between the particle axis and the field, for the particle of CrO_2 shown in the photograph: length ~1.5µm

Figure 8 shows that $H_r(0)$ increased only slightly until ψ was 50°, and it is probable that this is the value of ψ_0. Theoretical curves of H_c as a function of ψ for reversal by curling [4] show that this corresponds to a reduced radius S of 1.6, which is near to the lower figure shown in Table 1, of S=1.74. However, in calculating the theoretical curve the crystalline anisotropy was neglected, which is not a permissible approximation for CrO_2. Also, approximately the same values of S may be used to describe the relationship between H_c and ψ on the basis of fanning. Thus whilst the curve of Fig. 8 appears to be in accord with existing theories, it does not yield new information about the nature of the reversal process in CrO_2.

Metal Powder (iron)

The term metal powder is here used to refer to acicular particles of iron, stabilized by a layer of oxide [19].

Figure 9 shows a histogram of $H_r(0)$ for 100 particles of a very dilute lacquer, supplied by the tape manufacturer. An additional 7 particles were found which could not be reversed by the maximum available field of 1700 Oe (135 kA/m). As in Fig. 6, the differentiated remanence curve of the corresponding tape is also shown, which is displaced from the histogram by 220 Oe (17.5 kA/m).

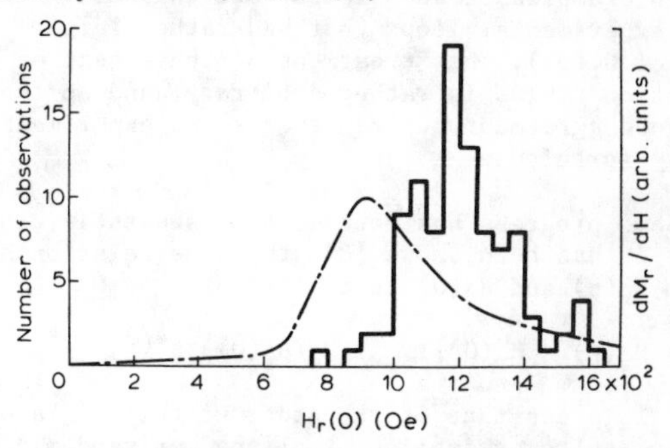

Fig. 9. A histogram of $H_r(0)$ for 100 particles of metal powder (iron). Also shown is the differentiated remanence curve of the corresponding loop.

The powder was prepared by reducing particles of α-FeOOH to iron and then partially oxidising them. This process caused some degradation of the morphology and this suggests that most reverse by fanning. For a chain of two spheres, $H_c(0)$ = 900 Oe (72 kA/m), and for three spheres then $H_c(0)$ = 1400 Oe (111 kA/m). For an infinite chain, then $H_c(0)$ = 1750 Oe (139 kA/m) [18]. These figures do not take the crystalline anisotropy field into account, which as shown in Table 1 is quite large. Like γ-Fe_2O_3, iron is of cubic symmetry, but the easy axis is [100]. Presumably the particles of oxide did not change their crystallographic orientation when they were reduced to iron, so the iron particles, like those of γ-Fe_2O_3, are expected to be polycrystalline with a [110] fibre texture. The average anisotropy field is then $-\frac{1}{2}K_1/M_s$ = -140 Oe (-11 kA/m), and it is supposed that the calculated values for $H_r(0)$ in Table 1 are smaller by this amount. For a particle of [100] orientation or texture, it is supposed that the calculated values are to be increased by $2K_1/M_s$ or 560 Oe (45 kA/m). Similarly, for a [111] orientation, they are to be decreased by 373 Oe (30 kA/m). In general, when the effect of the anisotropy is taken into account, albeit in a rather qualitative way, the fanning model accounts quite well for the experimental results.

It is to be remarked however, that Luborsky and Morelock [20] have reported measurements of $H_c(0)$ on individual iron whiskers. As these had a nearly ideal prismatic shape, it is probable that they reversed by curling. Their results, both theoretical and experimental, showed a very rapid variation of

$H_c(0)$ with particle diameter, which overlap the values reported here. Nevertheless, it is probably safe to say that the reversal processes are quite different in these two types of particle.

4. LOOP SYNTHESIS

The results of the measurements on particles of γ-Fe_2O_3 were used to calculate the remanent loop of the corresponding tape. The data required was:

a) The histogram of $H_r(0)$.
b) The variation of $H_r(0)$ with angle ψ.
c) The distribution function describing the orientation of the particles in the tape.

Examples of a) and b) have been given above: c) was determined by the work described elsewhere [21]. From a), b) and c) the remanent loop was calculated [2] for an assembly of widely separated particles, having the same orientation as that in the corresponding tape. To allow for the effect of packing the particles together, the scale of the field axis was multiplied by $(1 - p)$, where p is the packing factor. Then if (2) is valid, the remanent coercive force of the calculated loop should agree with that of the tape. Although examples of such loops were the same shape as the experimental loops (but had rather larger values of $H_r(p)$), this treatment of the effect of the interaction fields is rather arbitrary, and so the quite good agreement between theory and experiment is possibly fortuitous.

Some progress has been made subsequently, in as much as it has been shown [22] that the relationship between $H_r(p)$ and $H_r(0)$ is of the form:

$$H_r(p) = H_r(0)[1 - C(M_s/H_r(0))^2 p^{4/3}] \qquad (3)$$

where C is a constant of the order of unity. Values of C appropiate to assemblies of aligned or randomly orientated particles were determined by computer simulations. Very good agreement is found between the predictions of (3) and experimental results, for both 'tape' particles and elongated single domain particles of iron. It is hoped that extensions of this model will enable a better correlation to be made between the properties of the individual particles and those of the corresponding tape.

5. DISCUSSION

It has been shown that particles of the types commonly used in tape recording probably reverse by an interesting variety of mechanisms: γ-Fe_2O_3 by fanning in particles of poor morphology and buckling in particles of better morphology, CrO_2 by curling, and metal powder (iron) by fanning. Interpretation of the results, e.g. of the range in $H_r(0)$ found in a particular sample, is limited by defects in both theory and experimental technique.

The theory now available deals with rather idealized situations such as a cylindrical particle with uniaxial crystalline anisotropy aligned along the particle axis. The fanning model was devised to take some account of the effects of poor morphology, but neglects crystalline anisotropy. An actual particle is often of irregular shape, polycrystalline and with a cubic anisotropy. Such problems are probably too complicated to be dealt with by the techniques of micromagnetics, and the best approach would seem to be to develop numerical methods along the lines indicated by Matsumoto [23].

The experimental technique described here would seem to provide sufficient data to make progress in the long-term aim of being able to predict the properties of a tape from those of its constituent particles. A limitation of the technique is that the resolution of the optical microscope is totally inadequate to allow a correlation to be made between $H_r(0)$ of a particle and its morphology. To make a significant advance in this respect, it would seem to be essential to measure $H_r(0)$ in a transmission electron microscope, perhaps by sensing the direction of the magnetization in a particle by observing the interaction of a beam of low energy electrons with the external magnetic field of the particle. Whatever the method, the present work shows that it is highly desirable to make observations on a large number of particles.

ACKNOWLEDGEMENTS

The author is very grateful to those tape manufacturers and Mr P.J. Flanders who provided samples and data, some of which has been quoted in the text; and to Dr H.B. Haanstra who supplied Fig. 4.

REFERENCES

[1] J.E. Knowles, IEEE Trans. Magn. MAG-14, 858 (1978).
[2] J.E. Knowles, IEEE Trans. Magn. MAG-16, 62 (1980).
[3] J.E. Knowles, Proc. ICF 3 (Japan), in press.
[4] E. Kneller in Magnetism and Metallurgy edited by A.E. Berkowitz and E. Kneller (Acad. Press, New York, 1969) Vol. 1, Ch. 8.
[5] A.H. Morrish and S.P. Yu, Phys. Rev. 102, 670 (1956).
[6] P.C. Scholten, IEEE Trans. Magn. MAG-11, 5, 1400 (1975).
[7] R.K. Waring Jr., J. Appl. Phys. 38, 1005 (1967).
[8] H.B. Haanstra and A.A. Staals, unpublished.
[9] G. Bate in Ferromagnetic Materials edited by E.P. Wohlfarth (North-Holland, Amsterdam, 1980) Vol. 2, Ch. 7.
[10] D.F. Eagle and J.C. Mallinson, J. Appl. Phys. 38, 995 (1967).
[11] H. Takei and S. Chiba, J. Phys. Soc. Japan 21, 1255 (1966).
[12] Y. Imaoka, S. Umeki, Y. Kubota and Y. Tokuaba, IEEE Trans Magn. MAG-14, 649 (1978).
[13] P.J. Flanders and G. Kaganowicz, unpublished.
[14] J.E. Knowles, J. Magn. Magn. Mat. 22, 263 (1981).
[15] D.S. Rodbell, J. Phys. Soc. Japan 21, 1224 (1966).
[16] D.S. Rodbell, R.C. DeVries, W.D. Barber and R.W. DeBlois, J. Appl. Phys. 38, 4542 (1967).
[17] R. Gustard and H. Vriend, IEEE Trans. Magn. MAG-5, 326 (1969).
[18] I.S. Jacobs and C.P. Bean, Phys. Rev. 100, 1060 (1955).
[19] A.A. Van der Giessen, IEEE Trans. Magn. MAG-9, 191 (1973).
[20] F.E. Luborsky and C.R. Morelock, J. Appl. Phys. 35, 2055 (1964).
[21] J.E. Knowles, R.F. Pearson and A.D. Annis, IEEE Trans. Magn. MAG-16, 42 (1980).
[22] J.E. Knowles. To be published.
[23] M. Matsumoto, Proc. ICF 3 (Japan), in press.

High density recording characteristics of sputtered γ-Fe_2O_3 thin-film disks

S. Yoshii and O. Ishii

Ibaraki Electrical Communication Laboratory, N. T. T., Tokai 319—11, Japan

S. Hattori

Musashino Electrical Communication Laboratory, N. T. T., Musachino, Tokyo 180, Japan

T. Nakagawa and G. Ishida

Fujitsu Laboratories Limited, Kamikodanaka, Kawasaki 211, Japan

Sputtered γ-Fe_2O_3 thin films are attractive in practical application to high density magnetic recording. This paper presents basic processes of γ-Fe_2O_3 thin-film fabrication, magnetic properties, and high density recording characteristics of the films. The film is prepared by reactive sputtering of an iron alloy containing a small amount of Co, Ti, and/or Cu in an Ar-O_2 atmosphere on anodized Al alloy substrates and successive heat treatment. The film shows a high coercivity of 1000 Oe and a coercive squareness of 0.84 with the reduced thickness. The anodized Al alloy provides hard, smooth, and chemically stable substrates. The continuous γ-Fe_2O_3 thin films show high recording density with an excellent signal to noise ratio (SNR). First, the recording characteristics of the film employed in practical disk storage with the areal recording density of 24 000 bit/mm^2 are described. SNR improves as the average crystallite size decreases. It is shown that 35 dB can be obtained at the linear density of 1000 flux reversals/mm with the track width of 10 μm. Finally, the density D_{6dB} of 2600 flux reversals/mm is experimentally shown by using a Mn-Zn ferrite head with the small gap length of 2g = 0.15 μm operated at the heat-medium spacing of 0.1 μm.

PACS numbers: 85.70.Kh, 75.70.Dp

INTRODUCTION

The demand for ever increasing recording density in magnetic recording has stimulated introduction of thin films of continuous magnetic materials for the production of disks. With regard to application to rigid disk recording, continuous thin films of an iron oxide, γ-Fe_2O_3, are very attractive because of their high coercivity and high remanent magnetization as well as their resistance to corrosion and wear. In order to increase recording density, a magnetic recording medium must have increased coercivity and much reduced thickness. The high density recording with an extremely small head-medium spacing also required much improved surface smoothness and mechanical strength of the disks. It has been shown that γ-Fe_2O_3 thin films prepared by reactive sputtering and successive heat-treatment on adonized Al alloy substrates have superior magnetic and mechanical properties, and high density recording characteristics (1-4). The γ-Fe_2O_3 films in a range of 0.1-0.2 μm show the recording density over one thousand bit per mm. The anodized Al alloy substrate provides a smooth and mechanically hard disk surface. The films show excellent adhesion to the substrate. Recently, the first disk storage that employs the continuous γ-Fe_2O_3 thin film recording media has been developed, and it is scheduled to be put into practical use (5). Since the γ-Fe_2O_3 thin film disk has potential for further improvement in the recording density, it is one of the most effective components for high capacity disk storage in the coming age. In this paper, the basic processes of γ-Fe_2O_3 thin film fabrication and magnetic properties of the films will be summarized and then the high density recording characteristics will be presented.

SUBSTRATE

An Al-4wt.%Mg alloy was employed as the substrate. Disks are 360 mm and 210 mm in diameter and 1.9 mm in thickness. The substrates are anodized (6,7) to form an aluminum oxide (Al_2O_3) layer on the both sides. The anodized layer is produced in the 10% sulfuric acid solution. This Al_2O_3 layer improves the hardness of the substrate surfaces. Disk substrates should be extremely smooth to keep stably a small head-medium spacing, defect free to avoid signal errors, hard and chemically stable. A thick Al_2O_3 layer formed is polished to a 2 μm thickness. The surface is finished to an arithmetic average Ra less than 0.01 μm. Surface burnishing with a sapphire slider head is applied to eliminate protuberances prior to film preparation. Due to the thermal expansion coefficient difference between Al_2O_3 and the Al-Mg alloy, the oxidized layer is subjected to cracks when the substrate is exposed to higher temperature. The maximum possible range for the Al_2O_3 thickness and heat-treatment temperature was experimentally determined; for the heat-treatment at 350°C the thickness of Al_2O_3 is kept below 3 μm. The Al-4wt.%Mg alloy used in the conventional disk substrates, involves intermetallic compound precipitates

Reprinted with permission from *J. Appl. Phys.*, vol. 53, no. 3, pp. 2556-2560, Mar. 1982.

containing impurities, Fe, Si and Mn. These precipitates make defects on the substrate surface, when the substrate alloy is anodized. A purified Al-4wt.%Mg alloy is, therefore, used in order to reduce the surface defects. By using the high purity aluminum alloy for the substrates, signal errors can be reduced to a practical level.

THIN FILM FABRICATION PROCESS

Iron oxide films can be prepared by sputtering in a variety of processes (1-4,7). γ-Fe_2O_3 or Fe_3O_4 as well as Fe can be employed as a sputtering target in reactive or non-reactive (if necessary) sputtering (7). In this section, the most basic processes are summarized. The basic processes involve reactive sputtering and successive heat-treatment; the processes are I) deposition of an α-Fe_2O_3 film, followed by reduction of the film to Fe_3O_4 and then transformation into γ-Fe_2O_3 (1,2), and II) direct formation of an Fe_3O_4 film and transformation into γ-Fe_2O_3 (3,9,10).

The deposition rate of the films in reactive sputtering of an Fe target in an Ar-O_2 atmosphere shows a hysteresis loop in a decreasing and increasing cycle of oxygen partial pressure P in the order of 10^{-4}-10^{-3} Torr (9,10). The deposition rate has the pronounced steps at the critical values P* and P** (P*<P**). Below P* or P**, the increased rate of deposition are observed. This is due to a difference in the sputtering rates between iron and iron oxide; iron has a higher sputtering rate than α-Fe_2O_3. In the direction of decreasing the oxygen partial pressure, an γ-Fe_2O_3 film is deposited above P* and an Fe film is deposited below P*. Above P* the target surface is oxidized to α-Fe_2O_3 in the reactive sputtering, and hence α-Fe_2O_3 is sputtered and deposited. Below P* the sputtering rate of iron exceeds the oxide formation rate. In the increasing direction of the oxygen partial pressure, however, between P* and P**, a deposited film consists of Fe and Fe_3O_4, or only Fe_3O_4, and above P** an α-Fe_2O_3 film is deposited. Below P* the Fe film is deposited. Between P* and P**, the target surface is of iron , and oxidation process of the deposited film into Fe_3O_4 proceeds during and just after the successive deposition. The Fe_3O_4 single phase film is formed in a higher oxygen partial pressure side. The phenomena below P* and above P** are the same as the former case. The values of the critical oxygen partial pressure, P* and P**, depend on rf-sputtering power as well as Ar partial pressure and target temperature. Using the stages described above, α-Fe_2O_3 and Fe_3O_4 films can be prepared by the reactive sputtering. γ-Fe_2O_3 films are fabricated as follows.

I) An α-Fe_2O_3 film is deposited in the Ar-O_2 atmosphere with the total pressure of 2×10^{-2}-4×10^{-2} Torr. The α-Fe_2O_3 film is reduced to Fe_3O_4 at 300-330°C for 2-4 hours in the wet H_2 atmosphere. By subsequent heat-treatment the Fe_3O_4 film is oxidized to γ-Fe_2O_3 at 300-330°C for 2-4 hours in air.

II) An Fe_3O_4 single phase film is directly formed by the reactive sputtering of an iron target in the Ar-O_2 atmosphere with the total pressure of 2×10^{-2} Torr, where the substrate is heated at the temperatures up to 150°C. The Fe_3O_4 film is transformed into γ-Fe_2O_3 by the heat-treatment at 300-330°C for 2-4 hours in air. This process is attractive one, because it excludes the reduction heat-treatment in H_2.

A schematic view of our new sputtering apparatus is shown in Fig. 1. Simultaneous deposition on the both substrate surfaces is provided by a pair of targets. The target is equipped with a magnetron device to enhance the deposition rate. The substrate is rotated during the sputtering so as to make uniform deposition. About five substrates, 360 or 210 mm in diameter are loaded in a cassette box, and transferred between a cassette loading vacuum chamber and the sputtering chamber one by one. Cassettes which are set in the cassette box and to be set from the outside of the vacuum chamber can be exchanged without interrupting the sputtering by closing a vacuum valve between the cassette loading and sputtering chambers, so that continuous fabrication is ppssible (7).

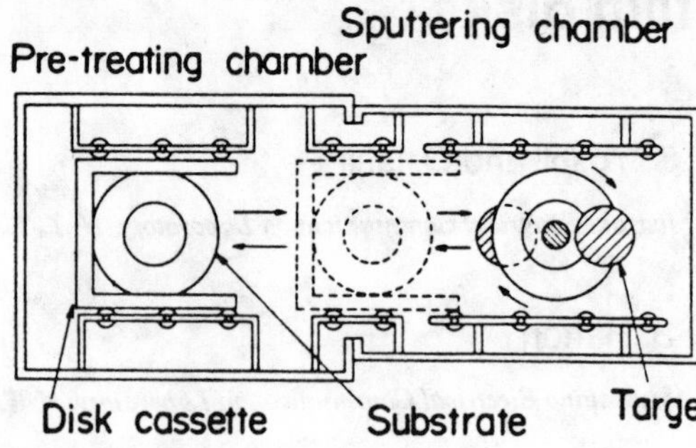

Sputtering chamber
Pre-treating chamber
Disk cassette
Substrate
Target

Fig. 1. Schematic view of sputtering apparatus.

MAGNETIC PROPERTIES

Structure and magnetic properties of the films were examined by X-ray and electron diffraction, electron microscopy, Mössbauer effect measurement, ESCA (electron spectroscopy for chemical analysis), and magnetic measurement with a vibrating sample magnetometer. Physical properties of the films (1.3.8.9.12) are in good agreement with the data of γ-Fe_2O_3 fine particles (13,14) and Fe_3O_4 (15,16).

Addition of Co increases coercivity of the films (1-4). Addition of Ti has an effect to widen the temperature range of reduction from α-Fe_2O_3 to Fe_3O_4 (1,2,8). The Ti addition improves squareness ratio of the γ-Fe_2O_3 films as well (1,8). The addition of Cu, on the other hand, lowers the reducing temperature. By simultaneous addition of Ti and Cu, a wider and lower temperature range is obtained as shown in Fig. 2. This makes it possible to obtain uniform magnetic properties.

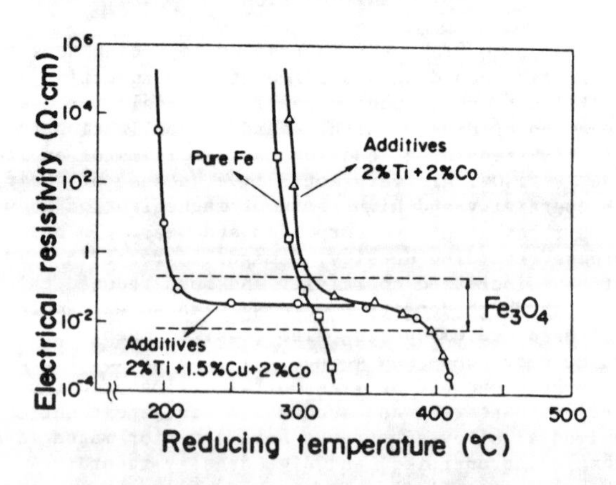

Fig. 2. Relation between electrical resistivity and reducing temperature. Heating duration is 2 hrs.

Variations in coercive force and squareness ratio with oxidation temperature are shown in Fig. 3. The Hc begins to increase above 150°C, and then shows a broad peak around 250°C. This peak is due to the magnetic interactions of vacancies and ferrous ions in the spinel lattice. The squareness ratio gradually decreases as the temperature increases to 250°C, and then increases to 0.8 when the temperature reaches 300°C. We confirmed that Fe_3O_4 is completely converted to γ-Fe_2O_3 above 290°C.

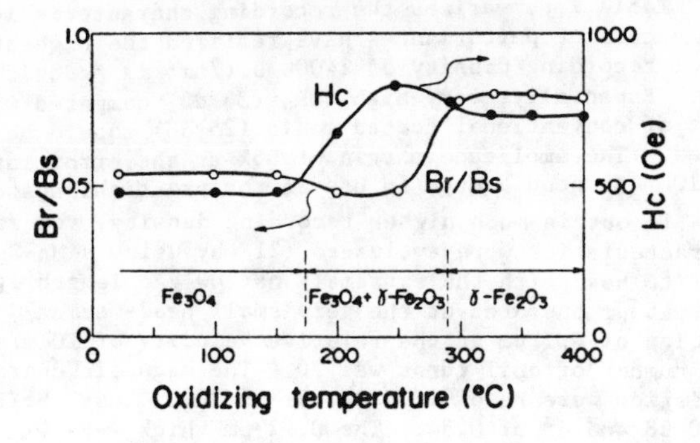

Fig. 3. Dependence of coercive force (Hc) and squareness ratio (Br/Bs) on oxidizing temperature.

Figure 4 shows transmission electron micrographs of the γ-Fe_2O_3 thin films with various additives. The film with the addition of 2 at. % Ti and 2 at. % Co has an average crystallite size of about 1000 Å. With the further addition of 1.5 at. % Cu, the average crystallite size decreases to 400 Å. Crystalline grain growth during the heat-treatment is suppressed by Cu addition; the improved signal to noise ratio at recorded state SNR_w is obtained by reduced crystalline grain size.

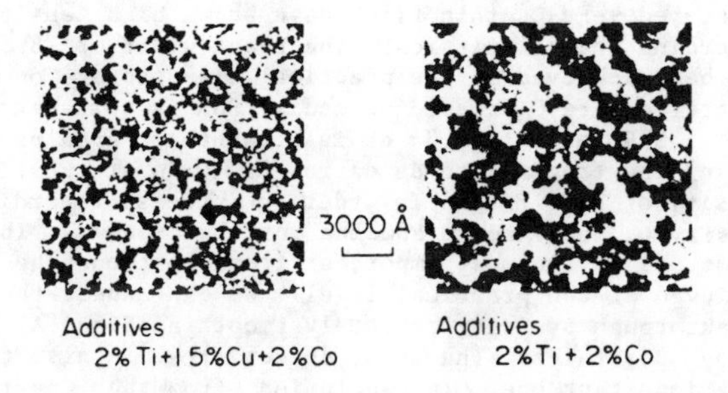

Fig. 4. Transmission electron micrographs with various additives.

0.12 µm thick γ-Fe_2O_3 films with simultaneous addition of Co, Ti and Cu were prepared on anodized Al-4wt.%Mg alloy substrates with the diameter of 210 mm. The contents of doped Co, Ti and Cu in the sputtering target were 2.0, 2.0 and 1.5 atomic percent, respectively. The films have the magnetic characteristics, Hc of 1000 Oe, Br of 3000 Gauss, Br/Bs of 0.88 and coercive squareness S* of 0.84.

HIGH DENSITY RECORDING CHARACTERISTICS

The recording density at 6 dB attenuation in the signal amplitude, D_{6dB}, of 1100 FRPM (flux reversals/mm) was demonstrated by γ-$(Fe_{0.955}Ti_{0.025}Co_{0.020})_2O_3$ (1,2), γ-$(Fe_{0.98}Co_{0.02})_2O_3$ (3), and double layer γ-$(Fe_{0.945}Ti_{0.025}Co_{0.03})_2O_3$/γ-$(Fe_{0.965}Ti_{0.025}Co_{0.01})_2O_3$ (18) films with the thickness of 0.14 µm prepared on the anodized 360 mm Al-Mg substrate using a Mn-Zn ferrite head with the head gap of 0.7 µm operated at the head-medium spacing of 0.2 µm. The coercive force was in the range 700-730 Oe. The excellent SNR was also shown (1-4,18).

Continuous γ-Fe_2O_3 thin film recording media prepared on the anodized 210 mm Al-4wt.%Mg substrate have successfully been employed in practical 3.2 GByte multi-device disk storage (5,19). The magnetic characteristics are Hc of 700 Oe, Br of 2700 Gauss and S* of 0.78. The thickness of the film is 0.17 µm.

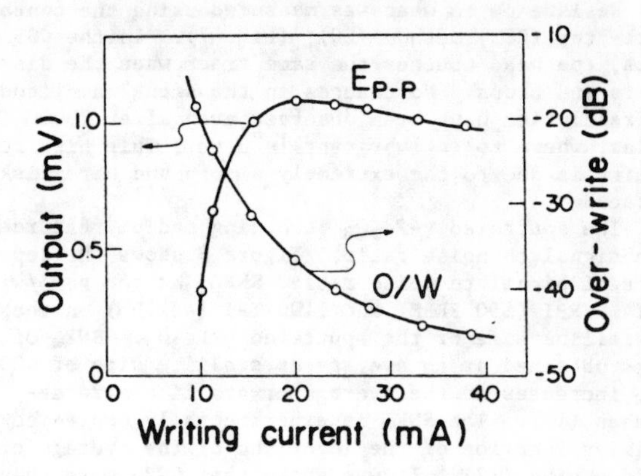

Fig. 5. Dependence of output voltage (E_{p-p}) and over-write (o/w) on writing current.

Figure 5 shows the dependence of output E_{p-p} and over-write on writing current, where the over-write is the ratio of 1F (5.4 MHz) and 2F (10.8 MHz) components after signals are recorded at 1F and then other signals are recorded at 2F on the same track. The E_{p-p} has the maximum value at a writing current of 20 mA, which shows the recording medium is magnetized to saturation. At a writing current of 25 mA, an over-write of -40 dB is obtained.

Figure 6 shows the dependence of output voltage on recording density. The recording density, D_{6dB}, is 22 kFRPI (880 FRPM).

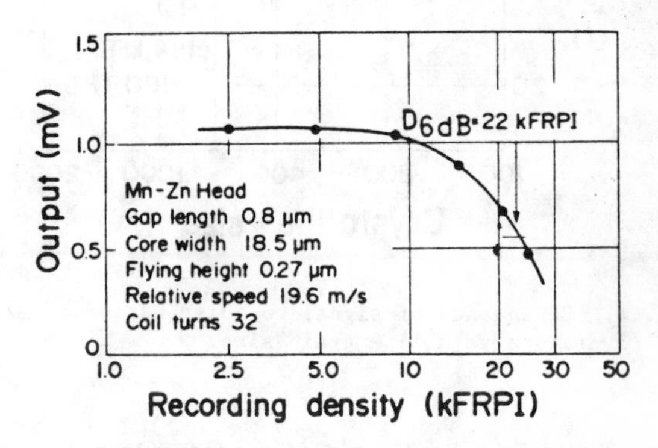

Fig. 6. Dependence of output voltage on recording density.

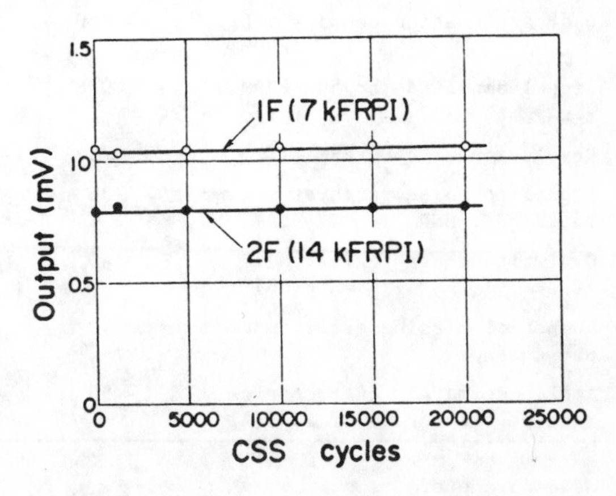

Fig. 7. Contact-start-stop (CSS) test.

Resistance to wear was measured using the contact-start-stop (CSS) method (20), (Fig. 7). In the CSS tests, the head touches the same track when the disk starts and stops. No changes in the signal amplitude and resolution have been observed even after 20000 CSS cycles, where solid lubricant is used. This high reliability is due to the extremely smooth and hard disk surfaces.

The sputtered γ-Fe_2O_3 recording medium features high signal to noise ratio. Figure 8 shows the dependence of signal to noise ratio, SNR_w, at the read/write of 14 kFRPI (550 FRPM) and 1100 TPI (43 TPM) on the crystalline size of the sputtered γ-Fe_2O_3. SNR_w of 36 dB is obtained at an average crystallite size of 400 Å. SNR_w increases as the average crystallite size decreases (17). The SNR_w is experimentally expressed as a linear function of the logarithm of the average crystallite size (17). It was shown that SNR_w more than 35 dB is attainable for the recording density of 1000 FRPM and the track width of 10 μm at the film thickness of 0.14 μm and the average crystallite size below 250 Å (17). The excellent SNR_w is attributed to smooth magnetic transition in a magnetization reversal region. Crystallite grain growth during the heat-treatment is suppressed by the Cu addition.

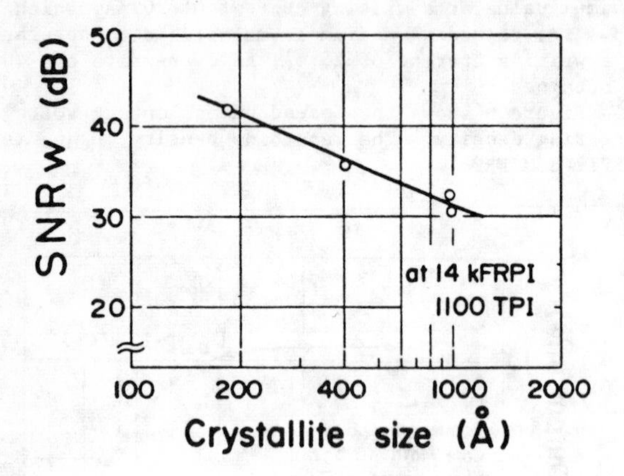

Fig. 8. Dependence of signal to noise ratio (SNR_w) on average crystalline grain size.

Table I γ-Fe_2O_3 thin film recording media performances

	Performances
6 dB attenuation density : D_{6dB} (FRPM)	880
Signal amplitude at 550 FRPM : E_{p-p} (mV)	0.8
Resolution at 550 FRPM (%)	78
Signal to noise ratio at 550 FRPM : SNR_w (dB)	36
Overwrite characteristics : 1F/2F (dB)	-40
Number of missing error spots per surface	30
Amplitude margin characteristics at error rate of 10^{-5}(%)	55

Head parameters;
Gap length (2g) 0.8 μm
Track width (Tw) 18.5 μm
Number of coil turns (N) 32
Head-medium spacing at 19.6 m/sec (d) 0.27 μm

Table I summarizes the recording characteristics. The excellent performances have realized the highest areal recording density of 24000 bit/mm^2 in practical use. Especially, very high SNR_w (36 dB) compared with that of conventional coated media (25 dB) should be noted. The amplitude margin of 55% at the error rate of 10^{-5} is good enough to use in the practical system.

To obtain much higher recording density, recording characteristics were evaluated (21) by using a Mn-Zn ferrite head with the extremely narrow gap length 2g of 0.15 μm operated at the very small head-medium spacing of 0.1 μm at the relative velocity of 10 m/sec. The number of coil turns was 20. The magnetic characteristics were Hc of 1000 Oe, Br of 3000 Gauss, Br/Bs of 0.88 and S* of 0.84. The 0.12 μm thick γ-Fe_2O_3 film was prepared on the 210 mm anodized substrate. The very high density, D_{6dB} of 2600 FRPM, was obtained. The extrapolation of Talke and Tseng's calculations was in good agreement with the above experimental value. The signal amplitude had the maximum at the write current of 20 mA. The head core was not saturated during the write process and the recording medium was sufficiently magnetized even for the high coercive force (1000 Oe). This was due to the high coercive squareness.

CONCLUSIONS

The fabrication processes for the sputtered γ-Fe_2O_3 thin film have been established. The physical properties show that the films are a stable γ-Fe_2O_3 compound. The films have excellent magnetic properties, i.e. high coercivity and high coercive squareness with much reduced thickness. Owing mainly to these, the sputtered γ-Fe_2O_3 thin films have shown high density recording characteristics. The track width of 18.5 μm has been achieved in the practical disk storage on the sputtered γ-Fe_2O_3 recording media. SNR is quite excellent. SNR_w more than 35 dB is attainable. even postulating the track width as narrow as 10 μm at the linear density of 1000 FRPM. In order to increase recording density, a head-medium spacing must be reduced. This seems one of the most important items that must be achieved at the practical level. We can expect the breakthrough by using extremely smooth anodized Al alloy substrates. The item, defect free, is also the most important one. In conclusion, from the linear density, track width and SNR summarized here, we can expect a disk storage with the areal density in the range 60-100 Kbit/mm^2 by the sputtered γ-Fe_2O_3 thin film disk in accordance with the progress in head technology.

ACKNOWLEDGEMENTS

Many individuals have contributed to the research and development reviewed in this paper. The authors wish to express their thanks to these contributors, Y. Ishii, A. Tago, I. Sato, T. Nakanishi, S. Ohara, K. Kogure and R. Kaneko of Musashino Electrical Communication Laboratory, N.T.T., A. Terada, and S. Ohta of Ibaraki Electrical Communication Laboratory, N.T.T. and K. Makino, M. Shinohara and A. Ikari of Fujitsu Laboratories Ltd.

REFERENCES

(1) S. Hattori, Y. Ishii, M. Shinohara and T. Nakagawa, IEEE Trans. Mag., MAG-15, 1549 (1979).
(2) M. Shinohara, K. Makino and S. Hattori, Fujitsu Sci. Tech. J., 15, 99 (1979).
(3) Y. Ishii, A. Terada, O. Ishii, S. Ohta, S. Hattori and K. Makino, IEEE Trans. Mag., MAG-16, 1114 (1980).
(4) Y. Ishii, A. Terada, O. Ishii, S. Ohta, T. Nakagawa and M. Shinohara, Proc. 3rd释Internl'. Conf. Ferrites, (1981).
(5) R. Kaneko and S. Yoshii, Review of Elect. Comm. Lab. (Review of ECL), 30 to be published (1982).
(6) N. Inagaki, S. Hattori, Y. Ishii, A. Terada and

H. Katsuraki, IEEE Trans. Mag., MAG-12, 785 (1976).
(7) S. Hattori, A. Tago, Y. Ishii, A. Terada, O. Ishii and S. Ohta, Review of ECL, 30, to be published (1982).
(8) A. Terada, O. Ishii, M. Shinohara and T. Nakagawa, Trans. IECE Jpn., to be published.
(9) Y. Ishii, S. Ohta, O. Ishii, S. Hattori and K. Makino, Trans. IECE Jpn., J63-C, 609 (1980).
(10) Y. Ishii, S. Ohta and S. Hattori, Trans. IECE Jpn., J64-C, 483 (1981).
(11) Y. Ishii, S. Hattori and S. Ohta, 真空 , 24, 360 (1981).
(12) O. Ishii, A. Terada, S. Ohta and T. Nakagawa, Trans. IECE Jpn., to be published.
(13) B.J. Armstrong, A.H. Morrish and G.A. Sawatzky, Phys. Lett., 23, 414 (1966).
(14) N.S. McIntyre and D.G. Zetaruk, Analytic. Chem., 49, 1521 (1977).
(15) P.A. Miles, W.B. Westphal and A. von Hippel, Rev. Mod. Phys., 29, 279 (1957).
(16) B.J. Evans and S.S. Hafner, J. Appl. Phys., 40, 1411 (1969).
(17) A. Terada, O. Ishii, S. Ohta and T. Nakagawa, submitted to IEEE Trans. Mag.,
(18) T. Nakagawa, M. Shinohara and O. Ishii, IEEE Trans. Mag., to be published (1981).
(19) T. Nakanishi, Y. Koshimoto and S. Ohara, Review of ECL, to be published (1982).
(20) K. Kogure, T. Kita and S. Fukui, Review of ECL, to be published (1982).
(21) O. Ishii, S. Ohta and T. Nakagawa, Trans. IECE Jpn., Section E, to be published (1982).

MICROSTRUCTURE AND MAGNETIC PROPERTIES OF ELECTROLESS Co-P THIN FILMS GROWN ON AN ALUMINUM BASE DISK SUBSTRATE

Tu Chen, D. A. Rogowski and R. M. White
Xerox Palo Alto Research Center
Palo Alto, California 94304

ABSTRACT

The microstructure and magnetic properties of electroless Co-P films (~100 to 1000 Å) deposited on an aluminum base substrate are examined. The squareness and coercivity characterizing the magnetic hysteresis are found to correlate with the size and nature of the microcrystallites making up the film. It is argued that this correlation is due to coupling between the microcrystallites.

INTRODUCTION

Electroless Co-P films have been regarded as a promising digital recording medium since their discovery by Brenner and Riddell [1]. For high density recording one wants a thin medium (~400 Å) with a high coercivity (H_c ~500 Oe) and a rectangular hysteresis loop. In the case of Co-P films these properties, as well as the remanent magnetization (M_r), are strongly dependent upon the microstructure [2,3]. The microstructure is, in turn, governed by the plating conditions and the substrate. Most of the previous studies correlating magnetic properties with microstructure have been carried out on films grown on an activated glass or mylar substrate. For some recording applications, however, one would like to deposit the Co-P on a rigid aluminum substrate. The purpose of this study has been to investigate the magnetic characteristics of Co-P films grown on multilayer coated aluminum substrates.

EXPERIMENTAL TECHNIQUES

The chemistry of the plating baths and the operating conditions were similar to those of Brenner and Riddell [1]. The temperature of the baths was maintained at 76 ± 1°C and the total volume was 800 ml. Two bath parameters were varied: the phosphorous acid concentration, H_3PO_3, and the ammonium hydroxide concentration, NH_4OH, which adjusts the pH. The values of these parameters for the five baths used in this study are listed in Table I.

Bath	pH (Room Temperature Value)	H_3PO_3 (mole/liter)
I	8.5	0.01
II	8.5	0.10
III	9.5	0.01
IV	9.5	0.10
V	10.5	0.01

TABLE I

Characteristics of the five baths used in this study.

The substrates consisted of an aluminum alloy disk approximately 1 cm in diameter overcoated with a non-magnetic Ni-P layer followed by a layer of Cu.

The magnetic hysteresis loops were measured with a vibrating sample magnetometer and the microstructure was analyzed with a transmission electron microscope (TEM). For TEM analysis of the film the substrate was removed by a process involving chemical etching, electropolishing, and ion milling. The composition and compositional variation with depth was determined by ion scattering spectrometry using bulk polycrystalline $Co_{1.95}P$ as a standard.

RESULTS AND DISCUSSION

In Table II we list the results of the deposition rate, the phosphorous content, and the saturation magnetization of films prepared from the five different baths. We note that M_s increases with increasing phosphorous content. Other studies [4,5] show different dependences of M_s on phosphorous content. This is not surprising when we consider that the state of the phosphorous in these films may vary with the bath conditions. For example, it might be dissolved as a metastable solid solution with the Co, or exist as a pure phosphorous or phosphite phase between the Co grains. Thus, the magnetization is not a unique function of just the phosphorous content. Unfortunately it is very difficult to determine the chemical disposition of the phosphorous.

Bath	R [Å/sec]	at% P	M_s (emu/cm³)
I	4.3	7.8 ± 0.4	1025
II	5.6	7.6 ± 0.6	1000
III	7.5	8.1 ± 0.6	1056
IV	10.0	9.6 ± 0.6	1147
V	12.8	10.2 ± 0.7	1208

TABLE II

Deposition rate (R), atomic percent phosphorous and volume magnetization of Co-P films prepared from baths I thru V.

Table II shows that for a fixed H_3PO_3 concentration the deposition rate (R) increases with increasing pH. For a fixed pH, Table II also shows that R increases with increasing H_3PO_3 concentration.

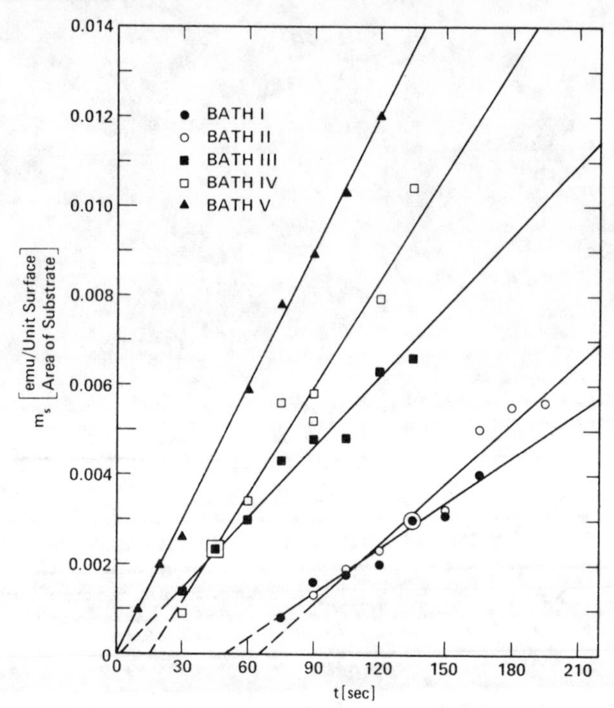

Fig. 1. Saturation magnetic moment vs. deposition time.

In Fig. 1 we plot the magnetic moment per unit surface area as a function of deposition time. The fact that these results are highly linear suggests that the deposition rate is constant and that the magnetization is uniform with thickness. Independent thickness measurements show that this moment per unit area is indeed proportional to the film thickness. The intercept along the time axis is therefore a measure of the nucleation time, τ, or the time it takes a particle to fluctuate beyond the critical radius for stable growth. The results show that for a fixed H_3PO_3 concentration τ increases with decreasing pH, while for a fixed pH τ increases with increasing H_3PO_3 concentration.

The microstructure of these films may be distinguished by the size of the crystalline grains. We shall refer to structures having grains less than 100 Å as type I while those having grain sizes of the order of several hundred angstroms or larger as type II. "Well developed" type II grains are characterized by sharp edges and clearly identifiable stacking faults within the grains. The fine grain particles characterizing the type I structure generally coagulate into globular lumps separated by channels. The electron diffraction rings of films containing a large amount of type I

Reprinted with permission from *J. Appl. Phys.*, vol. 49, no. 3, pp. 1816–1818, Mar. 1978.

crystallites are broadened due to the small crystallite size in the manner reported by Aspland, et al. [6].

Fig. 2 shows some typical microstructures. Fig. 2a corresponds to a film plated from bath V. It shows a type II grain in the center surrounded by type I structure.

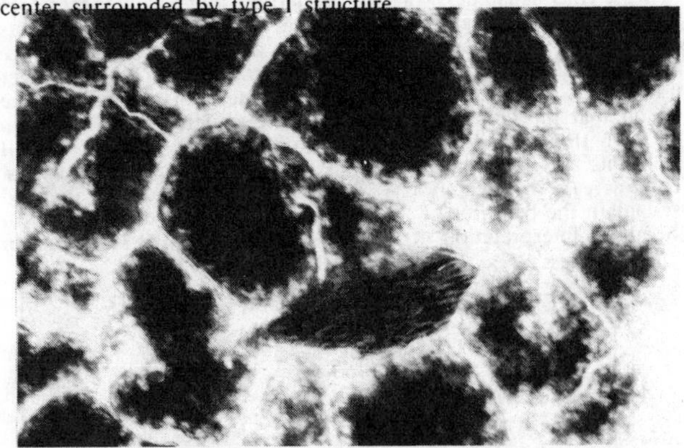

Fig. 2a. Micrograph of film with type I and type II structures. 225,000X

Fig. 2b. Micrograph of film with poorly developed type II structure. 225,000X

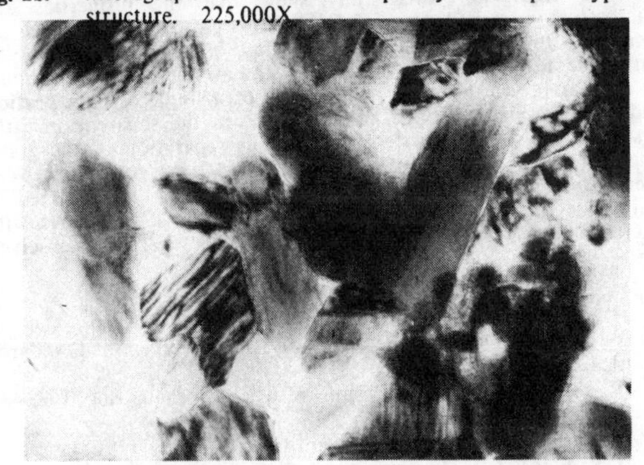

Fig. 2c. Micrograph of film with well developed type II structure. 225,000X

Films plated from baths I to IV exhibit only type II microstructure. However, depending upon the plating bath, the crystallites have other distinguising features besides a difference in grain size. These are the definition of the grain, the separation between grains and the evidence of stacking faults. Two extreme cases which illustrate these features are shown in Figs. 2b and 2c. In Fig. 2b, the grains are poorly developed as indicated by the poorly defined grain boundary, whereas in Fig. 2c the grains are well developed and the stacking faults have a better definition. Also by comparing films with these two morphologies of grain, it is noted that the relative separation between the grains is greater for the case of well developed rather than poorly developed grains.

In general the films plated from the four baths have a microstructure ranging between these two cases. For film thickness $\leq$ 600 Å, films plated from baths I and II exhibited well developed grains, films plated from bath III exhibited poorly developed grains, *but very uniform in size* and films plated from bath IV exhibited partially developed grains. In all the cases, *the grain definition improves with increasing film thickness*, and the separation between grains is decreased.

In all the films investigated the *average grain size also increases with increasing film thickness.* Comparing the nucleation time, τ, and the deposition rate with the grain size for the films plated from baths I to IV shows that the change in grain size in the films follows the change in nucleation period rather than the deposition rate. This result suggests that the nucleation rate may be at least as important, if not more important, than the deposition rate in determining the microstructure of these films.

For plating bath V, which has a pH of 10.5, the solution is relatively highly activated as evidenced by the high deposition rate (12.8 Å/sec) and by the fact that the solution shows no initial nucleation period. The high deposition rate and easy nucleation characterizing this bath may be the reason why films from this bath exhibit the fine grain (type I) microstructure.

Electron diffraction patterns of the films from baths I to IV indicate that the crystallites have a hcp structure and exhibit some preferred orientation. For all the films examined, the (100) reflection ring was generally stronger than for the case of a random orientation of crystallites, and the (002) ring was suppressed. The suppression of the (002) ring was observed to increase as the grain definition improved. A weak (002) reflection implies that the c-axis prefers to be perpendicular to the plane of the film. Thus films from bath III, having the least developed grains, also have less of a preferred orientation.

As was mentioned above, the magnetization (moment per unit volume) is independent of film thickness. The squareness of the hysteresis loop and the coercive force H_c, however, do depend upon the film thickness (as well as plating conditions). Fig. 3 shows the dependence of H_c upon thickness for films from the different plating baths. The coercivities of films from bath V varied from 50 to 200 Oe in an irreproducible manner and are therefore not included.

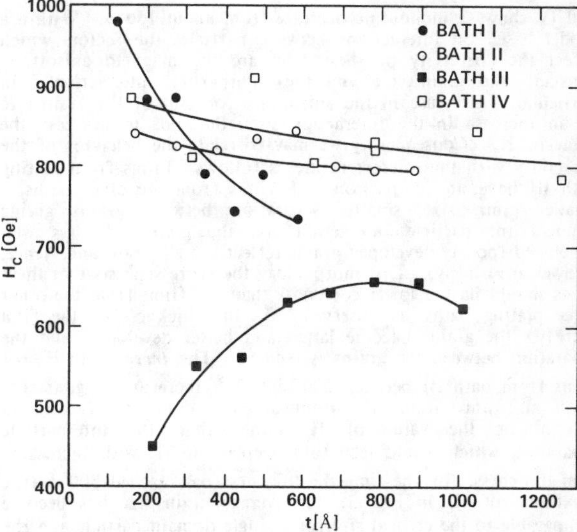

Fig. 3. H_c as a function of film thickness.

As we see from Fig. 3, within the range of film thicknesses investigated, the coercivities of the films plated from baths I, II and IV show a monotonic decrease from about 900 to 700 Oe as the film thickness increases. In contrast, H_c of the films plated from bath III increases with increasing film thickness and reaches a maximum at ~800 Å. Also their values of H_c are generally smaller (ranging between 450-650 Oe).

The hysteresis loops of all the films fall into three categories. For films plated from baths I, II, III and IV, the loops are quite similar and have a high coercivity, as illustrated by Fig. 4 a. However, for films plated from bath V, the hysteresis loops vary from very narrow loops with a small coercivity (Fig. 4b) to constricted loops (Fig. 4c). The extent of the constriction varies irregularly from film to film. We suggest these constricted shapes are a direct result of the presence of both type I and type II microstructures in these films. That is, the hysteresis loop reflects the presence of two different magnetization processes. In such a case it is not possible to define a meaningful coercivity.

For the reproducible loops obtained from baths I to IV, it is convenient to characterize their squareness by the parameter S* introduced by Williams and Comstock [7]. This is defined in terms of the remanent magnetization, M_r, and the slope of the major hysteresis curve at $H = -H_c$ according to $S^* = [H_c(dM/dH) - M_r]/H_c(dM/dH)$. An ideal square loop would have $S^* = 1$. Fig. 5 shows that S* increases with increasing film thickness. An alternative squareness, defined by M_r/M_s, varies from 0.7 to 0.9 with films from bath III having the highest values.

Let us now consider how these hysteresis loops correlate with microstructure. When a film shows a pure type I behavior (very fine grains), the hysteresis loop is very narrow as in Fig. 4b with $H_c < 100$ Oe. Low coercivities in fine grain materials are generally attributed to superparamagnetism. However, the reduction in H_c based on this mechanism is also accompanied by a reduction in the remanence. Since the remanence in type I films is not appreciably reduced, the grains are not superparamagnetic. In Fig. 2d we saw that those fine grains tend to coagulate. The grains within these globules are very likely strongly coupled. As a result, the magnetization process can occur via domain wall motion. Similar results have been observed in films grown on glass [6,8].

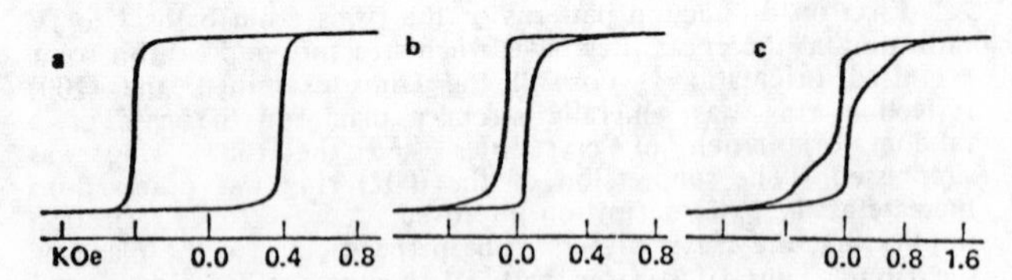

Fig. 4. Typical hysteresis loop: **a)** normal high H_c loop, **b)** normal low H_c loop, **c)** constricted loop.

For films with only type II microstucture the coercivity is relatively high, and has a wide range of values depending on the grain size, grain separation and how well developed the grains are. As we saw in Fig. 3, films from bath III have lower coercivities than those from baths I, II, or IV. Also the thickness dependence of H_c for films from bath III shows a maximum at about 800 Å whereas the thickness depndence of H_c for films from bath I, II and IV shows a monotonic decrease from about 150 Å. Within a model based on interactions between particles the factors which affect the coercivity of these film are the magneto anisotropy constant, the grain size and the interparticle interaction. In particular, a decrease in the anisotropy constant or the grain size or an increase in the interaction strength tends to decrease the value of H_c. Consequently we may correlate the behavior of the coercivity with the microstructure as follows. Films from plating bath III have, in comparison with films from the other baths, a smaller grain size, smaller separation between grains giving stronger interparticle interaction and the grains are less well developed (poorly developed grains reflect a high strain and, hence, a lower anisotropy). This morphology therefore suggests that these films should have a lower coercivity than the films from the other three plating baths, as observed. As the thickness of the film increases the grains become larger and better developed, and the separation between the grains is reduced. The *increase* in H_c for films from bath III between 200 to 800 Å therefore suggests that grain size and grain definition are the dominant factors in determining the value of H_c rather than the interparticle separation which would lead to a decrease in H_c with increasing film thickness. By the time the thickness has reached 800 Å (the maximum of H_c in Fig. 3), the average grain size has become comparable to the critical size for a single domain particle and the grains are well developed. As the film thickness and, hence, the grain size increases further we have the possibility of multidomain formation within single grains. The formation of multidomain grains and the continued increase in interparticle interaction is presumably responsible for the observed decrease in coercivity as film thickness increases.

For films from bath I, II and IV, with thicknesses between 100 to 200 Å, the grains are already well developed and the average grain size is near the critical size for single domain particle. Also since the separation of the grains in these films at this thickness is large, the interparticle interactions are small making the coercivity relatively high. Increase in film thickness of these films leads to mulidomain particles and also causes an increase in interparticle interaction because of the smaller particle separation. Consequently, increasing the thickness causes a reduction in the coercivity of the film as we observed in Fig. 3. In these films, a maximum in the coercivity similar to that observed in films from bath III would presumably occur at thicknesses less than those studied.

Finally, let us consider the squareness, S*. As we mentioned above, the c-axis of the grains in films from bath III are more randomly oriented. Therefore in the absence of magnetic interactions between grains, the Sonter-Wohlfarth model would predict that films from bath III should show *less* squareness. This is the opposite of what Fig. 5 shows. In films deposited on glass or mylar, it is observed that domain motion occurs from grain to grain [8] suggesting they are magnetically coupled. Reimer [9] has studied the behavior of 294 permanent magnets consisting of 100 mm cubes of barrium ferrite mounted 15 mm apart. He finds that the magnetic interactions between these magnets does indeed increase the squareness of the hysteresis loops. We therefore suggest that the intergrain coupling is also governing the squareness of the loops in our samples. Films with a more uniform grain size and a smaller separation between the grains will have stronger interactions and therefore a higher squareness. This could be the reason that the films from bath III have a higher S*. Increasing the film thickness decreases the separation between grains for all the films and this explains why S* increases with film thickness. This increase in interaction with thickness, which we have invoked to explain the behavior of S* with thickness, is also consistent with the above interpretation of the decrease in H_c with the increase in film thickness.

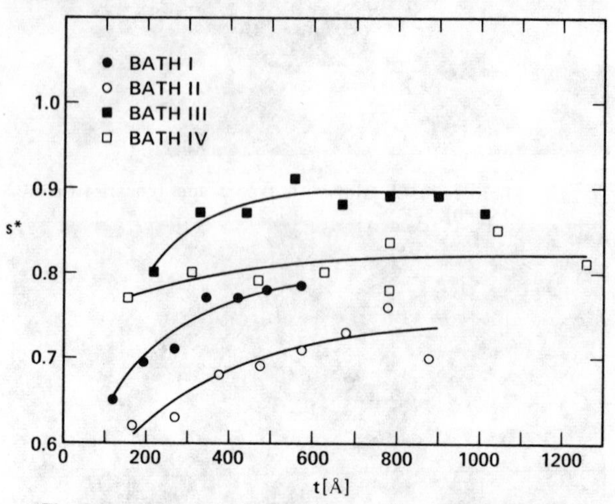

Fig. 5. Hysteresis loop squareness S* vs. film thickness.

CONCLUSION

The results of this study show that the properties and microstructural correlations of Co-P films grown on aluminum are similar to those grown on activated glass or mylar substrates. In particular, to obtain a high coercivity the average grain size should be comparable to the *critical* grain size of a single domain particle and have a well developed grain. However, the squareness of the hysteresis loop seems to depend upon the separation of the grain. Therefore, the optimum film represents a compromise between coercivity and squareness. In this paper we have also tried to emphasize the importance of correlating magnetic properties with microstructure rather than the plating bath conditions themselves.

REFERENCES

1. A. Brenner and G. E. Riddell, J. Research Nat'l Bur. Standards 3, 31 (1946).
2. R. D. Fisher and W. H. Chilton, J. Electrochem. Soc. **109**, 485 (1962).
3. Y. Movadzadeh, J. Electrochem. Soc. **112**, 891 (1965).
4. J. S. Judge, J. R. Morrison and D. E. Speliotis, J. Electrochem. Soc. **113**, 547 (1966).
5. J. R. Depew, J. Electrochem. Soc. **120**, 1187 (1973).
6. M. Aspland, G. A. Jones and B. K. Middleton, IEEE Trans. MAG-5, 314 (1966).
7. M. L. Williams and R. L. Comstock, AIP Conf. Proc. on Magnetism and Magnetic Mat. No. 5, Rept. 1, 738 (1971).
8. G. A. Jones and A. Farnsworth, Phys. Stat. Sol **(a)15**, 454 (1973).
9. L. Reimer, Z. Angew Phys. **17**, 196 (1964).

Co-Cr RECORDING FILMS WITH PERPENDICULAR MAGNETIC ANISOTROPY

Shun-ichi Iwasaki and Kazuhiro Ouchi*

ABSTRACT

For a new perpendicular magnetic recording system, a Co-Cr recording film with perpendicular anisotropy has been developed by an RF sputtering. The Co-Cr films are found to show some suitable properties for high density recording such as perpendicular anisotropy, a rectangular M-H loop, and fine grain structure. An extremely high recording density of 100,000 bits/inch was realized by using the Co-Cr film. The crystal and microscopic structure of the films are also discussed, and the perpendicular anisotropy of the Co-Cr films is mainly originated from the uniaxial magneto-crystalline anisotropy.

INTRODUCTION

We have recently proposed a new perpendicular magnetic recording system, and confirmed that a flat response curve of the output voltage can be realized in the high density region.[1] This system must consist of a recording head which produces a pure perpendicular magnetic field, and a medium which has an easy axis of magnetization perpendicular to the medium plane.

For the perpendicular recording medium, it has been pointed out[1] that a high saturation magnetization and a high coercive force as well as perpendicular anisotropy are necessary to obtain high output voltage and high recording resolution. Furthermore, the mechanical and the chemical stability of the medium and the productivity are also desired.

Taking into account these properties, the authors have prepared the Co-Cr perpendicular anisotropy film.[2]

Cobalt has a large magneto-crystalline uniaxial anisotropy energy, hence it can be used to develop the perpendicular anisotropy film. The films must have the anisotropy field H_K surpassing the maximum demagnetizing field $4\pi Ms$. Therefore, it is necessary to add other metals to reduce Ms, keeping the c-axis oriented perpendicularly to the film surface. We have chosen chromium as an additional metal, because Co-Cr alloy has a relatively stable hcp phase at a lower content of Cr, and at the same time, the saturation magnetization is expected to decrease when a small amount of Cr is added.

To prepare the film of Co-Cr alloy, RF sputtering was used, since it is suitable to prepare the films of a high melting point Co-Cr alloy, and superior to the other methods for the adhesion of the deposited magnetic layer to the substrate, and also for reproducibility. In addition, it is also convenient that Cr has the same sputtering yield as Co in the RF sputtering process.

It was found that the RF sputtered Co-Cr film has a large perpendicular magnetic anisotropy, a high coercive force, and also other favorable properties for the high density magnetic recording.

In this paper, the magnetic properties of the Co-Cr films, and the origin of the perpendicular anisotropy of the films are discussed.

Manuscript received March 14, 1978.

* Research Institute of Electrical Communication, Tohoku University, Sendai, 980, Japan.

UNIAXIAL PERPENDICULAR ANISOTROPY OF THE Co-Cr FILMS

Co and Cr were co-deposited by an RF sputtering on the Polyimide film, from the cobalt target on which a number of electrolytic Cr pellets were placed at regular intervals in a grid pattern. The composition of the film was controlled by changing the surface area of the Cr pellets. An alloy target of Co-Cr was also successfully used for RF sputtering.

The RF sputtering was done in an Argon gas atmosphere after baking the vacuum chamber and the substrate holder at about 300°C. The back ground pressure reaches below 2×10^{-7} Torr.

The thickness of the film was controlled by the sputtering time. The deposition rate is mainly influenced by the RF power density and the Argon pressure. In this study, we have chosen such a sputtering condition as deposition rate of 0.33 μm/1hr., the Argon pressure is 0.01 Torr and the RF power density is 0.44 watt/cm^2.

The most influential factor on the magnetic properties was found to be Cr content of the film. The saturation magnetization Ms of the film decreases almost linearly with an increase of the Cr content, as shown in Fig.1. The change in Ms with the Cr content agrees with that of bulk Co-Cr[3], which is depicted by a dotted line in Fig.1.

Fig.2 shows the M-H loops of the films of different Cr contents, measured parallel (//) and perpendicular (⊥) to the film surface. The measurements were taken of disk samples (5mmϕ) of a 0.8 μm-thick film. In the figure, no compensation for demagnetization is made for the perpendicular M-H loops. Therefore, it is supposed that an intrinsic M-H loop (⊥), when compensated for the demagnetization, has a rectangular shape with an almost infinite slope. On the contrary, the M-H loop (//) is isotropic in the film plane and has a very small hysteresis loss.

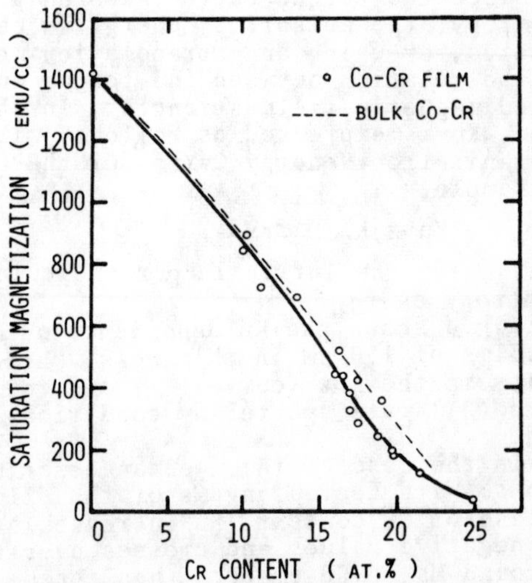

Fig.1 Ms vs. Cr content for Co-Cr film.

Reprinted from *IEEE Trans. Magn.*, vol. MAG-14, pp. 849-851, Sept. 1978.

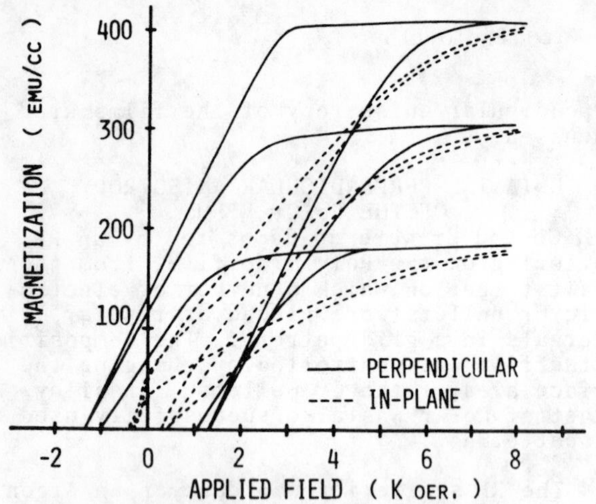

Fig.2 M-H loops of Co-Cr films.

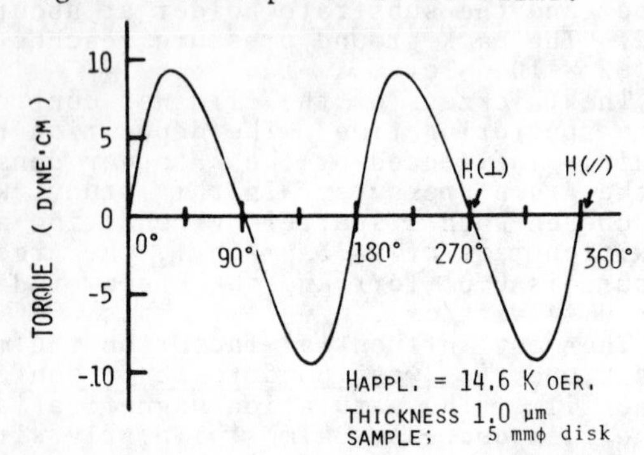

Fig.3 Torque curve of Co-Cr film.

From the result, it is safe to conclude that the Co-Cr film has an easy axis of magnetization in the normal of the film plane and a hard axis lying in the film plane.

In order to confirm the anisotropy, we have measured the torque curve of the Co-Cr film of Ms = 300 emu/cc in the normal plane of the disk sample. The measured torque curve, as shown in Fig.3, is a slightly distorted sine wave with a period of 180°. The polarity of the curve, together with its period, suggests that the film has the uniaxial anisotropy whose easy axis lies along the film normal.

Since the torque curve shows uniaxial anisotropy, the anisotropy energy Ku can be evaluated, by using an extrapolation method[4], from the relation between the torque and the applied magnetic field strength. In the method, Ku is expressed as follows, with the shape anisotropy energy $2\pi Ms^2$ for the circular disk sample,

$$Ku = K_{\perp} - 2\pi Ms^2 \quad (1)$$

where, $K_{\perp}$ is the intrinsic perpendicular anisotropy energy.

Fig.4 shows the Ku dependence on Ms for the films of 1.0 μm in thickness, where the results in the low (curve (A)) and the high (curve (B)) rate sputtering conditions are shown.

In the case of (A), we can describe as follows: With decreasing Ms of the film by an increase of Cr content, Ku increases steeply from negative values and crosses over the zero Ku around Ms = 700 emu/cc, then reaches to the maximum value of Ku = 4.9×10^5 erg/cc at Ms = 300 emu/cc. The positive values of Ku in

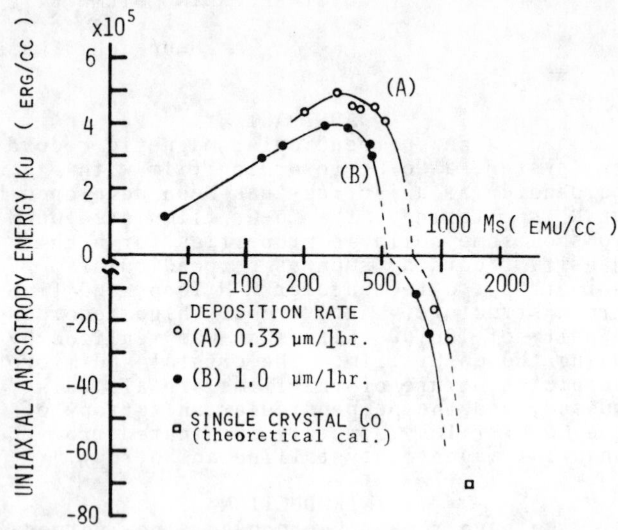

Fig.4 Ku vs. Ms in Co-Cr film for deposition rate (A) and (B).

Fig4 mean that $K_{\perp}$ surpasses $2\pi Ms^2$, then the thin film disk is magnetizable along the normal of the film surface. $K_{\perp}$, calculated by the equation(1), is positive in the whole Ms region. Typically, $K_{\perp} = 1.0 \times 10^6$ erg/cc was obtained for a Ms of 300 emu/cc. Pure Co film has the highest $K_{\perp}$, which decreases monotonically when the Cr content is incresed. The $2\pi Ms^2$, however, decreases more rapidly than $K_{\perp}$ with the increasing Cr content, hence the relation, $K_{\perp} > 2\pi Ms^2$, holds for films of Cr content greater than 13 at.%. Comparing the curves (A) and (B) in Fig.4, we found that the lower deposition rate yields films of higher perpendicular anisotropy.

CRYSTAL STRUCTURE OF THE Co-Cr FILMS

From the X-ray analysis, the Co-Cr film was found to have a hcp structure, and neither σ phase nor bcc phase of Cr. The lattice constant C is 4.055 Å for the film of 20 at.% Cr, which is slightly smaller than the lattice constant of bulk Co (4.069 Å). An X-ray diffraction pattern in Fig.5(a) shows only the (002) line, suggesting that the c-plane of hcp structure is oriented parallel to the film plane. Therefore, the c-axis of a hcp Co-Cr solid solution, or the easy axis of the uniaxial magneto-crystalline anisotropy, lies mainly along the normal of the film plane.

To investigate the c-axis dispersion of the crystallites, a rocking curve was

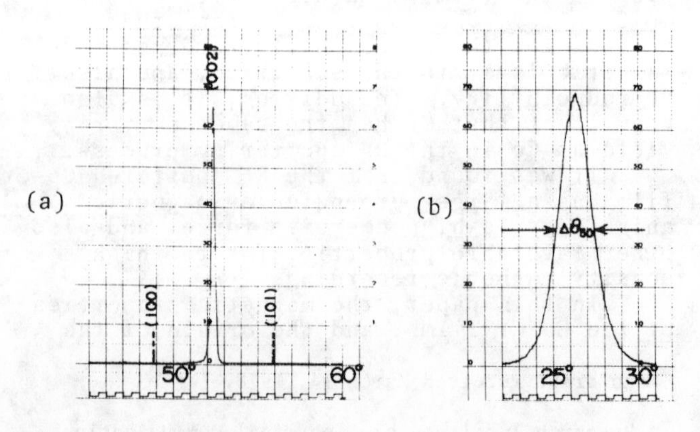

Fig.5 (a) X-ray diffraction pattern
(b) Rocking curve of (002) plane.

measured as shown in Fig.5(b). The rocking curve represents the angle distribution of the intensity of the X-ray diffracted from the (002) plane, and consequently, expresses the c-axis dispersion of the crystallites around the normal of the film. As shown in Fig.5(b), we defined the half angle width $\Delta\theta_{50}$ as the degree of orientation of the c-axis. The values of $\Delta\theta_{50}$ are 7.5° for the Co film, and 2.5° for the films of 13 at.% Cr. Therefore, the Co film, as well as the Co-Cr films, is found to have the c-axis of hcp structure which is oriented perpendicularly to the film plane. It can further be said that the c-axis dispersion of the Co-Cr film is narrower than that of the Co film. In the aforementioned high rate sputtering, the $\Delta\theta_{50}$'s are about two times as large as in the low rate sputtering, in the whole Ms range. The large values of $\Delta\theta_{50}$'s mean the broader angle dispersion of the c-axis. Combined with the result of Fig.4, the above result shows that the higher perpendicular anisotropy clearly corresponds to the narrower dispersion of the c-axis. Consequently, it is concluded that $K_\perp$ is primarily caused by the c-axis orientation of the hcp structure, and the dispersion of the c-axis must be narrowed to obtain the higher $K_\perp$.

A transmission electron microscopy revealed that the Co-Cr film is composed of fine grains as shown in Fig.6. The crystallites are of uniform size and as small as 0.04 μm in diameter for the film of a thickness of 0.34 μm. The average grain size was calculated to be one-fourth and one-ninth of the film thickness for thickness 0.1 μm and 0.5 μm, respectively. Since the grains grow at a very slower rate than the thickness, it is supposed that the crystallites in the film grow into a rod like structure which elongates in the direction of the film thickness. The abovementioned micro-structure of the film is thought to contribute to the perpendicular anisotropy.

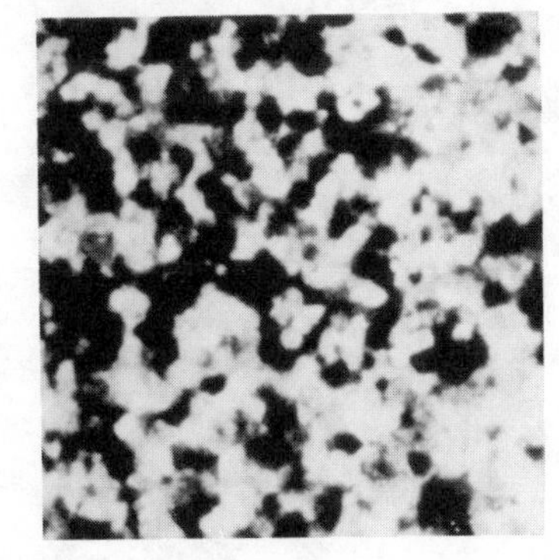

0.1 μm

Fig.6 Transmission image by electron microscopy of Co-Cr film of thickness 0.34 μm.

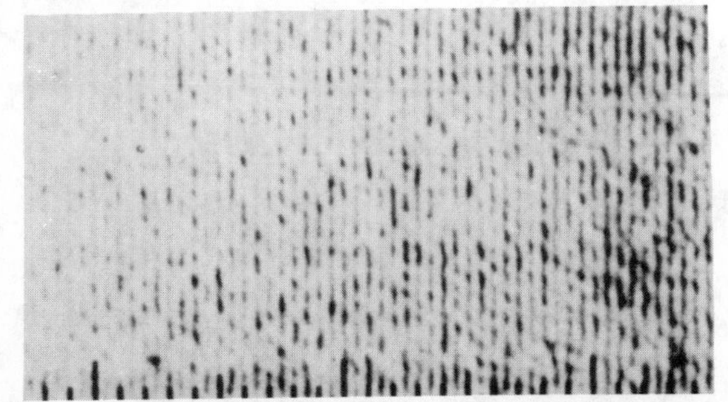

Fig.7 Bitter pattern of recorded signal on Co-Cr film of thickness 0.7 μm. One division of scale is 1.0 μm.

As described above, the Co-Cr sputtered film shows some suitable properties for the perpendicular recording medium, such as perpendicular anisotropy, fine grain structure, and rectangular M-H loop. We recorded a signal on a Co-Cr film of 0.7 μm thickness by a single pole head and realized an extremely high density recording. Fig.7, shows the recorded signal of 100,000 bits/inch developed by a Bitter technique. Since the direct observation is very difficult at the remarkably high density, the Bitter pattern was made by applying the perpendicular dc field to the film after recording[5], hence, the stripe's spacing is two times as large as the recorded bit's interval of 0.25 μm. By this fact, we confirmed that the signal of 100,000 bits/inch has been recorded on a relatively thick film.

CONCLUSIONS

The Co-Cr alloy films with perpendicular anisotropy have been successfully developed by an RF sputtering technique. To obtain the higher perpendicular anisotropy, the lower deposition rate was found to be preferable. It has also been confirmed that, as a high density recording medium, the Co-Cr film has some desirable properties such as a high recording resolution and a high output in the new perpendicular recording system. It was confirmed that Co-Cr sputtered films have sufficient reliability to record signals in high densities. The origin of the perpendicular anisotropy of the films was found to lie in the magneto-crystalline anisotropy and the shape effect of the rod like structure. The quantitative evaluation of the each origin of the anisotropy has not yet been done, and the reason why the addition of the Cr causes the high orientation of the c-axis is not yet known.

ACKNOWLEDGMENTS

This work was sponsored by the foundation of the scientific research of the Ministry of Education, Science and Culture, and by the Japan Society for the Promotion of Science.

The authors wish to thank Professor J. Shimoiizaka for providing magnetic fluid and K. Takemura for observing the Bitter Pattern.

REFERENCES

[1] S.Iwasaki and Y.Nakamura; "An analysis for the magnetization mode for high density magnetic recording," IEEE Trans. Magn., vol.MAG-13, no.5, pp.1272-1277, 1977.

[2] S.Iwasaki and H.Yamazaki; "Co-Cr sputtered films with perpendicular magnetic anisotropy," 7th Ann. Conf. on Magnetics, Japan, 4pA-7, 1975, (in Japanese).

[3] R.M.Bozorth; "Ferromagnetism," p.289, D. Van Nostrand Company, Inc., 1951.

[4] H.Miyajima, K.Sato and T.Matsuda; "New method for measuring the perpendicular anisotropy by using torquemeter," 7th Ann. Conf. on Magnetics, Japan, 5pB-13, 1975, (in Japanese).

[5] S.Iwasaki and K.Takemura; "An analysis for the circular mode of magnetization in short wavelength recording," IEEE Trans. Magn., vol.MAG-11, no.5, pp.1173-1175, 1975.

Part III
Head Fields

STUDIES ON MAGNETIC RECORDING *)

by W. K. WESTMIJZE 621.395.625.3

II. FIELD CONFIGURATION AROUND THE GAP AND THE GAP-LENGTH FORMULA

1. Types of head to be discussed

We are interested in the configuration of the magnetic field around the gap for two reasons: (1) on the recording side it informs us about the magnetic fieldstrengths the tape traverses in passing the gap, while (2) on the reproducing side it enables us, making use of the reciprocity theorem, to calculate the flux through the coil of the reproducing head due to the presence of a sinusoidally magnetized tape. This may be shown as follows.

The field distribution being known, the flux Φ through an arbitrary cross-section of the tape caused by a current I through the coil is also known. Now the reciprocity theorem states that, on the other hand, a current I round this cross-section of the tape excites the same flux Φ through the coil of the head. Replacing the tape by a series of currents of appropriate strength round the tape, the resulting flux through the coil of the head may be found by summing the contributions of all these currents.

It is easily seen that if the recorded wavelength is great compared with the length of the gap and small compared with the length of the head, the flux through the coil equals the flux in the tape in front of the gap. Deviations occur if the wavelength is of the order of the length of the head or of the order of the gap length.

We are chiefly interested in the latter case. Here the contributions to the flux are mainly due to elements of the tape in the neighbourhood of the gap. So it makes no difference if we suppose the head to be infinitely extended in the longitudinal direction of the tape.

A further simplifying assumption, which makes the potential problem a two-dimensional one, is that both the tape and the head are of infinite width. Finally we suppose the permeability of the tape to be unity, and that of the head to be infinite.

We shall discuss in this section three types of head (fig. 4). In the first one (fig. 4a) the gap is formed by two parallel planes $x = -l/2$ and $x = +l/2$. The tape is moved parallel to the x-axis. Although this type of head is impracticable, for the tape would have to cross an infinitely small slit in the walls of the gap, it is of some theoretical interest. The fieldstrength in this gap is easily calculated, and it is for this simple model that the well-known gap-loss formula $\sin(\pi l/\lambda)/(\pi l/\lambda)$ holds.

The second type (fig. 4b) is formed by two thin sheets, extending in the plane $y = 0$, one from $x = -\infty$ to $x = -l/2$, and the other from $x = l/2$ to $x = \infty$. This model resembles more or less some heads used in practice. The field distribution as well as the gap-length formula can be calculated.

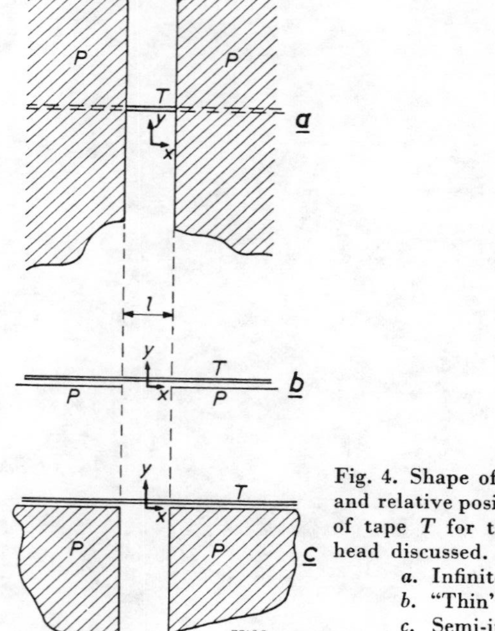

Fig. 4. Shape of the pole pieces P and relative position of an element of tape T for the three types of head discussed.
a. Infinite gap;
b. "Thin" gap;
c. Semi-infinite gap.

In the third type (fig. 4c) the left pole piece is bounded by the plane $y = 0$ from $x = -\infty$ to $x = -l/2$ and by the plane $x = -l/2$ from $y = 0$ to $y = -\infty$. The right pole piece is symmetrical to the left with respect to the plane $x = 0$. This type bears close resemblance to practical heads, but the calculations are rather difficult to carry out.

Seen in the direction normal to the tape the gap is infinitely extended in both the positive and the negative direction in the first type; in the

*) Continued from Philips Res. Rep. **8**, 148-157, 1953.

Reprinted with permission from *Philips Res. Rep.*, Part II, vol. 8, no. 3, pp. 161–183, June 1953.

second it is infinitely small, while in the third type it is infinitely extended only in the negative direction. Thus the latter type is intermediate between the other two.

In all three cases the permeability of the head material is supposed to be infinite, so that the pole surfaces are magnetic equipotentials. The coil of the head, consisting of one turn, may be supposed to be wound round a connecting element of the two pole pieces. A current I through this coil results in a magnetic potential difference I between the pole pieces. Thus the potential function $V(x, y)$ solving the problem has to satisfy the boundary condition $V = \frac{1}{2} I$ and $= -\frac{1}{2} I$ respectively for the two pole pieces. From the symmetry it is clear that $V = 0$ for $x = 0$.

A current I through the head excites in an element of the tape of width b and thickness dy a flux

$$d\Phi = -\mu_0 \frac{\partial V}{\partial x} b \, dy .$$

According to the reciprocity theorem a current I round the element bdy excites the same flux in the coil. If the tape is magnetized according to $M_x = M_0 \cos kx$, $(k = 2\pi/\lambda)$, $M_y = M_z = 0$ (longitudinal magnetization), the magnetic moment of an element of length dx will be equivalent to a current $I' = (M_0/\mu_0) \cos kx \, dx$ round the element. So the flux in the coil caused by this layer of the tape is

$$d\Phi = -\frac{M_0}{I} b \, dy \int_{-\infty}^{+\infty} \frac{\partial V}{\partial x} \cos kx \, dx . \tag{1}$$

Here $\partial V/\partial x$ is taken along this layer. For a tape of finite thickness the total flux is found by integration over y.

For reasons of symmetry it is obvious that $\partial V/\partial x$ is an even function. Thus a tape magnetized according to $M_x = M_0 \sin kx$ will induce no flux in the coil, $\partial V/\partial x \sin kx$ being odd and therefore integration from $-\infty$ to $+\infty$ giving zero.

2. First type, infinite gap

In this simple one-dimensional problem the potential is $V = Ix/l$, and the fieldstrength $H = I/l$.

The flux coming from a cross-section bdy of a tape magnetized according to $M_x = M_0 \cos kx$ is (eq. (1))

$$d\Phi = -\frac{M_0}{I} b \, dy \int_{-l/2}^{+l/2} \frac{I}{l} \cos kx \, dx = M_0 \, b \, dy \frac{\sin(kl/2)}{kl/2} .$$

The contribution of a tape of thickness d and width b is

$$\Phi = M_0 \, bd \frac{\sin(kl/2)}{kl/2} = \Phi' \, G(\pi l/\lambda) , \tag{2}$$

where $\Phi' = M_0 \, bd$ and $G(x) = \sin x/x$.

This is the well-known gap-loss formula, derived by Lübeck [10]) for a head of type c. The assumption Lübeck makes in the derivation is that the lines of force leaving the tape are distributed in inverse proportion to the distance to the pole pieces. This condition is fulfilled in the case discussed here, but not, as we shall see, for a head of type b or c.

This gap-loss formula bears close resemblance to the formula for the loss due to the finite width of the light-slit in optical recording and reproduction. In the latter case, however, the amplitude is proportional to the width of the light-slit, while in the case discussed here for long wavelengths the flux is independent of the gap length (as long as the reluctance of the air gap remains great compared with that of the path through the coil).

3. Second type, "thin" gap

In order to discuss case b we consider the transformation

$$z = (il/2) \sinh(\pi W/I) \tag{3}$$

giving a conformal mapping of the W-plane $(W = U + iV)$ on the z-plane $(z = x + iy)$. The lines $U =$ const. and $V =$ const. in the W-plane are represented in the z-plane by confocal ellipses and hyperbolas respectively. The line segments AB and CD (fig. 5) are special cases of the confocal hyperbolas, obtained for $V = \frac{1}{2} I$ and $V = -\frac{1}{2} I$ respectively. So the potential function $V(x, y)$ satisfying the boundary conditions of our problem may be found by separating in eq. (3) real and imaginary parts and eliminating U.

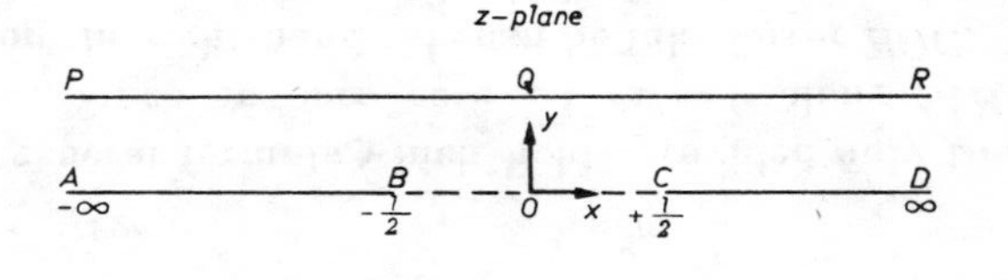

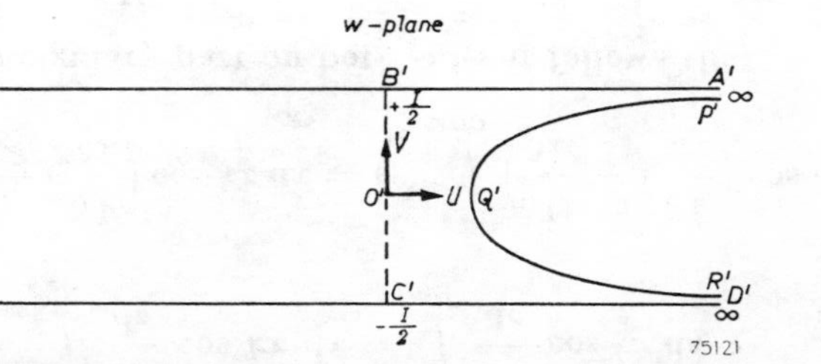

Fig. 5. Mapping of the z-plane $(z = x + iy)$ on the W-plane $(W = U + iV)$ for the case b.

Since
$$\frac{dW}{dz} = \frac{\partial U}{\partial x} + i\frac{\partial V}{\partial x},$$
or, with the Cauchy-Riemann relation $\partial U/\partial x = \partial V/\partial y$,
$$\frac{dW}{dz} = \frac{\partial V}{\partial y} + i\frac{\partial V}{\partial x},$$
it follows that the absolute value of the fieldstrength is given by
$$\sqrt{\left(\frac{\partial V}{\partial x}\right)^2 + \left(\frac{\partial V}{\partial y}\right)^2} = \left|\frac{dW}{dz}\right|.$$
Thus in our case the fieldstrength $H_n = -\partial V/\partial z$ normal to the planes $V=$ constant is found by taking the absolute value of
$$\frac{dW}{dz} = -i\frac{2}{\pi}\frac{I}{l}\{1-(2z/l)^2\}^{-1/2},$$
therefore
$$H_n = \frac{2}{\pi}\frac{I}{l}|\{1-(2z/l)^2\}^{-1/2}|.$$
For the special case $x = 0$ we have
$$H_x = \frac{2}{\pi}\frac{I}{l}\{1+(2y/l)^2\}^{-1/2}.$$

To obtain the flux induced in the coil of the reproducing head by a magnetized tape in a plane $y =$ constant $\int_{-\infty}^{+\infty} \partial V/\partial x \cos kx \, dx$ has to be calculated, where $\partial V/\partial x$ is taken along the line PQR in the z-plane.

Since $|dW/dz| \to 0$ for $z \to \infty$, integration of $(dW/dz)\exp(ikz)$ along the semi-circles at infinity in the upper half of the z-plane gives zero. Thus for any line $y =$ const. $\geqslant 0$
$$\int_{x=-\infty}^{+\infty} \frac{dW}{dz} e^{ikz}\, dz = 0$$
and therefore
$$\int_{-\infty}^{+\infty} \frac{dW}{dz} e^{ikx}\, dx = 0. \tag{4}$$

On the other hand integration of $(dW/dz)\exp(-ikz)$ along the contour $AODRQPA$ gives zero and thus
$$e^{ky}\int_{PQR} (dW/dz)\, e^{-ikx}\, dx = \int_{AOD} (dW/dz)\, e^{-ikx}\, dx. \tag{5}$$
Therefore with eq. (4)
$$e^{ky}\int_{PQR} \frac{dW}{dz}\cos kx \, dx = \int_{AOD} \frac{dW}{dz}\cos kx\, dx$$
or
$$\int_{PQR}\left(\frac{\partial U}{\partial x} + i\frac{\partial V}{\partial x}\right)\cos kx\, dx = e^{-ky}\int_{AOD}\left(\frac{\partial U}{\partial x} + i\frac{\partial V}{\partial x}\right)\cos kx\, dx.$$
Taking the imaginary part on both sides it follows that
$$\int_{PQR}\frac{\partial V}{\partial x}\cos kx\, dx = e^{-ky}\int_{AOD}\frac{\partial V}{\partial x}\cos kx\, dx. \tag{6}$$

This is a general formula which holds provided only that $|dW/dz| \to 0$ for $|z| \to \infty$. Since in our case $\partial V/\partial x = 0$ along AB and CD the integration on the right-hand side can be taken over BOC.

To evaluate the integral on the right-hand side of eq. (6) we have to find the dependence of $x(V)$ on V for the x-axis.

From eq. (3) it follows by separating the real and imaginary parts that
$$x = -\frac{l}{2}\cosh\frac{\pi U}{I}\sin\frac{\pi V}{I}$$
$$y = \frac{l}{2}\sinh\frac{\pi U}{I}\cos\frac{\pi V}{I}.$$
Thus for $y = 0$ it follows that $U = 0$, and therefore $x = -(l/2)\sin(\pi V/I)$. Eq. (6) now becomes
$$\int_{PQR}\frac{\partial V}{\partial x}\cos kx\, dx = e^{-ky}\int_{+I/2}^{-I/2}\cos\left\{k\left(-\frac{l}{2}\sin\frac{\pi V}{I}\right)\right\}dV =$$
$$= -\frac{I}{\pi}e^{-ky}\int_{-\pi/2}^{+\pi/2}\cos\left(\frac{kl}{2}\sin\tau\right)d\tau = -I\, e^{-ky} J_0\left(\frac{kl}{2}\right).$$

So we obtain for the flux contribution of an element bdy at a distance y from the head
$$d\Phi = M_0 b\, J_0(kl/2)\, e^{-ky}\, dy.$$
Integration over y from a to $a+d$ gives for the flux from a tape of thickness d on which is recorded a sinusoidal magnetization with an amplitude M_0 and a wavelength λ $(= 2\pi/k)$, and with a space a between head and tape,
$$\Phi = \Phi' \frac{1-e^{-2\pi d/\lambda}}{2\pi d/\lambda}\, e^{-2\pi a/\lambda}\, J_0(\pi l/\lambda) \tag{7}$$
where $\Phi' = M_0 bd$.

From a tape magnetized according to $\Phi' \cos\{2\pi(x-x')/\lambda\}$ only the even term $\Phi' \cos(2\pi x/\lambda)\cos(2\pi x'/\lambda)$ contributes to the flux in the coil. If the tape is moved with a speed v along the head ($x' = vt$) the flux in the head is

$$\Phi' \frac{1-e^{-2\pi d/\lambda}}{2\pi d/\lambda} e^{-2\pi a/\lambda} J_0(\pi l/\lambda) \cos \omega t;\ (\omega = 2\pi v/\lambda)$$

and hence eq. (7) represents the amplitude of the flux variation in the head if a sinusoidally magnetized tape is moved along the head.

The three factors describing the influence of tape thickness d, space between head and tape a, and gap length l are occurring separately. Thus the gap loss is given by the Bessel function $J_0(\pi l/\lambda)$, independent of tape thickness and distance head-to-tape. In fig. 9 this function is plotted against l/λ. Comparison with the gap-loss function $\sin(\pi l/\lambda)/(\pi l/\lambda)$ for a head of type a shows that the decrease of the amplitude of the maxima and minima is slower. This is also shown by the first term of the asymptotic expansion of the Bessel function

$$J_0(\pi l/\lambda) \approx \frac{\sin(\pi l/\lambda + \pi/4)}{\pi\sqrt{l/2\lambda}}.$$

The above solution of the potential problem also holds if, instead of the boundary discussed, the pole pieces are bounded by two blades of hyperbolas with focal points B and C (fig. 5). However, a tape parallel to the x-axis is impossible in this case for it has to cross the pole pieces. When, however, this cross point occurs for x large as compared with the gap-length it makes a negligible difference on the flux in the coil if for larger x the tape follows the surface of the head instead of crossing it.

4. Third type, semi-infinite gap

In this case the potential problem can be solved with the Christoffel-Schwarz method. Because of the symmetry it is sufficient to consider the problem where the potential of ABC (fig. 6) is $\frac{1}{2}I$, and that of DE is zero. Application of the theorem of Schwarz and Christoffel gives as the equation from which the transformation of the contour $ABCDE$ in the z-plane into the ξ-axis of the ζ-plane ($\zeta = \xi + i\eta$) is found

$$\frac{dz}{d\zeta} = C\frac{\sqrt{\zeta}}{\zeta - 1},$$

which on integration leads to

$$z = 2C\left(\sqrt{\zeta} + \tfrac{1}{2}\ln\frac{\sqrt{\zeta}-1}{\sqrt{\zeta}+1}\right) + C'.$$

If $\sqrt{\zeta}$ is defined such that $\sqrt{\zeta}$ lies in the first quadrant for $\mathrm{Im}(\zeta) > 0$, and the logarithm such that for $\zeta \to \infty$, $\ln\{(\sqrt{\zeta}-1)/(\sqrt{\zeta}+1)\} \to 0$ the values of the constants C and C' are found to be $C = il/2\pi$ and $C' = 0$. The required transformation now becomes

$$z = \frac{il}{\pi}\left(\sqrt{\zeta} + \tfrac{1}{2}\ln\frac{\sqrt{\zeta}-1}{\sqrt{\zeta}+1}\right). \qquad (8)$$

On the other hand the equation

$$W = \frac{I}{2\pi}\ln(\zeta - 1);\ (W = U + iV) \qquad (9)$$

transforms the W-plane into the ζ-plane such that, if the logarithm is real for $\zeta > 1$, $V = 0$ is mapped on $D'E'$, and $V = \frac{1}{2}I$ on $A'C'$. In general a line $V = V_0$ in the W-plane is mapped in the ζ-plane on a straight line through the point C' (1,0) making an angle $2\pi V_0/I$ with the positive ξ-axis.

Elimination of ξ from (8) and (9) gives the desired transformation, where V satisfies the potential equation and the boundary condition.

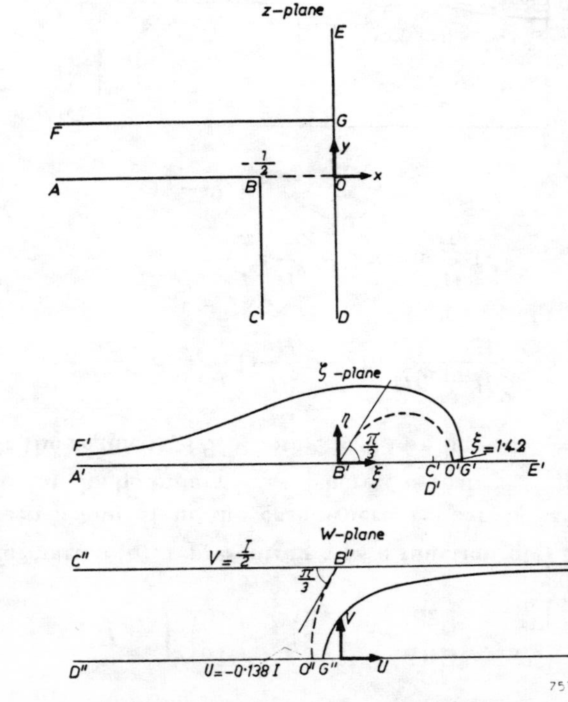

Fig. 6. Mapping of the z-plane ($z = x + iy$) and the W-plane ($W = U + iV$) on the ζ-plane ($\zeta = \xi + i\eta$) for the case c.

An analogous solution has been given recently by Booth [13]).

It is impossible to give $V(x, y)$ in explicit form for the general case; this is only possible in the limiting cases $U/I \ll -1$ and $U/I \gg 1$.

For $U/I \ll -1$: $\quad |\zeta - 1| = |e^{2\pi W/I}| \ll 1$,

$$z \approx \frac{il}{\pi}(1 - \ln 2 + \pi W/I),$$

so that
$$V \approx -\frac{I}{l}x.$$

Because $y \approx lU/I$, and $U/I \ll -1$ this gives the solution deep into the gap. Here the equipotentials are straight lines parallel to the gap wall. The fieldstrength $H = I/l$.

For $U/I \gg 1$: $\quad \zeta \approx e^{2\pi W/I}, \quad z \approx \frac{il}{\pi}\sqrt{\zeta}$,

whence

$$x + iy \approx i\frac{l}{\pi}e^{\pi U/I}\{\cos(\pi V/I) + i\sin(\pi V/I)\},$$

$$V \approx -\frac{I}{\pi}\arctan\frac{x}{y}.$$

At great distance from the gap the equipotentials are radial and the lines of force circular. This is the same solution as for an infinitely short gap. The radial component of the fieldstrength is zero, while the tangential $H_t = I/\pi r = (l/\pi r)H_0$, where H_0 is the fieldstrength deep into the gap.

For intermediate values of U/I, that means in the neighbourhood of the edge of the gap, even an approximate solution cannot be given. Fig.7 gives some equipotentials (V = constant) and lines of force (U = constant), calculated in a graphical way. The dashed lines are lines of constant fieldstrength, the value of which is expressed in the ratio to the fieldstrength H_0 deep into the gap, $H_0 = I/l$.

As may be expected the fieldstrength rises indefinitely towards the sharp edge of the gap. For magnetic recording, however, this has no real significance, firstly because fig. 7 shows that even for the sharp edge a tape passing at a distance greater than about 1/10 of the gap length does not experience a fieldstrength greater than that in the gap, and secondly because the gap edge of an actual head is never completely sharp.

Although a general solution for the potential $V(x, y)$ and for the fieldstrength $H(x, y)$ cannot be given, an expression may be derived for the fieldstrength along the boundary. From $dz/d\zeta = (il/2\pi)\sqrt{\zeta}/(\zeta - 1)$ and $dW/d\zeta = (I/2\pi)/(\zeta - 1)$ it follows that

$$\frac{dW}{dz} = -i\frac{I}{l}\frac{1}{\sqrt{\zeta}} = -i\frac{H_0}{\sqrt{\zeta}}. \tag{10}$$

Since
$$H = \sqrt{\left(\frac{\partial V}{\partial x}\right)^2 + \left(\frac{\partial V}{\partial y}\right)^2} = \left|\frac{dW}{dz}\right| = \frac{H_0}{|\sqrt{\zeta}|},$$

the transformation (8), representing z as a function of $\sqrt{\zeta}$, provides a relation between z and H in the case where $\sqrt{\zeta}$ can be expressed in $|\sqrt{\zeta}|$. This is true for the boundary lines, where ζ is real.

Thus for the segment AB, where $\zeta < 0$, we have $\sqrt{\zeta} = i|\sqrt{\zeta}| = iH_0/H$. Therefore

$$z = \frac{il}{\pi}\left(i\frac{H_0}{H} + \tfrac{1}{2}\ln\frac{iH_0 - H}{iH_0 + H}\right)$$

whence

$$x = -\frac{l}{\pi}\left(\frac{H_0}{H} + \arctan\frac{H}{H_0}\right),$$

$$y = 0.$$

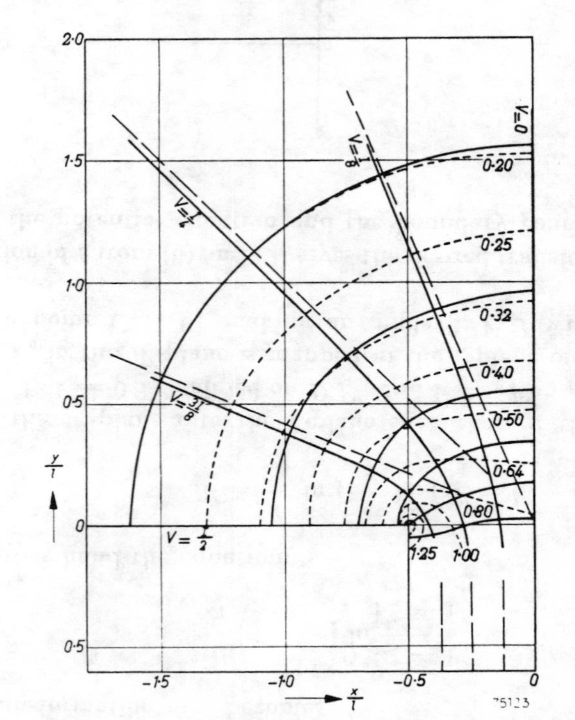

Fig. 7. Equipotentials, lines of force and lines of constant fieldstrength (dashed curves) for the case c.

For the segments BC and DE, $\zeta > 0$ and thus $\sqrt{\zeta} = |\sqrt{\zeta}| = H_0/H$. Hence for BC

$$x = -\,l/2\,,$$

$$y = \frac{l}{\pi}\left(\frac{H_0}{H} + \tfrac{1}{2}\ln\frac{1 - H_0/H}{1 + H_0/H}\right),$$

and for DE

$$x = 0\,,$$

$$y = \frac{l}{\pi}\left(\frac{H_0}{H} + \tfrac{1}{2}\ln\frac{H_0/H - 1}{H_0/H + 1}\right).$$

Fig. 8 gives the fieldstrength along the boundary computed with these formulae.

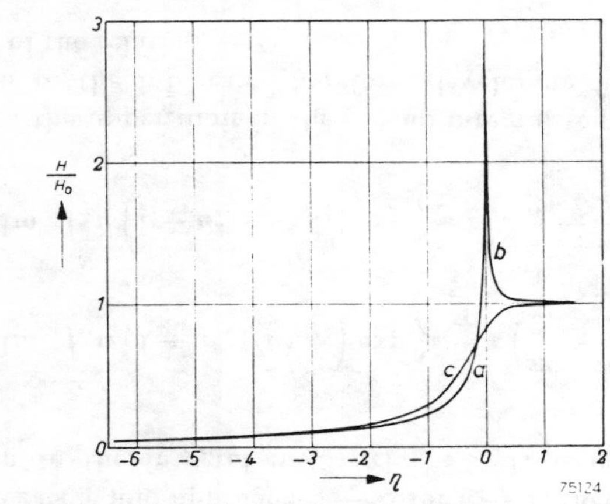

Fig. 8. Calculated fieldstrength along the boundary $ABCDE$ (fig. 6).
(a) H_y/H_0 along AB as a function of $\eta = x/l + \frac{1}{2}$,
(b) H_x/H_0 ,, BC ,, ,, ,, ,, $\eta = y/l$,
(c) H_x/H_0 ,, DE ,, ,, ,, ,, $\eta = y/l$.

In order to calculate $\int_{-\infty}^{0} \frac{\partial V}{\partial x} \cos kx \,\mathrm{d}x$ along the line FG in fig. 6, needed for the evaluation of the flux through the coil, we consider

$$\oint \frac{\mathrm{d}W}{\mathrm{d}z} e^{ikz}\,\mathrm{d}z \quad\text{and}\quad \oint \frac{\mathrm{d}W}{\mathrm{d}z} e^{-ikz}\,\mathrm{d}z\,.$$

The first integral is zero along $FGEF$ which gives with (8) and (9):

$$e^{-ky}\int_{-\infty}^{0} \frac{\partial W}{\partial x} e^{ikx}\,\mathrm{d}x + \frac{I}{2\pi}\int_{G'}^{\infty} \frac{1}{\zeta - 1}\exp\left\{-\frac{kl}{\pi}\left(\sqrt{\zeta} + \tfrac{1}{2}\ln\frac{\sqrt{\zeta}-1}{\sqrt{\zeta}+1}\right)\right\}\mathrm{d}\zeta = 0\,.$$

Since ζ is real and greater than 1 the latter integral is real. Thus, taking the imaginary part, it follows that

$$\int_{-\infty}^{0} \frac{\partial U}{\partial x} \sin kx\,\mathrm{d}x + \int_{-\infty}^{0} \frac{\partial V}{\partial x}\cos kx\,\mathrm{d}x = 0\,. \qquad (11)$$

The second integral is zero along $ABCDGFA$ which gives

$$e^{ky}\int_{-\infty}^{0} \frac{\partial W}{\partial x} e^{-ikx}\,\mathrm{d}x = \frac{I}{2\pi}\int_{-\infty}^{G'} \frac{1}{\zeta - 1}\exp\left\{\frac{kl}{\pi}\left(\sqrt{\zeta} + \tfrac{1}{2}\ln\frac{\sqrt{\zeta}-1}{\sqrt{\zeta}+1}\right)\right\}\mathrm{d}\zeta\,.$$

Taking again the imaginary part:

$$e^{ky}\left\{-\int_{-\infty}^{0} \frac{\partial U}{\partial x}\sin kx\,\mathrm{d}x + \int_{-\infty}^{0}\frac{\partial V}{\partial x}\cos kx\,\mathrm{d}x\right\} =$$

$$= \frac{I}{2\pi}\,\mathrm{Im}\left[\int_{-\infty}^{1} \frac{1}{\zeta - 1}\exp\left\{\frac{kl}{\pi}\left(\sqrt{\zeta} + \tfrac{1}{2}\ln\frac{\sqrt{\zeta}-1}{\sqrt{\zeta}+1}\right)\right\}\mathrm{d}\zeta\right]$$

or with (11)

$$\int_{-\infty}^{0} \frac{\partial V}{\partial x}\cos kx\,\mathrm{d}x = -\,\tfrac{1}{2}\,I\,e^{-ky}\,S\left(\frac{kl}{2}\right),$$

where

$$S(\tau) = \mathrm{Im}\left[\frac{1}{2\pi}\int_{-\infty}^{1} \frac{1}{1-\zeta}\exp\left\{\frac{2}{\pi}\,\tau\left(\sqrt{\zeta} + \tfrac{1}{2}\ln\frac{\sqrt{\zeta}-1}{\sqrt{\zeta}+1}\right)\right\}\mathrm{d}\zeta\right].$$

Following eq. (1) the flux from an element b dy is

$$\mathrm{d}\Phi = M_0 b\,\mathrm{d}y\,e^{-ky}\,S(kl/2)\,.$$

Thence a tape of thickness d and width b at a distance a from a head of type c gives a flux

$$\Phi = \Phi'\,\frac{1 - e^{-2\pi d/\lambda}}{2\pi d/\lambda}\,e^{-2\pi a/\lambda}\,S(\pi l/\lambda)\,. \qquad (13)$$

The gap loss is given by $S(\tau)$, with $\tau = kl/2 = \pi l/\lambda$. For $S(\tau)$ we may write

$$S(\tau) = \mathrm{Im}\,\frac{1}{\pi}\int_{0}^{\infty} \frac{u}{1+u^2}\exp\left\{i\,\frac{2\tau}{\pi}\left(u - \arctan u + \frac{\pi}{2}\right)\right\}\mathrm{d}u \;+$$

$$+\,\mathrm{Im}\,\frac{1}{\pi}\int_{0}^{1} \frac{u}{1-u^2}\exp\left\{\frac{2\tau}{\pi}\left(u + \tfrac{1}{2}\ln\frac{1-u}{1+u} + i\,\frac{\pi}{2}\right)\right\}\mathrm{d}u\,, \qquad (14)$$

or

$$S(\tau) = \frac{1}{\pi}\int_0^{\pi/2} \tan\Phi \sin\left\{\frac{2}{\pi}\tau\left(\tan\Phi - \Phi + \frac{\pi}{2}\right)\right\} d\Phi +$$

$$+\frac{1}{\pi}\sin\tau \int_0^1 \frac{u}{1-u^2}\exp\left\{\frac{2}{\pi}\tau\left(u + \tfrac{1}{2}\ln\frac{1-u}{1+u}\right)\right\} du. \qquad (15)$$

For large values of τ the main contribution to these integrals comes from small values of the argument. In the first integral of (15) this is true because $\sin\{(2\tau/\pi)(\tan\Phi - \Phi + \pi/2)\}$ becomes a rapidly oscillating function if $(2\tau/\pi)\tan\Phi > \pi/2$ thus giving no contribution to the integral, while in the second integral

$$u + \tfrac{1}{2}\ln\frac{1-u}{1+u} = -\frac{u^3}{3} - \frac{u^5}{5} - \dots$$

is always negative and approaches $-\infty$ for $u \to 1$, therefore also giving no contribution to the integral if $(2\tau/\pi)\,u^3/3 \gg 1$. Hence eq.(14) may be written

$$S(\tau) = \frac{1}{\pi}\,\mathrm{Im}\int_0^\infty u\left(1 - u^2 + u^4 \dots\right)\exp\left\{i\frac{2}{\pi}\tau\left(\frac{\pi}{2} - \frac{u^3}{3} - \frac{u^5}{5} - \dots\right)\right\} du$$

$$+\frac{1}{\pi}\,\mathrm{Im}\int_0^1 u\left(1 + u^2 + u^4 \dots\right)\exp\left\{\frac{2}{\pi}\tau\left(i\frac{\pi}{2} - \frac{u^3}{3} - \frac{u^5}{5} - \dots\right)\right\} du.$$

Extension of the upper limit in the second integral to ∞ gives a negligible contribution to the integral. Therefore, developing for powers of u and making use of the relation

$$\int_0^\infty x^p \exp(-w\,x^q)\,dx = \frac{1}{q}\,\frac{\Gamma\{(p+1)/q\}}{w^{(p+1)/q}}$$

the asymptotic expansion

$$S(\tau) \sim \frac{3^{1/2}\Gamma(2/3)}{\pi(2\tau/\pi)^{2/3}}\sin(\tau + \pi/6) + \frac{3^{5/6}\Gamma(4/3)}{\pi(2\tau/\pi)^{4/3}}\sin(\tau - \pi/6) + O(\tau^{-8/3}) \quad (15a)^{*)}$$

is found.

The above expansion is obtained by developing eq. (14) for small values of u, and therefore of ζ. Since the region round $\zeta = 0$ is mapped on the edge of the gap this means physically that only elements of the tape close to this edge are considered. Therefore the leading term of the expansion (15a) can also be arrived at by expressing $\partial V/\partial x$ in x for small values of ζ and then according to eq. (6) evaluating

$$\int_{-l/2}^{0}\frac{\partial V}{\partial x}\cos kx\,dx.$$

Eq. (8) gives in first approximation

$$z = -\frac{l}{2} - \frac{il}{3\pi}\zeta^{3/2}.$$

In order to move along BO the argument of $\zeta = re^{i\varphi}$ has to be taken $\varphi = \pi/3$ which means that in fig. 6 the representation on the ζ-plane of the line OA in the z-plane makes an angle $\pi/3$ with the ξ-axis. Thus $x = -l/2 + (l/3\pi)r^{3/2}$.

Since $\dfrac{dW}{dz} = -i(I/l)/\sqrt{\zeta}$ (eq. (10)) it follows that

$$\frac{\partial V}{\partial x} = -\frac{I}{l}\,\mathrm{Re}\,\frac{1}{\sqrt{\zeta}} = -\frac{I}{l}r^{-1/2}\cos\frac{\pi}{6} = -\frac{I}{l^{2/3}}\,\frac{3^{1/6}}{2\pi^{1/3}}\left(x + \frac{l}{2}\right)^{-1/3}$$

and

$$\int_{-l/2}^{0}\frac{\partial V}{\partial x}\cos kx\,dx = -\frac{I}{l^{2/3}}\,\frac{3^{1/6}}{2\pi^{1/3}}\int_{-l/2}^{0}\frac{\cos kx}{(x + l/2)^{1/3}}\,dx,$$

where

$$\int_{-l/2}^{0}\frac{\cos kx}{(x + l/2)^{1/3}}\,dx \approx \int_0^\infty \cos\{k(x - l/2)\}\,x^{-1/3}\,dx = \mathrm{Re}\int_0^\infty e^{ikx - ikl/2}\,x^{-1/3}\,dx =$$

$$= \mathrm{Re}\,e^{-ikl/2}\int_0^\infty e^{-ky}\,(iy)^{-1/3}\,i\,dy = \frac{\Gamma(2/3)}{k^{2/3}}\sin\left(\frac{kl}{2} + \frac{\pi}{6}\right),$$

it follows that

$$S\left(\frac{kl}{2}\right) = -\frac{2}{I}\int_{-l/2}^{0}\frac{\partial V}{\partial x}\cos kx\,dx = \frac{3^{1/6}\Gamma(2/3)}{\pi^{1/3}}\,\frac{\sin(kl/2 + \pi/6)}{(kl)^{2/3}}$$

which is the leading term of (15a).

For smaller values of τ, $S(\tau)$ can be computed numerically. In the next table the values of $S(\tau)$ as computed by The Mathematical Centre at Amsterdam are given for $0 < \tau < 5\pi$. Column 3 gives the values obtained by using the first two terms of eq. (15a), and column 4 those of the first term.

---

*) In the course of this work The Mathematical Centre at Amsterdam also verified eq. (15a) by a more rigorous analysis.

TABLE I

Values of $S(\tau)$ for $0 \leqslant \tau \leqslant 5$, as computed according to different formulae.

τ/π	$S(\tau)$	1st and 2nd term of (15a)	1st term of (15a)
0·000	1		
0·125	0·969		
0·250	0·880		
0·375	0·740		
0·50	0·565	0·571	
0·75	0·180	0·182	
1·00	—0·135	0·135	
1·25	—0·282	—0·282	
1·50	—0·244	—0·244	
1·75	—0·084	—0·084	
2·00	0·091	0·092	
2·25	0·188	0·188	
2·50	0·167	0·167	
2·75	0·057	0·057	
3·00	—0·072	—0·071	—0·078
3·25		—0·147	—0·144
3·50		—0·132	—0·123
3·75		—0·044	—0·035
4·00		0·061	0·065
4·25		0·122	0·120
4·50		0·110	0·103
4·75		0·037	0·030
5·00		—0·053	—0·056

It is seen from the table that from $\tau = 0{\cdot}5\pi$ onwards the approximation by the first two terms is very satisfactory, while from $\tau = 5\pi$ onwards, that is past the fifth zero, the approximation by the first term is reasonable.

The gap loss $S(\pi l/\lambda)$ as a function of l/λ is represented in fig. 9. It is seen that, as could be expected on physical grounds, the gap-loss function for a head of type c is intermediate between those for types a and b.

Equations (2), (7) and (13) give the output for the three types of head in the case of a longitudinally magnetized tape. For perpendicular magnetization a head of type a gives no response as may be seen from the symmetry, whereas for the other heads the output is still given by equations (7) and (13). We proceed to prove this.

In order to apply the reciprocity theorem to perpendicular magnetization we have first to ask for the flux through an element $b\mathrm{d}x$, lying in the plane of the tape. This flux is given by $-\mu_0(\partial V/\partial y)\, b\, \mathrm{d}x$. Thus, applying the reciprocity theorem in the same way as above, we find for the flux through the coil of a reproducing head from a tape magnetized according to $M_y = M_0 \sin kx$, $M_x = M_z = 0$

$$\mathrm{d}\Phi = \frac{-M_0}{I}\, b\, \mathrm{d}y \int_{-\infty}^{+\infty} \frac{\partial V}{\partial y} \sin kx\, \mathrm{d}x\,.$$

Here the contributions from a cosine term in the magnetization cancel each other out.

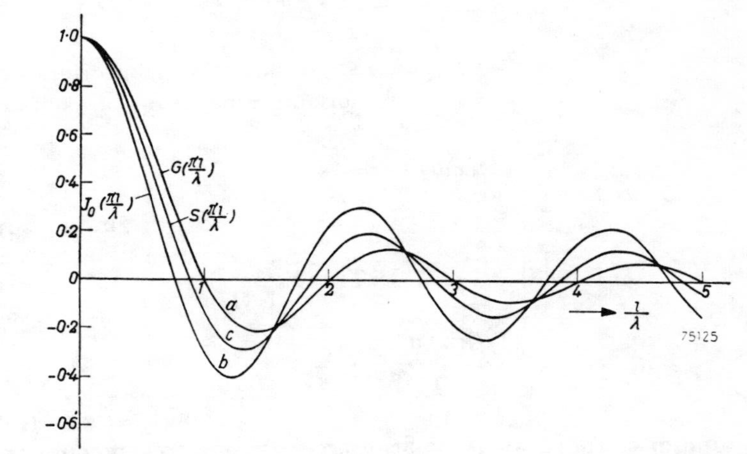

Fig. 9. The three gap-loss functions $G(\pi l/\lambda)$, $J_0(\pi l/\lambda)$ and $S(\pi l/\lambda)$ as a function of l/λ for heads of types a, b and c respectively.

Since $\partial V/\partial y = 0$ in the gap of a head of type a the output is zero for this type. For types b and c, however, it follows from the fact that V is an analytical function that $\partial V/\partial y = \partial U/\partial x$. Putting in eq.(4) $\mathrm{d}W/\mathrm{d}z = \partial U/\partial x + i\, \partial V/\partial x$ and taking the imaginary part yields

$$\int_{-\infty}^{+\infty} \frac{\partial U}{\partial x} \sin kx\, \mathrm{d}x = \int_{-\infty}^{+\infty} \frac{\partial V}{\partial x} \cos kx\, \mathrm{d}x\,,$$

whence

$$\mathrm{d}\Phi = \frac{M_0}{I}\, b\, \mathrm{d}y \int_{-\infty}^{+\infty} \frac{\partial V}{\partial x} \cos kx\, \mathrm{d}x\,.$$

This is eq.(1) with opposite sign.

5. Reproduction at long wavelengths

Until now we have assumed that the reproducing head is long compared with the wavelength to be reproduced. In practice this will not always be so, and, as a consequence, deviations from the predicted frequency characteristic may occur at low frequencies.

In order to determine the influence of the finite length of the head we can proceed along the same lines as above for the influence of the finite gap length. First we determine the field distribution around an energized head of finite length and then calculate the flux from a magnetized tape by application of the reciprocity theorem. This in fact was the method followed by Clark and Merrill [14]) who determined the field distribution by direct measurement.

The field measurements being difficult to carry out with sufficient accuracy, a calculation of the field may be useful. On the other hand such calculations become very difficult without a number of simplifying assumptions. Perhaps the most significant of these is that the problem can still be treated as a two-dimensional one, although in reality the width of the head is of the same order of magnitude as the wavelength and sometimes even smaller.

We shall first carry out the calculation for the simple model of a head consisting of two thin magnetically conducting sheets of finite length, AO and OB in fig.10*a*, in close proximity over which the tape is transported. This is the model discussed in section 3 where, however, the length of the head was assumed to be infinite. To simplify matters we shall here assume the gap to be infinitely short.

The potential problem we have to solve here is that in the z-plane AO be at a potential $I/2$ and OB at $-I/2$. Carrying out an inversion with centre in the origin, and inversion radius $L/2$, $z'z = (L/2)^2$, gives in the

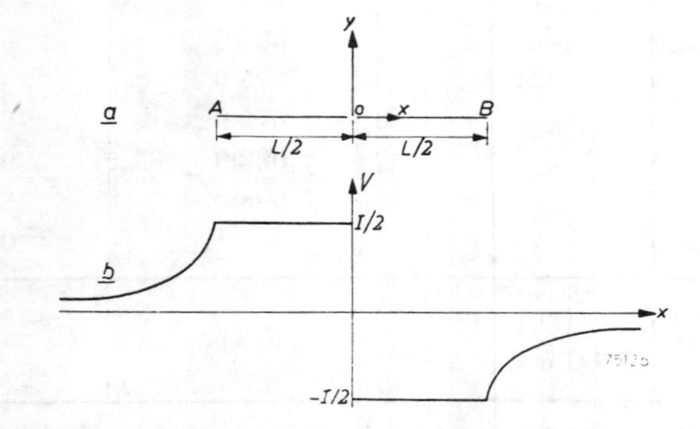

Fig. 10. *a*. Schematization of a thin head of finite dimensions. *b*. Distribution of the potential along the x-axis.

z'-plane the boundary condition that the potential V be $I/2$ from $x' = -\infty$ to $L/2$ and $-I/2$ from $x' = L/2$ to $+\infty$. The solution to this problem is given by eq.(3), whence the transformation satisfying our problem is found to be

$$z = -i \frac{L}{2} \frac{1}{\sinh(\pi W/I)}.$$

In order to calculate the flux through a coil, wound round a connecting element of the two pole pieces, we can apply eq.(1) and calculate therefore

$$\int \frac{\partial V}{\partial x} \cos kx \, dx$$

along the x-axis. The potential along this axis is sketched in fig. 10*b*. Integration along AO and OB gives no contribution since here $\partial V/\partial x = 0$. Therefore the contributions to this integral are coming from the line segments outside AB, and from the origin, where $\partial V/\partial x$ is infinite. Outside AB, $U = 0$, so that

$$x = -\frac{L}{2 \sin(\pi V/I)}.$$

Hence

$$\frac{\partial V}{\partial x} = \frac{IL}{2\pi} \frac{1}{x^2} (1 - L^2/4x^2)^{-1/2} \text{ for } |x| \geqslant L/2.$$

In the origin

$$\frac{\partial V}{\partial x} = -I\delta(x),$$

where δ is the Dirac delta function.

Therefore, the flux in the coil is found to be

$$\Phi = \Phi_0 \left(1 - \frac{2}{\pi} \int_1^\infty \frac{\cos \beta\tau \, d\tau}{\sqrt{\tau^2 - 1}\, \tau}\right), \tag{16}$$

where $\beta = kL/2 = \pi L/\lambda$ and $\Phi_0 = M_0\, bd$ is the amplitude of the flux variations in the tape. The integral can be transformed by differentiating with respect to β, which gives

$$-\int_1^\infty \frac{\sin \beta\tau}{\sqrt{\tau^2 - 1}} d\tau = -\frac{\pi}{2} J_0(\beta). \text{ [15])}$$

Hence

$$\frac{2}{\pi} \int_1^\infty \frac{\cos \beta\tau \, d\tau}{\sqrt{\tau^2 - 1}\, \tau} = -\int_0^\beta J_0(x)\, dx + c.$$

Since for $\beta = 0$ the first integral of this equation is 1 and and the second 0 it follows that $c = 1$, and therefore

$$\Phi/\Phi_0 = \int_0^{\pi L/\lambda} J_0(x)\,\mathrm{d}x\,, \tag{17}$$

which is a tabulated function [16]).

An approximation of this result can be obtained by noting that in eq. (16) the main contribution to the integral comes from $\tau \approx 1$. This is especially true for large values of β since then the contributions to the integral coming from larger values of τ cancel each other out owing to the fluctuating character of $\cos \beta\tau$. Therefore

$$\int_1^{\infty} \frac{\cos \beta\tau}{\sqrt{\tau^2 - 1}} \frac{\mathrm{d}\tau}{\tau} \approx \int_0^{\infty} \frac{\cos \{\beta\,(1+k)\}}{2^{1/2}\,k^{1/2}}\,\mathrm{d}k = \frac{\sqrt{\pi}}{\beta^{1/2}} \cos\,(\beta + \pi/4)$$

and hence

$$\Phi/\Phi_0 \approx 1 - \frac{2^{1/2}}{\pi(L/\lambda)^{1/2}} \cos \{\pi(L/\lambda + \tfrac{1}{4})\}\,. \tag{17a}$$

From fig.11, giving Φ/Φ_0 computed from eq.(17) (full line) and from (17a) (dashed line) it is seen that the approximation is excellent for values of $L/\lambda > 1$.

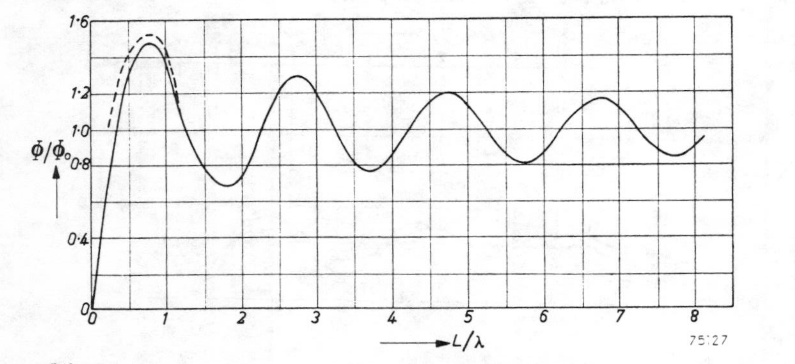

Fig. 11. $\int_0^{\pi L/\lambda} J_0(x)\mathrm{d}x$ as a function of L/λ (full line) and the approximation of this integral by $1 - 0{\cdot}45 \dfrac{\cos\,[\pi(L/\lambda + 1/4)]}{(L/\lambda)^{1/2}}$ (dashed line).

We can now proceed to calculate in the same approximation the more realistic case that the edges of the pole pieces where the tape approaches and leaves the head, A and B in fig. 12, are 90° and not 180° as in the preceding case.

Application of the theorem of Christoffel and Schwarz leads to the equation transforming the contour $CAOBD$ into the ξ-axis of a ζ-plane:

$$z = \frac{L}{\pi} \{\zeta \sqrt{1 - \zeta^2} + \arcsin \zeta\}\,.$$

Here the principal value of the arcsin has to be taken, and that value of the root which is positive in the origin.

The edges A and B are mapped on $\zeta = -1$ and $\zeta = +1$ in the ζ-plane respectively.

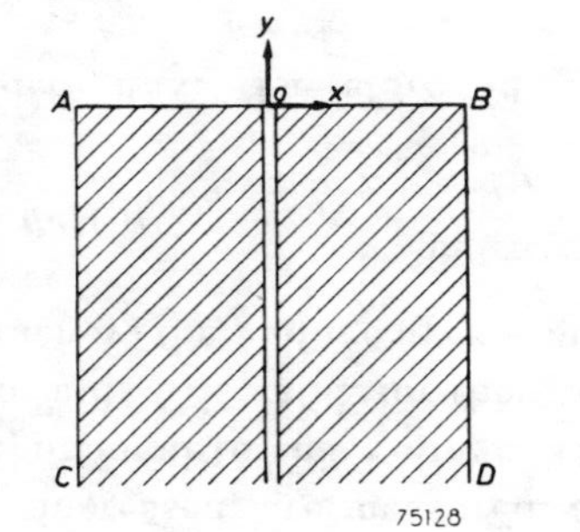

Fig. 12. Schematization of a finite head with 90° edges.

On the other hand, the equation

$$W = \frac{I}{\pi} \ln \zeta - \frac{iI}{2}$$

transforms a W-plane ($W = U + iV$) on the ζ-plane such that $V = +I/2$ is mapped on the negative and $V = -I/2$ on the positive ξ-axis.

From these equations it is found that the fieldstrength at the edge B is in first approximation

$$-\frac{\partial V}{\partial x} \approx -\frac{I}{L^{2/3}} \frac{3^{1/6}}{2^{5/3}\pi^{1/3}} \frac{1}{(x - L/2)^{1/3}}\,,$$

so that the required integral is

$$\int_{L/2}^{\infty} \frac{\partial V}{\partial x} \cos kx\,\mathrm{d}x \approx I \frac{3^{1/6}}{2^{5/3}\pi^{1/3}} \frac{\Gamma(2/3)}{(kL)^{2/3}} \cos\left(\frac{kL}{2} + \frac{\pi}{6}\right).$$

Hence the total flux through the head resulting from the contributions of the gap and the two edges is

$$\frac{\Phi}{\Phi_0} \approx 1 - \frac{3^{1/6}}{\pi 2^{1/3}} \frac{\Gamma(2/3)}{(L/\lambda)^{2/3}} \cos\left\{\pi\left(\frac{L}{\lambda}+\frac{1}{6}\right)\right\} = 1 - 0{\cdot}205 \frac{\cos\{\pi(L/\lambda+1/6)\}}{(L/\lambda)^{2/3}}. \quad (18)$$

In fig. 13 Φ/Φ_0 according to (18) is plotted (full line) together with two measured response curves. For the first of these (dashed line) the tape was led immediately over the edges of the head. Apart from the fact that the zeros are shifted to lower values of L/λ there is good agreement with the theoretical curve. The difference may be due to the non-validity of the assumption that the head is broad compared with the wavelength.

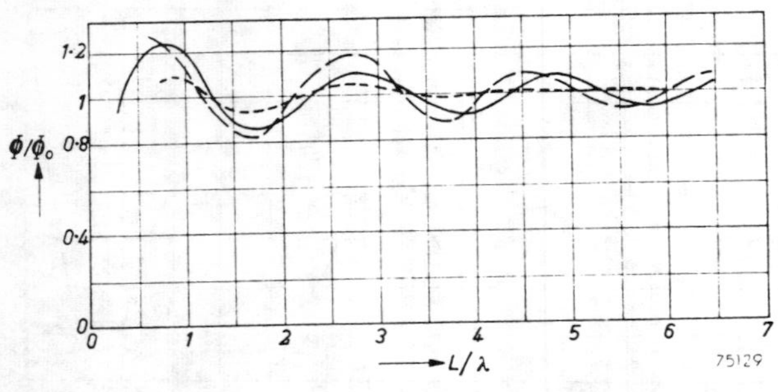

Fig. 13. Calculated output for a finite head with 90° edges (full line) and measured curve for the case that the tape is led immediately over the edges (dashed line) and at a distance of 2 mm from the edges (dotted line).

In the other case (dotted line) the tape was led in a straight line in front of the head, touching the head only at the place of the gap. Owing to the receding angles of the front planes the tape now passes the edges of the head at a certain distance, the influence of which may be described in first approximation by addition of a factor $\exp(-2\pi a/\lambda)$ to the last term of eq. (18), where a is the distance between tape and edge. This effects an increased damping of the oscillations in conformity with the measurements.

6. Discussion

In our analysis we supposed the permeability of the tape to be unity. A general solution for the case of a permeability other than unity cannot be given. For a special case, viz. $\mu = 3$ and a gap length equal to the thickness of the tape, a graphical solution was given by Begun [17]).

Our calculations show that for the three types of head discussed the factors describing the influence of tape thickness, space between tape and head, and gap length are occurring separately. For the "open" heads, types b and c of fig. 4, the influence of the first two is for longitudinal and perpendicular magnetization given by

$$\frac{1-e^{-2\pi d/\lambda}}{2\pi d/\lambda} e^{-2\pi a/\lambda}.$$

The same expression was deduced by Wallace [12]) in a different way. In both derivations, however, the assumption is made that the permeability of the tape equals unity. We shall see in Part IV that for a tape of higher permeability the expression becomes more complicated.

The gap-loss formula $\sin(\pi l/\lambda)/(\pi l/\lambda)$ is valid only in the theoretical case of a head of type a. For the more realistic types b and c the general behaviour, i.e. an oscillating function with decreasing amplitude, is the same. But the existing differences have important consequences for some frequently used procedures in magnetic recording. Thus it is common practice to deduce the gap length from tape velocity and frequency of the first zero, on the supposition that here $l = \lambda$. Fig. 9 shows that, for a type-c head, $l = 0{\cdot}9\lambda$ for the first zero, so that the actual gap is 10% smaller than the calculated gap with $l = \lambda$. This is confirmed by the measurements of Lübeck [10]) who concluded that the magnetic gap is 10% longer than the mechanic gap. If the length of the gap is comparable with the depth (wide-gap measurements) the difference may become even greater; for then a head of type b is approached.

It has been proposed [18]) to use wide-gap measurements as a means to determine the actual flux in the tape. Then the successive maxima of the voltage across the open terminals of a reproducing head, which occur if the frequency is gradually increased, are measured, and the recorded flux in the maxima is put proportional to this voltage. The latter is only true, however, if the gap-loss formula G is valid, for then, since the open voltage is proportional to the frequency, the voltage over a one-turn coil is given by

$$E = \omega \Phi G(\pi l/\lambda) = 2\pi \frac{v}{\lambda} \frac{\sin(\pi l/\lambda)}{\pi l/\lambda} \Phi,$$

where Φ is the recorded flux. Therefore in the extremes, where $|\sin(\pi l/\lambda)| = 1$,

$$E_m = \frac{2v}{l} \Phi,$$

independent of the wavelength.

If, however, the measurements are carried out with a head of type c, as is normally done, the gap-loss function $S(\pi l/\lambda)$ has to be used. Then the voltage induced by a thin tape in close contact with the head is given approximately by

$$E'_m \approx \frac{3^{1/6}\Gamma(2/3)}{2^{2/3}} \left(\frac{l}{\lambda}\right)^{1/3} \frac{2v}{l} \Phi \approx 1{\cdot}023 \left(\frac{l}{\lambda}\right)^{1/3} \frac{2v}{l} \Phi,$$

rising, therefore, with respect to the former formula as $(l/\lambda)^{1/3} \sim \omega^{1/3}$, or 2 dB per octave. The calculated dependence of the voltage on l/λ for a constant amplitude of the flux in the tape is given in fig. 14.

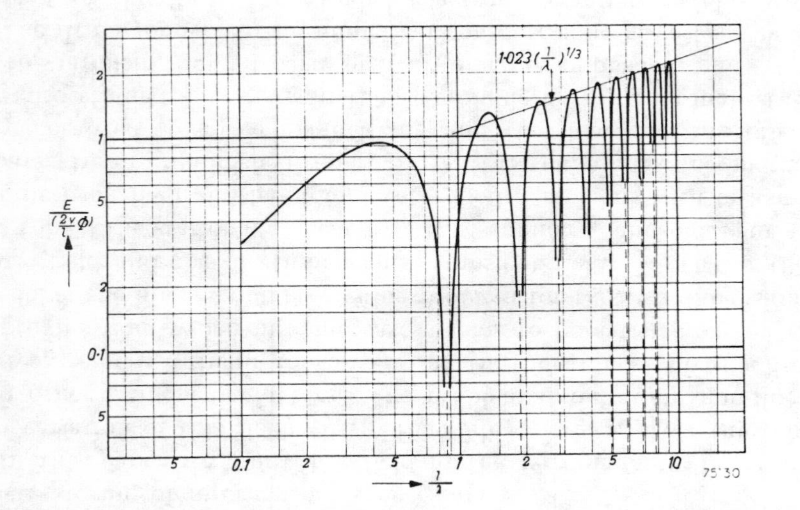

Fig. 14. Calculated output for wide-gap measurements with a semi-infinite gap.

Further it must be remembered that, although in wide-gap measurements the gap is long compared with the wavelength, the dependence of the output voltage on the space between head and tape is the same as in the case of a narrow gap, being given in both cases by the formula $\exp(-2\pi a/\lambda)$.

Recently Schmidbauer [10]) has also derived the 2dB/octave rise. He arrives at an approximate formula for the gap loss analogous in behaviour to our function $S(\pi l/\lambda)$. An expression is also given for the long-wave deviations of the frequency characteristic due to the finite length of the reproducing head, which qualitatively explains the experiments.

In another recent publication Mankin [20]) discussed these deviations, under the name of interference effects, by comparing the head with a multiplicity of gaps each acting as a source of flux. The edges of the head act as such sources. This is in conformity with our formula (1) since at the edges $\partial V/\partial x$ is large and will, therefore, give a large contribution to the integral and thus to the reproduced flux.

CALCULATION OF THE MAGNETIC FIELD IN THE FERROMAGNETIC LAYER OF A MAGNETIC DRUM

BY OLLE KARLQVIST

1. Introduction

A magnetic drum is used to store information. The information may come from an electronic computer or from a telephone dial. The magnetic drum consists of cylinder on the surface of which a ferromagnetic material is coated. It is equipped with reading and recording heads, which enables binary data to be recorded in the form of magnetized elements on its surface when the drum rotates. When a pulse is read in, it is stored as an elementary magnet with a north and a south pole. The length of this dipole is of the order of magnitude of 3 millimetres. The drum contains a number of channels, the distance between them being usually a few millimetres.

The drum has been proved to be a reliable store and it is used in most electronic computers. Drums have recently found applications in telephony (MALTHANER and VAUGHAN, 1953) and other branches where storing of information is necessary.

In this paper we study the field in the ferromagnetic layer and the variation of this field with permeability, airgap, layer thickness and other influencing factors. The problem is definitely non-linear and extremely difficult to solve. But the linear case gives a first approximation, which in some cases seems to be satisfactory. Here we solve the linear boundary value problem for the two-dimensional static field and the one-dimensional transient field. WALLACE 1953 has solved the stationary linear case by assuming a sinusoidal surface magnetizing field and has found very good agreement with measurements. In an unpublished report by D. HUNTER, WALLACE's method is generalized to space magnetizing and the demagnitizing effects are also studied. WESTMIJZE 1953 has made a very thorough study of the problem, but all parts of it are not yet published.

The drum for which the numerical computations have been made is that of the Besk, the Swedish electronic computer. This work was, however, done after this drum was designed. The pulse frequency is assumed low enough to neglect eddy current losses in the head and layer, that are made of a spinel material.

2. Main results

Below are given the analytical expressions for the magnetic field in three different cases. The numerical values of the magnetic field for special values of the parameters can be found in sections nr 6 and 7.
A recording head of normal construction is shown in fig. 1. Usually the most interesting region is around the gap and this is shown in fig. 2.

The three cases are:

1. The permeability in the layer is 1.
2. The permeability in the layer is greater than 1, but the layer is infinitely thick.
3. As 2, but the layer is finite.

The notations are

d = layer thickness
μ = layer permeability
N = half the pole distance
b = distance head — layer
B_0 = induction in the pole gap measured in voltsecond per square metre

$B_0 = \mu_0 V/N$, where V is the magnetic potential of the head.
$\mu_0 = 4\pi \cdot 10^{-7}$ in the $MKSA$ system.

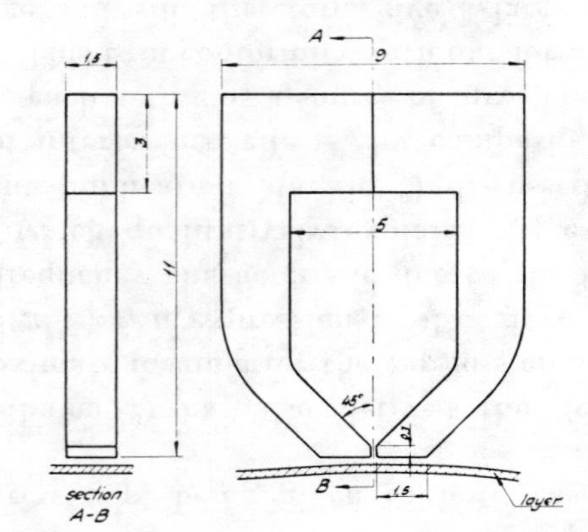

Fig. 1. Kernel to a recording head used in the Besk. The airgaps are exaggerated. Measurements in millimetres.

Reprinted with permission from *Trans. Roy. Inst. Technol. Stockholm*, no. 86, 1954.

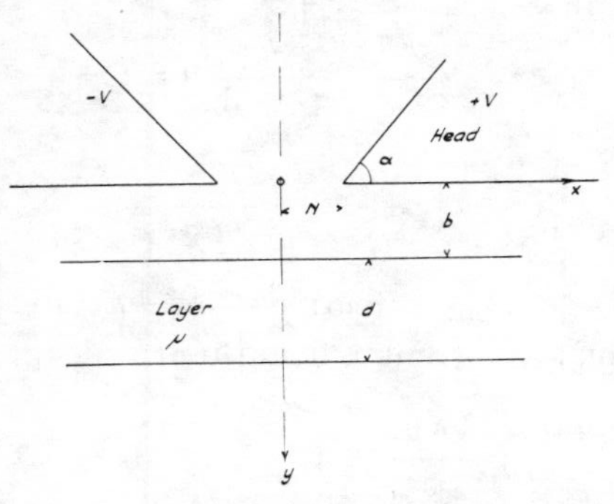

Fig. 2. The recording head in the neigbourhood of the airgaps.

The expressions in the three cases for the magnetic flux $B(x, y) = (B_x, B_y)$ at $y = b$ (inside the layer) are:

1. $$B_x = -\frac{1}{\pi} B_0 \left(\operatorname{arctg} \frac{N+x}{y} + \operatorname{arctg} \frac{N-x}{y} \right)$$

$$B_y = \frac{1}{2\pi} B_0 \log \frac{y^2 + (N+x)^2}{y^2 + (N-x)^2}$$

2. $$B_x = -\frac{2\mu}{\pi} B_0 \int_0^\infty \frac{dt}{t} \frac{\cos \frac{tx}{b} \sin \frac{tN}{b}}{\mu \sinh t + \cosh t}$$

$$B_y = \frac{2\mu}{\pi} B_0 \int_0^\infty \frac{dt}{t} \frac{\sin \frac{tx}{b} \sin \frac{tN}{b}}{\mu \sinh t + \cosh t}$$

Asymptotic expansions (see appendix 1 for definition) for the B-fields are:

$$B_x = -B_0 \frac{2Nb\mu^2}{\pi x^2} \left[1 - \frac{b^2}{x^2} \left(6\mu^2 - \frac{N^2}{b^2} - 5 \right) + \dots \right]$$

$$B_y = B_0 \frac{2N\mu}{\pi x} \left[1 - \frac{b^2}{x^2} \left(2\mu^2 - 1 - \frac{N^2}{3b^2} \right) + \dots \right]$$

3. $$B_x = -B_0 \frac{4\mu}{\pi} \int_0^\infty \frac{dt}{t \cdot K} \cos \frac{tx}{b} \sin \frac{tN}{b} \left(\sinh \frac{td}{b} + \mu \cosh \frac{td}{b} \right)$$

$$B_y = B_0 \frac{4\mu}{\pi} \int_0^\infty \frac{dt}{t \cdot K} \sin \frac{tx}{b} \sin \frac{tN}{b} \left(\mu \sinh \frac{td}{b} + \cosh \frac{td}{b} \right)$$

$$K = (\mu+1) \sinh t \left(1 + \frac{d}{b}\right) + (\mu - 1) \sinh t \left(1 - \frac{d}{b}\right) +$$

$$+ \mu(\mu+1) \cosh t \left(1 + \frac{d}{b}\right) - \mu(\mu-1) \cosh t \left(1 - \frac{d}{b}\right)$$

Asymptotic expansions are:

$$B_x = -B_0 \frac{2bN\mu}{\pi x^2} + \dots$$

$$B_y = B_0 \frac{2N}{\pi x} - \dots$$

The expressions in the three cases are derived by assuming linear magnetic potential along $y = 0$. Thus

$$v = -V \qquad x < -N$$

$$v = V \cdot x/N \qquad -N < x < N$$

$$v = +V \qquad x > N$$

This linear potential between the corners is the result of investigations made for the cases $\mu = 1$ and $\mu = \infty$ treated with conformal mapping.

The results may have their greatest interest when estimating the effect on the field due to the drum eccentricity, layer thickness, airgap, layer permeability and so on. In the manufacturing of drums this may be of interest in order to keep mechanical tolerances below certain values.

3. Idealization of the problem

The first approximation is to regard the drum surface as a plane. The variation of the distance b (fig. 2) due to the curved surface is about 10 % for the interval $0 < x < 10\,N$. The quotient b/N is usually between 0,5 and 2. The drum diameter is 120 millimetres and the gap is about 0,02 millimetres. The length of the head is about 100 times the gap width, so we assume that the head has infinite length.

The width of the head is also about 100 times the gap width and this shows that it is satisfactory to treat the two-dimensional problem only. An investigation shows that the field along a generatrice is very flat under the head.

The permeability of the head is about 1000 for the frequencies encounted, and this means that the lines of force leave the head nearly perpendicularly. The magnetic potential of the head is therefore assumed constant, and is $+V$ on the right half and $-V$ on the left half of the head.

In order to investigate how the pole length (0,3 mm in fig. 1) influences the field in the layer, the angle α is introduced. If $\alpha = 0$ this length is zero, and if $\alpha = 90°$ the length is infinite. This has already been studied by BOOTH 1952 and is included here only because we study the potential between the corners of the head.

4. The boundary value problem

The problem is to find the magnetic potential $v(x, y)$ in the region $y > 0$, $-\infty < x < \infty$ (fig. 2) when the potential along $y = 0$ is prescribed. The magnetizing vector is then

$$H = -\operatorname{grad} v(x, y) \tag{1}$$

The potential satisfies the equation:

$$\frac{\partial}{\partial x}\mu\frac{\partial v}{\partial x} + \frac{\partial}{\partial y}\mu\frac{\partial v}{\partial y} = 0 \tag{2}$$

If μ, the permeability of the layer, is a constant, we get Laplace's equation:

$$\Delta v = 0 \qquad \Delta = \frac{\partial^2}{\partial x^2} + \frac{\partial^2}{\partial y^2} \tag{3}$$

Usually equation (2) is a non-linear equation. The boundary conditions along $y = b$ and $y = b + d$ read:

$$\frac{\partial v_1}{\partial y} = \mu\frac{\partial v_2}{\partial y} \tag{4}$$

$$\mu\frac{\partial v_3}{\partial y} = \frac{\partial v_4}{\partial y} \tag{5}$$

where

v_1 = the potential above the layer $(y = b - 0)$
v_2 = the potential in the layer $(y = b + 0)$
v_3 = the potential in the layer $(y = c - 0)$
v_4 = the potential below the layer $(y = c + 0)$
μ = the permeability of the layer

The non-stationary one-dimensional field can be computed from the equation (μ is a constant):

$$\frac{\partial^2 H}{\partial x^2} = \sigma\mu\mu_0\frac{\partial H}{\partial t} \tag{6}$$

σ = conductivity of the layer

Equation (6) is a parabolic equation, while equations (2) and (3) are elliptic equations. In the special cases $\mu = 1$ and $\mu = \infty$ equation (3) is solved by means of conformal mapping. The general case is solved by means of Fourier transforms. The non-stationary case is solved by means of Laplace transforms.

5. The special cases $\mu = 1$ and $\mu = \infty$

The four most interesting cases are treated by conformal mapping (WEBER 1950). We use the notations

$$H_0 = V/N$$

$$H_1 = V/b$$

1. $\mu = \infty, \alpha = 90°.$

Because of the symmetry it is sufficient to consider the part ABCDEF in fig. 3.

The Schwarz-Christoffels integral gives the following expressions for the mapping function:

Fig. 3. The case $\mu = \infty$, $\alpha = 90°$.

$$\bar{z} = x - j\,y = \frac{2\,N}{\pi}\,\text{arctg}\left(\frac{N}{b}\,\text{tgh}\,p\right) + \frac{2\,b}{\pi}\,p \tag{7}$$

$$w = -\frac{v}{\pi}\log\,(1 - t) \tag{8}$$

$$H = H_x + j\,H_y = j\,H_1\,\text{tgh}\,p \tag{9}$$

The parameter p is defined by

$$t = \frac{a\,N^2\,\text{tgh}^2\,p}{b^2 + N^2\,\text{tgh}^2\,p} \tag{10}$$

The field along BC has been computed for $b/N = 0{,}5$, 1 and 2 and is plotted in fig. 4.

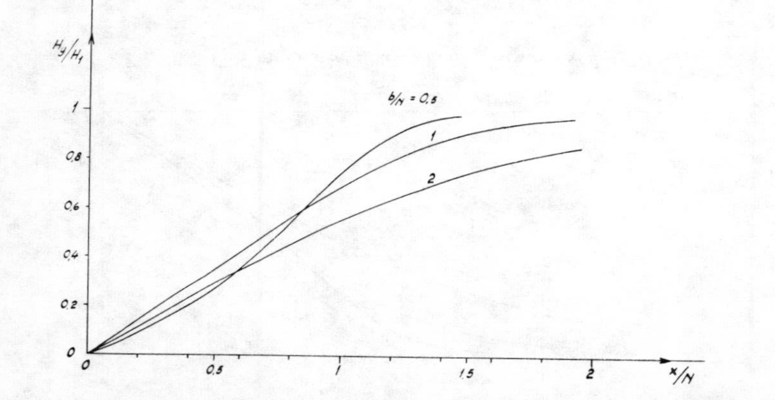

Fig. 4. The magnetic field on the surface of the layer. $\mu = \infty$, $\alpha = 90°$.

2. $\mu = \infty$, $\alpha = 0°$

The mapping function is:

$$\bar{z} = \frac{2\,b}{\pi}\left(\frac{a}{1-a}\,\text{tgh}\,p + p\right) \tag{11}$$

$$w = -\frac{V}{\pi}\log\,(1 - \text{tgh}^2\,p) \tag{12}$$

$$H = j\,H_1\,\frac{\sinh 2\,p}{\cosh 2\,p + \dfrac{1+a}{1-a}} \tag{13}$$

Here,

$$t = a \cdot \text{tgh}^2\,p \tag{14}$$

The parameter a is to be solved from a transcendental equation.

The following values are obtained:

b/N	a
0,5	0,62056
1	0,34511
2	0,12821

The field along BC is shown in fig. 6.

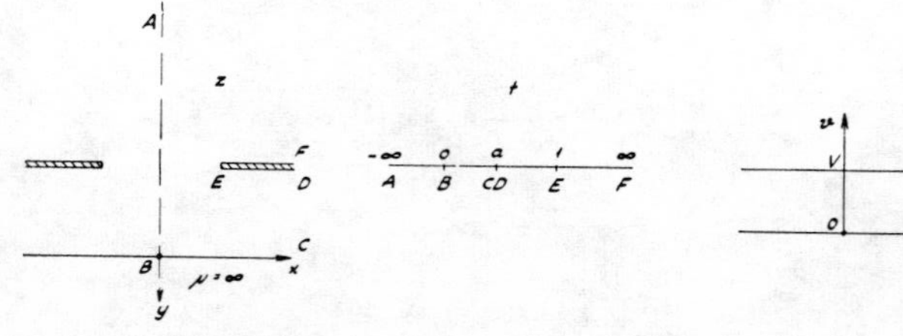

Fig. 5. The case $\mu = \infty$, $\alpha = 0°$.

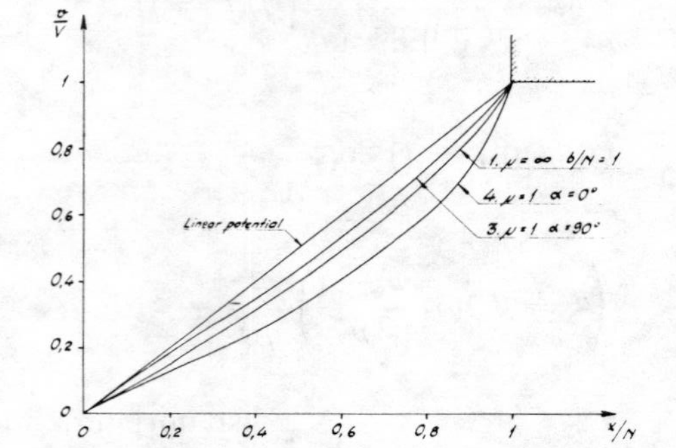

Fig. 6. The magnetic field when $\mu = \infty$, $\alpha = 0°$.

3. $\mu = 1, \alpha = 90°$

The mapping function is:

$$z = \frac{2N}{\pi} (\coth p - p) \tag{15}$$

$$w = -\frac{2V}{\pi} \log (\sinh p) \tag{16}$$

$$H = -H_0 \operatorname{tgh} p \tag{17}$$

$$t = \coth^2 p \tag{18}$$

The mapping function has been used in order to calculate the potential between the two corners in fig. 7. To compute the field for different b/N as in the cases above is a tremendous task. It is much easier to compute the field due to linear potential between the corners and then compute the difference field due to the difference between the linear potential and the actual potential between the corners.

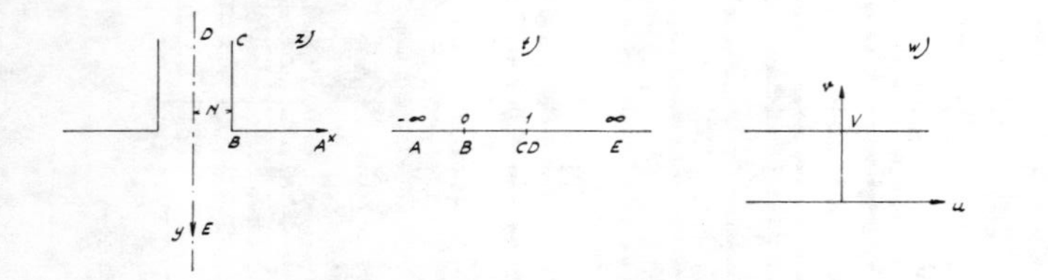

Fig. 7. The case $\mu = 1$, $\alpha = 90°$.

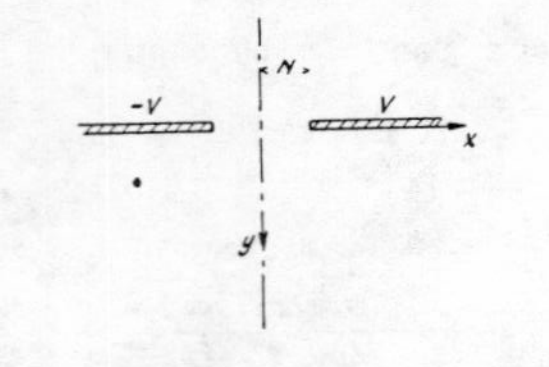

Fig. 8. The case $\mu = 1$, $\alpha = 0°$.

4. $\mu = 1, \alpha = 0°$

The field can be explicitely expressed in z:

$$H = -\frac{2H_0}{\pi \sqrt{1 - \frac{z^2}{N^2}}} \tag{19}$$

The variation of the potential between the corners of the head is computed for the four cases and is plotted in fig. 9. The linear potential is also drawn, and the question is whether this linear potential can be a good approximation for computation of the field. We have for the linear potential

$$\begin{aligned} v &= -V & x &< -N \\ v &= V \cdot x/N & -N &< x < N \\ v &= V & x &> N \end{aligned} \tag{20}$$

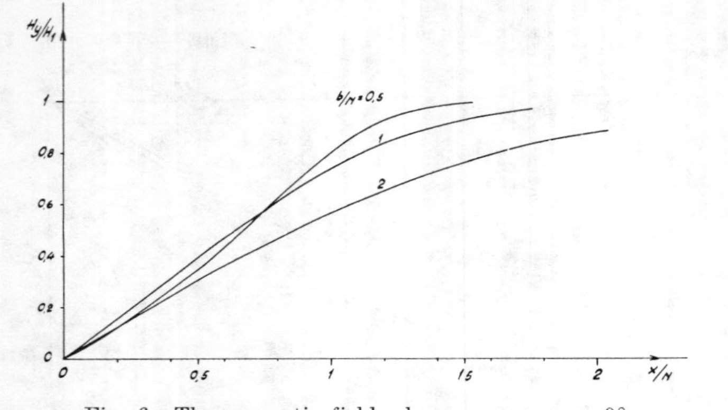

Fig. 9. The potential between corners for different cases.

The potential implies that we get a discontinuity in H_x instead of a singularly at the point $(0, N)$. H_y has still a singularity at that point. In practice we have no singularities because the corners are rounded off. The influence of rounded corners has been studied by COCKROFT 1928.

The potential $v(x, y)$ due to a known potential $f(x)$ along the line $y = 0$ can be computed from:

$$v(x, y) = \frac{1}{2b} \sin \frac{\pi y}{b} \int_{-\infty}^{\infty} f(t) \frac{dt}{\cosh \frac{\pi (t - x)}{b} - \cos \frac{\pi y}{b}} \tag{21}$$

If b is infinite we use the formula:

$$v(x, y) = \frac{y}{\pi} \int_{-\infty}^{\infty} f(t) \frac{dt}{y^2 + (t - x)^2} \tag{22}$$

Evaluating the integral (21) for the function defined by equation (20) we get the field for $\mu = \infty$.

$$H_y(x, b) = H_0 \frac{1}{\pi} \log \frac{\cosh \frac{\pi (x + N)}{2b}}{\cosh \frac{\pi (x - N)}{2b}} \tag{23}$$

The field for different b/N is shown in fig. 10. If we use equation (22) we get the field for $\mu = 1$, and this is shown in fig. 11. The field is always computed from equation (1).

The field is given by the eqs.

$$H_x(x, y) = -H_0 \frac{1}{\pi} \left(\operatorname{arctg} \frac{N + x}{y} + \operatorname{arctg} \frac{N - x}{y} \right) \tag{24}$$

$$H_y(x, y) = H_0 \frac{1}{2\pi} \log \frac{y^2 + (N + x)^2}{y^2 + (N - x)^2} \tag{25}$$

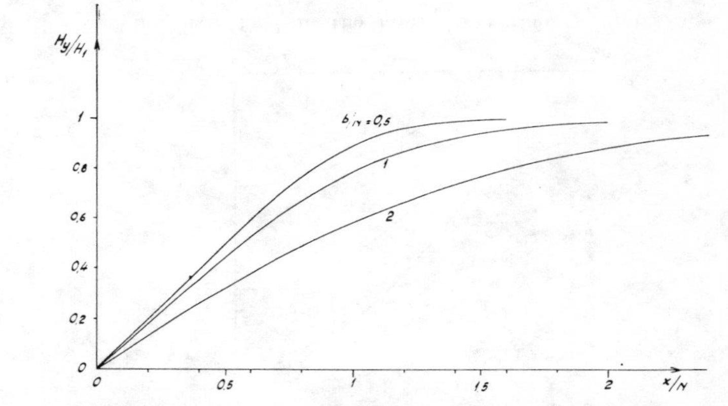

Fig. 10. The magnetic field when $\mu = \infty$ and linear potential between corners.

In the last three eqs. we have the field given explicity as simple functions of x and y, and it is very easy to compute an actual field. The approximation is found to be satisfactory for y-values greater than $0{,}5 \cdot N$. An estimation of the error involved can be done by using the integrals (21) or (22), where the function f is taken to be the difference between the linear potential and the actual potential between the corners. Another method is to solve Laplace's difference equation instead of the differential equation (KARLQVIST 1952) and as this is often favourable for numerical computation, it has been used here.

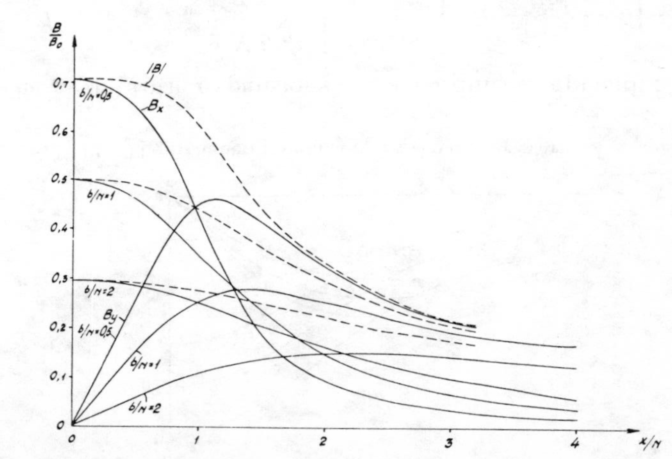

Fig. 11. The magnetic field when $\mu = 1$ and linear potential between corners.

6. The general case with finite μ but infinite layer thickness

In the last section it was shown that the linear potential is a good approximation to the special cases $\mu = 1$ and $\mu = \infty$, when $y > 0{,}5\,N$. Since the magnetic field is a continuous function of μ we obtain a good approximation also when μ is finite.

The equations of the general case are solved in appendix 2. In this section we consider the case when the layer thickness is infinite, which gives easier analytic expressions than having finite thickness. The equations for the field components along the layer surface $y = b$ are according to appendix 2:

$$B_{2x} = -\,B_0 \frac{2\,\mu}{\pi} \int\limits_0^\infty \frac{dt}{t} \frac{\cos\dfrac{t\,x}{b} \sin\dfrac{t\,N}{b}}{\mu \sinh t + \cosh t}$$

$$= -\,B_0 \frac{2\,\mu}{\pi\,(\mu+1)} \sum_0^\infty \left(\frac{\mu-1}{\mu+1}\right)^n \left(\operatorname{arctg}\frac{x+N}{b\,(2\,n+1)} - \operatorname{arctg}\frac{x-N}{b\,(2\,n+1)}\right) \qquad (26)$$

$$B_{2y} = B_0 \frac{2\,\mu}{\pi} \int\limits_0^\infty \frac{dt}{t} \frac{\sin\dfrac{t\,x}{b} \sin\dfrac{t\,N}{b}}{\mu \sinh t + \cosh t}$$

$$= B_0 \frac{\mu}{\pi\,(\mu+1)} \sum_0^\infty \left(\frac{\mu-1}{\mu+1}\right)^n \log\frac{(N+x)^2 + b^2\,(2\,n+1)^2}{(N-x)^2 + b^2\,(2\,n+1)^2} \qquad (27)$$

$$B_0 = \mu_0\,H_0$$

The field B_{2y} is measured inside the layer.

Figs. 12 and 13 show the fields for different μ and b/N.

The asymptotic expansions can be used to compute the tails of the fields and this gives information of the interaction of the pulses on the drum. The total field is

$$|B| = \sqrt{B_x^2 + B_y^2} \qquad (28)$$

and are the dotted lines in fig. 10, where the fields for the case $\mu = 1$ are plotted.

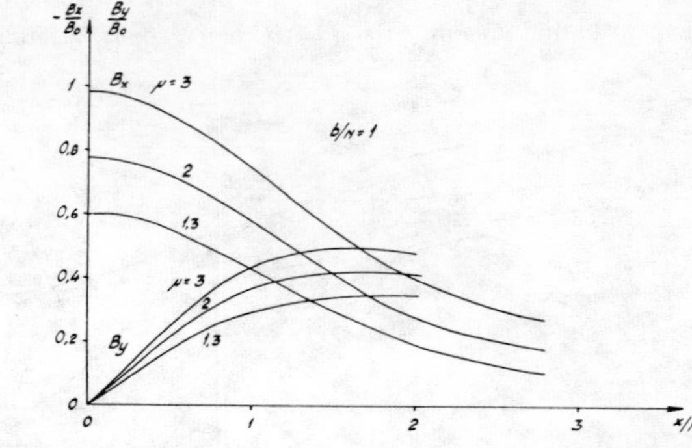

Fig. 12. The magnetic field in the layer for various μ and $b = N$.

The asymptotic expansions are according to appendix 2:

$$B_{2x} = -\,B_0 \frac{2\,N\,b\,\mu^2}{\pi\,x^2} \left[1 - \frac{b^2}{x^2}\left(6\,\mu^2 - \frac{N^2}{b^2} - 5\right) + \ldots\right] \qquad (29)$$

$$B_{2y} = B_0 \frac{2\,N\,\mu}{\pi\,x} \left[1 - \frac{b^2}{x^2}\left(2\,\mu^2 - 1 - \frac{N^2}{3\,b^2}\right) + \ldots\right] \qquad (30)$$

At the point $x = 0$ the infinite sum in equation (26) can be put in closed form, namely:

$$B_{2x}\,(0, b) = -\,B_0 \frac{2\,N\,\mu}{\pi\,b\,\sqrt{\mu^2 - 1}} \operatorname{arcosh}\mu$$

which shows that the B-field along the line $x = 0$ $(y > b)$ approaches infinity in the same way as $\log\mu$. This means that the field approaches infinity very slowly with an increase in μ.

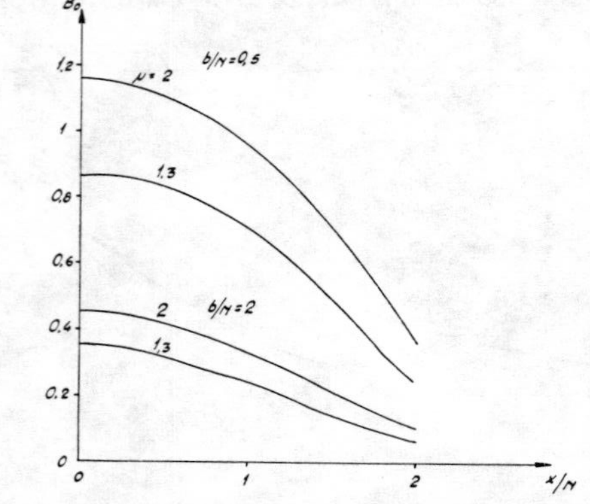

Fig. 13. The magnetic field in the layer for various μ and $b/N = 0{,}5$ and 2.

7. Finite μ and layer thickness

In the general case the field are given by:

$$B_{2x} = -B_0 \frac{4\mu}{\pi} \int_0^\infty \frac{dt}{t \cdot K} \cos \frac{tx}{b} \sin \frac{tN}{b} \left(\sinh \frac{td}{b} + \mu \cosh \frac{td}{b} \right) \quad (31)$$

$$B_{2y} = B_0 \frac{4\mu}{\pi} \int_0^\infty \frac{dt}{t \cdot K} \sin \frac{tx}{b} \sin \frac{tN}{b} \left(\mu \sinh \frac{td}{b} + \cosh \frac{td}{b} \right) \quad (32)$$

where B_{2x} and B_{2y} are the fields at $y = b + 0$, that is, the field inside the layer. At the other side of the layer $y = b + d - 0$ we have

$$B_{3x} = -B_0 \frac{4\mu^2}{\pi} \int_0^\infty \frac{dt}{t \cdot K} \cos \frac{tx}{b} \sin \frac{tN}{b} \quad (33)$$

$$B_{3y} = B_0 \frac{4\mu}{\pi} \int_0^\infty \frac{dt}{t \cdot K} \cdot \sin \frac{tx}{b} \sin \frac{tN}{b} \quad (34)$$

The function K is defined by:

$$K = (\mu + 1) \sinh t \left(1 + \frac{d}{b}\right) + (\mu - 1) \sinh t \left(1 - \frac{d}{b}\right) +$$
$$+ \mu(\mu + 1) \cosh t \left(1 + \frac{d}{b}\right) - \mu(\mu - 1) \cosh t \left(1 - \frac{d}{b}\right)$$

The corresponding asymptotic expansions are:

$$B_{2x} = -B_0 \frac{2bN\mu}{\pi x^2} + \dots \quad (35)$$

$$B_{2y} = B_0 \frac{2N}{\pi x} - \dots \quad (36)$$

$$B_{3x} = -B_0 \frac{2bN\left(\mu + \frac{d}{b}\right)}{\pi x^2} + \dots \quad (37)$$

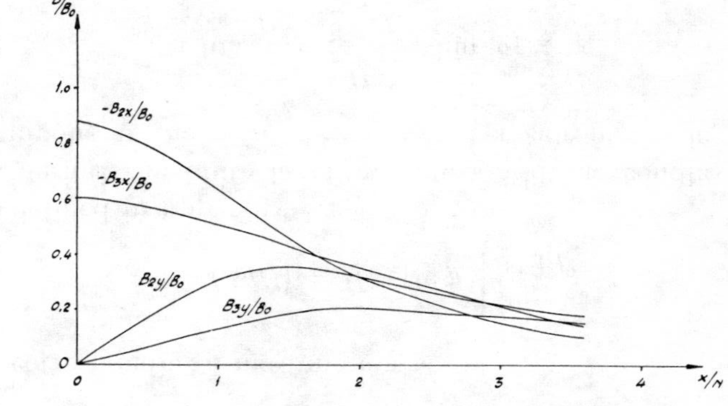

Fig. 14. The magnetic field for $\mu = 2$ and $d = b$.

$$B_{3y} = B_0 \frac{2N}{\pi x} - \dots \quad (38)$$

The next term in the asymptotic expansions is found in appendix 2. In fig. 14 are plotted the fields when $b/N = 1$ and $d/N = 1$, $\mu = 2$.

The deviation from the case when $\mu = 1$ is not very large. If the field is to be computed for many values of b/N it is convenient to write the factor

$$\cos \frac{tx}{b} \cdot \sin \frac{tN}{b}$$

in the form

$$\frac{1}{2} \left(\sin \frac{(N + x)t}{b} + \sin \frac{(N - x)t}{b} \right)$$

At a large distance from the origin the field is determined by B_{2y} eq. (36). This formula shows that the field is independent of μ and depends of x/N only. The same conclusion applies to B_{3y} and thus to the whole layer.

The linear case where μ is a constant can be used as a first approximation to the nonlinear case. The hysteresis loop for the material is shown in fig. 15 and if we use the initial magnetizing curve, we find that the μ-values vary from 1,1 at the origin to 2 at $x = 4N$. The induction B_0 was then 2500 gauss. The material is thus saturated for $x = \pm 4N$.

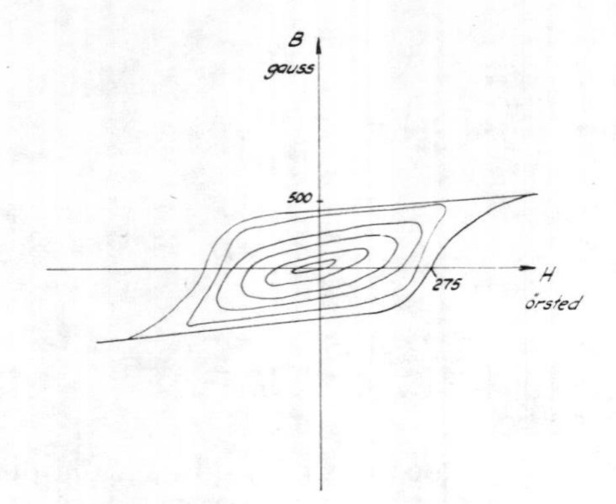

Fig. 15. The hysteresis loop for the layer material.

8. Pole and layer inductance

The magnetic flux is divided into two parts: the first part containing the flux going through the layer and the second part going through the pole gap (0,3 mm in fig. 1). The corresponding inductances are called L_l and L_p.

By integrating round the head we readily found:

$$L_p = \frac{n^2 A \mu_0 \mu_h}{l_h + 2 N \mu_h}$$

where

A = the area of the pole = 0,45 mm² for the head of fig. 1.
n = number of turns
l_h = the mean circumference round the head
μ_h = the permeability of the head.

The inductance L_l can be computed from eq. (24) if μ is small. Numerical computations shows that the contribution to this inductance due to the layer is very small and can normally be neglected.

We have

$$L_l = \frac{n^2 aN \mu_0 \mu_h}{\pi l_h + 2 \pi N \mu_h} \ln \left(1 + \frac{b^2}{N^2}\right) \frac{N^2 + (N+b)^2}{N^2 + (d+b)^2}$$

a = the width of the head (1,5 mm in fig. 1)

Actual values gives the result

$$L_p = 3 \text{ mH}$$
$$L_l = 0,3 \text{ mH}$$

This shows that we could save power if we keep the pole inductance low. The total inductance was 4,2 mH, that also includes the inductances from the sides of the head.

In order to verify the result we measured the small contribution to the inductance L_l when layer is present and not present. The voltage from the head was put in a bridge that was balanced when layer was not present. The unbalance when layer was present was found by THEVENIN's theorem to correspond to an inductance change of 0,04 mH. The computed value according to the formulas for the field was 0,036 mH. The quotient b/N was 1. Measurements was also made for greater values of this quotient and the result was fairly satisfactory.

9. The transient field

In this section we solve equation (6) for infinite layer and for finite layer. The x-coordinate in (6) is to be replaced by y.

We assume that a polarized electro-magnetic wave with the components H_x and E_z comes perpendicular to the layer. The wave is applied suddenly at $t = 0$ and the airgap b is assumed to be zero. The initial value problem is for infinite layer:

$$H(0, t) = H_0; \quad H(\infty, t) = 0; \quad H(y, 0) = 0;$$

Then the Laplace transform of the equation (6) can be written:

$$h = \frac{1}{s} \cdot H_0 e^{-k y \sqrt{s}}$$

$$k^2 = \sigma \mu \mu_0$$

The corresponding time function is

$$H(y, t) = H_0 \operatorname{erfc} y \sqrt{\frac{\sigma \mu \mu_0}{4 t}}$$

erfc is defined in appendix 1.

Considering the finite layer, we must add the condition that E_z is continuous at the point $y = d$ (= other side of the layer).

$$\lim_{y \to d-0} \sigma_2 \cdot \frac{\partial H}{\partial y} = \lim_{y \to d+0} \sigma_1 \frac{\partial H}{\partial y}$$

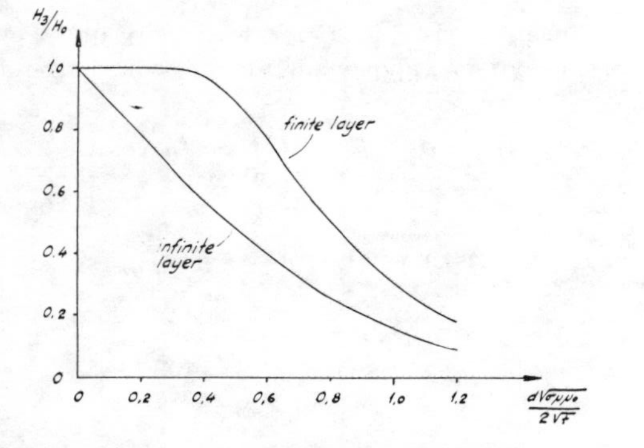

Fig. 16. The magnetic field in the transient case.

where σ_1 and σ_2 represent the conductivity on each side of the boundary. If the conductivity σ_1 of the layer is zero (spinel material) compared with the drum material (usually brass) we have the Laplace transform

$$h = H_0 \frac{e^{-(y-d)\,k\sqrt{s}}}{\cosh k\,d\sqrt{s}}$$

and the time function

$$H(d, t) = 2\,H_0 \sum_{n=0}^{\infty} (-1)^n \operatorname{erfc}\left[(2\,n+1)\,d\sqrt{\frac{\sigma\,\mu\,\mu_0}{4\,t}}\right]$$

Fig. 16 shows the two cases, infinite layer and finite layer, from which one can compute the transient time for a given material.

10. Summary

The magnetic field in the ferromagnetic layer on a magnetic drum is calculated having finite airgaps and finite permeability in the layer. The layer is assumed to be a spinel material for which the permeability is low. The special cases when the permeability is one and infinite is treated by conformal mapping. The results from this investigation suggest a linear potential distribution between the corners of the recording head. This approximation gives explicit expressions for the field, and the method is generalized to finite permeability. Expressions are given for the field on each side of the layer, and asymptotic expressions are given for the field at a large distance from the pole gap of the recording head. The inductance of the head is calculated, and measurements of the inductance change have been made when the permeability is increased from 1 to μ. The transient field is computed for the one-dimensional case, assuming the resistivity of the layer to be very large. The results can be used to analyze the influence on the field from permeability and geometric shape of the head.

11. Acknowledgements

The writer wishes to express his thanks to Messrs G. DAHLQUIST and E. STEMME of the Swedish Board for Computing Machinery, and to Mr D. WILLIS Cambridge, England, who has made many suggestions for solving the problem. He is also indebted to the Swedish Board for Computing Machinery for running some of the computations on the Swedish relay machine Bark. Professor T. LAURENT at the Royal Institute of Technology, Stockholm, has in many respects stimulated the writer through his interest in recent developments in the application of magnetic drums to automatic telephony.

12. Appendix 1

The sine-transform $g(x)$ of the function $G(t)$ is defined by

$$g(x) = \int_0^\infty \sin x\,t\,G(t)\,dt$$

Its inverse is

$$G(t) = \frac{2}{\pi}\int_0^\infty \sin t\,x\,g(x)\,dx$$

The Laplace-transform $f(s)$ of the function $F(t)$ is defined by

$$f(s) = \int_0^\infty e^{-s\,t}\,F(t)\,dt$$

The error integral is defined by

$$\operatorname{erf} x = 1 - \operatorname{erfc} x = \frac{2}{\sqrt{\pi}} \int_0^x e^{-t^2} \, dt$$

An asymptotic series diverges for all x, but can be used in numerical computation if a finite number of termes is included. Cf. WHITTAKER and WATSON 1927.

13. Appendix 2. Derivation of the field formulas in the general case

The boundary value problem defined by the equations (3), (4) and (5) is solved in the following way[1].

Let $g(x)$ be the linear potential (20). Along the lines $y = b$ and $y = c$ (each side of the layer) the H_y field makes a jump according to equations (4) and (5).

Put

$$g(x) = \int_0^\infty \sin xt \cdot G(t) \, dt$$

where

$$G(t) = \frac{2 V \sin N t}{\pi N t^2}$$

We want to determine three layers each with boundary value potentials $f_1(x)$, $f_2(x)$ and $f_3(x)$ at the lines $y = 0$, $y = b$ and $y = c$. Their potentials should be equal to the function $v(x, y)$. Set

$$f_1(x) = \int_0^\infty F_1(t) \sin xt \, dt$$

$$f_2(x) = \int_0^\infty F_2(t) \sin xt \, dt$$

$$f_3(x) = \int_0^\infty F_3(t) \sin xt \, dt$$

$$v(x, y) = \int_0^\infty V(t, y) \sin xt \, dt$$

[1]) The original proof assumed Cesàro summability of the sine transform of $g(x)$. This derivation due to Mr G. DAHLQUIST, avoids this difficulty.

Now we consider the function $u(x, y)$, defined by

$$u(x, y) = \int_0^\infty e^{-t|y|} \sin xt \, F_1(t) \, dt$$

This function satisfies Laplace's equation

$$\Delta u = 0$$

This shows that a layer with the boundary potential $f_1(x)$ at the line $y = 0$ gives a potential that is the sine transform of

$$e^{-t|y|} \cdot F_1(t)$$

and similar for F_2 and F_3

Then we have

$$V(t, y) = F_1(t) e^{-t|y|} + F_2(t) e^{-t|b-y|} + F_3(t) e^{-t|c-y|}$$

Now we express the boundary conditions for the transforms, which gives a linear system with three eqs. Solving this system we get

$$F_1(t) = (r^2 - e^{-2t(c-b)}) \frac{G}{D}$$

$$F_2(t) = (e^{-t(2c-b)} - r e^{-tb}) \frac{G}{D}$$

$$F_3(t) = e^{-tc} (r - 1) \frac{G}{D}$$

where

$$r = \frac{\mu - 1}{\mu + 1}$$

$$D = r^2 - r(e^{-2tb} - e^{-2tc}) - e^{-2t(c-b)}$$

We can now form the various derivatives of $v(x, y)$ and easily compute the field components given by the eqs. (26), (27) (c infinite), (31), (32), (33) and (34) (c finite).

To prove the asymptotic formulas, we observe that the integral

$$f(x) = \int_0^\infty F(t) \sin x t \, dt$$

can be expanded by repeated partial integration in a series of $1/x$:

$$f(x) = \frac{F(0)}{x} - \frac{F''(0)}{x^3} + \dots$$

In the same manner we expand the integral

$$g(x) = \int_0^\infty G(t) \cos x t \, dt$$

$$g(x) = -\frac{G'(0)}{x^2} + \frac{G'''(0)}{x^4} - \dots$$

In our case the series are divergent for all x but are still useful for large x (cf. appendix 1).
The derivatives are computed recurrently. For example, the integrand of the formula (26) is S/T, where

$$S = \frac{1}{t} \sin \frac{t N}{b}$$

$$T = \mu \sin h t + \cos h t$$

Writing

$$GT = S$$

we have

$$G'T + GT' = S'$$

and so on.

Proceeding in this way, we get the asymptotic expansions (29), (30) (c infinite) and (35), (36), (37) and (38) (c finite).

The two first terms in the last four expansions are:

$$B_{2x} = -B_0 \left[\frac{2 b N \mu}{\pi x^2} + \frac{2 b^3 N \mu}{\pi x^4} \left(12 \mu q + p^2 - 1 + 6 q^2 - \right. \right.$$
$$\left. \left. - \frac{6 q (2 \mu + r)}{\mu^2} \right) + \dots \right]$$

$$B_{2y} = B_0 \left[\frac{2 N}{\pi x} - \frac{2 b^2 N}{\pi x^3 \mu^2} (\mu^2 + 4 q \mu - 2 q^2 \mu^2 - p^2 \mu^2/3 - \right.$$
$$\left. - 4 q \mu^3 + 2 q^2) + \dots \right]$$

$$B_{3x} = -B_0 \left[\frac{2 b N (\mu + q)}{\pi x^2} + \frac{4 b^4 \mu}{\pi x^4} G'''(0) + \dots \right]$$

$$B_{3y} = B_0 \left[\frac{2 N}{\pi x} - \frac{2 b^2 N}{3 \pi x^3 \mu^2} (3 \mu^2 - p^2 \mu^2 + 12 q \mu + 6 q^2 - \right.$$
$$\left. - 3 q^2 \mu^2 - 6 q \mu^3) + \dots \right]$$

$$p = N/b$$

$$q = d/b$$

$$G'''(0) = \frac{p}{2 \mu^4} \left[12 q \mu^4 + 15 q^2 \mu^3 + (p^2 - 1) \mu^3 + p^2 q \mu^2 - \right.$$
$$\left. - 5 q^3 \mu^2 - 15 q \mu^2 - 18 q \mu - 6 q^3 \right]$$

14. References

BEGUN, S. J. 1949. Magnetic recording. — Murray Hill, New York.

BIERENS DE HAAN, D. 1867. Nouvelles Table d'Integrales Definies. — Reprint Stechert & Co New York 1939.

BOOTH, A. D. 1952. On two problems in potential theory and their application to the design of magnetic recording heads for digital computers. — British J. Appl. Ph. vol. 3 (1952), 307—308.

BOZORTH, R. M. 1951. Ferromagnetism. — Van Nostrand, New York.

BUTLER, O. I. and SARMA B. E. 1951. Relaxation methods to the problem of a. c. magnetization of ferromagnetic laminae. — J. I. E. E. paper nr 1026, (1951).

COCKCROFT, J. 1928. The effect of curved boundaries on the distribution of electrical stress round conductors. — J. I. E. E. vol. 66 (1928), 385—409.

DOETSCH, G. 1943. Laplace-transformation. — Dover, New York.

GUCKENBERG, W. 1950. Die Wechselbeziehungen zwischen Magnettonband und Ringkopf bei der Wiedergabe. — Funk und Ton vol. 4 nr 24 (1950).

KARLQVIST, O. 1952. Numerical solution of elliptic difference equations by matrix methods. — Tellus vol. 4 (1952), 374—384.

LATIMER, K. E. and MACDONALD, H. B. 1950. A survey of the possible applications of ferrites. — Comm. News vol. 11 (1950), 76—90.

LINDMAN, C. F. 1891. Examen des Nouvelles Tables d'Integrales Definies de M. Bierens de Haan. Stockholm 1891. — Reprint Stechert & Co, New York 1944.

MALTHANER, W. A. and VAUGHAN, H. E. 1953. An automatic telephone system employing drum memory. — P. I. R. E. vol. 41 (1953), 1341—1347.

WALLACE, R. L. 1951. The reproduction of magnetically recorded signals. — Bell System Tech. J. vol. 30 (1951), 1145—1173.

WEBER, E. 1950. Electromagnetic fields. I. — Chapman & Hall 1950.

WESTMIJZE, W. K. 1953. Studies on magnetic recording. — Philips Research Reports. vol. 8. April 1953.

WHITTAKER, E. T. and WATSON, G. N. 1927. A course of modern analysis. — Cambridge 1927.

ON RECORDING HEAD FIELD THEORY

J. C. Mallinson

ABSTRACT

Several well-known theoretical treatments of the properties of the fringing field from a magnetic recording head are shown to yield a non-zero dc flux response, which is physically impossible. This deficiency is due to the approximations in the models rather than mathematical error. Plausible long wavelength spectra may be derived from better models, but they are usually of great mathematical or computational difficulty. In this paper, a simple approximation is suggested which yields correctly the qualitative features of the whole spectrum. In conclusion, the topical case of narrow pole-tip-length heads is discussed.

INTRODUCTION

A full understanding of the properties of the fringing field from a recording head is a central problem in the theory of magnetic recording. Several studies have been made and different approaches to the calculation of these properties have been published. In this paper we distinguish between local models, yielding correctly the field adjacent to the gap, and total models which give the field everywhere along the tape path. Similarities and differences between several well-known local models are discussed. When the Fourier transform of the fringing field is considered, several mathematical simplifications occur which illuminate some of its general, and frequently overlooked, properties. It will be shown that the local models violate physical reality and, incorrectly, yield a non-zero dc flux response. The total models avoid this difficulty in principle but are of such mathematical or computational difficulty as to be almost beyond practical utility. The simple formulation suggested in this paper, although not of great accuracy, shows properly the long wavelength behavior of the flux response. The topical case of narrow pole-tip-length heads is discussed as an illustration of the utility of this simple approach. All discussion is limited to the usual case of two dimensional heads and tapes of unit permeability.

LOCAL MODELS

In well-known studies, both Booth[1] and Westmijze[2] used the Schwarz-Christoffel method of conformal transformation to deduce exactly the fringing field adjacent to the gap of an infinitely permeable head of semi-infinite dimensions. Unfortunately, this technique does not lead to readily comprehensible formulae and some type of graphical representation of the results is, therefore, mandatory. Later, Fan[3] solved the identical problem by harmonic analysis of Laplace's equations in the regions above and in the gap. His results, which are completely equivalent to those of Booth and Westmijze, are cast in the form of an infinite series of integrals. The first integral in the series for the longitudinal field component above the gap is,

$$h_x(x,y) = \frac{2h_o}{\pi}\int_0^{\infty} \sin kg \cos kx \cdot \frac{e^{-ky}}{k}\, dk = \frac{h_o}{\pi}\, \mathrm{TAN}^{-1}\left(\frac{2yg}{x^2+y^2-g^2}\right) \qquad (1)$$

and it dominates at all points except close to the gap edges. Here h_o is the deep-gap field, $2g$ is the gap length, x and y are the longitudinal and perpendicular coordinates.

The arctangent term is identical to that given by Karlqvist[4] who solved the same problem with one added simplification. He assumed that the magnetic scalar potential, Ψ, had a constant gradient, $d\Psi/dx$, across the top of the gap; in reality, of course, the constant gradient only occurs deep in the gap. It has been shown[5] that an arctangent expression may be derived also from uniform magnetic pole sheets on the gap faces which would occur if the magnetization within the head were longitudinal and uniform. Although true, this observation has caused a great deal of confusion which may best be resolved by the series of statements below;

a) The Booth, Westmijze and Fan results are all correct and yield the arctangent form as an approximation in the region above the gap,
b) In a real head, the arctangent form does not hold, even approximately, in the gap,
c) In a real head, the pole density on the gap faces is only uniform at points deep in the gap; a considerable increase in pole density occurs at the top of the gap and there are also poles on the top surfaces adjacent to the gap,
d) If the pole density were uniform, the arctangent form would hold exactly both above and in the gap,
e) In order to make the uniform pole model match Eq. 1 above the gap, a pole density just twice that occurring deep in the gap of a real head must be assumed. This is because, for a given deep gap field, the uniform pole model yields both surface fields and magnetic potentials which are half in magnitude and of the same form as in the Karlqvist simplification.[6]

FREQUENCY DOMAIN REPRESENTATION

The reciprocity expression[3] for the flux $\phi(x',y)$ in a head with fringing field vector $\bar{h}(x,y)$ due to a thin lamina of tape magnetized according to the vector $\bar{m}(x,y)$, is a convolution integral,

$$\phi(x',y) = \int_{-\infty}^{\infty} \bar{m}(x,y)\cdot \bar{h}(x'-x,y)\, dx \qquad (2)$$

and its Fourier transform (upper case symbols), is a sum of products,

$$\Phi(k,y) = H_x(k,y)\, M_x(k,y) + H_y(k,y)\, M_y(k,y), \qquad (3)$$

where k is the wavenumber (2π times the reciprocal wavelength). It is clear, since the transform of a pure sinusoid consists of Dirac delta functions only, that the transform, $H_x(k,y)$, of longitudinal fringing

Manuscript received April 6, 1974; Mr. Mallinson is affiliated with Ampex Corporation, 401 Broadway, Redwood City, California 94063.

Reprinted from *IEEE Trans. Magn.*, vol. MAG-10, pp. 773-775, Sept. 1974.

field, $h_x(x,y)$, is simply the reproduce head flux spectrum for longitudinal magnetization in the lamina and, analogously, $H_y(k,y)$ is the spectrum for perpendicular tape magnetization. We note that, since spatial differentiation is equivalent to multiplication by jk in the frequency domain, the corresponding voltage spectra are $jk\,H_x(k,y)$ and $jk\,H_y(k,y)$ respectively.

Now regardless of the head geometry or the magnetization patterns in the head, all physically realizable fringing fields must obey Laplace's equation in the region above the gap, and, therefore,

$$\nabla^2 h = \nabla^2 h_x = \nabla^2 h_y = 0 . \qquad (4)$$

It is possible to prove then that

$$H_x(k,y) = H_x(k,o)\, e^{-|k|y} , \qquad (5)$$

where $H_x(k,o)$ is $H_x(k,y)$ evaluated on the top surface of the head where $y=o$, and that

$$H_y(k,y) = j\,H_x(k,y) . \qquad (6)$$

This pair of equations, which are of the utmost generality, lead to several interesting observations;

a) By considering frequency domain representations, a partial separation of the variables, k and y, is always possible,
b) All heads must display the same $e^{-|k|y}$ type ($-55\,y/\lambda$ dB) of spacing loss behavior,
c) The spectrum of a head is determined completely by the fringing field spectrum on the top surface of the head, and
d) The spectrum for perpendicular tape magnetization is, apart from the j factor ($\pi/2$ change in phase angle), always identical to that for longitudinal magnetization.

Further insight may be gained upon considering the spectrum of the Karlqvist form (Eq. 1) which is

$$H_x(k,y) = 2g\,h_o \frac{\sin kg}{kg} e^{-|k|y} . \qquad (7)$$

At short wavelengths, $\lambda = 2g$, the spectrum vanishes giving the "first gap" null. At long wavelengths, $\lambda = \infty$, the spectrum is equal to $2g\,H_o$ which clearly violates physical reality. The dc value of the spectrum is

$$H_x(o,y) = \left[\int_{-\infty}^{\infty} h_x(x,y)\, e^{-ikx}\, dx \right]_{k=o} = \int_{-\infty}^{\infty} h_x(x,y)\, dx \qquad (8)$$

and the value of this line integral obviously must be zero for any path above a real head since no current loop is threaded. The failing is not due to the Karlqvist approximation; all of the local models discussed above incorrectly yield a non-zero dc response equal to the potential drop across the gap. The deficiency is due entirely to the unreality of the models and does not result from mathematical error.

It may be shown that any model which has either pole-pieces of infinite dimensions or an incomplete electrical coil will yield implausible long λ spectra. In all the cases discussed above, the pole-pieces extended infinitely in both $\pm x$ and $-y$ directions and the erroneous dc spectrum equals the potential drop. For pole-pieces which extend to infinity only in the $-y$ direction the dc spectrum turns out to be just one half the potential drop. When the return conductors in a head coil are not specified or included in the model, the spurious result is again just one half the potential drop.

TOTAL MODELS

In order to obtain physically reasonable spectra the computation must include characteristics of the entire head; this must include the whole of the pole-pieces and/or the complete electrical coil. When the pole-pieces are large compared with the coil dimensions it is permissible to suppose that the coil is of negligible size. On the other hand when the pole-pieces are small the real coil dimensions must be used.

Total model calculations have been done by Westmijze[2], Elabd[7], Potter et al[8], and Mallinson and Steele[9] in numerical studies of great complexity. An appreciation for the difficulty of these calculations may be inferred from Stavn's[10] recent paper on narrow pole-tip length heads. Here, despite extensive computations, the results appear flawed in at least two respects; the published spectra show dc response and differences between $|H_x(k)|$ and $|H_y(k)|$.

The fringing field found in properly performed total model analyses has negative lobes adjacent to the outer edges of the pole tip as well as the usual form adjacent to the gap. (see Fig. 1)

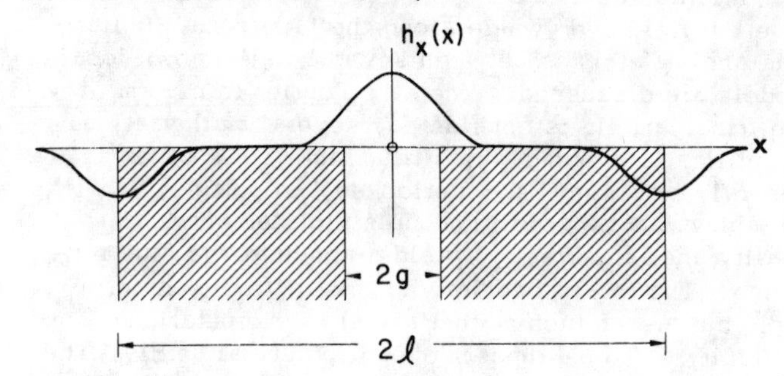

Fig. 1 Longitudinal component of the fringing field showing the mandatory negative lobes. (schematic only)

The algebraic sum of the areas underneath the h_x versus x plot is zero; this forces the dc value of the spectrum to be zero. It is important to realize that these lobes are not mere theoretical curiosities; they determine the long wavelength limit in all recording systems. Just as the shortest λ usable is approximately equal to the gap length, 2g, so the longest λ usable is given, within a factor of two, by the pole-tip-length 2ℓ. Bandwidths of 1000:1 or 10 octaves are typical of current head designs.

In the frequency domain the negative field lobes usually give rise to an oscillatory behavior at long wavelengths; these oscillations are the so-called "head bumps" which plague the system designer.

Simple expressions for the long wavelength spectra are rare. For infinitely permeable heads Westmijze suggests

$$H_x(k) = \int_0^{k\ell} J_o(u)\, du , \qquad (9)$$

a tabulated function which is qualitatively similar (see Fig. 2) to

$$H_x(k) = 1 - \frac{\sin k\ell}{k\ell} \qquad (10)$$

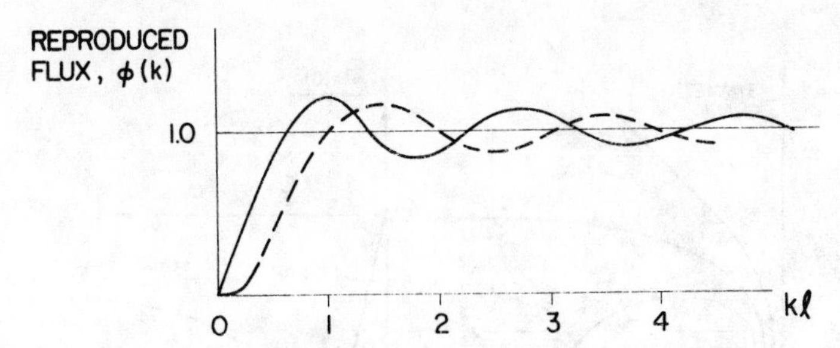

Fig. 2 Reproduce flux at long wavelengths showing

a) $\int_0^{k\ell} J_0(u)\,du$ (solid) and

b) $1 - \frac{\sin k\ell}{k\ell}$ (dotted) .

This similarity suggests that a reasonable and simple approximation for the flux spectrum of a real head is

$$H_x(k,y) = \left[\frac{\sin kg}{kg} - \frac{\sin k\ell}{k\ell}\right] e^{-|k|y} . \quad (11)$$

This spectrum is clearly an extension of Karlqvist's approximation of constant potential gradients. The magnetic scalar potentials on the top surface of the head which are consistent with the extended Karlqvist form and with the assumption of infinite permeability are shown in Fig. 3. The differences shown there correspond, of course, with the spectral variations occurring

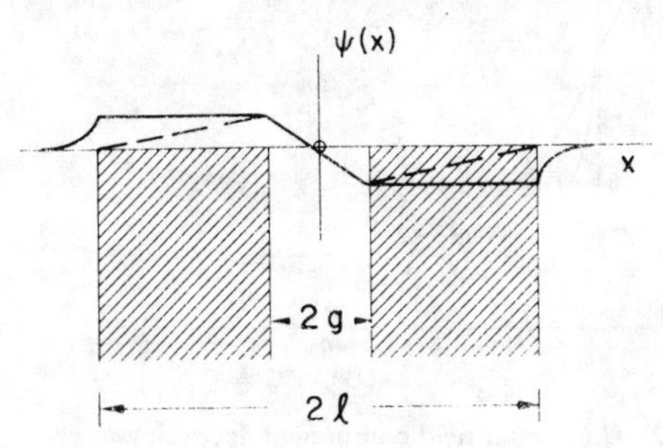

Fig. 3 Magnetic potential on top surface of head for infinite permeability case (solid) and constant gradient approximation (dotted).

in Fig. 2. It seems likely that, in real heads of finite permeability, the actual potential may be closer to the constant gradient case than it appears from Fig. 3.

We note that this spectrum (Eq. 11) always has a dc response of zero and that when $\ell \gg g$ (wide pole-tip-length) the first gap-null falls at $\lambda = 2g$. When the pole-tip-length, 2ℓ, is comparable to the gap-length, 2g, however, the long and short wavelength spectral oscil-

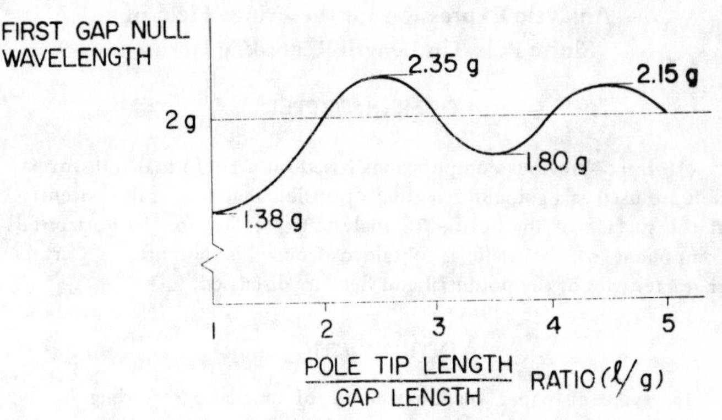

Fig. 4 Wavelength of first gap-null versus pole-tip-length to gap-length ratio (schematic only).

lations interfere and the first gap-null need no longer occur at $\lambda = 2g$. Dependent upon the value of the pole-tip-length, it may be shown that the first gap-null falls in the range 1.38g to 2.35g (see Fig. 4).

These observations are of special relevance in the field of digital recording where there is currently great interest in narrow pole-tip-length heads. In digital recording, the improved short wavelength response attainable with narrow pole-tip-length heads narrows the isolated output voltage pulse. From Fig. 4 we see that the occurrence of pulse narrowing may depend critically upon the value of pole-tip-length to gap-length ratio. Since this pulse narrowing always occurs at the expense of the long wavelength response of the head, it is by no means obvious that the effect is in fact advantageous. From the standpoint of communications theory, the poor long wavelength response is one of the most serious limitations in magnetic recording and this deficiency can be only aggravated by the use of narrow pole-tip-length heads.

REFERENCES

1. A.D.Booth, Brit. J. Applied Physics, 3, 307 (1952).
2. W.K.Westmijze, Philips Res. Reports,8,161 (1953).
3. G.Fan, IBM J. Res. Develop., 5, 321 (1961).
4. O.Karlqvist, Royal Inst. Tech., Stockholm Trans., 86, 1 (1954).
5. R.O.McCary, IEEE Trans Mag-7, 1, 4 (1971).
6. H.N.Bertram, Private Communication.
7. I. Elabd, IEEE Trans Audio, 11, 21 (1963).
8. R. Potter et al, IEEE Trans Mag-7, 689 (1971).
9. J.Mallinson and C.Steele, IEEE Trans Mag-8, 503 (1972).
10. M.J.Stavn, IEEE Trans Mag-9, 698 (1973).

Analytic Expression for the Fringe Field of Finite Pole-Tip Length Recording Heads

ROBERT I. POTTER

Abstract—Previous computations based on a conformal transformation are used as a guide in selecting a plausible magnetic scalar potential at the surface of the head. An analytic expression for the horizontal component of the field is obtained from this potential. Certain characteristics of the potential and field are discussed.

INTRODUCTION

In a recent paper on the theory of magnetic recording heads, Mallinson [1] considers the horizontal field component $H_x(x,y)$ for the finite pole-tip length head. He argues that

$$\int_{-\infty}^{\infty} H_x(x,y)\, dx = 0 \tag{1}$$

for all real heads, which is true. He also gives an approximate expression for the Fourier transform of $H_x(x,y)$ based on a saw-tooth-like magnetic scalar potential at $y = 0$ that may be characteristic of ferrite heads.

The purpose of this letter is to suggest an analytic approximation to the exact two-dimensional field component $H_x(x,y)$ for the finite pole-tip length head. This expression is derived from a scalar potential that is different from Mallinson's, and is applicable to heads with finite-permeability single-film permalloy pole-pieces. The pole-tip geometry, notation, and coordinate system shown in Fig. 1(a) are identical to those of Potter, Schmulian, and Hartman [2] (PSH).

DISCUSSION OF EQUATION (1)

Equation (1) is not a consequence of the lack of currents in the upper half plane, because in two dimensions the contribution from the usual semicircular contour at infinity does not vanish. The integral in (1) vanishes in two dimensions only if the total current is zero; that is, only if the return current outside the yoke is included in the problem. Even though the return current path for a single-turn film head could be located at $x = 0$, $y = -l$ with l essentially infinite compared to the x and y dimensions of the yoke, its influence on $H_x(x,y)$ must be included when the integral in (1) is evaluated. This can be done in an approximate way, for large l and finite p, by adding

$$H_x^{(r)}(x,y) = \frac{-I}{2\pi}\frac{y}{x^2+(y+l)^2} \tag{2}$$

to $H_x(x,y)$ as computed with the return current neglected. If this current is neglected, then

$$\int_{-\infty}^{\infty} H_x(x,y) = \begin{cases} I, & \text{for infinite } p. \\ I/2, & \text{for finite } p. \end{cases} \tag{3}$$

The return current is neglected in the following.

ANALYTIC EXPRESSION FOR $H_x(x,y)$

Fig. 3 of PSH, schematically reproduced here as Fig. 1(a), provides guidance in choosing the approximate antisymmetric scalar potential $\phi(x,0)$ shown in Fig. 1(b) for $x \geq 0$ only. The choice $\phi(x,0) = C_0 + C_1 x^{-\alpha}$ with $\alpha = 1$ for $x \geq p + g/2$ is one of several possible choices, although Fig. 3 of PSH suggests α should be somewhat greater than unity. Having chosen this inverse x dependence, the constants C_0 and C_1 are completely determined by the continuity of ϕ and the condition $\phi \to -I/4$ as $x \to \infty$. The choice $\phi(x,0) = -I/2$ for $g/2 \leq x \leq p + g/2$ is an excellent approximation for finite-permeability film heads with $p/g \simeq 1$, since for any reasonable permeability the potential drop across the pole-tip is only a few percent of that across the gap. The choice $\phi(x,0) = -Ix/g$ for $0 \leq x \leq g/2$ is standard. This potential must be scaled by the head efficiency factor in the usual way.

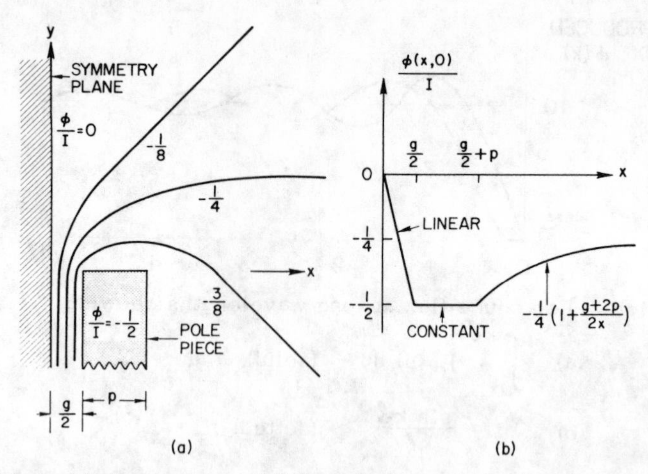

Fig. 1. (a) Cross section of finite pole-tip length head with several equipotential lines schematically shown. Refer to [2, Fig. 3] for an accurate drawing. (b) Assumed magnetic scalar potential at $y = 0$ is shown.

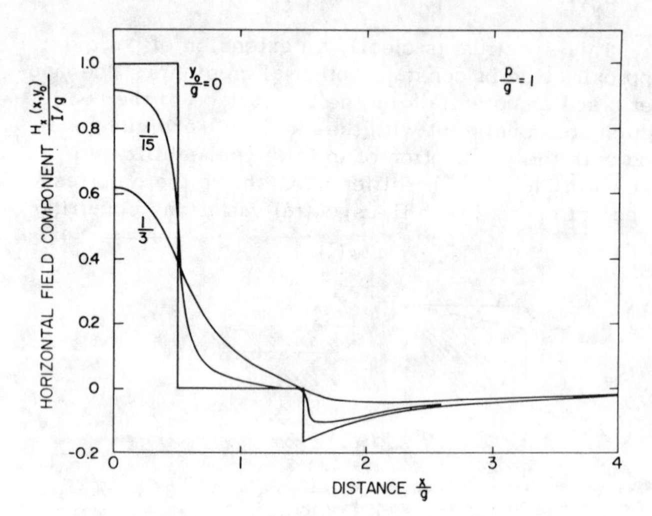

Fig. 2. Horizontal field component according to (7).

Having chosen $\phi(x,0)$ to approximate the conformal mapping result, standard theory [3] gives in CGS units

$$H_x(x,y) = -\frac{\partial \phi}{\partial x} \tag{4}$$

where

$$\phi(x,y) = \frac{-1}{4\pi}\int_{-\infty}^{\infty} \phi(x',0)\left.\frac{\partial G(x,y,x',y')}{\partial y'}\right|_{y'=0} dx' \tag{5}$$

and where

$$G(x,y,x',y') = \ln \frac{(x-x')^2+(y-y')^2}{(x-x')^2+(y+y')^2}. \tag{6}$$

Manuscript received September 9, 1974.

The author is with the IBM Thomas J. Watson Research Center, Yorktown Heights, N.Y. 10598, on leave from IBM Research, San Jose, Calif. 95193.

Reprinted from *IEEE Trans. Magn.*, vol. MAG-11, pp. 80–81, Jan. 1975.

The result is

$$H_x(x,y) = \frac{I}{\pi g}\left(\tan^{-1}\frac{x+g/2}{y} - \tan^{-1}\frac{x-g/2}{y}\right) - \frac{I}{2\pi}\frac{y}{x^2+y^2}$$
$$+ \frac{I}{4\pi}\frac{p+g/2}{x^2+y^2}\left[\frac{x^2-y^2}{x^2+y^2}\left(\tan^{-1}\frac{x+p+g/2}{y}\right.\right.$$
$$\left.- \tan^{-1}\frac{x-p-g/2}{y} - \pi\right) + \frac{xy}{x^2+y^2}$$
$$\left.\ln\frac{(x+p+g/2)^2+y^2}{(x-p-g/2)^2+y^2}\right]. \qquad (7)$$

This function is plotted in Fig. 2 such that it is directly comparable with certain curves in Figs. 4 and 6 of PSH.

REFERENCES

[1] J. C. Mallinson, "On Recording Head Field Theory", presented at the 1974 INTERMAG Conference, Toronto, Canada, May 14–17, Paper 23.4; also in *IEEE Trans. Magn.*, vol. MAG-10, pp. 773–775, Sept. 1974.

[2] R. I. Potter, R. J. Schmulian, and K. Hartman, "Fringe field and readback voltage computations for finite pole-tip length recording heads," *IEEE Trans. Magn.*, vol. MAG-7, pp. 689–695, Sept. 1971.

[3] See, for instance, J. D. Jackson, *Classical Electrodynamics.* New York: Wiley, 1962, pp. 18–19.

Analytic Expressions for Field Components of Nonsymmetrical Finite Pole Tip Length Magnetic Head Based on Measurements on Large-Scale Model

THEODORE J. SZCZECH

Abstract—An analytic expression for the x component of the field of a finite pole tip length magnetic head (thin film type) has been given by Potter [1]. Potter described the boundary potential extending beyond the pole tip edge with the expression $V(x, 0) = V_0(C_0 + C_1/x)$, where V_0 is the pole tip potential. Computations based on conformal transformations of pole tips of semi-infinite height were the basis for Potter's selection of boundary potential. The slightly modified expression $V(x, 0) = V_0[C_0 + C_1/(x + C_2)]$, as an approximation for the boundary potential beyond the tip edge, leads to expressions for the field components which agree well with those measured on a large scale model. C_0, C_1, and C_2 were determined from a consideration of experimental field components for various experimental conditions and found to be a function of head geometry. The boundary potential used in these calculations has a slight discontinuity. This results from approximating the actual potential by a $1/x$ function to simplify the problem. However, the purpose is to give analytical expressions for the field components agreeing with experimental measurements rather than to restrict the boundary potential to meet certain conditions as done by Potter.

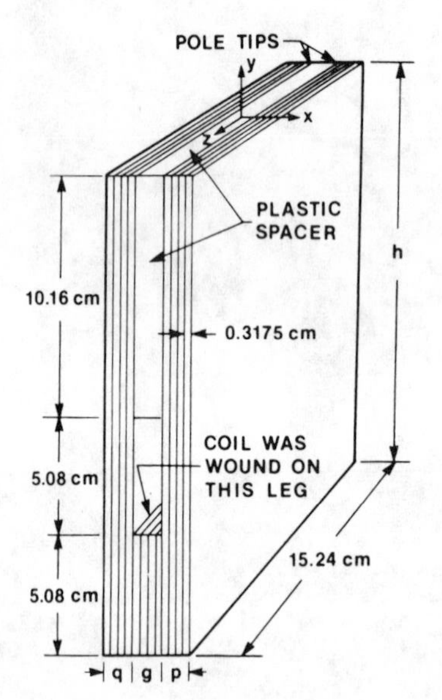

Fig. 1. Large scale model of thin film head.

I. INTRODUCTION

ANALYTIC EXPRESSIONS for the field components of a magnetic head are extremely useful as shown by the wide-spread use of Karlqvist's equations [2]. The main reason for the popularity of Karlqvist's equations is their simplicity. Unfortunately, the expressions given in this paper are somewhat more complicated than those given by Karlqvist. Yet, the expressions are simple enough for calculation with a pocket calculator. The calculations for this paper were done with a TI 59 programmable calculator. The calculator programs are available to anyone interested and should be easily adaptable to any comparable calculator. Since current theories of digital recording (see [3]-[5] as examples) rely upon use of a digital computer, the application of the expressions given in this paper for extending those theories to thin film heads is relatively straightforward.

To obtain the boundary potential experimentally, the x component of field was measured on the large scale model (Fig. 1) at $y = 0.26$ cm, plotted, and graphically integrated. From this procedure, it was determined that $V(x, 0) = V_0[C_0 + C_1/(x + C_2)]$ is a good approximation to the boundary potential beyond the pole tip edge. The boundary value problem was solved for H_x and the C were determined by matching calculated and experimental plots of H_x versus x for various experimental conditions. The resulting boundary potential differs from Potter's as shown in Fig. 2 for the example, $q = g = p = 1$. Neither of the potentials in Fig. 2 is strictly correct as x approaches infinity since they are two-dimensional approximations and would have to be modified at distances large compared to head height or width.

Manuscript received March 2, 1979; revised May 25, 1979.

The author is with the Data Recording Products Division, 3M Company, 3M Center, St. Paul, MN 55101.

II. THEORY

The problem was solved for pole pieces of unequal length. The solution to the two-dimensional problem in the region $-\infty < x < \infty$, $y > 0$ is well known [1], [2] and given below as (6). The boundary potential for the problem is described by the following identities:

$$V(t, 0) = V_0\left(\frac{C_3}{t + C_4} - C_5\right); \quad -\infty < t \leqslant -(q + g/2) \tag{1}$$

$$V(t, 0) = -V_0; \quad -(q + g/2) \leqslant t \leqslant -g/2 \tag{2}$$

$$V(t, 0) = 2V_0 t/g; \quad -g/2 \leqslant t \leqslant g/2 \tag{3}$$

$$V(t, 0) = V_0; \quad g/2 \leqslant t \leqslant p + g/2 \tag{4}$$

$$V(t, 0) = V_0\left(\frac{C_1}{t - C_2} + C_0\right); \quad p + g/2 \leqslant t < \infty. \tag{5}$$

Reprinted from *IEEE Trans. Magn.*, vol. MAG-15, pp. 1319–1322, Sept. 1979.

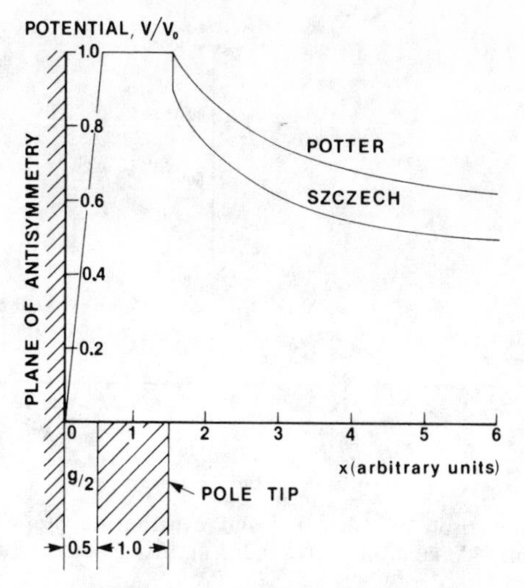

Fig. 2. Boundary potentials assumed by Potter and author. Shown for $q = g = p = 1$. See Fig. 1 for definitions of q, g, and p.

The potential at any point (x, y) is given by

$$V(x,y) = \frac{y}{\pi}\int_{-\infty}^{\infty} \frac{V(t,0)\,dt}{y^2 + (x-t)^2} \tag{6}$$

and the field components are

$$H_x = -\partial V/\partial x \tag{7}$$

and

$$H_y = -\partial V/\partial y. \tag{8}$$

The gap field H_g is defined as the horizontal component of the field at $y = 0$, $-g/2 \leqslant x \leqslant g/2$ and is given by

$$H_g \equiv -\partial V/\partial t = -2V_0/g. \tag{9}$$

Evaluating the steps described by (1)-(9) gives

$$\begin{aligned}
H_x = {} & \frac{H_g}{\pi}\left[\tan^{-1}\left(\frac{g/2+x}{y}\right) + \tan^{-1}\left(\frac{g/2-x}{y}\right)\right] \\
& - \frac{gH_g y}{2\pi}\left[\frac{(1-C_5)}{y^2+(q+g/2+x)^2} + \frac{(1-C_0)}{y^2+(p+g/2-x)^2}\right] \\
& + \frac{gH_gC_3}{2\pi}\left\{\frac{(x+C_4)^2-y^2}{[(x+C_4)^2+y^2]^2}\left[\tan^{-1}\left(\frac{q+g/2+x}{y}\right)-\frac{\pi}{2}\right]\right. \\
& - \frac{(q+g/2+2x+C_4)\,y}{[(x+C_4)^2+y^2]\,[y^2+(q+g/2+x)^2]} \\
& \left. - \frac{y(x+C_4)}{[(x+C_4)^2+y^2]^2}\ln\left[\frac{(q+g/2-C_4)^2}{y^2+(q+g/2+x)^2}\right]\right\} \\
& + \frac{gH_gC_1}{2\pi}\left\{\frac{(x-C_2)^2-y^2}{[(x-C_2)^2+y^2]^2}\left[\tan^{-1}\frac{(p+g/2-x)}{y}-\frac{\pi}{2}\right]\right. \\
& - \frac{(p+g/2-2x+C_2)\,y}{[(x-C_2)^2+y^2]\,[y^2+(p+g/2-x)^2]} \\
& \left. + \frac{y(x-C_2)}{[(x-C_2)^2+y^2]^2}\ln\left[\frac{(p+g/2-C_2)^2}{y^2+(p+g/2-x)^2}\right]\right\}
\end{aligned} \tag{10}$$

and

$$\begin{aligned}
H_y = {} & \frac{H_g}{2\pi}\ln\left[\frac{y^2+(g/2-x)^2}{y^2+(g/2+x)^2}\right] \\
& + \frac{gH_g}{2\pi}\left[\frac{(q+g/2+x)(1-C_5)}{y^2+(q+g/2+x)^2} - \frac{(p+g/2-x)(1-C_0)}{y^2+(p+g/2-x)^2}\right] \\
& + \frac{gH_gC_3}{4\pi}\left\{\frac{1}{y^2+(x+C_4)^2} - \frac{2y^2}{[y^2+(x+C_4)^2]^2}\right\} \\
& \cdot\left\{\ln\left[\frac{(q+g/2-C_4)^2}{y^2+(q+g/2+x)^2}\right]\right. \\
& \left. - \frac{2(x+C_4)}{y}\left[\tan^{-1}\frac{(q+g/2+x)}{y}-\frac{\pi}{2}\right]\right\} \\
& - \frac{gH_gC_1}{4\pi}\left\{\frac{1}{y^2+(x-C_2)^2} - \frac{2y^2}{[y^2+(x-C_2)^2]^2}\right\} \\
& \cdot\left\{\ln\left[\frac{(p+g/2-C_2)^2}{y^2+(p+g/2-x)^2}\right]\right. \\
& \left. + \frac{2(x-C_2)}{y}\left[\tan^{-1}\frac{(p+g/2-x)}{y}-\frac{\pi}{2}\right]\right\} \\
& + \frac{gH_gC_3}{2\pi[y^2+(x+C_4)^2]}\left\{\left(\frac{x+C_4}{y}\right)\right. \\
& \left. \cdot\left[\tan^{-1}\left(\frac{q+g/2+x}{y}\right)-\frac{\pi}{2}\right] + \frac{(x+C_4)(q+g/2+x)-y^2}{y^2+(q+g/2+x)^2}\right\} \\
& + \frac{gH_gC_1}{2\pi[y^2+(x-C_2)^2]}\left\{\left(\frac{x-C_2}{y}\right)\right. \\
& \left. \cdot\left[\tan^{-1}\left(\frac{p+g/2-x}{y}\right)-\frac{\pi}{2}\right] + \frac{(x-C_2)(p+g/2-x)+y^2}{y^2+(p+g/2-x)^2}\right\}.
\end{aligned} \tag{11}$$

III. Experimental Procedures and Results

As mentioned previously, the C_n in (10) and (11) were determined experimentally by employing a large scale model. The large scale thin film head was constructed of 0.3175 cm thick plates of low carbon iron; the gap material was 1.27 cm thick plexiglass. Several extra iron plates were fabricated so that the pole tip lengths could be increased. The coil was 200 turns of 22 gauge enameled wire wound around the back leg well removed from the pole tips (see Fig. 1). The resulting equations describe the field produced by both the coil and pole tips. The use of a large-scale model has several advantages in that anisotropy and frequency effects are avoided, and the field components can be easily measured.

The field components were measured at the center of the head track (i.e., $z = 0$ in Fig. 1) with a Bell model 240 gaussmeter having an A-2401 axial probe and a T-2401 transverse probe. The probes were translated with a device capable of measuring probe movement to 0.1 mm. Since each probe samples a finite volume, the measurements represent a value which has been averaged over that volume. However, the head

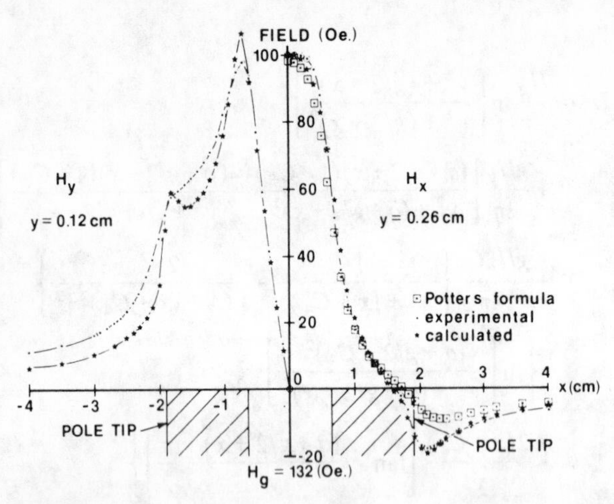

Fig. 3. Comparison of calculated and experimental field components. Axial probe at $y = 0.26$ cm and transverse probe at $y = 0.12$ cm. Symmetric head, $p = q$.

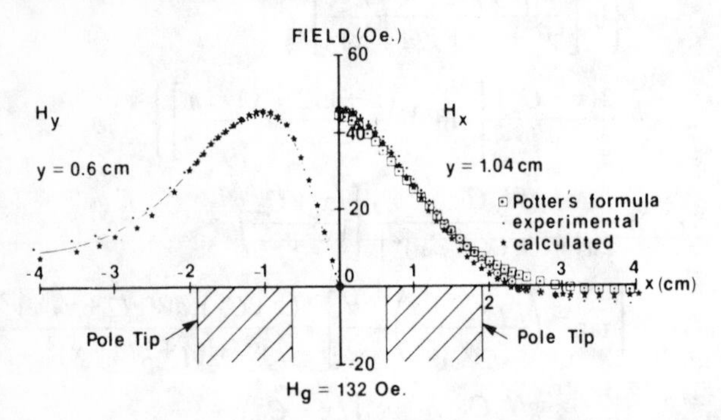

Fig. 4. Comparison of calculated and experimental field components. Axial probe at $y = 1.04$ cm and transverse probe at $y = 0.6$ cm. Symmetric head, $p = q$.

dimensions compared to the probe dimensions are such that this error is negligible in most cases. The values given for y in Figs. 3-5 are the distance from the head surface to the center of the probe element. Since the boundary potential was specified along $y = 0$, H_g is the field along $y = 0$. This value was measured with the transverse probe in a hole drilled in the plexiglass spacer. The measured value was 132 Oe as denoted in Figs. 3-5. This value was also used in the calculations.

In Fig. 3 is shown H_g versus x at an axial probe distance of 0.26 cm and H_y versus x at a transverse probe distance of 0.12 cm. This was as close as each probe could approach the head surface. The calculated values are also shown. Only half of each curve is given since the H_x curve is symmetrical about the origin and the H_y curve is antisymmetrical about the origin. The C_n values were determined by a consideration of the agreement between calculated and experimental values for H_x for various experimental conditions and were found to be $C_0 = C_5 = 0.41$, $C_1 = p/2$, $C_2 = C_4 = g/2$, and $C_3 = q/2$. For gap length, an effective length of $g = 1.1g_m$, where g_m is the measured gap length, gives a better fit than the measured value. This is because the potential in the gap is not quite linear and is better approximated by a straight line if $g = 1.1g_m$. In Fig. 3, it is seen that the agreement between calculated and

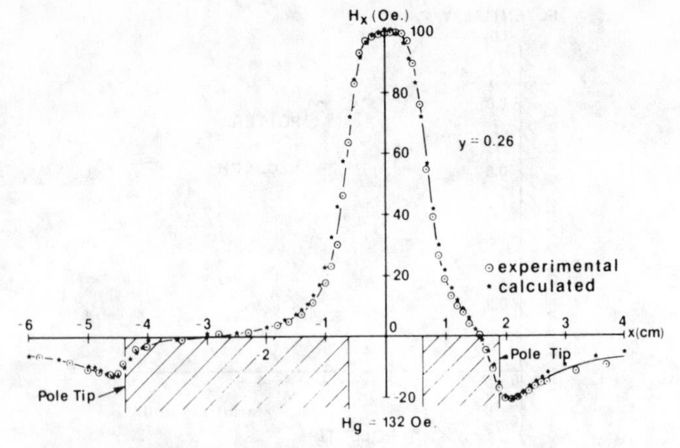

Fig. 5. Comparison of calculated and experimental horizontal field components. Axial probe at $y = 0.26$ cm. Asymmetric head, $q = 3p$.

experimental values is excellent for H_x. For H_y, the agreement is not quite as good. In the region $-q - g/2 < x < -g/2$ much of the disagreement in H_y is caused by the averaging effect of the probe and the fact that the measurement was taken extremely close to the head. In the region $x < -q - g/2$, error is introduced by an inaccurate boundary potential. As pointed out by Potter [1], $V(x) \sim 1/x^\alpha$ where $\alpha > 1$. It is believed that this form of boundary potential would result in a solution for H_x and H_y giving better agreement than $V(x) \sim 1/x$. Unfortunately, (6) is not easily solved for this type boundary potential. In Fig. 3 is also plotted Potter's formula for the x component of field. Notice the better agreement to the experimental field component using (10).

As y increases, the agreement between calculated and experimental values improves as illustrated in Fig. 4. Again the differences shown are due to the slightly inaccurate assumptions of a linear boundary potential for $-g/2 \leq x \leq g/2$ and a $1/x$ boundary potential for $x > p + g/2$ and $x < -q - g/2$. Potter's formula is again plotted here and does not agree as well as (10). Equation (10) was verified for $y/g \cong 0.2$ whereas (11) was verified for $y/g \cong 0.1$. Since (10) and (11) are derived from the same potential function, it is reasonable to assume that (10) is as accurate as (11) at $y/g \cong 0.1$. The author does not recommend use of these equations for smaller values of y than $y/g = 0.1$.

The author feels that (10) is of more interest than (11) since most magnetic recording media are horizontally oriented and most theories consider only the x component. Thus the calculated values of the field components for heads having pole tips of unequal length was verified only for the x component. The geometry of the large scale model head was changed so that $g = p = q/3$ as shown in Fig. 5. The axial probe was again positioned to measure as close to the head surface as possible ($y = 0.26$ cm). As seen from Fig. 5, the agreement between calculated and experimental values is excellent. For these calculations the values $C_0 = 0.40$ and $C_5 = 0.42$ give the best fit where $C_0 = C_5 = 0.41$ previously. The slight change is probably due to experimental error. Thus, although there had been a considerable geometry change, little change in these coefficients resulted. It is reasonable to assume that C_0 and C_5

may be a function of h and head width. It was not possible to determine that dependency, if any, in this investigation.

IV. Conclusions

Analytic expressions for the field components of a thin film head are given. Although formidable looking, the expressions can be programmed for computation with a TI 59 pocket calculator. Applying the expression to recent digital recording theories to extend these theories to thin film heads should be relatively straightforward, since digital computers are commonly employed for such calculations. The expressions given in (10) and (11) have undefined coefficients C_n which were determined from measurements on a large scale model thin film head. The values were found to be $C_0 = C_5 = 0.41$, $C_1 = p/2$, $C_2 = C_4 = g/2$, and $C_3 = q/2$. These coefficients, particularly C_0 and C_5, may change if h and head width are changed significantly. In extending these expressions to actual thin film heads, anisotropy and frequency effects in the thin film heads may introduce inaccuracies.

Acknowledgment

The author would like to thank W. A. Bernett, R. E. Fayling, E. T. Gorman, D. B. Pendergrass, D. B. Richards, and J. E. Ross of San Diego State, and Prof. W. F. Brown, Jr., of University of Minnesota, for their assistance.

References

[1] R. I. Potter, "Analytic expression for the fringe field of finite pole-tip length recording heads," *IEEE Trans. Magn.*, vol. MAG-11, pp. 80–81, Jan. 1975.

[2] O. Karlqvist, "Calculation of the magnetic field in the ferromagnetic layer of a magnetic drum," *Trans. Royal Inst. Tech.* (Stockholm, Sweden), no. 86, pp. 1-27, 1954.

[3] R. I. Potter, "Analysis of saturation magnetic recording based on arctangent magnetization transitions," *J. Appl. Phys.*, vol. 41, no. 4, pp. 1647-1651, Mar. 1970.

[4] M. L. Williams and R. L. Comstock, "An analytical model of the write process in digital magnetic recording," in *AIP Conf. Proc. on Mag. and Magn. Matls.*, pp. 738-742, 1971.

[5] C. S. Chi and D. E. Speliotis, "Dynamic self-consistent iterative simulation of high bit density digital magnetic recording," *IEEE Trans. Magn.*, vol. MAG-10, pp. 765-768, Sept. 1974.

ACCURATE FIELD COMPUTATION WITH THE BOUNDARY ELEMENT METHOD

M.H. Lean and A. Wexler

Abstract - Interface problems in magnetostatics are formulated as boundary integral equations of the second kind involving the appropriate scalar (no current sources) or vector potentials. The boundary element method (BEM), which employs parametric representation of surfaces and sources, is used to solve some two-dimensional examples by way of illustration. A novel approach, automated to address Green's functions singularities over arbitrarily-shaped geometries, is introduced.

INTRODUCTION

The accurate solution of boundary value problems posed in integral form, require both faithful representation of problem geometry and source variation, and precise approximation of the integral operation. Traditional modes of solution fall into the class of moment methods [1] with the most common involving pulse expansion and point-matching. This technique uses planar collocation sections over which the source is assumed constant. Boundary conditions are relaxed by 'matching' only at discrete 'points' which are the geometric centers of collocation sections. The evaluation of integrals for matrix fill is usually accomplished analytically especially in the vicinity of kernel singularities. One immediate consequence is the loss in geometrical fidelity when curved boundaries are encountered.

The BEM is a composite methodology that may be viewed as a refinement on the preceding moment method scheme. It has the capability to model arbitrarily-shaped boundaries through a piecewise assembly of parametric, non-planar boundary elements. Sub-domain basis (expansion) functions of appropriate order, derived from Lagrange interpolation considerations, are used in conjunction with specified node-point coordinates to attain very precise geometrical representation. Intra-element source variation is handled by the same type of basis functions so that use of the same order results in an 'isoparametric' scheme. Each element in n-dimensional 'global' space is linked by a mapping to a standard simplex in 'local' n-1 space. This feature leads to algorithmic convenience since matrix accumulation is performed on a per-element basis. Another useful consequence is the reduction in overhead as expansion functions and quadrature data need to be specified only once on the simplex.

Problem discretization is via the Rayleigh-Ritz procedure on the variational functional which can be shown to result in a form identical to that resulting from a direct application of Galerkin's method. In the accumulation of matrix entries, Gauss quadratures [2] of appropriate order and form are solicited for precise integration over each element. In particular, singularities introduced through the use of Green's functions and their derivatives are handled accurately by a fully automated numerical scheme. By tailoring Gauss quadratures for specific applications, required precision can be attained with minimal sampling. A welcomed consequence of these innovative treatments is that matrix diagonal strength is enhanced, thus further ensuring the well-conditioning of the integral equation-generated matrix. Further details on the BEM are published elsewhere [3], [4].

M.H. Lean recently completed his Ph.D. degree at the University of Manitoba and is now with the Xerox Corporation, 141 Webber Avenue, North Tarrytown, NY 10591. A. Wexler is professor of Electrical Engineering at the University of Manitoba, Winnipeg, Canada R3T 2N2. Financial assistance from the Natural Sciences and Engineering Research Council of Canada and the Xerox Corporation are gratefully acknowledged.

SINGULAR KERNELS

The main difficulty encountered in integral equation solution lies in the numerical approximation to the integration process for singular kernels. This singularity is a direct consequence of Green's function when the observer and source locations coincide. Conventional methods of addressing this crucial issue include analytic integration over 'flat' intervals, and the evaluation of Cauchy principal values. Being analytic, these techniques are unavoidably problem geometry dependent, thus restricting their widespread application. With few exceptions, their usage incur tedious manipulations that contribute to both core length and overhead.

The BEM addresses this issue with a two-step procedure: first, by choosing a weight function that will uniquely specify a Gauss quadrature formula; and second, by sectioning the element about the singular location and ensuring that the quadrature rule is oriented in the appropriate directions to collectively handle the limiting behaviour. Singularities of Green's functions for static fields in two- and three-dimensions (logarithmic and r^{-1} behaviour), are treated by essentially the same sectioning philosophy. To illustrate, consider a line element in two-dimensions that is sectioned into two parts about the singular location. Since the form of the kernel singularity is logarithmic a quadrature scheme with weight $w(\zeta) = -\ell n\zeta$ is chosen for this application. This quadrature data is then operated on by two linear transformations subject to the constraint that the behaviour of the weight function be preserved. In effect, this maneuver positions the quadrature so that the singular location is approached on both sides in logarithmic fashion. The net result is the construction of a special set of data that will integrate a logarithmic singularity within the line element. The inherent merit of this scheme is that it is defined on the simplex so that the form of the singularity is preserved regardless of the actual size or shape of the 'global' element. Extensions of this scheme to three-dimensions and to time-harmonic Green's functions are straightforward [3], [5].

SCALAR POTENTIAL

A typical example of an interface problem is that of the perturbation of a uniform magnetostatic field by a permeable body (Fig. 1). Let $\hat{n}_i$ and $\hat{n}_e$ be unit

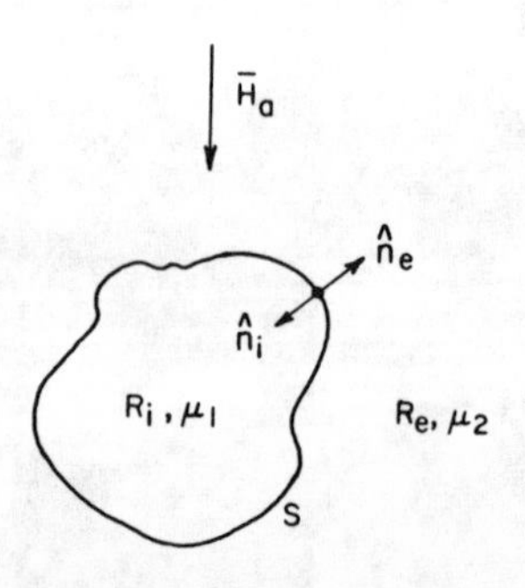

Fig.1. Interface problem geometry

Reprinted from *IEEE Trans. Magn.*, vol. MAG-18, pp. 331-335, Mar. 1982.

normals directed into and out of the region R_i. Also, let material constants be μ_1 and μ_2 for the interior and exterior regions, respectively. From linearity of Maxwell's equations, one may consider an equivalent problem posed by the superposition of the applied field in R_e, and the perturbation field due to a polarization source distribution on S. This source distribution is a direct consequence of the interface condition and vanishes when $\mu_1 = \mu_2$. Thus, signifying $\bar{H}$, $\bar{H}_a$ and $\bar{H}_m$ to be the total, applied and perturbation fields respectively, then

$$\bar{H} = \bar{H}_a + \bar{H}_m \tag{1}$$

and in particular if ϕ_a and ϕ_m are Laplacian in all R, then the total potential $\tilde{\phi}$ will be given by the algebraic summation of the corresponding scalar potentials, or

$$\tilde{\phi}(\bar{r}) = \phi_a(\bar{r}) + \phi_m(\bar{r}) \tag{2}$$

where $\bar{H}$ is given by

$$\bar{H} = -\nabla\tilde{\phi} \tag{3}$$

The following subsections discuss the origins and relative merits of three formulations that may be used to address this problem. For convenience, material permeabilities are set to be $\mu_2 = \mu_0$ for free-space, and $\mu_1 = \mu\mu_0$ where μ is the relative constant.

<u>Simple-Layer Kernel Formulation</u>

This formulation is derived from the use of a distribution of simple-layer polarization sources $\sigma(s)$ on S. The perturbation potential ϕ_m everywhere in R is given by

$$\phi_m(\bar{r}) = \frac{1}{\mu_0}\int G[\bar{r}|s']\,\sigma(s')ds' \tag{4}$$

where G is Green's function which is singular on S thus requiring special handling. Boundary conditions on S require that both

$$\tilde{\phi}_1 = \tilde{\phi}_2 \tag{5}$$

and

$$\mu_1\frac{\partial\tilde{\phi}_1}{\partial n_i} + \mu_2\frac{\partial\tilde{\phi}_2}{\partial n_e} = 0 \tag{6}$$

be satisfied. Since ϕ_m is harmonic in R and continuous across the interface, condition (5) is satisfied when $\phi_{a_1} = \phi_{a_2}$ on S. For flux continuity, consider the Neumann formulations of (4) taken along $\hat{n}_i$ and $\hat{n}_e$ such that

$$\frac{\partial\phi_{m_1}(s)}{\partial n_i} = \frac{1}{\mu_0}\int\frac{\partial G[s|s']}{\partial n_i}\sigma(s')ds' - \frac{\sigma(s)}{2\mu_0} \tag{7}$$

and

$$\frac{\partial\phi_{m_2}(s)}{\partial n_e} = \frac{1}{\mu_0}\int\frac{\partial G[s|s']}{\partial n_e}\sigma(s')ds' - \frac{\sigma(s)}{2\mu_0} \tag{8}$$

on S where the last term of each of the above expressions represent the jump in flux in crossing the boundary. Expanding (6) into component form, dividing throughout by μ_0, and recognizing that the normal derivative of ϕ_{a_1} and ϕ_{a_2} are identical, the result is the simplified interface condition

$$\mu\frac{\partial\phi_{m_1}}{\partial n_i} + \frac{\partial\phi_{m_2}}{\partial n_e} = (\mu - 1)\frac{\partial\phi_a}{\partial n_e} \tag{9}$$

Substituting (7) and (8) into (9) and noting that the normal derivative of G in the $\hat{n}_i$ and $\hat{n}_e$ directions sum to zero, the expression

$$\frac{1}{\mu_0}\int\frac{\partial G[s|s']}{\partial n}\sigma(s')ds' + \frac{(\mu+1)}{(\mu-1)}\frac{\sigma(s)}{2\mu_0} = H_{an} \tag{10}$$

is obtained where subscripts have been dropped with the understanding that normals are directed along $\hat{n}_e$. The right-hand side of (10) represents the normal component of the applied field and is computed from

$$H_{an} = -\hat{n}\cdot\nabla\phi_a = -\frac{\partial\phi_a}{\partial n} \tag{11}$$

Once $\sigma(s)$ is know, ϕ_m everywhere may be computed from (4) and $\tilde{\phi}$ from (2).

<u>Double-Layer Kernel Formulations</u>

An alternative to (10) is to express perturbation potential ϕ_m everywhere in R as a function of its values on S, i.e.

$$\int\frac{\partial G[s|s']}{\partial n'}\phi_m(s')ds' + \frac{(\mu+1)}{(\mu-1)}\frac{\phi_m(s)}{2} = \int G[s|s']H_{an}ds' \tag{12}$$

which is seen to retain the same form except for the right-hand side which is considerably more involved. Details of the derivation are reported elsewhere [3]. Computation of ϕ_m subsequent to the solution of (12) requires the determination of ϕ_m' on S. This requirement however, could be removed by eliminating the flux term, a procedure justified by the fact that $\phi_m(s)$ alone is sufficient to define ϕ_m everywhere in R. Hence, through a series of algebraic manipulations, the total potential $\tilde{\phi}$ is obtainable as

$$\tilde{\phi}(\bar{r}) = \begin{cases}\phi_a(\bar{r}) + \phi_m(\bar{r}) & ;\ \bar{r}\in R_e\\ \phi_a(s) + \phi_m(s) & ;\ \bar{r}\in S\\ \phi_a(\bar{r}) + \frac{1}{\mu}\phi_m(\bar{r}) & ;\ \bar{r}\in R_i\end{cases} \tag{13}$$

where ϕ_m for $\bar{r}\notin S$ is given by

$$\phi_m(\bar{r}) = (\mu-1)\int\{G[\bar{r}|s']H_{an}(s') - \phi_m(s')\frac{\partial G[\bar{r}|s']}{\partial n'}\}ds' \tag{14}$$

A more compact expression than that of (12) is provided by

$$\int\frac{\partial G[s|s']}{\partial n'}\tilde{\phi}(s')ds' + \frac{(\mu+1)}{(\mu-1)}\frac{\tilde{\phi}(s)}{2} = \frac{1}{(\mu-1)}\phi_a(s) \tag{15}$$

which is a formulation in terms of total potential $\tilde{\phi}$ on S. Corresponding $\tilde{\phi}$ everywhere is calculated from

$$\tilde{\phi}(\bar{r}) = \begin{cases}\phi_a(\bar{r}) + \phi_m(\bar{r}) & ;\ \bar{r}\in R_e\\ \frac{1}{\mu}\phi_a(\bar{r}) + \frac{1}{\mu}\phi_m(\bar{r}) & ;\ \bar{r}\in R_i\end{cases} \tag{16}$$

where ϕ_m is now given by

$$\phi_m(\bar{r}) = -(\mu-1)\int\frac{\partial G[\bar{r}|s']}{\partial n'}\tilde{\phi}(s')ds' \tag{17}$$

The expressions (10), (12) and (15) are three formulations of the same problem in terms of second kind Fredholm integrals with fairly well-behaved kernels. By replacing σ/μ_0 by M_n, the normal component of equivalent magnetization sources, the matrix of (10) is seen to be the transpose of (12) and (15). The most noticeable difference is in the form of the excitation function where (10) and (15) are simple and compact. In the case of (12), a double-surface integral has to be evaluated. Besides additional time required in matrix fill, the kernel also requires special treatment since G is singular on S. Consequently, this formulation is not a viable alternative. The main difference between (10) and (15) lies in the computation of $\tilde{\phi}$ once the equations have been solved for boundary sources. With the former, the expression for ϕ_m is valid everywhere without reservation; but the same cannot be said for the latter. As such, numerical inconsistencies with respect to sign changes may arise for $\bar{r}$ close to the interface. The formulation in terms of ϕ_m on S has another disadvantage when it comes to field determination. From inspection of (14), $\phi_m(\bar{r})$ is dependent on the product of μ with the difference between two small values. In the limit as $\mu \to \infty$, the error in ϕ_m computation is magnified especially for the exterior fields. The interior expression is more stable due to the added factor of μ^{-1}.

The choice of which of (10) or (15) to implement really depends on the parameter of interest. If interface potentials are important, then (15) would be preferred since $\tilde{\phi}$ is the unknown variable. But if field definition within R is deemed important, then (10) would be a better choice due to the continuity of ϕ across S.

Permeable Square Cylinder Problem

This example entails a two-dimensional analysis of a square, permeable cylinder oriented as shown in Fig. 2. The uniform $\bar{H}_a$ - field is directed along the positive y - axis by setting $\phi_a = -y$. Using quarter plane symmetry, only the fourth quadrant need be addressed. Galerkin's method is applied individually to (10), (12) and (15) to solve this problem.

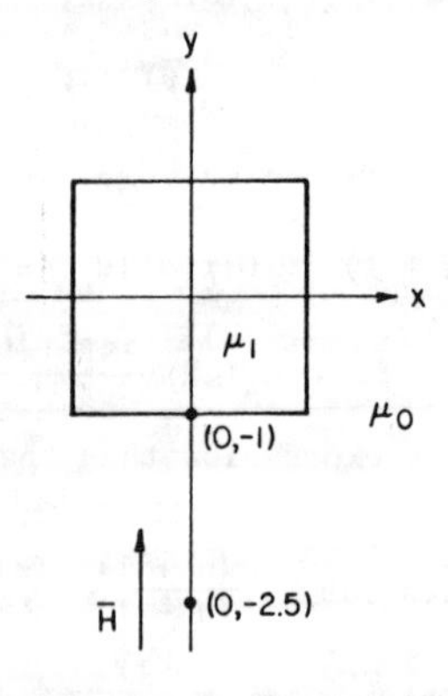

Fig. 2. Permeable square cylinder in uniform H-field

Each of these equations is seen to be of the second kind with well-behave kernels. In particular, the normal derivative of Green's function is

$$\frac{\partial G}{\partial n} = \hat{n} \cdot \nabla G = -\frac{1}{2\pi} \frac{\cos(\hat{n}, \bar{r} - \bar{r}')}{|\bar{r} - \bar{r}'|} \qquad (18)$$

which is finite in the limit as $\bar{r} \to \bar{r}'$ since the argument of the cosine function tends to $\pi/2$. As a result, only the Gauss-Legendre weight $w(x) = 1$ need be used for integration over the source variable.

The BEM model consists of 4 cubic elements involving 13 unknowns. Denoting the formulations of (10), (12) and (15) as F_1, F_2 and F_3 respectively, the potential $\tilde{\phi}$ computed at coordinate positions (0, -1) and (0, -2.5) are compared in Table 1. for various values of μ.

Table 1. Computed potentials $\tilde{\phi}$

	$-\tilde{\phi}(0, -1)$			$-\tilde{\phi}(0, -2.5)$		
μ	F_1	F_2	F_3	F_1	F_2	F_3
1	1.0	1.0	1.0	2.5	2.5	2.5
2	.649	.649	.648	2.332	2.332	2.332
10	.175	.174	.173	2.074	2.070	2.074
10^2	$.208 \times 10^{-1}$	$.199 \times 10^{-1}$	$.188 \times 10^{-1}$	1.979	1.929	1.979
10^3	$.383 \times 10^{-2}$	$.293 \times 10^{-2}$	$.190 \times 10^{-2}$	1.968	1.457	1.968

For the situation $\mu = 1$, the interface does not exist, i.e. F_1 and F_2 have only the trivial solution for M_n and ϕ_m respectively. F_3 has the solution $\tilde{\phi} = \phi_a$ as expected. Therefore, the exact values of $\tilde{\phi}(0,-1)$ and $\tilde{\phi}(0, -2.5)$ are returned in the first row. As μ is increased, the results for F_2 in the exterior region, deteriorates in comparison to F_1 and F_3. Also, $\tilde{\phi}$ computed close to the boundary, have negative signs expecially for large μ. This discrepancy is due to the formulation (14) used to recover ϕ_m. As $\mu \to \infty$, the $\bar{H}$-field in the cylinder is vanishing so that for (14) to hold true, very precise field cancellation is required between the applied and the magnetization components.

Potentials on the surface are more aptly computed by F_3 since $\tilde{\phi}$ is the unknown. F_2 requires the summation of ϕ_a and ϕ_m on S. The difference in the results of F_2 and F_3 is attributed to the inaccuracy of solution of ϕ_m - the error being introduced in the computation of the excitation vector. Of the three, F_1 is the least accurate in terms of surface potentials due to the crude recovery scheme used. Theorectically (4) is singular on S so that special attention is necessary. In comparing the accuracy of computed exterior ϕ, no difference can be detected between F_1 and F_3. Thus, this observation reiterates the theorectical claim that F_1 is more viable in terms of computational economy. Equipotential plots of $\tilde{\phi}$ computed using F_1 are shown in Fig. 3.

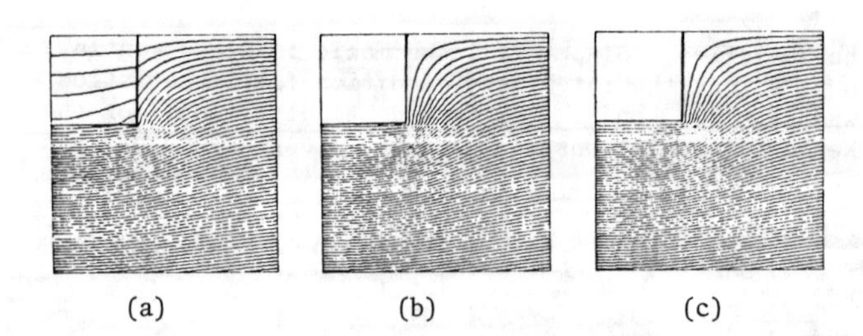

Fig. 3. Contours of equipotentials $\tilde{\phi}$:
(a) $\mu = 10$; (b) $\mu = 100$; (c) $\mu = 1000$

VECTOR POTENTIAL

A vector formulated problem is posed by a current coil radiating through a permeable body into free-space. In the case of a magnetic recording head where the length is much greater than cross-section dimensions, a two-dimensional analysis is sufficient. As such, the following derivation is for the z-component of magnetic vector potential $\bar{A}$. Again, citing the linearity of Maxwell's equations, the equivalent problem to be considered is the superposition of the effects of a current conductor alone in air; and a distribution of magnetic sources on the head cross-section. The component of vector potential $\bar{A}$ due to a z-directed current density is give by

$$A_z^J(\bar{r}) = \mu_0 \int G[\bar{r}|R'] \, J_z(R')dR' \quad ; \quad \bar{r} \in R_1 + R_2 \qquad (19)$$

where R', R_1 and R_2 denote conductor cross-section, head cross-section, and free-space respectively. Using a simple-layer kernel formulation for the sources, the following equation may be written separately for R_1 and R_2

$$A_z^M(\bar{r}) = \int G[\bar{r}|s'] \, M_t(s')ds' \qquad (20)$$

where M_t denotes the tangential component of the magnetization on S caused by B_{tan} due to the current source. The total potential $\tilde{A}_z$ is then

$$\tilde{A}_z(\bar{r}) = A_z^J(\bar{r}) + A_z^M(\bar{r}) \qquad (21)$$

Interface conditions on S require continuity of normal component of $\bar{B}$ and tangential component of $\bar{H}$, or

$$\tilde{A}_{z_1} = \tilde{A}_{z_2} \qquad (22)$$

and

$$\frac{1}{\mu_1}\frac{\partial \tilde{A}_{z_1}}{\partial n} = \frac{1}{\mu_2}\frac{\partial \tilde{A}_{z_2}}{\partial n} \qquad (23)$$

Enforcement of (23) with $\mu = \frac{\mu_1}{\mu_2}$ leads to the reduced interface condition

$$\frac{\partial A_{z_1}^M}{\partial n_i} + \mu \frac{\partial A_{z_2}^M}{\partial n_e} = (\mu - 1)\frac{\partial A_z^J}{\partial n_e} \qquad (24)$$

Substituting the appropriate normal derivatives of (20) into the above and rearranging, the result is

$$\int \frac{\partial G[s|s']}{\partial n} M_t(s')ds' + \frac{(1+\mu)}{(1-\mu)}\frac{M_t(s)}{2} = -\frac{\partial A_z^J(s)}{\partial n} \qquad (25)$$

where the right-hand side is computed from

$$\frac{\partial A_z^J(s)}{\partial n} = \mu_0 \int \frac{\partial G[s|R']}{\partial n} J_z(R')dR' \quad ; \quad s \in S \qquad (26)$$

Once (25) is solved, $\tilde{A}_z$ everywhere is calculated from (19), (20) and (21). In practice, the parameters of interest are the downtrack (B_x) and vertical (B_y) fields which are give by

$$B_x(\bar{r}) = \frac{\partial A_z^M(\bar{r})}{\partial y} + \frac{\partial A_z^J(\bar{r})}{\partial y} \qquad (27)$$

and

$$B_y(\bar{r}) = -\frac{\partial A_z^M(\bar{r})}{\partial x} - \frac{\partial A_z^J(\bar{r})}{\partial x} \qquad (28)$$

These derivatives may be carried within the integral since they are with respect to the unprimed variables, thus giving rise to a neat and compact algorithm.

A Recording Head Problem

Galerkin's method is used to solve (25). The Green's function used is given by

$$G[\bar{r}|\bar{r}'] = -\frac{1}{2\pi}\ln|\bar{r} - \bar{r}'| + \frac{1}{2\pi}\ln|\bar{r}_R - \bar{r}'| \qquad (29)$$

where the last term on the right-hand side is included to make the integral operator positive-definite. The reference point $\bar{r}_R$ is chosen quite far away to make this term approximately constant. The physical effect of this manipulation is to regularize the potential behaviour so that it vanishes logarithmically as $|\bar{r}| \longrightarrow |\bar{r}_R|$. Problem dimensions are 2 X 10 microns for the conductor, and 6 X 13 microns for the head. The BEM models of the conductor and head cross-sections are shown in Fig. 4. A constant current density of .03A/micron^2 flows in the conductor which is modelled by 4 linear triangular elements. A basic configuration for the head is the 8 element model of Fig. 4(c), with the facility to increase the degree of interpolation from a linear (8 nodes) to a quartic (32 nodes) approximation.

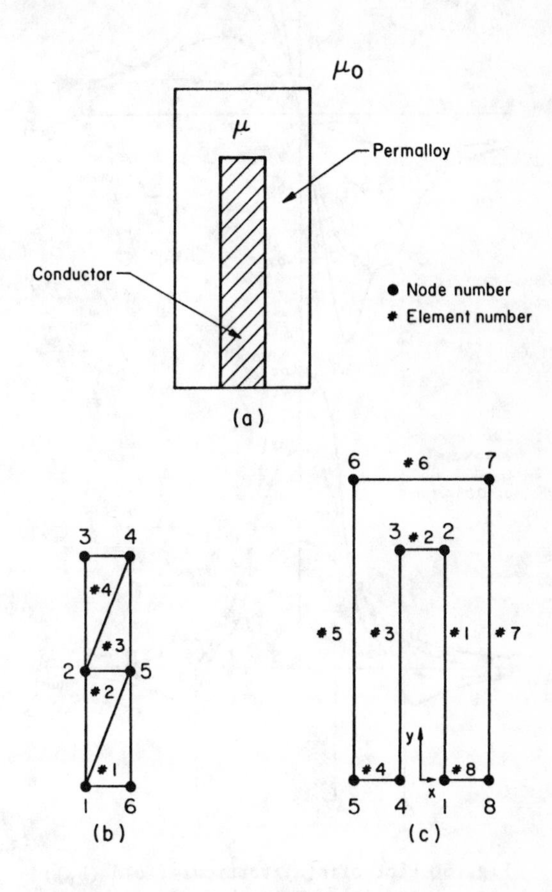

Fig. 4. Magnetic recording head: (a) cross-section; (b) BEM model of conductor; and (c) BEM model of head cross-section

Fig. 5 shows the vertical and downtrack fields at distances of .5, 1, 2 and 4 microns below the head. The 'bumps' in each curve at approximately micron are not observed until a cubic interpolation scheme is used. Evidently, the rapid undulation of the field demands higher-order interpolation of the boundary sources.

CONCLUSION

The BEM offers high fidelity in geometrical representation in addition to the theorectical guarantee for convergence through the use of the variational/Galerkin method. Solution accuracy is made economical by the tailoring of quadrature formulas to obtain maximum precision with minimal sampling. The technique for addressing Green's function singularities allows for automation in the treatment of arbitrary geometries [5]. In addition, the problem of source singularities due to geometry-at edges or corners may also be handled by incorporating the form of the singularity into the trial function set [3]. Application of the BEM to vector, three-dimensional problems as in electromagnetic scattering, are published elsewhere [3].

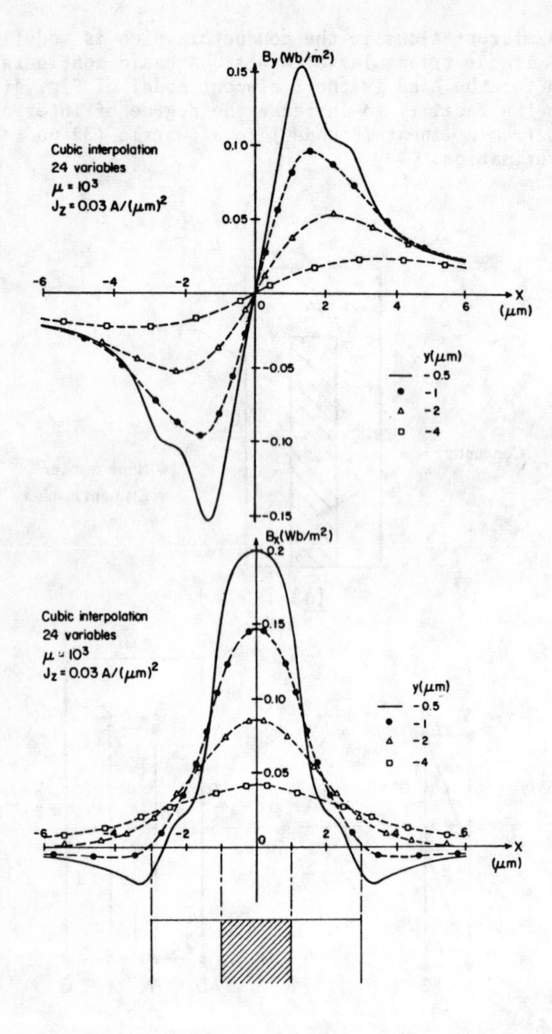

Fig. 5 Plot of: (a) vertical field (B_y); and (b) downtrack field (B_x)

ACKNOWLEDGEMENT

The authors are indebted to Dr. Dan Bloomberg of Advanced Development Laboratory, Xerox Corporation for many helpful discussions.

REFERENCES

[1] R.F. Harrington, Field Computation by Moment Methods. New York: Macmillan, 1968.

[2] A.H. Stroud and D. Secrest, Gaussian Quadrature Formulas. New Jersey: Prentice-Hall, 1966.

[3] M.H. Lean, Electromagnetic Field Solution with the Boundary Element Method. Ph.D. Dissertation, University of Manitoba, Winnipeg, Canada R3T 2N2, 1981.

[4] M.H. Lean, M. Friedman and A. Wexler, "Advances in Application of the Boundary Element Method in Electrical Engineering Problems, "in Developments in Boundary Element Methods - I, P.K. Banerjee and R. Butterfield, Eds. Barking: Applied Science Publishers, 1979, pages 207 - 250.

[5] M.H. Lean and A. Wexler, "Automated Handling of Green's Function Singularity on Arbitrarily-Shaped Boundaries." To be submitted for publication.

DISCUSSION

W. R. Hodgkins: It was stated that the method handles singularities at corners and edges. Is there any special feature of singularities at corners different from those at edges?

M. Lean: Yes. In the order of singularity given by $r^{-\nu}$ where $\nu = 1 - \frac{\pi}{\delta}$ and δ is the re-entrant angle. The two-dimensional edge corresponds to $\delta = 2\pi$. A 'corner' has δ in the range: $\pi < \delta < 2\pi$.

F. Moon: Have you made any progress in applying B. EM. to nonlinear magnetostatic problems.

M. Lean: No, not yet. We have concentrated so far on developing the tools necessary for building up the BEM code. However, this problem area will definitely be investigated in the very near future.

Part IV
Thin-Film Heads

Analysis of the Efficiency of Thin-Film Magnetic Recording Heads

Andrew Paton
IBM United Kingdom Laboratories Ltd., Hursley, Hampshire, England
(Received 18 January 1971; in final form 22 July 1971)

Equations for the flux, magnetic field, and current distributions in a single-turn planar magnetic recording head are derived making the assumption that the magnetic material has constant permeability. These equations are solved in the frequency domain and an expression is given for the head efficiency in terms of the head dimensions, permeability, and frequency.

Recently there have been reports[1,2] of attempts to produce thin-film magnetic recording heads. In this letter we report a theoretical analysis of a single-turn planar recording head. A typical geometry and some definitions are given in Fig. 1. The analysis reported here is concerned with the part that the dimensions t_1, t_2, and l play in determining the efficiency of the head. As an approximation we assume that the fringing field pattern ($x < 0$) does not vary with t_1, t_2, or l, but that its magnitude is proportional to the gap field at $x = 0$. Our main assumption is that the permeability μ of the magnetic films is constant and that $\mu \gg 1$. For simplicity we assume that quantities do not vary in the y direction.

Since $\mu \gg 1$ it is clear that the flux $\phi(x)$ in the magnetic films will be essentially parallel to the films and that the magnetic field $H(x)$ in the copper-filled gap will be almost perpendicular to the films.

We assume that the frequency of operation and the thickness of the films are such that we may neglect variations in the current density in the z direction, so that the current density J is a function of x alone. With these definitions and assumptions we now apply the integral forms of Maxwell's equations to an element of the head. This is illustrated in Fig. 2.

Calculating the integral of the magnetic field around the rectangle ABCD, we get

$$-\frac{\partial H}{\partial x} t_2 + 2H_{\text{int}} = J(x)t_2, \tag{1}$$

where $H_{\text{int}}(x)$ is the internal magnetic field of the films.

Applying the conservation of magnetic flux to the element gives

$$\frac{\partial \phi}{\partial x} = \mu_0 H W. \tag{2}$$

Calculating the integral of the electric field around the rectangle BEFC, we get

$$\frac{\partial J}{\partial x} = \mu_0 \sigma \frac{\partial H}{\sigma t}, \tag{3}$$

where σ is the conductivity of the gap material.

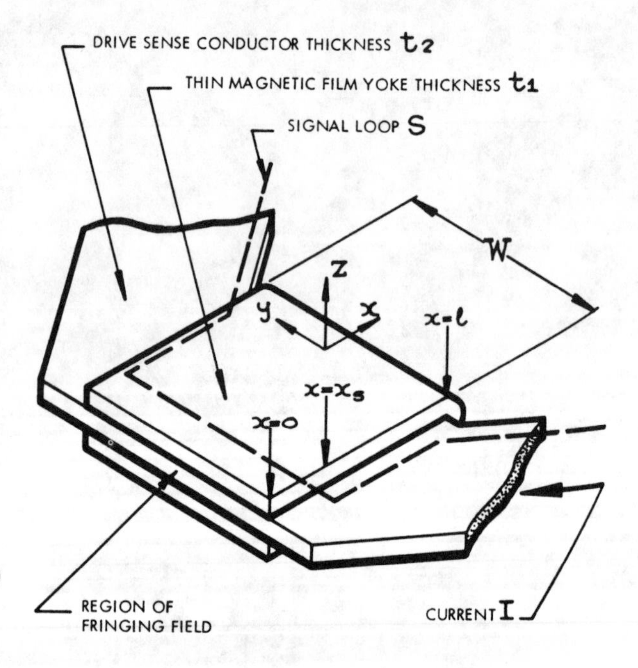

FIG. 1. Typical thin-film magnetic recording head showing coordinate conventions used.

Reprinted with permission from *J. Appl. Phys.*, vol. 42, no. 13, pp. 5868–5870, Dec. 1971.

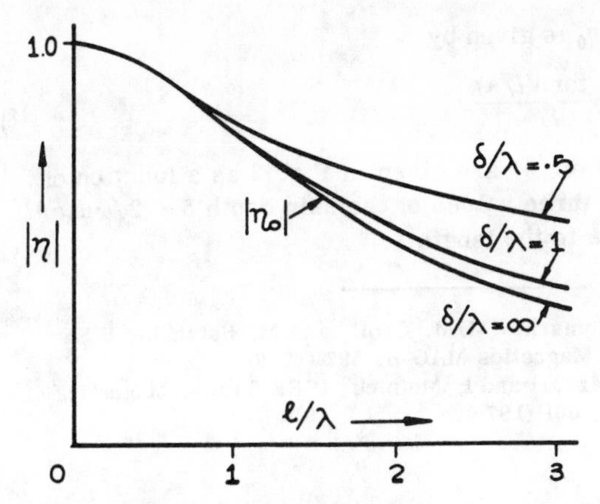

FIG. 2. An element of the head in the x direction. The integral forms of Maxwell's equations are applied to the loops ABCD and BEFC.

The internal field H_{int} and the internal flux ϕ are related by

$$\phi = \mu\mu_0 W t_1 H_{\text{int}}. \tag{4}$$

Using Eq. (4) we may write Eq. (1) in the form

$$\frac{\partial H}{\partial x} = \frac{2\phi}{W t_1 t_2 \mu\mu_0} - J(x). \tag{5}$$

Differentiating Eq. (5) with respect to x and then using Eqs. (2) and (3) we get

$$\frac{\partial^2 H}{\partial x^2} = \frac{1}{\lambda^2} H - \mu_0 \sigma \frac{\partial H}{\partial t}, \tag{6}$$

where

$$\lambda = (\tfrac{1}{2}\mu t_1 t_2)^{1/2}. \tag{7}$$

Usually we know the total current I through the drive sense conductor, so that

$$t_2 \int_0^l J(x)\, dx = I. \tag{8}$$

Assuming now that all of the dependent variables have a time dependence of the form $e^{j\omega t}$ we can solve the equations for H, ϕ, and J in a straightforward way. We let $H = H_a e^{j\omega t}$ and similarly for the other variables.

The general solution to Eq. (6) is

$$H_a = A \sinh kx + B \cosh kx, \tag{9}$$

where

$$k^2 = (1/\lambda^2) - j\omega\mu_0\sigma. \tag{10}$$

Using this general solution for H_a in Eqs. (3) and (5) we can obtain the general solutions for $J_a(x)$ and $\phi_a(x)$, namely,

$$J_a = C + (j\omega\mu_0\sigma/k)(A \cosh kx + B \sinh kx), \tag{11}$$

$$\phi_a = \tfrac{1}{2} W t_1 t_2 \mu\mu_0 [A(k + jw\mu_0\sigma/k) \cosh kx + B(k + j\omega\mu_0\sigma/k) \sinh kx + C]. \tag{12}$$

We now select two sets of boundary conditions which represent the head—first in the write mode and second in the read mode. In the case of the write mode we assume first a nonzero current I_a in the conductor, second that the head is flux closed at $x = l$, i.e., $H_a(l) = 0$, third that the flux ϕ is zero at $x = 0$. This latter assumption implies that the flux of the fringing field is small in comparison to the total flux across the gap.

With these assumptions and some algebra we can determine the coefficients A, B, and C from Eqs. (8)–(12). In particular, we find

$$H_a(0) = B = \frac{I_a k^2 \lambda^2 \tanh kl}{t_2(kl - j\omega\mu_0\sigma\lambda^2 \tanh kl)}. \tag{13}$$

In the case of the read mode we assume that the total current through the drive sense conductor may be neglected, i.e., $I_a = 0$. As before, we assume that $H_a(l) = 0$. Finally, we assume an input flux ϕ_0 at $x = 0$. With these conditions and again some algebra, we can obtain the coefficients A, B, and C. There is one further step in calculating the signal from the flux distribution $\phi_a(x)$, and that is the question of how much of the flux contributes to the signal. To answer this, we must use Faraday's law:

$$\oint_S \vec{E} \cdot d\vec{l} = -\frac{\partial \phi_s}{\partial t}, \tag{14}$$

where S is a suitable loop and ϕ_S is the flux linking the loop S. In practice the simplest loop we can choose is probably the one illustrated by a dotted line in Fig. 1 in which the position of the portion of the loop passing through the gap is chosen such that the resistive contribution to the signal is zero, i.e.,

$$J(x_s) = 0. \tag{15}$$

The signal is therefore due only to the flux contribution which is clearly $\phi(x_s)$.

Using the previously determined coefficients A, B, and C and Eq. (15) we can show that the signal volt-

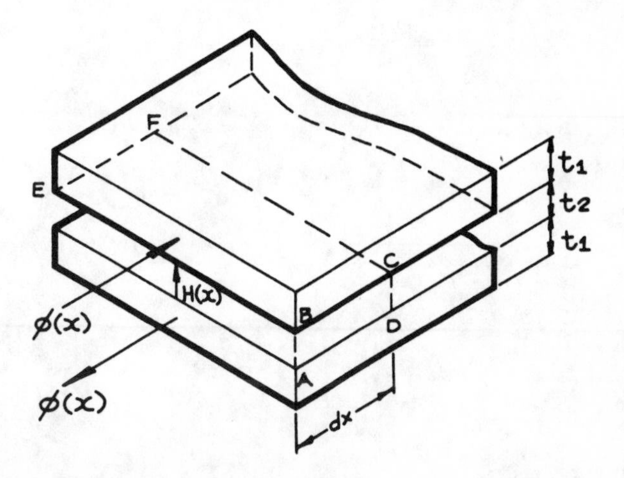

FIG. 3. Modulus of the efficiency η as a function of l/λ, with the skin depth $\delta = (2/\omega\mu_0\sigma)^{1/2}$ relative to $\lambda = (\frac{1}{2}\mu\, t_1 t_2)^{1/2}$ as a parameter.

age V_S is given by

$$V_S = -j\omega\phi(x_S) = \frac{-j\omega\phi_0 k^2\lambda^2 \tanh kl}{kl - j\omega\mu_0\sigma\lambda^2 \tanh kl}. \quad (16)$$

As we might expect from the reciprocity theorem, the efficiency factor η can be obtained either from Eq. (16) for the signal or Eq. (13) for the write field. We get

$$\eta = \frac{k^2\lambda^2 \tanh kl}{kl - j\omega\mu_0\sigma\lambda^2 \tanh kl}. \quad (17)$$

As $\omega \rightarrow 0$ we see that the low-frequency efficiency factor η_0 is given by

$$\eta_0 = \frac{\tanh(l/\lambda)}{l/\lambda}. \quad (18)$$

Graphs of $|\eta|$ are given in Fig. 3 as a function of l/λ for three values of the skin depth $\delta = (2/\omega\mu_0\sigma)^{1/2}$ relative to the length λ.

---

[1]L. T. Romankiw, I. M. Croll, and M. Hatzakis, IEEE Trans. Magnetics MAG-6, 597 (1970).

[2]J. D. Lazzari and I. Melnick, IEEE Trans. Magnetics MAG-6, 601 (1970).

Domain Effects in the Thin Film Head

R. E. Jones, Jr.
IBM Corporation
San Jose, CA 95193

ABSTRACT

An analysis is presented showing that delays in the response of thin film heads can lead to distortions in the shape of isolated pulses including decreases in the peak amplitude, increases in the isolated pulse half width, and, in extreme, the disappearance of the negative amplitude trailing edge of the pulse characteristic of the thin film head. Examples of thin film heads have been found exhibiting isolated pulse forms corresponding to variable delay times after near saturation of the head in writing. These examples suggest that variable domain structures in the head are associated with these delays. One specific model for domain wall motion in the head is considered in detail showing a relationship between the equations of motion of domain walls and the delay times associated with the portions of the head blocked by closure domains.

INTRODUCTION

Domains in thin film heads have been evoked as explanations of noise[1] and, more recently, read pulse shape irreproducibility following near saturation of the head in a write operation on a remote track. Examples of isolated pulses from heads exhibiting this phenomenon, are shown in Figures 1a and 1b. Several

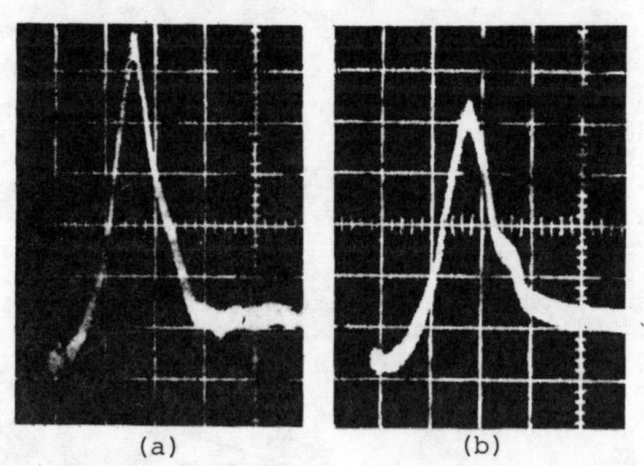

(a) (b)

Figures 1a and 1b. Thin film head isolated pulses before and after writing on a remote track.

features of the different pulse shapes roughly correspond to what would be expected if portions of the head were responding with a delay time constant, τ , which changes after the near saturation of the head. For example, the leading edges of the pulses show relatively small distortions, peak amplitudes are somewhat effected, while the trailing portions of the pulses are considerably distorted. Moreover, a more detailed analysis shows that the total area under the pulses is nearly identical for every case. In the next section the effects of introducing various delays to an ideal pulse will be analyzed and these general features demonstrated. It will also be shown that the complicated distortions in the trailing edges of the pulses (note Figure 1b) to a degree can be replicated by assuming that different portions of the track width respond with different delay times.

Although the details of thin film head domain structures are a result of many factors, some of which are accidental, such as film defects and stray fields, it is possible to postulate plausible models for domain structures which would exhibit delays corresponding to those observed in distorted pulses. One example of a domain structure is given by the Bitter pattern shown in Figure 2 for a single layer of a yoke structure. Here it can be seen that closure domains block a portion of the head track width so some flux traveling through the head can vary only by domain wall motion. In the third section we consider one idealized model for domain motion and calculate a time constant based on the equation of motion for domain walls.[2] The constants governing the equation of motion can be related to magnetic material properties, film thickness and the geometry of the domains. The results of these calculations show that the time constants observed, in fact, have a plausible explanation in terms of the known behavior of magnetic materials. As might be expected, the parameters which lead to small delays are generally those associated with small edge domains for the domain model considered.

Figure 2. Bitter pattern of domains in a thin film head pole piece.

EFFECTS OF MAGNETIC LAG

By considering the relationship between the flux, ϕ, linking the turns of the head and the magneto-motive force, u , across the recording gap, $u = R\phi$, a magnetic lag can be

Reprinted from *IEEE Trans. Magn.*, vol. MAG-15, pp. 1619-1621, Nov. 1979.

introduced by adding a term proportional to the time derivative of flux:

$$u = R(\phi + \tau\dot{\phi}) \qquad (1)$$

Since the time derivative of flux is proportional to the voltage sensed by the head, the derivative of Equation 1 implies:

$$V_O = V_\tau + \tau\dot{V}_\tau \qquad (2)$$

Where $V\tau$ is the response corresponding to a delay time τ and V_o, to the ideal signal for $\tau = 0$.

For an isolated pulse, the general solution to Equation 2 is given by

$$V_\tau(t) = \exp(-t/\tau)\int_0^t \exp(s/\tau)V_O\frac{ds}{\tau} \qquad (3)$$

Figure 3 shows the effects of various time constants on an ideal isolated pulse computed used this equation. The arbitrarily drawn pulse labeled $\tau = 0$ was deliberately made symmetrical with negative wings before and after the peak as expected for identical thin film pole tips[3] close to a thin media.

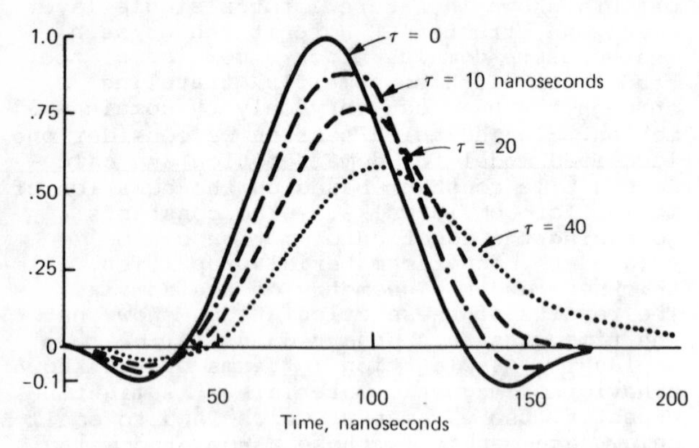

Figure 3. Isolated pulses from a thin film head subject to various delays.

In general, the effects of progressively long time constants are: (1) The peak amplitude decreases with increasing time constant such that the peak of the delayed pulse lies on the curve for V_O. This is a consequence of Equation 2. (2) The half height pulse width broadens slightly with increasing τ. (3) The negative swing in advance of the peak becomes more shallow with increasing τ; however, it never vanishes. (4) The negative swing trailing the peak vanishes for $\tau \geqq 20$ nanoseconds for the example shown in Figure 3. (5) However, from the integration of Equation 2 it can be shown that the total area under the pulse does not depend on τ.

Pulse forms such as those in Figures 1a and 1b can at least approximately be accounted for as members of a family of curves corresponding to different values of τ. For example, lower peak amplitudes are associated with larger trailing edge amplitudes, greater half widths, and more shallow leading edge undershoots. For several pulses in a series feature (5), the constant area under the curve, has been verified to within a few percent.[4]

Considering the superposition of a train of pulses such as those shown in Figure 3, it is clear that the intrinsic amplitude and resolution of the thin film head can be degraded by delays in the head response.

In detail, not all of the structure in the pulses can be accounted for as simple changes of τ. For example, none of the "wiggles" in the trailing edge of the peak in Figure 1b are present in Figure 3. However, features resembling these "wiggles" can be formed by superimposing two pulses with somewhat different τs. These two pulses might, for example, correspond to the response of the center and outer portions of the yoke of the head. Figure 4 shows such a pulse shape constructed from the curves in Figure 3 assuming that half the head responds with $\tau = 10$ nanoseconds and half with $\tau = 40$ nanoseconds.

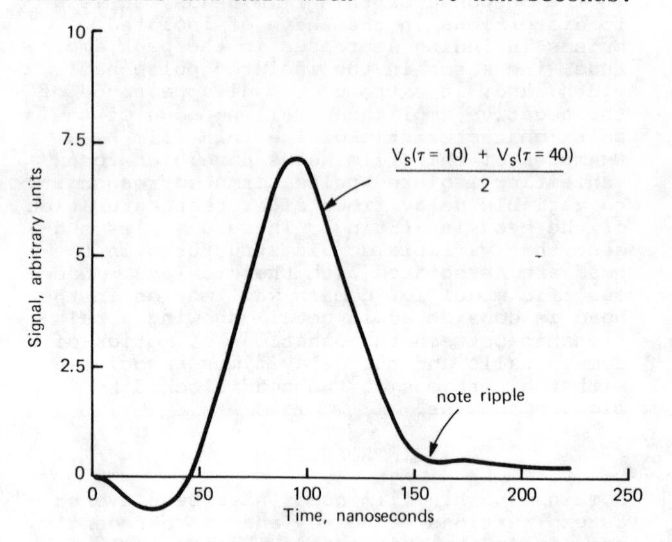

Figure 4. Superposition of two isolated pulses with delay times of 10 and 40 nanoseconds.

A MODEL FOR DOMAIN WALL MOTION

Assuming edge domains block a portion of the track width as shown in Figure 5, that the easy axis of the magnetic material is oriented across the track, and that the domain walls are pinned at the edges of the track for relatively small fields, an expression for the delay time, τ, can be developed. It will be assumed that the magnetostatic energy associated with magnetic poles appearing at the domain walls is large and that in equilibrium

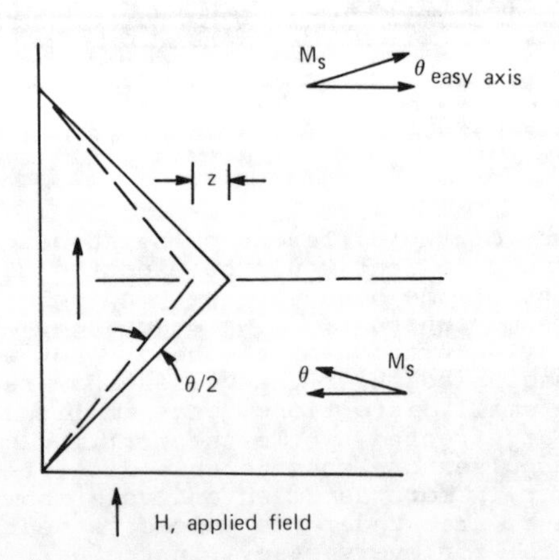

Figure 5. A closure domain pinned at the edge of the track in an applied field.

the walls will move to avoid creation of such magnetic poles. In equilibrium with a small applied field H and with no divergence of magnetization at the domain walls, the edge domain walls will rotate through an angle /2, where θ is the angle of rotation of magnetization in easy axis domains such that

$$H/H_k = \sin\theta \simeq \theta \quad (4)$$

This implies a motion z of the tips of the domains given by:

$$z \cong \theta D/2 \quad (5)$$

for small fields. Clearly, Equations 4 and 5 are somewhat arbitrary and valid only if the domain walls are pinned at the edges of the track. If the domains are freer to move, the distance moved by the tip of the domain will be greater than given by Equation 5. On the other hand, imperfections in the film could limit the motion even more.

For small reversible domain wall movements an equation of motion for domain walls can be written[5]:

$$\beta \frac{dz}{dt} + \alpha z = \Delta(\vec{H}\cdot\vec{M}) \quad (6)$$

where $\Delta(\vec{H}\cdot\vec{M})$ is the change in the $\vec{H}\cdot\vec{M}$ dot product from one side of the domain wall to the other. This is an approximate equation which neglects domain wall intertia, but includes a term for damping with coefficient β.

Considering the model for the case of equilibrium (dz/dt = 0) it can be shown from Equations (4) and (5) that the coefficient α in Equation (6) is

$$\alpha = \frac{2M_sH_k}{D} = \frac{4\times10^7}{D} \text{ erg cm}^{-4}, \quad (7)$$

for permalloy with the dimension D in micrometers.

The damping coefficient β is a sum of terms for two mechanisms, relaxation and eddy current damping. For permalloy films 1 micrometer or greater in thickness the latter should dominate. According to the eddy current theory for domain wall mobility given by Williams et al[6] the damping coefficient is

$$\beta = \frac{(1.052)\,32dM_s(4\pi M_s)_{\parallel}}{\pi^2c^2\rho} \quad (8)$$

where c is the speed of light, ρ is the film resistivity (20μΩ-cm for permalloy), $(4\pi M_s)_{\parallel}$ is the component of saturation induction parallel to the wall, and d is the film thickness. For permalloy:

$$\beta = 0.98\text{ d erg sec cm}^{-4} \quad (9)$$

for thicknesses d in micrometers.

For the fraction of the track width blocked by closure domains the flux increment in response to an applied field is proportional to z. It follows from equations 2 and 6 that the delay time associated with this fraction is given by

$$\tau = \beta/\alpha = 2.45\times10^{-8}\text{dD sec} \quad (10)$$

for permalloy with dimensions d and D in micrometers. This equation shows that for the model under consideration edge domains determine not only the fraction of the response subject to delay but also the delay time itself. It follows that the materials properties of the heads and the techniques used to fabricate it must be optimized to reduce these domains and thereby produce a more ideal, reproducible pulse forms.

REFERENCES

1. J. Lazzari and I. Melnick, IEEE Trans. Mag., MAG-7, 146 (1971).

2. For example, C. Kittel, Phys. Rev. 70, 965 (1946).

3. R. I. Potter, R. J. Schmulian, and K. Hartman, IEEE Trans. Mag., MAG-7, 689 (1971).

4. W. Nystrom, private communication

5. For example, C. Kittel and J. K. Galt, Solid State Physics 3, Academic Press, N. Y., p. 643

6. H. J. Williams, W. Shockley and C. Kittel, Phys. Rev. 80, 1090 (1950).

Magnetic behaviour of narrow track thin-film heads

W. F. Druyvesteyn, E. L. M. Raemaekers, R. D. J. Verhaar, and J. de Wilde

Philips Research Laboratories, 5600 MD Eindhoven, The Netherlands

J. H. J. Fluitman and J. P. J. Groenland

Twente University of Technology, Enschede, The Netherlands

The influence of the trackwidth on the performance of thin film heads has been tested. Results of experiments on the wafer have indicated an increase in the head efficiency with decreasing trackwidth. This was underlined by measurements of the head fringe field and tape recording experiments. A model which takes the domain structure into account has been developed to interpret this behaviour.

PACS numbers: 75.60.Ch, 85.70. – w

INTRODUCTION

Thin film heads are especially suited for use in recording systems with a high information density e.g. a narrow trackwidth. Therefore it is important to know the influence of the trackwidth w on its recording performance. The magnetic yoke of a thin film head has dimensions such that it will not be magnetised uniformly but will split up into domains, owing to the high demagnetising field. A study of this domain structure revealed [1] that it greatly depends on the trackwidth of the head.

(a) Physical aspects of this trackwidth dependence.

It is usual to have the easy axis (e.a.) of magnetisation parallel to the turn (fig.1). In this situation domains will occur with a period d which depends on w [6].

When a magnetic field H is applied perpendicular to the turn, domain wall displacement and rotation of the magnetisation will occur as illustrated in fig. 3. We assume that the coercive field is equal to zero so that domain wall displacement occurs at the lowest fields.

If $w > d$ domain wall displacement hardly contributes to the magnetisation in the direction of the field. In this case only rotation plays a role so that the permeability is determined by the ansiotropy field H_K, $\mu_0\mu_r = B_s/H_K$, where B_s is equal to the magnetic induction of the saturated strip. However, if w is reduced, wall displacement contributes to the permeability and $\mu_0\mu_r > B_s/H_K$.

If the e.a. is perpendicular to the turn only wall displacement governs the magnetisation process. Therefore no trackwidth dependence will be expected.

(b) Relation between efficiency and permeability.

For a thin film head the relation between the efficiency η and the permeability of a thin film head has been derived by Paton [2]. In general η increases with increasing μ_r, but in some heads the influence is very weak. In Table I we give some examples of heads with different poletip length p and height h.

Table I clearly shows that for some heads the influence of μ_r on η is quite strong (type I) while for other heads (type II) hardly an influence exists. In our experiments type I was chosen as we wanted to study the influence of the domain structure on η. However, in a recording system one mostly prefers to eliminate this effect and uses type II head.

TABLE I

	μ	500	800	1000	1500	2000
I	p = 1 μm g = 2 μm h = 20 μm	47	56	60	67	72
II	p = 4 μm g = 2 μm h = 5 μm	87	91	93	95	97

Efficiency η (in %) as a function of μ_r for two types of heads following [2].

MANUFACTURING OF THE HEADS

In order to measure the influence of w on the magnetic behaviour of a thin film head, special test circuits have been designed containing a series of identical single turn heads with different w. Heads have been processed with a symmetrical (two Permalloy shields) structure on oxydised silicon wafers. Non-magnetostrictive Permalloy films with a thickness of 1 μm are obtained by electrodeposition. A 0.06 μm thick sputtered NiFe layer serves as the plating base. Some typical magnetic parameters of this electrodeposited material are: $H_c = 80$ A/m; $H_K = 400$ A/m, while most films are made with a well-defined e.a.. From these layers the heads are formed by means of chemical etching or ion-milling. Sputtered SiO_2 is used as an insulator and sputtered MoAuMo as a conductor.

The heads are encapsulated and polished so that either the height h = 20 μm (type I) or h = 5 μm (type II).

WAFER TESTING

The magnetic properties of the heads with w ranging from 600 μm to 10 μm have been measured on the wafer.

In order to get a clear picture of the influence of w, it is necessary to measure a large number of identical heads. Therefore the electronic set-up was arranged around an Electroglas 900 wafer tester. The induced voltage v across the test turn (fig. 1) was measured as a function of the ac current i flowing in the writing turn. The frequency was 0.5 MHz. The effective permeability μ_r was found from the mutual inductance M by using the transmission line model [2].

In the heads with e.a. oriented parallel to the writing turn, all experiments indicate that μ_r increases with decreasing trackwidth. This is illustrated in fig. 1. Curve I represents the results of heads formed by means of chemical etching and curve II heads formed by ion-milling. No difference can be seen. If w is larger than 200 μm, only a slight increase can be detected. However, if $w < 50$ μm, especially in the circuits with w down to 10 μm, we see a dramatic increase in μ_r (fig. 1, curve III).

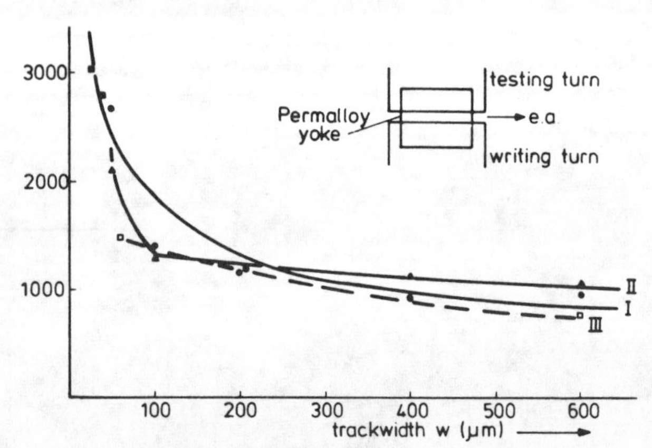

Fig. 1. The trackwidth dependence of μ_r as measured in wafer testing experiments.

On the other hand the heads with the e.a. perpendicular to the turn do not show this behaviour, but taking all heads on one wafer into account the standard deviation is very high.
The μ_r of these heads depends on the accidental occurrence of pinning points and structural imperfections of the Permalloy which was in agreement with the observed domain structures.

HEAD FRINGE FIELD

The head fringe field has been measured with a magnetoresistive transducer (MRT) and analysed in a way described earlier [3].
The fields on the energised heads were measured by placing them underneath the magnetoresistor at an accurately adjusted distance, and moving them laterally while keeping the separation at a constant magnitude. The thickness of the MRT was 0.03 μm and its height 2.8 μm. Since the latter value is large compared with the field inhomogeneities, the analysis of the field tracks takes the transducer response into account [4]. In this way 'field tracks' were produced as a function of head/transducer separation and as a function of writing current. The orientation of the transducer is such that the perpendicular field component is detected. Measurements were performed on heads with a gap length of 2.6 μm and pole tips of 1.15 and 0.9 μm respectively.

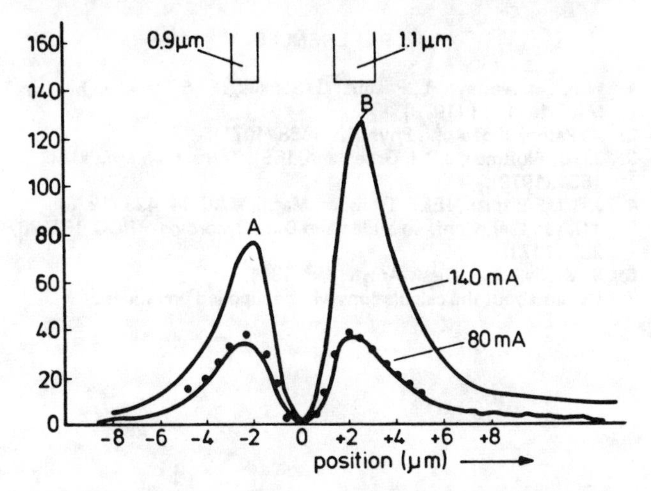

Fig. 2. Typical sensor output for head fringe measurements at a head/transducer distance of 1.0 μm. The solid curve is experimental, while the dots represent computed results. In the curve at I = 140 mA the thinnest Permalloy shield is saturated (peak A).

Occasionally these calculations must include the field components introduced by the transducer mirror image in the pole tips.
The relation between the writing current and the head field was found to be linear up to 100 mA. However, a small hysteretic effect was found indicating the presence of domains. At 100 mA saturation of the thinnest poletip starts. The current dependence of the peak amplitudes is illustrated in fig. 2. (Note that the response of the MRT is quadratic.)

The structure of the head field has been analysed by comparing the experimental numbers with the simulated results. The form of the theoretical field was calculated by means of the method of van Lier [5].
The curves were fitted by using the head efficiency as a parameter. The agreement between theoretical and experimental curves is good (cf. fig. 2).

TABLE II

w(μm)	wafer	field	recording 12 kHz	recording 18 kHz
100	62	63	59	55
200	55	55	59	51
400	45	45	46	49
600	45	45	45	45

Head efficiency (in %) from different experiments.

RECORDING EXPERIMENTS

Recording experiments on tape have been carried out with the same heads as used in the field measurements.

The reproducing efficiency was measured on video tape ($H_c = 55 \times 10^3$ A/m). A harmonic signal was recorded with a ferrite head. The tape velocity was 19 cm/s and the recorded frequencies were 12 kHz and 18 kHz respectively. These signals were reproduced by means of the thin film heads. The detected levels were compared with the signal from a conventional ferrite head with known efficiency. The resulting efficiency η has a function of the trackwidth of the head as been given in Table II. It is clear that all numbers in this table show the same increase with decreasing trackwidth.

We tried to do writing experiments with the same heads, although the low efficiency and the thin Permalloy yoke (saturation) limited the head field. Therefore audio tape ($H_c = 26 \times 10^3$ A/m) was chosen. The written information was reproduced with a ferrite head. Again the same track dependence was found: the non-linear behaviour of the recording process even emphasised this effect.

The same recording experiments have been performed with type II heads (cf. Table I).
Indeed when w was varied between 50 μm and 600 μm the head efficiency was constant both in the writing and reading measurements.

CALCULATIONS

A model has been developed for calculating the influence of w on μ_r. To characterise the magnetic behaviour of the head material, we consider it as an infinite strip with width w placed in an external magnetic field parallel to this strip (fig. 3). In the yoke of a thin film head, flux circulates in such a way that the direction of the magnetisation in the upper and lower shield are opposite to each other and are closed through the air gap and by the magnetic material in the back of the head. A small amount of the total flux will be closed outside the head (head fringe field) but for all the heads we used it can be shown that this flux is negligible compared with the total flux running in the head. Therefore the demagnetising field at the front of the head will be very small.
If the easy axis of the magnetisation is perpendicular to the strip, domains will occur with a period d depending on w [6].

The permeability is calculated by minimising the total magnetic energy with respect to the rotation angle α and the geometry of the closure domains e.g. ℓ_1 (or ℓ_2). Three energy terms are considered: the anisotropy energy, the wall energy and the magnetic field energy.

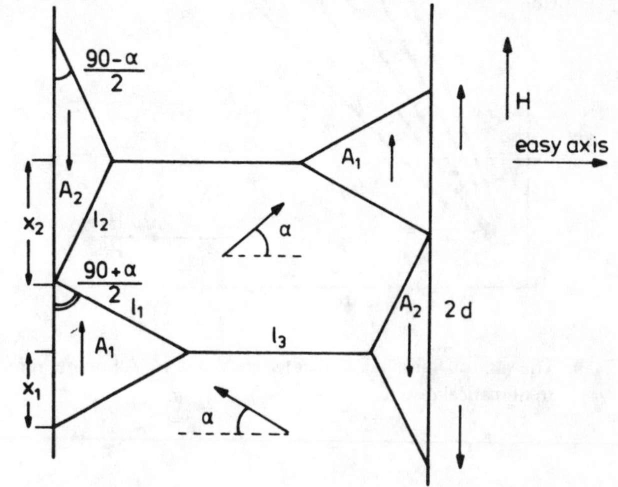

Fig. 3. Domain configurations as used in the calculations.

For the total energy per unit length E' we write:

$$E' = \frac{Kd}{2t}\,[(\ell_1{}^2 + \ell_2{}^2)\cos\alpha + (2wd - \ell_1{}^2\cos\alpha - \ell_2{}^2\cos\alpha)\sin^2\alpha]$$
$$+ \frac{\gamma d}{t}[2\ell_1 \sin^2(45 - \frac{\alpha}{2}) + 2\ell_2 \sin^2(45 + \frac{\alpha}{2}) + \Big[w - \ell_1 \sin(45 + \frac{\alpha}{2})$$
$$- \ell_2 \sin(45 - \frac{\alpha}{2})\Big]\cos^2\alpha] + \frac{M_s Hd}{t}\,[-(wd - \frac{\ell_1{}^2 + \ell_2{}^2}{2}\cos\alpha)\times$$
$$\sin\alpha - \tfrac{1}{2}\ell_1{}^2\cos\alpha + \tfrac{1}{2}\ell_2{}^2\cos\alpha] \qquad (1)$$

where t is the thickness of the films, K the uniaxial anisotropy constant, γ the wall energy of an 180° Bloch wall, M_s the saturation magnetisation of the strip.
When applying a field H, the number of domains remains constant, i.e. we assume that this d is fixed:

$$d = \sqrt{\frac{\gamma w}{K}} = \ell_1 \cos(45 + \frac{\alpha}{2}) + \ell_2 \cos(45 - \frac{\alpha}{2}) \qquad (2)$$

The energy thus depends on two variables, e.g. α and ℓ_1. For a stable domain configuration the first derivations of E' with respect to α and ℓ_1 should be zero. Solving α and ℓ_1 from these equations is straightforward, therefore we shall not reproduce this here [7]. The results in terms of M/M_s as a function of H for different values of w are given in fig. 4. In this model calculation the parameters were: $H_K = 400$ A/m, $\gamma = 2\times10^{-3}$ Jm^{-2}, K = 2 Jm^{-3} and the trackwidth varied between 40 µm and 100,000 µm.
The relative permeability deduced from these curves increases with decreasing trackwidth, even in this simple model.

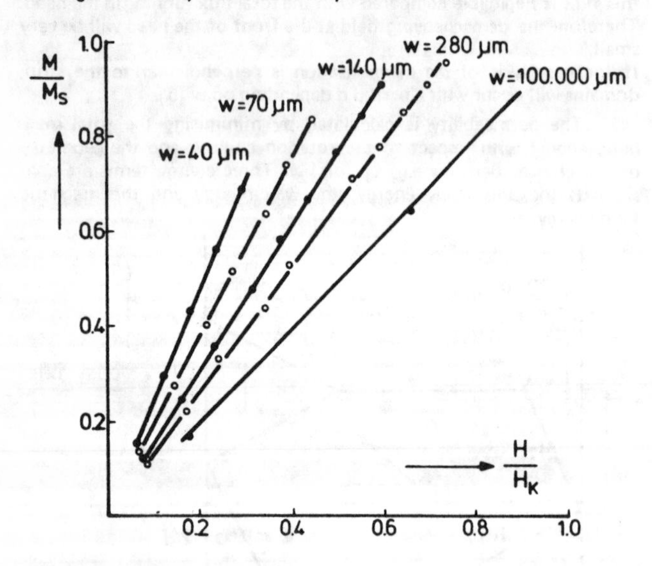

Fig. 4. The almost linear relation between M and H. All points refer to numerical results.

CONCLUSIONS

In the theoretical analysis [2] of the magnetic behaviour of thin film heads homogeneous magnetisation is assumed whereas the relation between B and H is characterised by the material constant μ_r. However, it is known that the NiFe yoke of the head is not uniformly magnetised but is split up into domains. We have shown that nevertheless several experimental results can be described with μ_r as an adjustable parameter. The permeability depends on the trackwidth of the head, if the easy axis of the magnetisation is parallel to the turn (fig. 1). From wafer testing experiments we conclude that the efficiency of a narrow track head should increase with decreasing trackwidth. Head fringe field measurements and tape recording experiments are in agreement with this fact and have shown the reliability of wafer test measurements.
The fringe field of the head is in agreement with the model calculations of Van Lier [5]. The dead layer in the Permalloy yoke, if present, is within the experimental error of 0.1 µm. The efficiency values of the heads achieved in the several experiments are in reasonable agreement with each other (Table II), and the model calculation which takes the domain structure into account, shows the same increase in the μ_r with decreasing trackwidth.

REFERENCES

1. W.F. Druyvesteyn, L. Postma, G. Somers, IEEE Trans. on Magn., MAG-15, 1613 (1979).
2. A. Paton, J. of Appl. Phys. **42**, 5868 (1971).
3. J.H.J. Fluitman, J.P.J. Groenland, IEEE Trans. on Magn., MAG-15, 1634 (1979).
4. J.H.J. Fluitman, IEEE Trans. on Magn., MAG-14, 433 (1978).
5. J.C. van Lier, Conf. on Video and Data Recording, PROC. IRE **26**, 285 (1973).
6. S. Middelhoek, Thesis Amsterdam 1961.
7. Details about the calculations will be supplied on request.

DIGITAL MAGNETIC RECORDING THEORY

Robert I. Potter*

ABSTRACT

A brief review of digital magnetic recording theory is given with emphasis on the analysis of ferrite, inductive thin film, and magnetoresistive heads. The utility of several approaches is demonstrated by three calculations: (1) resolution and signal amplitude of a shielded magnetoresistive head, (2) readback properties of a single sheet of permalloy with adjacent current conductor, and (3) analysis of spurious pulses generated by outside corners of ferrite heads. A discussion of high density digital recording is given, and it is concluded that the magnetoresistive head offers the greatest possible areal density of information. This density is conservatively estimated to be 1.6×10^7 flux reversals per cm^2 (10^8 per in^2) at a head-medium separation of 0.125μm. Experimental data in support of this conclusion are given.

I. INTRODUCTION

The technical goal of digital recording is to store and reliably recover as many magnetization reversals or "bits" per unit area of storage medium as possible. This inherently nonlinear process is sometimes called saturation magnetic recording because the record current is switched from one extreme to the other in an attempt to cause the remanent magnetization of the storage medium throughout all or part of its depth to do the same.

The purpose of this paper is threefold. The first is to give a brief review of digital magnetic recording theory with emphasis on methods of approach rather than detailed results. A recent review of a more general nature and from a somewhat different viewpoint is given by Mallinson [1]. Magnetic recording models are considered in Section II and head field calculations in Section III. The second is to illustrate several of these methods by three examples: resolution of the shielded magnetoresistive (MR) head (Section IV; see that section for meaning of "shield"), fringe field and readback properties of a single sheet of permalloy with adjacent conductor (Section V), and spurious pulses generated by outside corners of ferrite heads (Section VI).

The discussion of high density digital magnetic recording in Section VII is the third purpose. The focus is on the shielded MR head, which offers greater areal density than either ferrite or thin metallic film heads.

II. DIGITAL MAGNETIC RECORDING MODELS

The current state of digital recording theory is schematically indicated in Fig. 1. The head fringing field $\vec{H}$ is the backbone of all models. It is the means by which the data are recorded and is useful in calculating the readback waveform $e(\bar{x})$ via the principle of reciprocity. When the head field per unit coil current and the storage medium magnetization are known, then the reciprocity principle provides the formula [2]

$$e(\bar{x}) = \mu_o vW \int_{-\infty}^{\infty} dx \int_{d}^{d+\delta} dy \frac{\partial \vec{M}(x-\bar{x},y)}{\partial \bar{x}} \cdot \vec{H}(x,y) \quad (1)$$

where $\bar{x} \equiv vt$ is the location with respect to the head of

Manuscript received by IEEE April 22, 1974.

* IBM Thomas J. Watson Research Center, Yorktown Heights, New York 10598.

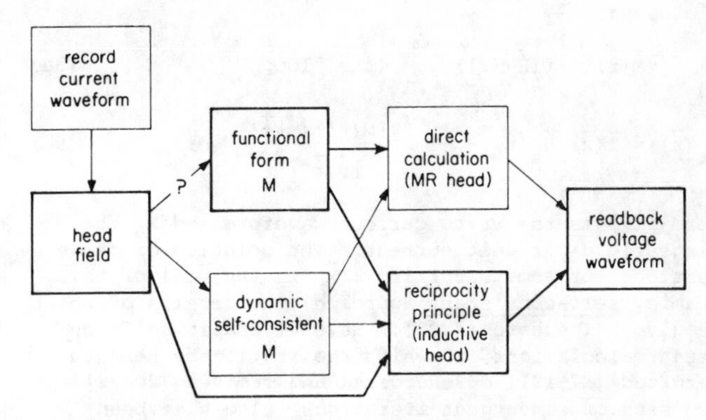

Fig. 1. Schematic of the present state of magnetic recording theory. Readback voltage calculations with storage medium magnetization $\vec{M}$ specified dominate the literature (heavy outlines).

a coordinate system in the medium, W the track width, d the head-medium separation, δ the medium thickness, and $\vec{M}$ the magnetization in the presence of the high permeability read head, which causes a slight remagnetization [3,4] from the demagnetized state. This formula indicates that the isolated digital pulse is a "smeared out" rendition of the appropriate head field component and provides considerable insight into the readback process. For example, if

$$M_x(x) = \frac{-2M_r}{\pi} \tan^{-1} \frac{x}{a}, \quad M_y = 0 \quad (2)$$

and δ is negligibly small, then for <u>any</u> $\vec{H}$ the output voltage is [5]

$$e(\bar{x}) \propto M_r \delta H_x(\bar{x}, d + a). \quad (3)$$

The most difficult problem in magnetic recording theory is determining $\vec{M}(\mathbf{r})$. There are basically two ways to proceed. The <u>functional form</u> theories [6-13] postulate the basic character of $\vec{M}$ but include one or more adjustable parameters. The most popular and convenient choice for digital recording is

$$M_x = \frac{2M_r}{\pi} \sum_{n=1}^{N} (-1)^n \tan^{-1} \frac{x-x_n}{a} + \frac{M_r}{2} [(-1)^N + 1], \; M_y = 0, \quad (4)$$

where x_n (n=1, 2, ... N) specifies the location of each transition and a is a transition length parameter that is chosen by fitting isolated pulses, by making demagnetization arguments [14,15], or by considering [13] both the demagnetizing field and the maximum head field gradient. Writing $\vec{M}$ as a sum of arctangents is equivalent to superimposing isolated arctangent-magnetization pulses.

An alternate approach is to Fourier transform the essentially digital magnetization pattern, multiply by the Wallace factors [6], and take the inverse transform. Such an approach may be useful when additional wavelength and/or frequency dependent factors are to be considered.

Functional form theories are not capable of predicting the subtle yet important effects arising from the recording process, such as the asymmetric shift of two adjacent pulses when the isolated pulse is symmetrical. Such phenomena require at least a <u>dynamic</u> approach [16], dynamic in the sense that the relative motion between head and medium is considered but not in the sense that time derivatives in Maxwell's equations are included. The dynamic models follow the magnetization at each point in the recording medium as it passes by the record head.

Reprinted from *IEEE Trans. Magn.*, vol. MAG-10, pp. 502-508, Sept. 1974.

An additional and enormous complication to the dynamic approach is due to the demagnetizing field, which causes the magnetization and total field to be related via

$$\vec{M}(\vec{r}) = f(\vec{H}_t(\vec{r})) \qquad \text{(the "loops")} \tag{5a}$$

and

$$\vec{H}_t(\vec{r}) = I(\bar{x})\,\vec{H}_a(\vec{r}) - \frac{1}{4\pi}\int \frac{\nabla\cdot\vec{M}(\vec{r}\,')(\vec{r}-\vec{r}\,')}{|\vec{r}-\vec{r}\,'|^3}\,d\vec{r}\,' \tag{5b}$$

where $I(\bar{x})$ is the write current waveform and $\vec{H}_a$ the fringe field per unit current. The solution of these equations for some specified $I(\bar{x})$ is the goal of this dynamic, self-consistent approach, the details of which are given elsewhere [4,23]. Several static self-consistent calculations, in which the stationary head is energized [17-19], or energized and removed [20-22] (two sets of convergent iterations) also have been reported. The necessity of including the demagnetizing field is shown in Fig. 2 where results from the Potter and Schmulian [4] model for thin metallic media are given with the demagnetizing field alternately present and suppressed.

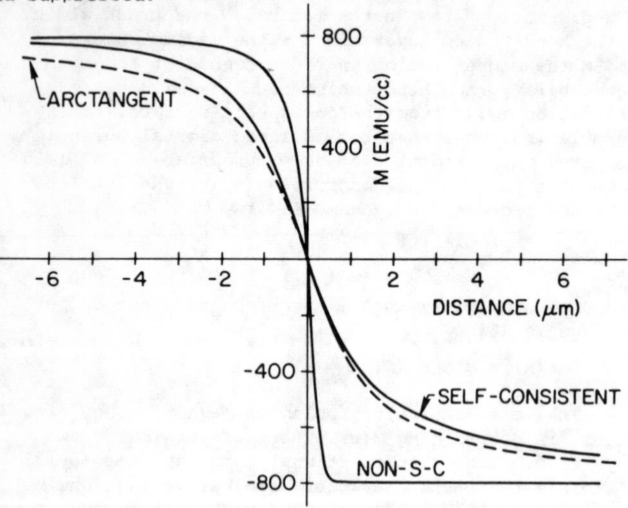

Fig. 2. Comparison of self-consistently and non-self-consistently computed magnetization transitions in thin media using the Potter-Schmulian model. Here, the image term in the demagnetizing field (and consequently head removal and replacement) is neglected for simplicity. Squareness parameter M_r/M_s=0.9, $I(\bar{x})|_{max}$ = 196.8 ma; see reference [4] for other details. Arctangent transition with parameter a=1.13 μm is shown dashed for comparison. Curves are displaced so that M(O)=0.

The situation for thick media is not so clear, because of the large y component of magnetization and the y dependence of both components. The recorded magnetization is quite complicated as the scaled-up experimental results of Tjaden [23] indicate. A satisfactory theoretical treatment of the recording process in thick media has not yet been given.

There exists a third class of models [24,25] in which it is assumed very localized magnetization transitions are written which subsequently broaden when beyond the influence of the head. In one case [25] each initial transition is assumed trapezoidal and then the entire medium is allowed to demagnetize with the final state computed by harmonic analysis. In these models it is assumed that the proximity of the record head reduces the demagnetizing field to a negligible value, an assumption that the self-consistent calculations (with separation parameter d comparable to gap length) do not support [4].

III. HEAD FIELD CALCULATIONS

The objective of a head field calculation is to determine the magnetic intensity surrounding a structure that is usually topologically equivalent to a toroid with a small gap, given the current density (the coil) and the relationship $\vec{M} = \overline{\overline{\chi}} \cdot \vec{H}_t$ between head magnetization $\vec{M}$ and total internal field $\vec{H}_t$. The elements of χ may be complex. The imaginary part of $\overline{\overline{\chi}}$ (denoted by $Im\{\overline{\overline{\chi}}\}$) has little effect on the spatial character of the fringe field when head efficiency is in a useable range [26,27], and the finite risetime caused by $Im\{\overline{\overline{\chi}}\}$ (or any other mechanism) has only slight effect on the length of the recorded magnetization transition [4]. Moreover, for thin film heads $Im\{\overline{\overline{\chi}}\}$ is negligible.

The aspect ratio of most heads is such that two-dimensional calculations in a plane normal to the track-width direction suffice. Thin film heads are fabricated with the easy axis of magnetization normal to this plane, and for such heads as well as polycrystalline ferrite heads $\overline{\overline{\chi}}$ is isotropic within it. This reduction in dimensionality is not possible, for example, when treating the "side reading" problem, which requires knowledge of at least one component of the head field as a function of x, y, and z.

A thorough discussion of the mathematical approaches available for computing head fields is beyond the scope of this paper. We discuss, therefore, only those that seem particularly useful, and restrict the number of dimensions to two.

Scalar Potential Specified

A magnetic scalar potential satisfying Laplace's equation may be defined in simply-connected current-free regions. This differential equation may be solved numerically or analytically once the scalar potential is specified on the boundary. The approach is most useful when the permeability can be assumed infinite, for then the pole-pieces become equipotentials with potential difference $\alpha I_o/g$, where α is the head efficiency (an ad hoc addition to the problem), I_o the total current, and g the gap length.

Karlquist's formula [28] is a famous example of this approach. The approximation of Karlquist can be stated several ways, but one of the more useful is: Replace the actual head boundary by the plane y=0. On this plane, let the potential be (with α=1 for simplicity)

$$\phi(x,o) = \begin{cases} I_o/(2g) & (x \le -g/2) \\ (I_o/g)\,x & (-g/2 \le x \le g/2) \\ -I_o/(2g) & (g/2 \le x)\ . \end{cases} \tag{6}$$

The expression

$$H_x(x,y) = \frac{I_o}{\pi g}\left[\tan^{-1}\frac{x+g/2}{y} - \tan^{-1}\frac{x-g/2}{y}\right] \tag{7}$$

follows from these two assumptions via standard boundary-value problem theory [29].

Karlquist's approximation is generalized to multi-gap structures in an obvious way. It is not necessary to recalculate $\vec{H}$ from ϕ for these cases; simply sum suitably displaced and weighted expressions of the Eq. (7) type. The technique works because potentials on a common boundary (the plane y=o) can be superimposed, and works only within the spirit of the original Karlquist approximation. The shielded magnetoresistive head is treated in Section IV using this method.

Conformal transformations are useful for several geometries. They can be used when current conductors are present [5], and in principle even when the pole pieces are not equipotentials. But the complexity of the transformations is overwhelming for all but the most idealized of structures.

Current Density and Susceptibility Specified

Maxwell's equations must be solved numerically when the head boundary cannot be simplified to the point where analytic techniques are useful. Once this threshold is crossed the goal should be a computer program of maximum flexibility and usefulness. To this end the integral form of the equations is probably more suitable than the differential form. The problem superficially resembles the self-consistent recording process calculations described above, but with applied field generated by the coil and with total field and head magnetization related by $\vec{M} = \chi(H)\vec{H}$. The problem is vastly easier, however, and for constant χ is equivalent to a matrix diagonalization.

Suppose the head consists of N uniformly magnetized elementary volume elements, which need not be of uniform size or shape. Let the field intensity at the center of the ith element due to the jth element be

$$H_x(i) = C_{xx}(i,j)\ M_x(j) + C_{xy}(i,j)\ M_y(j) \quad (8a)$$

$$H_y(i) = C_{yx}(i,j)\ M_x(j) + C_{yy}(i,j)\ M_y(j) \quad (8b)$$

where the C's are easily calculable analytic functions of the vectors $\vec{r}_i$ and $\vec{r}_j$, which are denoted by i and j for brevity. They also depend on the size, shape, and orientation of the jth element, which may, for instance, be square, rectangular, or triangular. They need to be evaluated only once per calculation. The field intensity at the ith element is also

$$H_x(i) = \frac{1}{\chi} M_x(i) - H_x^a(i) \quad (9a)$$

$$H_y(i) = \frac{1}{\chi} M_y(i) - H_y^a(i) \quad (9b)$$

where $\vec{H}^a$ is the known field produced by the coil. Therefore the problem is equivalent to solving the matrix equation

$$\begin{pmatrix} [C_{xx}(1,1) - \frac{1}{\chi}] & C_{xy}(1,1) & C_{xx}(1,2) & C_{xy}(1,2) & \cdots \\ C_{yx}(1,1) & [C_{yy}(1,1) - \frac{1}{\chi}] & C_{yx}(1,2) & C_{yy}(1,2) & \cdots \\ C_{xx}(2,1) & C_{xy}(2,1) & [C_{xx}(2,2) - \frac{1}{\chi}] & C_{xy}(2,2) & \cdots \\ C_{yx}(2,1) & C_{yy}(2,1) & C_{yx}(2,2) & [C_{yy}(2,2) - \frac{1}{\chi}] & \cdots \\ \vdots & \vdots & \vdots & \vdots & \end{pmatrix} \begin{pmatrix} M_x(1) \\ M_y(1) \\ M_x(2) \\ M_y(2) \\ \vdots \end{pmatrix} = - \begin{pmatrix} H_x^a(1) \\ H_y^a(1) \\ H_x^a(2) \\ H_y^a(2) \\ \vdots \end{pmatrix} \quad (10)$$

The field in a plane adjacent to the pole faces (or elsewhere) is readily calculable from the 2N magnetization components using the same functions C_{xx} and C_{xy}. When the field dependence of χ is to be included an interative solution is required, but this, too, can be done. The integral approach is flexible and can be generalized in several ways. Some sample calculations are given below.

IV. SHIELDED MAGNETORESISTIVE HEAD

The theory of the unshielded magnetoresistive (MR) head is given by Hunt [30]. The geometry of the shielded MR head is shown in Fig. 3. The purpose of the shields is not to reduce electromagnetic interference from external sources; rather, it is to shield the MR stripe from the approaching transition until the last possible moment and thereby increase the resolving ability of the head. In this section, expressions for the readback voltage of the shielded MR head are derived for both sinusoidal and arctangent storage medium magnetization.

Suppose the head material is highly permeable and imagine an energized coil wound around the MR element as shown in Fig. 3. The gap field near the shield faces would be essentially constant and the magnetic scalar potential would be as shown in that figure.

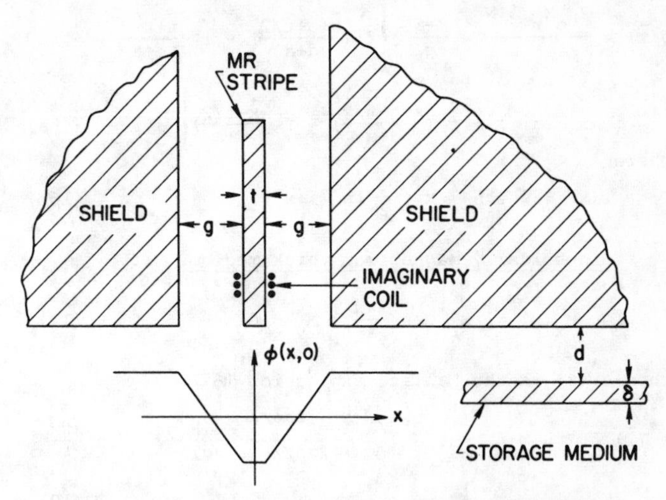

Fig. 3. Cross-section of shielded magnetoresistive head, showing coil that is imagined to exist for the purpose of calculating flux entering MR stripe from storage medium using reciprocity principle. Scalar potential at y=0 that would exist were this coil energized is shown according to the generalized Karlquist approximation (see text).

This is an example of the generalized Karlquist approximation. If the imaginary coil is located on the center of the MR element then the gap field is $H_g = I_o/2g$ where I_o is the total current. But if the coil is fairly close to the shield faces and if the MR element height L is great (L/g>>1) then H_g approaches I_o/g. The fringe field per unit current is

$$H_x(x,y) = H_x^*(x + \frac{g+t}{2}, y) - H_x^*(x - \frac{g+t}{2}, y) \quad (11)$$

where g is the separation between stripe and shields, t the stripe thickness, and H_x^* the Karlquist expression, Eq. (7), with I_o=1.

The flux entering the MR element from the storage medium is equal to the flux through the coil wound around its end and is calculable by reciprocity:

$$\phi(\bar{x}) = \mu_o W \int_{-\infty}^{\infty} dx \int_{d}^{d+\delta} dy\ \vec{M}(x-\bar{x},y) \cdot \vec{H}(x,y). \quad (12)$$

It is not necessary to perform these integrations because the flux is also obtainable from

$$\phi(\bar{x}) = \frac{1}{v} \int_{-\infty}^{\bar{x}} [e^*(\bar{x} + \frac{g+t}{2}) - e^*(\bar{x} - \frac{g+t}{2})]\ d\bar{x} \quad (13)$$

where $e^*(\bar{x})$ is the output voltage for the single-gap inductive head. This latter approach saves some effort provided e* is known. Although Eq. (13) can be integrated for a wide variety of cases, we restrict our attention to two of greatest interest.

Arctangent transition in thin medium. If

$$M_x(x) = -\frac{2M_r}{\pi} \tan^{-1}\frac{x}{a},\ M_y = 0 \quad (14)$$

then the inductive-head output in MKS units is [10-12]

$$e^*(\bar{x}) = \frac{2\mu_o M_r \delta W}{\pi} \frac{v}{g} [\tan^{-1}\frac{\bar{x}+g/2}{d+a} - \tan^{-1}\frac{\bar{x}-g/2}{d+a}] \quad (15)$$

and according to Eq. (13) the flux entering the MR element is

$$\phi(\bar{x}) = \frac{2\mu_o M_r \delta W}{\pi} \frac{d+a}{g} \left\{ f\left(\frac{\bar{x}+g+t/2}{d+a}\right) - f\left(\frac{\bar{x}+t/2}{d+a}\right) + f\left(\frac{\bar{x}-g-t/2}{d+a}\right) - f\left(\frac{\bar{x}-t/2}{d+a}\right) \right\} \qquad (16a)$$

where

$$f(x) \equiv x \tan^{-1} x - \frac{1}{2} \ln (1+x^2). \qquad (16b)$$

Sinusoidally magnetized thick medium. If

$$M_x(x) = M_r \operatorname{Cos} \frac{2\pi x}{\lambda}, \; M_y = 0 \qquad (17)$$

then $e^*(\bar{x})$ is the Wallace expression [6]

$$e^*(\bar{x}) = 2\pi\mu_o M_r \delta W \frac{v}{\lambda} e^{-2\pi d/\lambda} \frac{1-e^{-2\pi\delta/\lambda}}{2\pi\delta/\lambda} \frac{\sin \pi g/\lambda}{\pi g/\lambda} \sin \frac{2\pi\bar{x}}{\lambda} \qquad (18)$$

and

$$\phi(\bar{x}) = 2\mu_o M_r W\delta \sin \frac{\pi(g+t)}{\lambda} e^{-2\pi d/\lambda} \frac{1-e^{-2\pi\delta/\lambda}}{2\pi\delta/\lambda} \times \frac{\sin \pi g/\lambda}{\pi g/\lambda} \sin \frac{2\pi\bar{x}}{\lambda}. \qquad (19)$$

These expressions for $\phi(\bar{x})$ are related to the output signal as follows. The resistivity of ferromagnetic alloys (e.g. permalloy) is anisotropic and depends on the angle θ between magnetization and current density according to [31]

$$\rho = \rho_o + \Delta\rho_{max} \cos^2\theta = \rho + \Delta\rho_{max} \left[1 - \left(\frac{M_y}{M_s}\right)^2\right], \qquad (20)$$

where M_s is the saturation magnetization of the stripe and y is measured in the stripe-height direction, normal to the recording medium. This quadratic response is usually linearized by applying a bias field so that $\theta \simeq \pi/4$ in the absence of a signal field. Then, and also because within the stripe H << M,

$$\rho(y) \simeq \rho_o + \frac{1}{2} \Delta\rho_{max} - \sqrt{2}\, \Delta\rho_{max} \frac{\phi(y;\bar{x})}{\mu_o M_s tW} \qquad (21)$$

where $\phi(y;\bar{x})$ is the signal flux within the stripe at a vertical distance y from the face of the head when the storage medium coordinate system is at $x = \bar{x}$.

The flux within the stripe is not uniform because of leakage to the shields. This aspect of the problem can be treated via the one-dimensional "transmission line" theory proposed by Paton [32] for thin film heads. Let L be the vertical height of the stripe. The boundary conditions are $\phi(o;\bar{x})=\phi(\bar{x})$ [Eq. (16) or (19)] and $\phi(L;\bar{x})=0$. The characteristic length is $\beta = (tg\mu_r/2)^{1/2}$. When $L/\beta \leq 1$ the flux within the stripe decreases from $\phi(\bar{x})$ to zero in essentially linear fashion, and the average flux is $1/2\ \phi(\bar{x})$. The expression for arbitrary L/β is easily derived but not particularly useful since the criterion for maximum output consistent with tractable fabrication is $L/\beta \simeq 1$. When this condition is satisfied the shielded MR head output signal is

$$e(x) = I_o \Delta R(\bar{x}) \simeq JW \langle\Delta\rho(y;\bar{x})\rangle_{av}$$

$$\simeq -JW \Delta\rho_{max} \frac{\phi(\bar{x})}{\sqrt{2}\,\mu_o M_s tW} \qquad (22)$$

where J is the stripe current density and $\phi(\bar{x})$ is the flux entering the stripe, Eq. (16) or (19) as appropriate. The ratio $\sqrt{2}\ \phi_{max}/\mu_o M_s tW$ could be as high as one-quarter without causing serious distortion of digital pulses, because the resistivity of real films does not abruptly approach ρ_o as predicted by Eq. (20) and because the magnetic bias point could be adjusted for optimum symmetry of positive and negative pulses. Moreover, only the stripe edge nearest the storage medium experiences large signal fields. When this criterion is satisfied the peak-to-peak amplitude of isolated digital pulses is

$$e_{pp} = \frac{1}{4} JW \Delta\rho_{max} \simeq 90 \text{ volts per meter track width} \qquad (23)$$

assuming $J = 5\times10^{10}$ A/m^2 and $\Delta\rho_{max} = 7\times10^{-9}$ Ω-m for 200Å thick 81 Ni - 19 Fe permalloy [33]. Reliability tests [34] indicate that provided oxidation does not occur the mean time to failure for 200Å films at 85°C and $J \leq 10^{11}$ A/m^2 is of order 10^3 years.

The resolving ability of the shielded MR head is perhaps best indicated by Eqs. (22) and (19). In addition to the Wallace separation, thickness, and gap-length loss factors, Eq. (19) contains $\sin[\pi(g+t)/\lambda]$ in place of v/λ. This is an additional geometry-dependent loss factor that is peculiar to the two-gap head. Aside from this difference, the wavelength dependence of the shielded MR head is indistinguishable from that of the standard inductive head. In particular, the first gap null occurs when $\lambda \simeq g$ and effective gap-length is roughly one-half the shield-to-shield spacing since t is negligibly small (∿ 200Å).

It also is apparent from Eq. (16) that the shapes of the isolated pulse from the shielded MR and inductive heads are nearly the same, and that they become identical in the limit as g and t approach zero. The point need not be labored because the readback process is linear and thus whatever conclusions are drawn from the sine wave approach are also valid in the realm of digital recording and vice versa.

The output amplitude of the MR head is velocity independent but not frequency independent if velocity is held constant. The output depends on wavelength as it must because the source of the signal is the divergence of the storage medium magnetization. The amplitude is also limited by saturation and, unlike for inductive heads this must be considered in choosing the recording medium. The flux available, $2\mu_o M_r \delta W$, sometimes is more than sufficient to saturate the MR stripe, even after accounting for readback losses and especially when thin high-moment storage media are considered. This allows a reduction in $M_r\delta$ that is beneficial to the linear storage-density capacity of the medium and also guarantees that the condition $\sqrt{2}\phi_{max}/\mu_o M_s tW = 0.25$ can be met. In fact, this condition (with possibly an adjustment in the numerical value) provides the design criterion for the storage medium $M_r\delta$ product. For example, if g = d = 0.5 μm and a = 1.13 μm then with the aid of Eq. (16) the relationship $M_r\delta = 0.9\, M_s t$ is obtained.* Further, if $M_r = M_s \simeq 800$ EMU/cm^3 and t = 200Å, then an appropriate storage medium thickness would be about 180Å.

V. PERMALLOY SHEET WITH ADJACENT CONDUCTOR

The example in this section is chosen to demonstrate the integral approach to head field calculations and the saving in labor and insight afforded by the principle of reciprocity. The structure can be interpreted as a single-turn thin-film head without the second permalloy layer. The objective is to calculate the response of this structure to a step transition in a thin medium separated by d = 0.5 μm (or an arctangent

---

* This is not a precise argument because a is dependent on Mδ. Choice a=1.13 μm is discussed in Section VII.

transition with d+a = 0.5 μm), and to compare it to the response of the completed head. The structure and surrounding field is shown in Fig. 4 with the conductor energized (left) and with an isolated transition as the field source (right).

A direct calculation of the readback signal would require a large number of computations with different values of the transition location $\bar{x}$. For each computation, the flux linking the single-turn coil would have to be computed, taking into consideration the finite extent of the conductor. The formula for the signal voltage is

$$e(\bar{x}) = \mu_o vW \frac{\partial}{\partial \bar{x}} \left\{ \int_a^\infty dy\, H_x(y,\bar{x}) - \frac{1}{b-a} \int_a^b dy \int_a^y dy'\, H_x(y',\bar{x}) \right\} \quad (24)$$

where H_x is the field component (Fig. 4, right) perpendicular to the center plane of the coil located between y=a and y=b. This formula is derivable from the integral form of Faraday's law assuming (b-a)/W<<1.

The reciprocity principle gives

$$e(\bar{x}) = 2\mu_o vW\delta M_r H_x(\bar{x},-d) \quad (25)$$

where $2\delta M_r$ is the strength of the transition and H_x is the horizontal field component per unit current at y=-d (Fig. 4, left). This is compared with the signal from the completed head in Fig. 5. The signal from the completed head is slightly asymmetrical because the second permalloy layer bulges over the conductor.

Fig. 4. Permalloy sheet with adjacent conductor, or single-turn film head with second permalloy layer omitted. Left: field intensity with conductor energized. Right: field intensity with storage medium as source. Arrow length is proportional to $C_o + \ln(1+H)$ where C_o is a convenient constant.

VI. SPURIOUS PULSE GENERATED BY OUTSIDE CORNER OF READBACK HEAD

The problem considered in this section is the spurious readback pulse generated as an isolated magnetic transition passes under the sharp outside corner of a ferrite head. Conceptually, the problem is no different from that presented by the finite pole-tip length head [5,35,36] provided head permeability is

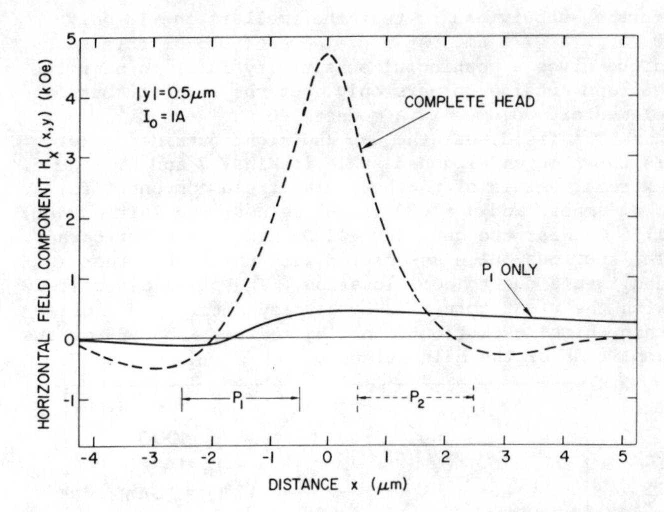

Fig. 5. Horizontal field component for permalloy sheet and conductor (solid curve) and complete head (dashed curve). Output voltage for arctangent transition in thin medium with d+a = 0.5 μm is proportional to these curves.

assumed infinite. The approach is to calculate the fringe field, where here the term is used in the broader sense to mean the entire magnetic field beneath the pole-pieces. According to the reciprocity principle all readback characteristics, including the spurious pulses, are calculable from this field.

We are concerned here with a secondary effect that takes place on the order of 10^2 gap lengths away from the readback gap. The infinite permeability approximation is questionable in this case, and coil location is important. The problem is best attacked from a numerical approach using the method described in Section III.

The problem is difficult from a numerical viewpoint because of the great disparity in distances: the magnetic field is needed at a distance of only a fraction of a micron from an object $\sim 10^3$ microns in length. The volume element size must be small compared to distances over which the magnetization appreciably changes; clearly, the number of volume elements can be held to a reasonable number only by employing a non-uniform grid. The geometry of the head and the grid chosen is shown in Fig. 6. Four of the major volume elements are

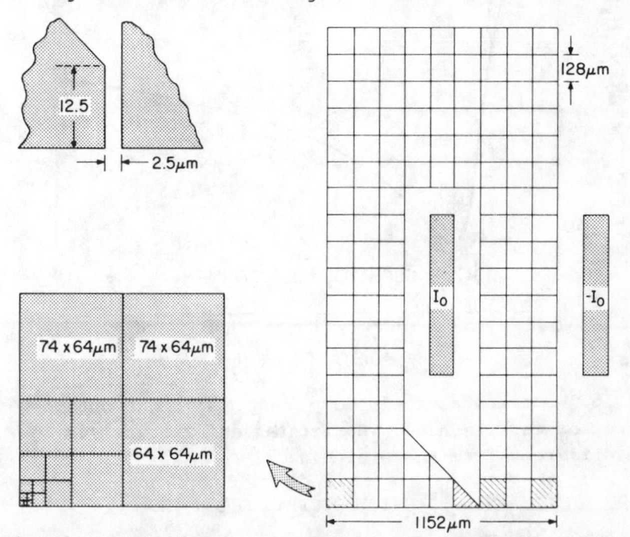

Fig. 6. Cross-section of ferrite head and coil, showing coarse grid and four major volume elements (cross-hatched) that are further subdivided in the manner indicated at lower left. Gap geometry is indicated at upper left.

further subdivided so that the smallest one is only 0.25 μm x 0.25 μm. At a distance of 0.5 μm this technique gives a meaningful sensitivity function near the gap and outside corners only, but the total number of elementary volumes is a modest 246.

The field near the gap and right outside corner is shown on an expanded scale in Figs. 7 and 8. The extremal values of the horizonal field component for I_o=1 ampere and μ_r=1000 are -7 Oe near the left corner, 1156 Oe near the gap, and -42 Oe near the right corner. The spurious pulse associated with the left corner is negligible due to coil location. The pulse associated with the right corner is highly asymmetric and, for a thin, high-resolution recording medium, about 3% of the amplitude of the main pulse.

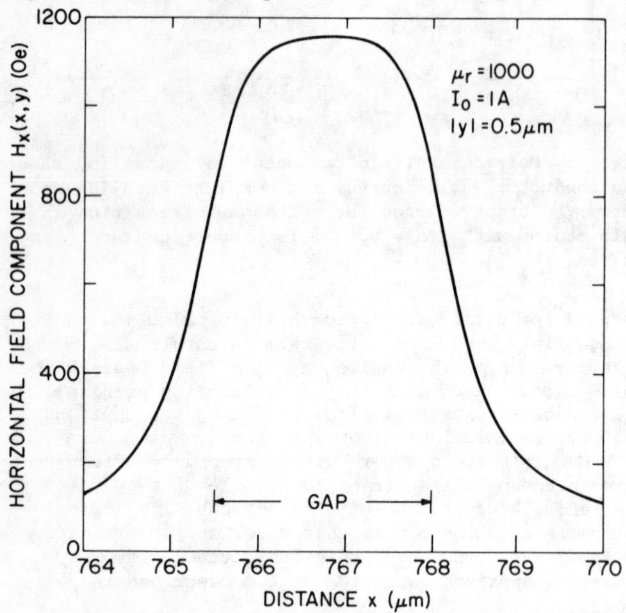

Fig. 7. Horizontal field component in vicinity of ferrite-head gap. Horizontal distance x is measured from left outside corner; left pole-tip length is 765.5 μm.

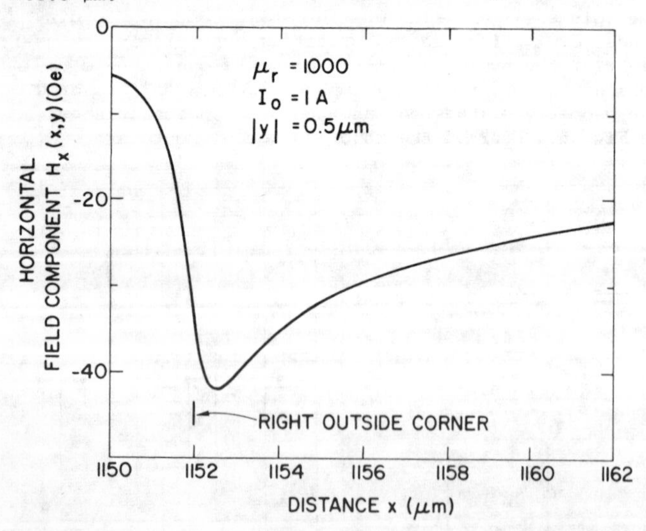

Fig. 8. Horizontal field component near right outside corner of ferrite head. Horizontal and vertical scales are different from those of Fig. 7.

VII. HIGH DENSITY DIGITAL RECORDING

A continuum, classical, electromagnetic theory predicts no limit to the achievable linear density since the equations involved scale. If track width and intrinsic material properties such as M_r and H_c are fixed while all remaining distances and drive currents are reduced by a factor β, then signal amplitudes remain fixed and linear density increases by the factor 1/β. The practical limitation is the head-medium separation d.

An estimate of maximum linear density, which we define as the -6 dB point on the all-ones roll-off curve, can be made with the arctangent model when d is specified. The transition length parameter, a, is chosen so that the pulse half-width, P_{50}, agrees with the self-consistently computed [4] half-width; a=1.13μm if d=0.5μm, g_w=1μm, and δM_r=0.1μm x 800 EMU/cm^3 where g_w is the write gap length. Scaling then provides the relationship a=2.26 d. The pulse half-width for thin media is [10]

$$P_{50} = \sqrt{g^2 + 4(d+a)^2} \ . \qquad (26)$$

Now g can be made negligibly small, but with head efficiency in mind and for the sake of simplicity, let g=d. The pulse shape is nearly Lorentzian in this case and $D_{-6} \simeq 1.39/P_{50}$, a result [37] that is exact when g=δ=0. Linear density as a function of head-medium separation, which is limited by surface roughness, is then given by the simple expression

$$D_{-6} = \frac{13{,}900}{d\sqrt{1+4(1+2.26)^2}} = \frac{2107}{d} \ \frac{\text{flux reversals}}{\text{cm}} , \qquad (27)$$

where d is in microns.

Track density is determined by the minimum tolerable signal-to-noise ratio and the mechanical considerations of finding and following the track. A long, imprecise, and possibly (depending on what the mechanical limiations are) irrelevant argument is avoided by assuming the signal must be at least 500 μV peak to peak at low density [38]. This criterion implies that for the MR head W=6 μm, using Eq. (23). At this track width the particulate or granular nature of the storage medium is not a fundamental problem [1]. There are, however, other sources of noise that deserve consideration for very narrow thin film inductive or magnetoresistive heads. They are the switching characteristics (Barkhausen noise) of the films, and thermal noise [39] in MR heads caused by frictional heating upon contact with storage medium asperities.

The MR head has a clear track width advantage over inductive heads. The ratio of MR head to inductive head output, from Eqs. (18), (19), and (22) with $\mu_o M_s$ = 1 W/m^2, $\sin[\pi(g+t)/\lambda] \simeq \pi g/\lambda$ and assuming unity inductive-head efficiency, is

$$\frac{e_{MR}}{e_{ind.}} \simeq \frac{1}{\sqrt{2}} \frac{g}{t} \frac{J\,\Delta\rho_{max}}{Nv} \simeq 10 \qquad (28)$$

where J=5x10^{10} A/m^2 is the assumed current density; $\Delta\rho_{max}$ = 7x10^{-9} Ω-m, g/t is assumed 20, N is the number of inductive-head turns and v the velocity. For a wide variety of cases Nv ≃ 500 meter-turns/sec. [40] Moreover, the resolving ability of the shielded MR head is high. It is intermediate to that of the ferrite head (p/g>>1) and thin film head (p/g≃1) when the gap length of the inductive heads is equal to the shield-to-shield spacing. Head efficiency calculations indicate that a shielded MR head with narrow effective gap-length is easier to make than a narrow-gap inductive head.

We conclude that due to high output and high resolution the shielded MR head offers the highest areal density in digital magnetic recording on thin media. It is capable of resolving 4000, 8000, and 16,000 flux reversals per cm at head-medium separations of 0.5, 0.25, and 0.125 μm, respectively, with remaining parameters as given in Table I. Adequate signals are obtainable with 6 μm track width. The areal density

Table I. Shielded MR head linear density predictions

d(μm)	δ(Å)	g_w(μm)	a^a(μm)	g_r^b(μm)	D_{-6}^c(fr/cm)
0.5	1000	1	1.13	1	4,076
0.25	500	0.5	0.57	0.5	8,153
0.125	250	0.25	0.28	0.25	16,305

[a] Based on M_r=800 EMU/cc, H_c=300 Oe, with arctangent parameter fit to the results of Potter and Schmulian [ref. (4)]

[b] Shield-to-shield separation $\approx 2g_r$

[c] Based on $D_{-6}=1.39\ [g_r^2 + 4(d+a)^2]^{-1/2}$

corresponding to tracks centered on 10 μm intervals and to 0.125 μm head-medium separation is 1.6x10^8 flux reversals per square centimeter, or 10^9 reversals per square inch.

In support of these conclusions experimental roll-off curves for an MR head with 1 μm shield-to-shield spacing are given in Fig. 9. The $M_r\delta$ product of the thin disc is 1.9x10^{-3} EMU/cm^2, and the coercive force 480 Oe. The -6 dB point is 7200 fr/cm at d=0.3μm. Peak-to-peak isolated pulse output amplitude for this particular head and storage medium combination is 34 volts/meter at an estimated current density of 6x10^{10} A/m^2.

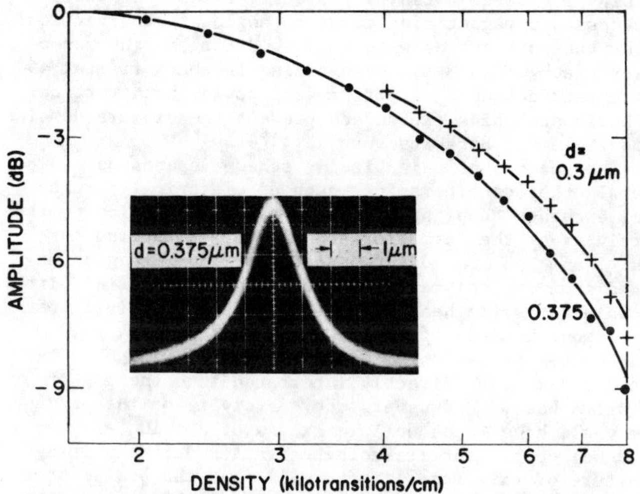

Fig. 9. Normalized amplitude of "all-ones" pulses versus density for shielded MR head and thin-film disc combination with 2 different head-disc separations. Oscilloscope photo of isolated pulse is shown in insert.

ACKNOWLEDGMENTS

Numerical computations were done in collaboration with R. W. Cole using a modification of a computer program developed by R. W. Cole, E. P. Valstyn, and R. I. Potter (unpublished). I thank Mr. Cole, to whom I am particularly indebted, and Dr. Valstyn for their permission to publish these results here. The MR head with which data were obtained was fabricated by L. T. Romankiw and co-workers. Experimental results were obtained with the technical assistance of D. C. Caldwell.

REFERENCES

1. J. C. Mallinson, AIP Conf. Proc. Magnetism and Magnetic Materials, 5, 743 (1971).
2. A. S. Hoagland, Digital Magnetic Recording (John Wiley and Sons, Inc., New York, 1963), pp. 57-60.
3. J. C. Mallinson, IEEE Trans. Mag., MAG-2, 233 (1966).
4. R. I. Potter and R. J. Schmulian, IEEE Trans. Magn., MAG-7, 873 (1971).
5. R. I. Potter, et. al., IEEE Trans. Magn., MAG-7, 689 (1971).
6. R. L. Wallace, Jr.; Bell System Tech. J., 30, 1146 (1951).
7. W. K. Westmijze, Phillips Res. Rep., 8, 148 (1953).
8. J. J. Miyata and R. R. Hartel. IRE Trans. Electronic Components, EC-8, 159 (1959).
9. D. E. Speliotis and J. R. Morrison, IBM J. Res. Develop., 10, 233 (1966).
10. P. I. Bonyhard, et. al., IEEE Trans. Magn., MAG-2, 1 (1966).
11. B. K. Middleton, IEEE Trans. Magn., MAG-2, 225 (1966).
12. R. I. Potter, J. Appl. Phys., 41, 1647 (1970).
13. M. L. Williams and R. L. Comstock, AIP Conf. Proc. Magnetism and Magnetic Materials, 5, 738 (1971).
14. D. W. Chapman, Proc. IEEE, 51, 394 (1963).
15. D. A. Lindholm, IEEE Trans. Magn., MAG-9, 3399 (1973).
16. J. R. Herbert and D. W. Patterson, IEEE Trans. Magn., MAG-1, 352 (1965).
17. G. C. Feth, AIEE Trans. Commun. Electron., 81, 267 (1962).
18. S. Iwasaki and T. Suzuki, IEEE Trans. Magn., MAG-4, 269 (1968). This calculation is a static one in the sense that head-medium motion is neglected.
19. N. Curland and D. E. Speliotis, J. Appl. Phys., 41, 1099 (1970).
20. ______, IEEE Trans. Magn., MAG-6, 640 (1970).
21. ______, IEEE Trans. Magn., MAG-7, 538 (1971).
22. D. J. George, et. al., IEEE Trans. Magn., MAG-7, 240 (1971).
23. D. L. A. Tjaden, IEEE Trans. Magn., MAG-9, 331 (1973).
24. C. W. Steele and J. C. Mallinson, IEEE Trans. Magn., MAG-4, 651 (1968).
25. B. Kostyshyn, IEEE Trans. Magn., MAG-7, 880 (1971).
26. W. K. Hodder and J. E. Monson, IEEE Trans. Magn., MAG-7, 686 (1971).
27. T. Suzuki and S. Iwasaki, IEEE Trans. Magn., MAG-8, 536 (1972).
28. O. Karlquist, Trans. Roy. Inst. Technol. Stockholm, 86, 1 (1954).
29. See, for instance, J. D. Jackson, Classical Electrodynamics, (John Wiley and Sons, Inc., New York, 1962), pp. 18-19.
30. R. P. Hunt, IEEE Trans. Magn., MAG-7, 150 (1971); U.S. Patent 3,493,694.
31. W. Döring, Ann. Physik, 32, 259 (1938).
32. A. Paton, J. Appl. Phys., 42, 5868 (1971).
33. S. Krongelb, J. Electron. Mater., 2, 227 (1973).
34. C. H. Bajorek and A. F. Mayadas, AIP Conf. Proc. Magnetism and Magnetic Materials, 10, 212 (1972).
35. L. W. Brownlow and C. King, IEEE Trans. Magn., MAG-8, 539 (1972).
36. M. J. Stavn, IEEE Trans. Magn., MAG-9, 698 (1973).
37. R. L. Comstock and M. L. Williams, IEEE Trans. Magn., MAG-9, 342 (1973). The factor 1.39 comes from recognizing that the root of sinh x=2x is x ≡ πp/(2s) = 2.18 (their notation).
38. R. Sykes, private communication.
39. R. D. Hempstead, to be published in IBM J. Res. Develop.
40. Based in part on W. B. Phillips and H. P. McGonough, IEEE Computer Soc. Int. Conf., pp. 101-104 (Feb. 1974).

AN INTEGRATED MAGNETORESISTIVE READ, INDUCTIVE WRITE HIGH DENSITY RECORDING HEAD

C. H. Bajorek, S. Krongelb, L. T. Romankiw and D. A. Thompson
IBM, Thomas J. Watson Research Center, Yorktown Heights, New York 10598

ABSTRACT

The design, fabrication and performance of an experimental thin film recording head is described which has 25 μm and 125 μm track width magnetoresistive (MR) read elements combined with a single turn thin film inductive write head. The MR read elements are shielded to provide high linear resolution with an equivalent read gap of 1/2 μm and are internally biased by means o a permanent magnet film in the gap. The best head performance with a 0.5 μm air bearing on a high performanc disk was: write current of 400 mA peak, isolated pulse readback signal of 40 mV peak per mm of track width at 10 mA sense current, with a 6 db density of 440 flux changes per mm (11 kfci).

INTRODUCTION

A substantial increase in storage density in magnetic recording systems is likely to require transducers with dimensions smaller than can be easily achieved with conventional recording heads. One alternative has been the thin film inductive head.[1]

A readback transducer using a magnetoresistive (MR) element is a thin film structure, is planar and can be integrated with a thin film inductive write head. The MR transducer has a significant track width advantage over inductive heads, since it can produce signals far larger than those of equivalent heads.[2,3]

We have therefore developed an experimental recording head consisting of a one turn inductive write head built directly on top of a shielded MR head. Novel features in the integrated transducer are the internal biasing of the magnetoresistor about its linear operating point by means of a permanent magnet film in the gap, th use of one magnetic layer for both a write pole tip and magnetoresistive shielding member, and a stepped conductor to improve the write head efficiency.

DESIGN CRITERIA

A schematic crossection of the integrated head is shown in Figure 1. It consists of a one turn vertical thin film inductive head directly on top of a vertical shielded MR head. The center shield is shared by both transducers. Several design requirements for both transducers have been described before.[1-8] We discuss only tl design features novel to this structure: Biasing the magnetoresistive element and provision of a stepped writ conductor. The latter allows for a narrow write gap g_w with lower write current density and power dissipation.

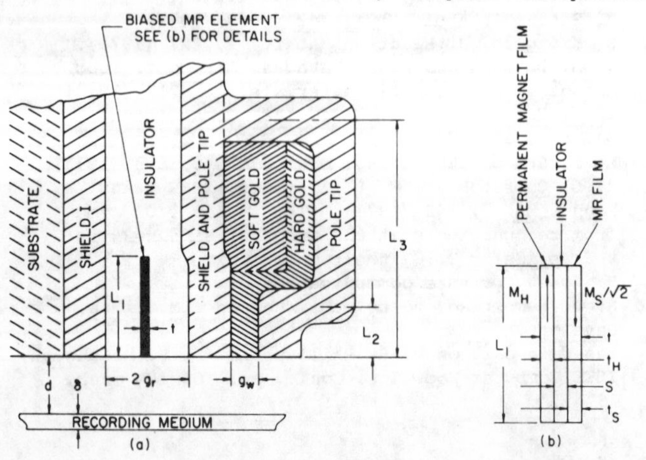

Fig. 1. Schematic Crossections of: (a) Integrated Head and (b) Biased MR film sandwich ($L_{1,2,3} >> g_{w,r}$).

The desirability of biasing an MR element, to maximize its sensitivity and linearize its response, has been described by Hunt[4] and others[5,6,8] for unshielded MR heads. For any high resolution application magnetic shields are required on both sides or the MR element; however these will also shield any fields originating from sources external to the heads. Thus a bias field source is required inside the shield structure.

We chose to implement internal biasing by using a permanent magnetic film.[3] Such a film in close flux coupling relationship ($t << L_1, g_r$) with a highly permeable (Permalloy MR) film will transfer most of its flux into the latter. In the simplest case where the track width (not shown in Fig. 1) is far larger than L_1 and the permanent magnet film is saturated transverse to the stripe, proper biasing of the magnetoresistor requires that

$$t_H M_H \simeq t_s M_s/\sqrt{2}, \qquad (1)$$

where t_H and t_s and M_H and M_s represent the permanent magnet film and Permalloy film thickness and saturation magnetization respectively. Satisfying Eq. 1 insures that the Permalloy film is magnetized at 45° to the stripe direction, corresponding to the inflection point of its magnetoresistance vs. vertical flux response.

The composite sensor and bias film stripe can be defined by one photolithographic step, a processing advantage. The bias field is independent of changes in L_1, making the biasing condition rather insensitive to wear. The bias may be adjusted by providing excess hard film thickness and magnetizing it at an angle off the vertical. In the case of devices with small width to height aspect ratio (track width $\approx L_1$) magnetizing the hard film at 45° to the vertical head direction can provide both vertical and horizontal bias fields and prevent the closure domains responsible for Barkhausen noise.[2,6]

The success of this biasing scheme depends on several other considerations such as electrical insulation, exchange coupling and preventing degradation of the coercivity of the Permalloy sense film as detailed in Refs. 3 and 9. Consideration of the field strengths expected from typical media as well as the stray field from the adjacent write head requires that the coercivity of the permanent magnet film be well in excess of 4×10^4 A/m (500 Oe) to avoid its demagnetization. The MR element could also be directly integrated into the gap of the write head. However, the coercivity of the permanent film would have to be well in excess of 2×10^5 A/m (2500 Oe) to prevent its demagnetization during writing. A feature of this read structure is that the bias-sensor film sandwich is considerably thinner (1,400 Å) than the other film thicknesses, resulting in a minimal step in the interhead region, and providing an adequate surface for subsequent fabrication of the inductive write yoke.

Write head design is dominated, not by the linear efficiency arguments of Paton,[7] but by more stringent requirements due to head saturation during writing. The design is suitable for fabrication over thickness ratios of at least a factor of two thicker or thinner, and for various media and flying heights, but is centered on 2 μm pole tips, a 1 μm gap, and a write field requirement of about 10^5 A/m (1900 Oe) in the medium. This is because twice H_c is required at the highest density to overcome the demagnetizing field and switching threshold of a perfectly square loop medium. For particulate media in the 30,000 to 50,000 A/m (400 to 600 Oe) range, with ordinary squareness ratios, this leads to the stated write field corresponding to a field in the internal gap of about 2.4 10^5 A/m (3000 Oe).

Copper and pure gold are good conductors, but too soft for satisfactory lapping and wear. The hard gold used for the gap has about the same resistivity as Per-

Reprinted with permission from *20th Annu. AIP Conf. Proc.*, no. 24, 1974, pp. 548–549.

malloy. Hence, nearly all the current in a simple head of uniform crossection would flow in the Permalloy, where its average efficiency is less than 50%. The thick soft gold conductor in the head provides a more magnetically efficient current flow. Saturation near the back of the yoke then limits the allowable pole tip length L_2 to 5μm for the required mmf at the pole tips of 240 mA-turns at a current of about 400 mA.

FABRICATION

The fabrication process uses sputter etching to define the MR stripe, electroplating through photoresist masks to form the conductors, and electroplating followed by chemical etching to make the Permalloy shields. Details and critical aspects of the individual process steps have already been given[10,11].

Shield 1 is a 1 to 2 μm layer of electroplated Permalloy. The plating base consists of 100 Å of Ti followed by 1000 Å of Permalloy evaporated on a thermally oxidized Si substrate. As an alternative, a polished ferrite slab may serve as both substrate and shield. The biasing and MR layers are deposited on a sputtered 2000 Å SiO_2 film which forms the first half of the gap. The permanent magnet film is a composite of 250 Å of α-Fe_2O_3 exchange coupled to a layer of 150 to 200 Å of Permalloy resulting in a combined coercivity greater than 180 Oe. Single layer films with even higher coercivities will be reported elsewhere.[9] The α-Fe_2O_3 film is formed first either by oxidizing an evaporated 150 Å Fe film in air (at 300°C for 15 minutes) or by reactively sputtering the Fe (at 200°C) in the presence of (10 μ) O_2. The exchange coupled Permalloy film, a 1000 Å separating layer of Schott glass and a 200 Å Permalloy MR film are then sequentially evaporated at 250°C.

The blanket MR film serves as a base for electroplating 2500 Å Au leads through a photoresist mask[10]. Sputter etching is then used to remove the unwanted deposits down to the SiO_2 as described in Ref. 10 with a photoresist mask designed to protect the gold leads as well as the sensor. The chemically non-selective sputter etch process is particularly advantageous for etching the multilayer structure.

A second layer of 3500 Å of sputtered SiO_2 completes the read gap resulting in $2\ g_r$ = 7500 Å. The second SiO_2 layer is etched to expose the MR leads for their full length to within about 0.1 mm of the sensor stripe. Thus subsequent metallizations used in fabricating the write head will deposit on the leads to increase their conductance. The openings in the SiO_2 are narrower than the leads so that some SiO_2 overlaps their edges as protection from subsequent etching steps.

The 2 μm shared shield is deposited over the read structure in the same manner as the first shield. The gold write conductor is now electroplated through a photoresist mask in a two stage process. The resist is first opened only in the throat and 2 μm of high conductivity gold (Selrex BDT-510) are plated. The gap and throat are both open during the second stage so that this plating defines the gap thickness of 1 μm and provides an additional 1 μm of conductor in the throat. A hard gold (Aurall 214, Knoop hardness 250 to 300) is used in the gap; the stepped structure compensates for the low conductivity inherent in this hard gold. If a positive resist is used for the first stage and plating done under safe-light conditions, the second plate-thru mask can be obtained by simply reexposing the first resist through a second photo mask. The outer shield is completed by electroplating 2 μm of Permalloy over the entire structure, followed by a 2 hour, 200°C stabilization treatment.

The shields are completed by spray etching in ferric chloride[11]. The resist mask protects both the shield and the leads so that the plated Permalloy will remain as part of the leads. The gold conductors must not be exposed during etching, since electrochemical couples would cause severe undercut and uncontrollable etching. An oversize etch mask ensures that the Permalloy in the final structure overlaps the gold on either side and encapsulates it. Etching is not critical since the front of the head will be defined by lapping after the head is mounted. The completed structure is shown in Figs. 2 and 3.

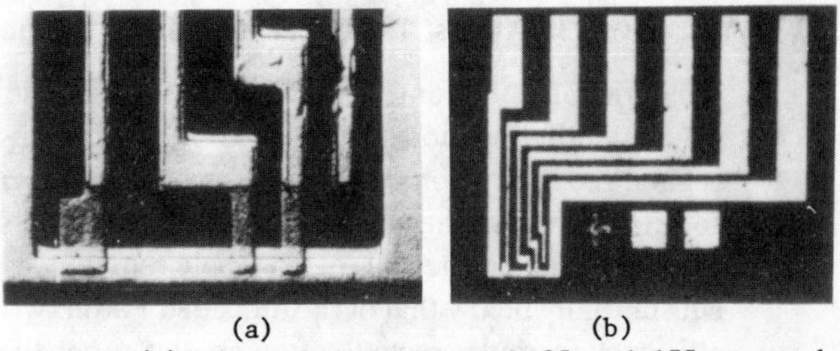

(a) (b)

Fig. 2. (a) Transducer area with 25 and 175 μm read track widths under a 375 μm wide write head and (b) overall planar view with contact pads.

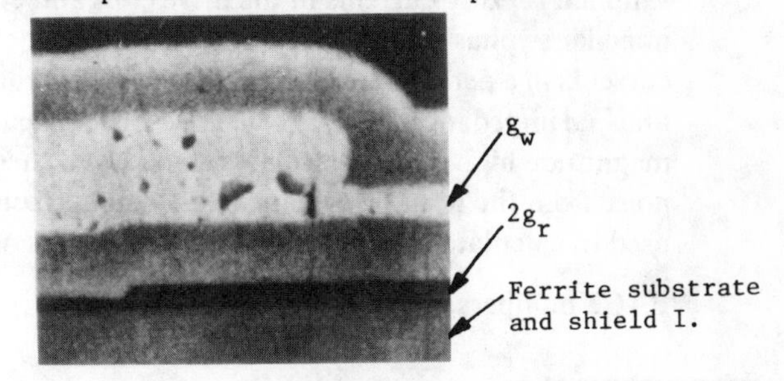

Fig. 3. SEM micrograph of lapped read/write yoke crossection (3000 X).

After dicing the processed substrate into individual elements, a layer of glass was epoxy bonded over the active portion of the head and the chip was then mounted on the end of an air bearing slider. The lower surface of the chip-slider assembly was lapped to a distance L_1 of 7 μm below the top edge of the MR stripe, with corresponding L_2 of about 5 μm and L_3 of 40 μm.

EXPERIMENTAL RESULTS

The transducer was tested with a conventional high performance particulate disk with a coercive force of approximately 480 Oe. The write current required was 400 mA peak and an isolated pulse readback signal of approximately 40 mV peak per mm of track width with a 10 mA sense current was achieved. The 6db resolution density was 440 flux changes per mm (11 kfci) at a 0.5 μm air bearing. Additional resolution data and a representative isolated transition pulse are found in Ref.2.

ACKNOWLEDGEMENTS

We gratefully acknowledge the technical assistance of N. Mazzeo, R. Anderson, D. Johnson, P. McCaffrey, E. Castellani, E. Harden, B. Stoeber and A. Pfeiffer.

REFERENCES

1. E. P. Valstyn, Advances in Magnetic Recording, Annals N.Y. Acad. Sci. 189, 21-51 (1972).
2. R. I. Potter, IEEE Trans. Mag., MAG-10, 502 (1974).
3. C. H. Bajorek, et al., U.S. Patent 3,840,898.
4. R. P. Hunt, IEEE Trans. Mag., MAG-7, 150 (1971); U.S. Patent 3,493,694.
5. R. L. Anderson,et al., AIP Conf. Proc.,10,1445(1973).
6. C. H. Bajorek,et al., IBM J. Res. Dev.,18,541,(1974).
7. A. Paton, J. Appl. Phys., 42, 5868 (1971).
8. F. W. Gorter,et. al.,IEEE Trans.Mag.,MAG-10,899(1974).
9. C. H. Bajorek and R. Hempstead, to be published.
10. L. T. Romankiw, et al., IEEE Trans. Mag., MAG-10, 828 (1974).
11. L. T. Romankiw and P. Simon, to be published in IEEE Trans. Mag.

Thin film recording head efficiency and noise

Gordon F. Hughes[a)]
Seagate Technology, Scotts Valley, California 95066

(Received 30 August 1982; accepted for publication 29 March 1983)

Analytic formulas for thin film recording head efficiency and thermal noise are presented. Paton's transmission line model is replaced by two-dimensional eddy currents in four regions: the coil region, the gap region, and two permeable pole regions. Predicted efficiencies are lower, and close to actual thin film heads. Basically, a four-region solution is needed because the gap region is a reluctive load on the coil region's pole reluctance, which lowers the predicted efficiency substantially below that of the unloaded Paton two region (coil and pole) transmission line. Paton efficiency formulas seem to often predict either too high an efficiency, or too low a permeability. For typical head dimensions, the four region efficiency formulas reduces to hand calculator simplicity. Eddy currents in the drive coils and conductive (permalloy) poles can cause the impedance phase angle θ of these heads to hover closer to 45° than to a purely inductive 90° (eddy currents in a permeable, plane slab have 45° average phase angle). Such eddy currents can give an unusual impedance $Z(f)$, and can raise the real part of the head impedance ($|Z|\cos\theta$) an order of magnitude higher than the dc resistance, and thereby similarly raise the mean-square thermal noise from the head. Magnetic field formulas from the four-region eddy current equations are used to calculate this impedance and the excess noise factor.

PACS numbers: 85.70.Kh, 89.80. + h

INTRODUCTION

Figure 1 shows the simplest thin film head geometry giving accurate efficiency predictions. Region "0" represents the write-read conductor(s) of permeability μ_0, skin depth δ_0, length h_0, and thickness $2g_0$. It does not appear to be necessary to model the individual conductors to get satisfactory efficiency formulas (as was done in the computer models of Jones,[1] Yeh,[2] and Katz[3]). Regions "2" and "3" are the pole layers, of thickness p. The model can easily be extended to allow different p values for regions "2" and "3". Region "1" is the (insulating) gap or throat area, of zero conductivity, permeability μ_0, length h_1 and thickness $2g_1$. Paton's two-region head[4] corresponds to regions "0" and "2."

Such a four-region head can be analyzed by a loaded two-transmission line model with mismatched impedances, as suggested by Thompson.[5] Miura[6] also uses a four-region model, based on Thompson.[5]

However, the eddy currents can substantially modify the head impedance, and so this paper will solve the full two-dimensional Maxwell equations, allowing eddy currents in all regions. Because of the presence of currents in the field regions, the vector potential form of the Maxwell equations will be used. For efficiency calculations and typical head dimensions, the equations can be simplified. If eddy currents are negligible, and the pole layers are sufficiently thin and close to each other, then the method is equivalent to a loaded two-transmission line model.

MATHEMATICS

All currents are in the z direction (perpendicular to the plane of Fig. 1), so the four vector potentials A_0, A_1, A_2, A_3 will represent z components. In Fig. 1, it simplifies the final formulas if we use two x coordinates: x_0 in regions "0" and "2," x_1 in regions "1" and "3." Only one y coordinate is used, measured from the head center symmetry plane.

Now, the Maxwell eddy current equation for a region is [Stoll,[7] Eq. (1.16)]

$$\nabla^2 A_z(x,y,t) = -\mu J_z(x,y,t) = \mu\sigma\partial A_z(x,y,t)/\partial t - \mu J_a(t), \quad (1)$$

where μ is the permeability; the drive current is

$$J_a(t) = \sigma\partial V_a(t)/\partial z, \quad (2)$$

and V_a is the scalar potential, which is the applied voltage across the head (driving function) in the z direction. Only region "0" has an applied voltage and therefore a nonzero J_a. It is not a function of (x,y) but the total J_z will be. Equation (1) expresses the fact that the electric field is the gradient of the scalar potential (driving function) plus the time derivative of the vector potential (induced eddy currents).

We assume that J_a is implicitly a steady-state sinusoidal driving function, viz., that the actual drive or writing current is $J_a e^{j\omega t}$ (ω being the write or read frequency). Then all fields B_{xi}, B_{yi}, and vector potentials A_i are complex amplitudes

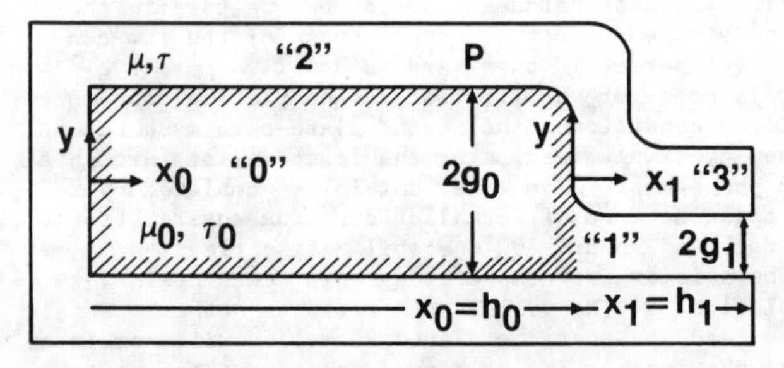

FIG. 1. Thin film head geometry and symbols.

[a)] Formerly with the Xerox Palo Alto Research Center.

Reprinted with permission from *J. Appl. Phys.*, vol. 54, no. 7, pp. 4168–4173, July 1983.

times $e^{j\omega t}$, where $i = 0,1,2,3$ is the region number. So Eq. (1) for the four regions is

$$\nabla^2 A_0(x_0,y) = \alpha_0^2 A_0(x_0,y) - \mu_0 J_a \quad \text{(coil region)}, \quad (3)$$

$$\nabla^2 A_1(x_1,y) = 0 \quad \text{(gap region)}, \quad (4)$$

$$\nabla^2 A_2(x_0,y) = \alpha^2 A_2(x_0,y) \quad \text{(poles over coil)}, \quad (5)$$

$$\nabla^2 A_3(x_1,y) = \alpha^2 A_3(x_1,y) \quad \text{(poles over gap)}, \quad (6)$$

where the (complex) propagation constants are defined by

$$\alpha_0^2 \doteq j\omega\mu_0\sigma_0, \quad (7)$$

$$\alpha_0 = (1+j)/\delta_0, \quad (8)$$

$$\alpha^2 \doteq j\omega\mu\sigma, \quad (9)$$

$$\alpha = (1+j)/\delta. \quad (10)$$

Skin depths δ_0 and δ are defined by

$$\delta_0 = (\omega\mu_0\sigma_0/2)^{-1/2} \quad \text{(coil conductor)}, \quad (11)$$

$$\delta = (\omega\mu\sigma/2)^{-1/2} \quad \text{(pole regions)}. \quad (12)$$

Note that unsubscripted parameters μ,σ,α,δ refer to the head pole material, regions "2" and "3." Boundary conditions for solving Eqs. (3)–(6) will be first stated, and then discussed [below Eq. (22)]. First, the y field is zero at the back coil flux closure:

$$B_{y1}(0,0) = 0. \quad (13)$$

Next, normal and tangential field components match across the boundaries "0"–"2" and "1"–"3":

$$\partial A_0(x_0,g_0)/\partial x_0 = \partial A_2(x_0,g_0)/\partial x_0, \quad 0 \leqslant x_0 \leqslant h_0, \quad (14)$$

$$\mu\partial A_0(x_0,g_0)/\partial y = \mu_0 \partial A_2(x_0,g_0)/\partial y, \quad 0 \leqslant x_0 \leqslant h_0, \quad (15)$$

$$\partial A_1(x_1,g_1)/\partial x_1 = \partial A_3(x_1,g_1)/\partial x_1 \quad 0 \leqslant x_1 \leqslant h_1, \quad (16)$$

$$\mu\partial A_1(x_1,g_1)/\partial y = \mu_0 \partial A_3(x_1,g_1)/\partial y \quad 0 \leqslant x_1 \leqslant h_1. \quad (17)$$

At the coil-gap pole discontinuity "2"–"3" equate the flux Φ_{x2} out of "2" with Φ_{x3} into "3"

$$\Phi_{x2} = A_2(h_0,g_0+p) - A_2(h_0,g_0)$$

$$= \Phi_{x3} \quad (18)$$

$$= A_3(0,g_1+p) - A_3(0,g_1), \quad (19)$$

and match the mmf between the poles:

$$\int_{-g_0}^{g_0} H_{y0}(h_0,y)dy = \int_{-g_1}^{g_1} H_{y1}(0,y)dy. \quad (20)$$

Next, require that the exterior flux leaving the head be small, except at the recording gap. Namely, assume that exterior air-path reluctances are large compared to circuit reluctances within the head. This requirement will be met if the fields normal to the exterior pole faces are zero; and this can be done by requiring that the vector potential be zero on these surfaces.

Finally, match the fringing flux out of the pole tips with the air reluctance R:

$$R = \left[\int_{-g_1}^{g_1} H_{y1}(h_1,y)dy\right] \Big/ [A_3(h_1,g_1+p) - A_3(h_1,g_1)]. \quad (21)$$

Yeh[3] [Eq. (10] gives an approximation for R based on assumed semicircular fringing flux paths from pole to pole. However, it is shown in the Appendix that we can calculate fringing flux by vector potential differences in Elabd's narrow pole tip head conformal transformation,[8] and get an exact, but still reasonably simple fringing reluctance formula:

$$R = \pi/\ln[1 + (p/g_1)(1 + \sqrt{1 + (p/2g_1)}]. \quad (22)$$

Reluctances calculated from Eq. (22) are less than half the semicircular flux assumption formula, for typical head dimensions. However, typical thin film heads are so lightly loaded by the fringing flux (see appendix) that it will be ignored (as Paton originally did). For substrate test heads[9] where the pole tips are closed, we will assume zero reluctance.

There is a certain amount of arbitrariness in choosing the precise boundary conditions to impose (and there is arbitrariness in choosing the details of the four-region geometry in Fig. 1). A considerable number of alternate models and boundary conditions were tried, and the ones presented here are the simplest set that give accurate efficiency predictions for thin film heads useful in present recording systems. It is felt that they are physically plausible also.

Solution of Eqs. (3)–(6) based on Eqs. (13)–(20) is detailed in the Appendix. The resulting overall efficiency is $\xi = \xi_0\xi_1$, where ξ_0 is the efficiency of the coil region "0":

$$\xi_0 \doteq \|H_{y0}(h_0,0)/(I_0/2g_0)\|,$$

$$= \frac{\|\mu g_0 g_1 k_0 k_1 p_e \tanh(k_0 h_0)\|}{\|h_0 g_1 k_1 + h_0 g_0 k_0 \tanh(k_0 h_0)\tanh(k_1 h_1) + \alpha_0^2 \mu g_0 g_1 (k_1/k_0) p_e \tanh(k_0 h_0)\|}. \quad (23)$$

In general, k_0 and k_1 are complex numbers, so Eq. (23) is an absolute value or modulus of a complex efficiency which includes the phase information.

The gap region efficiency ξ_1 is

$$\xi_1 = \|H_{y1}(h_1,0)/H_{y1}(0,0)\| = \|1/\cosh(k_1 h_1)\|. \quad (24)$$

In Eqs. (23)–(24), the wavenumbers k_0 and k_1 are implicitly defined by five equations in five unknowns $\lambda_0,\lambda_2,\lambda_3,k_0,k_1$:

$$\mu_r(\lambda_0/\lambda_2)\tan(\lambda_0 g_0)\tan(\lambda_2 p) = 1, \quad (25)$$

$$\mu_r(k_1/\lambda_3)\tan(k_1 g_1)\tan(\lambda_3 p) = 1, \quad (26)$$

$$k_0^2 - \lambda_0^2 - \alpha_0^2 = k_0^2 - \lambda_2^2 - \alpha^2 = k_1^2 - \lambda_3^2 - \alpha^2 = 0. \quad (27)$$

Here, the relative permeability $\mu_r \doteq \mu/\mu_0$ and the effective pole tip thickness p_e is defined to be $p_e \doteq \tanh(\alpha p)/\alpha$. For comparison, the Paton efficiency of region 0 is [Paton,[4] Eq. (17)]

$$\xi_p = \frac{\|\mu p g_0 k_0^2 \tanh(k_0 h_0)\|}{\|k_0 h_0 - \alpha_0^2 \mu p g_0 \tanh(k_0 h_0)\|}. \quad (28)$$

To get a Paton efficiency for a two region head, we will multiply the individual region Paton efficiencies ξ_p and ξ_1 [Eqs.

(28) and (24)]. It is admitted that there is no reason to believe that correct efficiencies will be obtained in this manner, because Paton's model was derived for a uniform gapped head. However, many workers in the recording community *do* use the Paton formula in just this way, for lack of a simple (non-computer) formula that would apply better. Notation similar to Paton has been used for ease in comparison (observe that k_0, k_1 will be *numerically* different because of the extended physical model used here).

Equations (25)–(27) can be solved in two or three Newton–Raphson iterations (e.g., by complex variable Fortran programming) (NBS,[10] Sec. 3.9.5). Iterate Eq. (25) to get a new λ_0, and get a new λ_2 from λ_2^2 equals $\lambda_0^2 + \alpha_0^2 - \alpha^2$. Iterate Eq. (26) to get a new k_1, and get a new λ_3 from λ_3^2 equals $k_1^2 - \alpha^2$.

For typical head dimensions, it will be shown that the magnitude of the complex numbers $\|k_0 g_0\|$, $\|k_1 g_1\|$, $\|\alpha_0 g_0\|$ are all $\ll 1$. Then Eqs. (25)–(27) simplify to "Paton Length" formulas:

$$\lambda_0^2 = 1/\mu g_0 p, \tag{29}$$

$$k_1^2 = 1/\mu g_1 p, \tag{30}$$

$$k_0^2 = \alpha_0^2 + 1/\mu g_0 p. \tag{31}$$

In the general case Eqs. (29) and (30) make good initial guesses for the Newton–Raphson iteration on Eqs. (25) and (26).

If the coil conductor skin depth δ_0 is also large compared to λ_0, then from Eq. (31)

$$k_0^2 = 1/\mu g_0 p. \tag{32}$$

Equations (23) and (24) then reduce to a reasonably simple formula for the overall efficiency ξ, which does not involve complex numbers, and can therefore be done on a hand calculator:

$$\xi = \frac{\tanh(k_0 h_0)}{k_0 h_0 \cosh(k_1 h_1) + k_1 h_0 \tanh(k_0 h_0) \sinh(k_1 h_1)}. \tag{33}$$

Use Eq. (30) for k_1 and Eq. (32) for k_0. Note that this simplified efficiency formula is not a function of frequency. (It will therefore be inaccurate at sufficiently high frequencies, where Eq. (23) must be used.)

Paton's efficiency[4] [Eq. (18)] is gotten from Eq. (33) by letting the gap reluctive load go to zero; e.g., set $k_1 h_1 = 0$.

Use the total head flux linkage Φ_0, the total drive coil current I_0 and the dc drive coil resistance R_0 to get the complex impedance $Z(\omega)$, [Stoll,[7] Eq. (2.25)].

$$Z(\omega) = R_0 + j\omega \Phi_0 / I_0. \tag{34}$$

Resistance R_0 is derived from the scalar potential [Eq. (2)], and Φ_0 is the total flux generated (or linking) total region "0" drive current I_0. Note that I_0 will be complex, since it includes the eddy current reaction onto the resistive current J_a. Also Φ_0 will be complex, and therefore $j\omega\Phi_0/I_0$ will have a real part, raising the real part of the impedance, and hence the thermal noise. Similarly, the inductance is the real part of Φ_0/I_0.

Current I_0 and total flux Φ_0 are derived in the Appendix [Eqs. (A6) and (A14)], and the normalized impedance becomes

$$Z/R_0 = 1 + \mu_r \alpha_0^2 g_{0e} p_e \frac{k_1 g_1 + k_0 g_0 \tanh(k_0 h_0)\tanh(k_1 h_1) - k_1 g_1 \tanh(k_0 h_0)/k_0 h_0}{k_1 g_1 + k_0 g_0 \tanh(k_0 h_0)\tanh(k_1 h_1) + \mu k_1 g_{0e} g_1 p_e \alpha_0^2 \tanh(k_0 h_0)/k_0 h_0}. \tag{35}$$

In Eq. (35), the effective coil gap g_{0e} is defined to be $\sinh(\alpha_0 g_0)/\alpha_0$, and the effective pole p_e is $\tanh(\alpha p)/\alpha$.

Now the mean-square thermal noise voltage is (Davenport and Root,[11] Sec. 9.4)

$$\langle e^2 \rangle = 2kT \int \mathrm{Re}[Z(f)] df. \tag{36}$$

So the real part of Eq. (35), numerically integrated over the bandpass of the read channel, is the factor of mean-square noise increase over the dc resistance alone.

NUMERICAL RESULTS

Efficiency, inductance, and noise results will be given for five typical digital disk thin film heads (after Jones[1,9]). For experimental comparisons, we will use Jones[1] results. Although his efficiency values are calculated from a dc theoretic model, they are assumed to be accurate at sufficiently low test frequencies. The inductance comparisons are to the Anderson and Jones[9] calculations; they present experimental inductance values at 1 MHz supporting their accuracy. Case (A) is from Jones,[1] Fig. 3; case (B) is from Jones,[1] p. 511; case (C), Anderson and Jones,[9] Fig. 2, open gap; case (D) is case (C) after throat height lapping (to $h_1 = 5$ μm); and case (E) is Anderson and Jones,[9] Fig. 2, closed gap.

For example, case (A) has "Paton efficiencies" of

$$\xi_0 = \tanh(\lambda_0 h_0)/(\lambda_0 h_0) = 79\%,$$

$$\xi_1 = 1/\cosh(k_1 h_1) = 99\%,$$

$$\xi \doteq \xi_0 \xi_1 = 79\%.$$

See Eq. (24) for the ξ_1 formula. This combined value of 79% is listed in Table I, case (A). Recall that this is done to see how serious is the error committed by applying the two-region Paton formula to a four-region head.

Simplified Eq. (33) gives $\xi = 69\%$. More accurate Eqs. (23) and (24) give 70%. Reference efficiencies are about 68% (Jones,[1] Fig. 3).

To see if simplified Eq. (33) is valid at 1 MHz, using inefficient head (B) from Table I,

$$\|\tan(\lambda_0 g_0)/\lambda_0 g_0\| = 1.000,$$

$$\|\tan(k_1 g_1)/k_1 g_1\| = 1.000,$$

$$\|\tan(\lambda_2 p)/\lambda_2 p\| = 0.997,$$

$$\|\tan(\lambda_3 p)/\lambda_3 p\| = 1.000,$$

which are all close to unity, as assumed in simplifying Eqs. (25) and (26). At 5 MHz, the worst ratio is 0.9873. At 1 MHz,

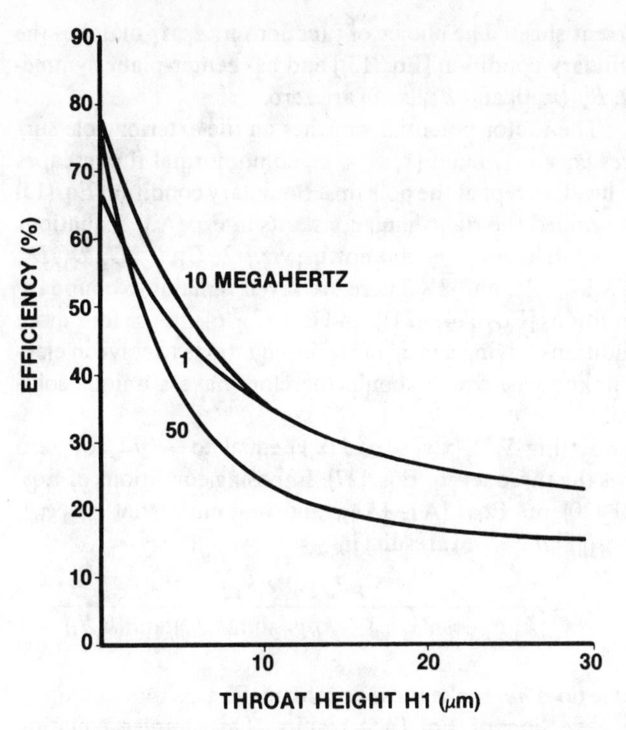

FIG. 2. Efficiency vs throat height h_1 for $p = 2$ μm, $2g_0 = 6.5$, $h_0 = 92.5$, $2g_1 = 0.8$, $h_1 = 5$, $\mu = 800$.

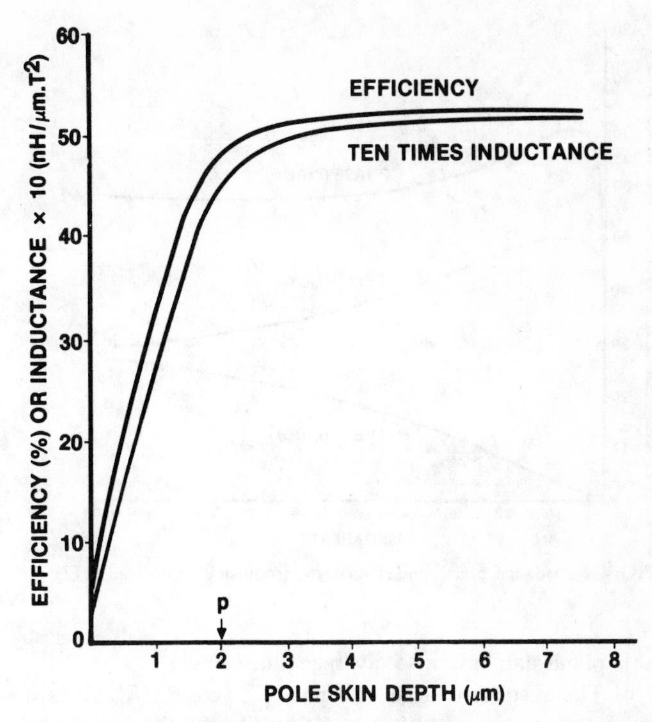

FIG. 3. Inductance and efficiency vs pole skin depth δ_1, same head as Fig. 2.

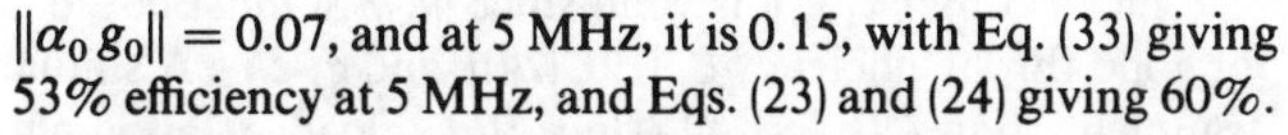
$\|\alpha_0 g_0\| = 0.07$, and at 5 MHz, it is 0.15, with Eq. (33) giving 53% efficiency at 5 MHz, and Eqs. (23) and (24) giving 60%.

Figure 2 shows the predicted efficiency [Eq. (23) times (24)] versus throat height h_1 at 1, 10, and 50 MHz for the head of case (B) but having the 0.8-μm gap $2g_1$ of case C. Note that the 1- or 10-MHz efficiency has fallen to 62% with 1-μm throat height h_1, and is only 35% at 10 μm. By comparison, conventional ferrite ring heads with 25-μm or 1-m in. throat heights, 800 permeability, and efficiency above 60% are common. Note that the 10- and 50-MHz efficiencies are larger at low throat heights than the 1-MHz values. This effect is due to coil region eddy currents, because it disappears if Eqs. (23) and (24) are evaluated for zero conductivity σ_0. It is believed that coil region eddy currents reduce the ac leakage flux from pole inside the region. Also, the coil currents tend to flow on the surfaces of the region, where they can couple more tightly to the poles.

Figure 3 shows that if the pole skin depth δ of this same head drops below the pole thickness p, the efficiency drops, and the inductance shuts off. The scale values represent ten times the inductance per μm of head width, and per turn squared.

Figure 4 shows the impedance real part and phase angle θ, and the efficiency, both versus frequency. As frequency rises, the phase angle begins to rise rapidly towards an inductive 90°, but the increasing eddy currents pull it back towards 45° (recall that one-dimensional eddy currents in a perme-

TABLE I. Typical thin film heads.

	Head example case:	(A)	(B)	(C)	(D)	(E)
p	pole thickness (μm)	2	2	2	2	2
$2g_0$	total coil thickness (pole to pole, μm)	6.5	6.5	8	8	8
h_0	coil length (μm)	92.5	92.5	80	80	80
$2g_1$	total gap dimension (μm)	1.5	1.5	0.8	0.8	0.8
h_1	throat length (μm)	5	5	30	5	30
τ_0	coil conductor resistivity (Cu, μΩ cm)	1.7	1.7	1.7	1.7	1.7
μ_0	coil permeability (cgs)	1.0	1.0	1.0	1.0	1.0
τ	pole resistivity (81% Ni, 19 Fe, μΩ cm)	20	20	20	20	20
μ	pole permeability (cgs)	1600	800	2000	2000	2000
ξ	Efficiency: Reference	68%	50%	...	...	0%
	Paton	79%	66%	75%	88%	75%
	Simplified Eq. (33)	69%	53%	32%	73%	0%
	Complex Eq. (23), (24)	70%	54%	32%	72%	0%
L	Inductance: Reference (ph/μm T^2)	7.5	...	19	...	22
	Calculated Eq. (35)	6.6	5.0	18	8.5	24
R_{NOISE}/R_0	0–12 MHz	2.7	2.0	6.4	3.2	9.1
	0–25 MHz	4.4	3.2	15	5.1	24

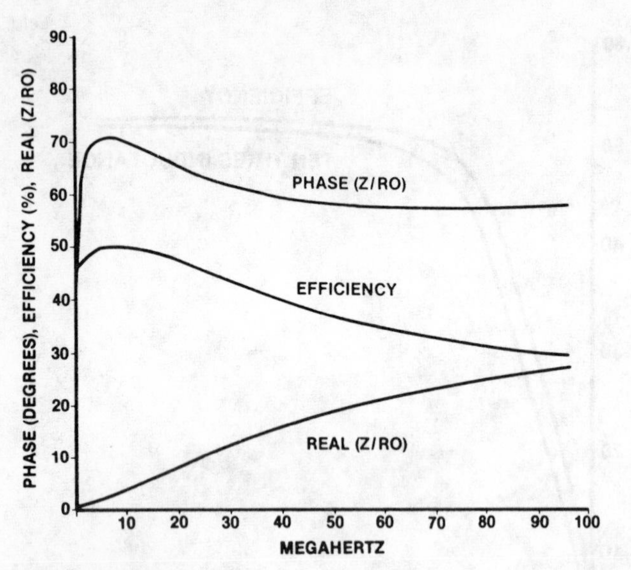

FIG. 4. Impedance, phase, and efficiency vs frequency, same head as Fig. 2.

able plane slab have a 45° average phase angle[7]).

The resultant large real part $\|Z\|\cos(\theta)/R_0$ gives a mean-square noise voltage 16.6 times larger than might be expected from the dc resistance R_0 alone (numerical integration was done by a complex variable Fortran program).

The efficiency curve in Fig. 4 first rises and then falls off at high frequencies, when the pole skin depth approaches the pole thickness p. It is felt that the initial rise in the efficiency is due to the conductor eddy currents resisting leakage of flux from pole to pole, across the conductor, as discussed above.

ACKNOWLEDGMENTS

This work was done while the author was with the Xerox Palo Alto Research Center. The author wishes to thank K. Norris for the experimental data suggesting excess noise and 45° phase in thin film heads. Also, thanks to D. Bloomberg and P. Kocsis for advice and help.

APPENDIX: EQUATION DERIVATIONS

Suitable forms of A_i solving Eqs. (3)–(6) are

$$A_0(x_0,y) = C_0 \cosh(k_0x_0)\cos(\lambda_0 y) + (\mu_0 J_a/\alpha_0^2)(1-\cosh \alpha_0 y), \quad \text{(A1)}$$

$$A_1(x_1,y) = [C_1 \cosh(k_1x_1) + D_1 \sinh(k_1x_1)]\cos(k_1y), \quad \text{(A2)}$$

$$A_2(x_0,y) = -C_2 \cosh(k_0x_0)\sin \lambda_2(y-g_0-p) - \mu J_a \sinh(\alpha_0 g_0)\sinh \alpha(y-g_0-p)/\alpha_0\alpha \cosh(\alpha p), \quad \text{(A3)}$$

$$A_3(x_1,y) = -[(C_3 \cosh(k_1x_1) + D_3 \sinh(k_1x_1)] \times \sin \lambda_3(y-g_1-p). \quad \text{(A4)}$$

Equation (3) which has the forcing function $J_a e^{j\omega t}$, has the right-most term of Eq. (A1) as a particular solution, plus the superimposed eddy current potential with amplitude C_0. If $\alpha_0 = 0$, (no eddy currents) the particular solution represents a x field linearly increasing with y; i.e., a symmetric uniform current sheet. The choice of functions in A_0,A_1 matches the boundary condition [Eq. (13)] and has center-plane symmetry: $B_{x0}(x_0,0)$ and $B_{x1}(x_1,0)$ are zero.

The vector potential vanishes on the exterior pole surfaces (x_0,g_0+p) and (x_1,g_1+p), so no normal flux escapes the head, except at the pole tips. Boundary condition Eq. (15) determined the right-hand constants in Eq. (A3). Equations (A1)–(A4) have eleven unknowns; viz., C_0, C_1, C_2, C_3, D_1, D_3, k_0, k_1, λ_0, λ_1, and λ_2. There are seven remaining boundary conditions [Eqs. (14)–(21)], and Eqs. (3)–(6) impose four more conditions, giving a total of eleven equations to solve in eleven unknowns, which should therefore have a unique solution.

Setting $\nabla^2 A_i(x,y) - \alpha_i^2 A(x,y)$ equal to $-\mu J_a$ or zero gives the three sets of Eqs. (27). Imposing conditions of Eqs. (14)–(20) on Eqs. (A1)–(A4), and assuming that $\|k_0 g_0\|$, $\|k_1 g_1\|$, $\|\alpha_0 g_0\| \ll 1$, results in

$$C_0 = \frac{-\mu J_a g_{0e} g_1 k_1 p_e}{k_1 g_1 \cosh(k_0h_0) + k_0 g_0 \sinh(k_0h_0)\tanh(k_1h_1)}. \quad \text{(A5)}$$

If one does *not* make these approximations, a more cumbersome version of Eq. (A5) results. The simpler equation shown here gives efficiencies accurate to 1% for the experimental heads (Table I), when compared to the exact version of Eq. (A5). Contact the author for details.

Neglecting $\|k_1 g_1\|$ is valid since $\|k_1h_1\|$ must be small for good throat efficiency ξ_1 [Eq. (24)], and $g_1 \ll h_1$. A similar argument shows $\|k_0 g_0\| \ll 1$, since the unloaded Paton efficiency [Eq. (28)] must be high if the loaded efficiency ξ_0 is to be reasonably high. Finally, g_0 is normally small compared to the conductor skin depth, so $\|\alpha_0 g_0\| \ll 1$, and so g_{0e}, defined to be $\sinh(\alpha_0 g_0)/\alpha_0$, is approximately g_0 (but g_{0e} has been left in the equation because it can differ from g_0 by 10% at 25 MHz). To restate this argument, thin film heads efficient enough to be usable in a recording system, normally have parameters which satisfy these approximations. Refer to the discussion under Eq. (22).

To get the net total drive current I_0, use Eq. (1) for the total current density $J_z(x,y)$:

$$I_0 = \int_{-g_0}^{g_0} dy \int_0^{h_0} [J_a - \alpha_0^2 A_0(x_0,y)]dx_0, \quad \text{(A6)}$$

$$= 2J_a h_0 g_0 - 2C_0\alpha_0^2 g_0 \sinh(k_0h_0)/k_0. \quad \text{(A7)}$$

From $B = \nabla \times A$, we can get the fields; e.g.,

$$H_{y0}(x_0,0) = -\partial A_0(x_0,0)/\partial x_0, = C_0k_0 \sinh(k_0x_0). \quad \text{(A8)}$$

Thereby the region "0" efficiency ξ_0 is

$$\xi_0 \doteq \|H_{y0}(h_0,0)/(I_0/2g_0)\| = \|k_0 \sinh(k_0h_0)/[J_0h_0/C_0 - \alpha_0^2 \sinh(k_0h_0)/k_0]\|. \quad \text{(A10)}$$

Equation (23) comes by putting C_0 from Eq. (A5). Equation (24) follows by differentiating Eq. (A2) similarly to Eq. (A8).

To get the total flux linkage Φ_0 for Eq. (34), we can take the y flux which crosses the region 0 symmetry plane $y=0$, from $x_0 = 0$ to h_0 plus the back pole flux $\Phi_{y0}(x_0=0)$, which is the same as $\Phi_{x2}(x_0=0)$. Take care of the varying flux linkage by assuming a distributed coil, viz., the turn of width dx_0 at x_0 links all the B_y flux from 0 to x_0 (the return turns are to the left of $x_0 = 0$ in Fig. 1).

$$\Phi_0 = -1/h_0 \int_0^{h_0} dx_0 \int_0^{x_0} B_{y0}(x_0',0)\,dx_0' - \int_{g_0}^{g_0+p} B_{x2}(0,y)dy \quad \text{(A11)}$$

$$= 1/h_0 \int_0^{h_0} [A_0(x_0,0) - A_0(0,0)]dx_0 - \int_{g_0}^{g_0+p} [\partial A_2(0,y)/\partial y]dy \quad \text{(A12)}$$

$$= (1/h_0) \int_0^{h_0} A_0(x_0,0)dx_0 - A_0(0,0) - A_2(0,g_0+p) + A_2(0,g_0) \quad \text{(A13)}$$

$$= C_0 \sinh(k_0 h_0)/(k_0 h_0) + \mu \sinh(\alpha_0 g_0)\tanh(\alpha p)/\alpha\alpha_0. \quad \text{(A14)}$$

Complex coefficient C_0 can be substituted from Eq. (A5), and Eq. (35) follows from Eq. (34). Equation (22) for the pole tip fringing flux reluctance is obtained from Potter,[12] by subtracting his vector potential at the right pole tip $\psi(w = \lambda)$, from $\psi(w = 1)$ on the left pole tip, giving the flux leaving each pole tip. Dividing this flux into the mmf $2g_1 H_{y1}(h_1,0)$, taking the real part, and taking g_1 as the half-gap gives Eq. (22).

[1] R. E. Jones, IEEE Trans. Magn. **MAG-14**, 509 (1978).
[2] Yeh, IEEE Trans. Magn. **MAG-18**, 233 (1982).
[3] E. R. Katz, IEEE Trans. Magn. **MAG-14**, 506 (1978).
[4] A. Paton, J. Appl. Phys. **42**, 5868 (1971).
[5] D. A. Thompson, AIP Conf. Proc. MMM **24**, 528 (1974).
[6] Y. Miura, IEEE Trans. Magn. **MAG-14**, 512 (1978).
[7] R. L. Stoll, *The Analysis of Eddy Currents* (Clarendon, Oxford, 1974).
[8] I. Elabd, IEEE Trans. Audio **11**, 21 (1963).
[9] N. C. Anderson and R. E. Jones, IEEE Trans. Magn. **MAG-17**, 2896 (1981).
[10] *Handbook of Mathematical Functions* (AMS 55, U. S. Govt. Printing Office, 1965).
[11] Davenport and Root, *Random Signals and Noise* (McGraw-Hill, New York, 1958).
[12] R. I. Potter, R. J. Schmulian, and K. Hartman, IEEE Trans. Magn. **MAG-7**, 689 (1971).

Part V
Written Magnetization

STUDIES ON MAGNETIC RECORDING *)

by W. K. WESTMIJZE 621.395.625.3

III. THE RECORDING PROCESS

1. Description of the d.c. and a.c. biasing methods

In recording sound the first demand is that there be a perfect linear relation between the output and the input signal. If the recording characteristic is not linear this becomes apparent by the creation of harmonics or combination tones to which the human ear is very sensitive, especially to the latter.

In principle it should be possible to obtain a linear recording even without the aid of a special biasing field. The normal magnetization curve of a virgin magnetic material in fact has a point of inflection in the origin; by making the amplitude of the signal sufficiently small a region around the origin can be found that is sufficiently linear. In this region, however, the magnetic processes are for the major part reversible, so that the slope of the remanent virgin curve is practically zero. Because of the weakness of the recorded signal the ratio of the signal to the unavoidable background noise becomes too unfavourable.

An extension of the linear portion of the recording characteristic can be obtained either by the direct-current or by the alternating-current biasing method.

In the d.c. method the tape is magnetized to saturation (A in fig. 15) by a separate head. This head acts at the same time as an erasing head, for by bringing the tape to saturation anything that might have been recorded on it is obliterated. Leaving this head the tape has reached the remanent magnetization M_R. During the recording a magnetic field is added to the signal field to be recorded. The polarity of this biasing field is opposite to the field that has previously brought the tape to magnetization, and the magnitude is of the order of the coercive force, thus bringing the magnetization to a point C when no signal field is applied. If the biasing field is chosen such that C coincides with a point of inflection of the hysteresis loop, a region around C is essentially linear and has a much steeper slope and, therefore, a much greater amplitude of the recorded magnetization than the linear part of the virgin magnetization curve around the origin. After leaving the recording head the magnetization reaches the M-axis along a branch of the minor hysteresis loop, for instance DD' or CC'. These branches being essentially parallel the remanent magnetization remains a linear function of the applied signal field.

In the alternating-current biasing method an a.c. field of an amplitude approximately equal to the coercive force and a frequency well above the highest to be recorded is superimposed on the signal field. It appears that in this way too a linearization of the recording characteristic is achieved. Compared with the d.c. method the a.c. method has two advantages.

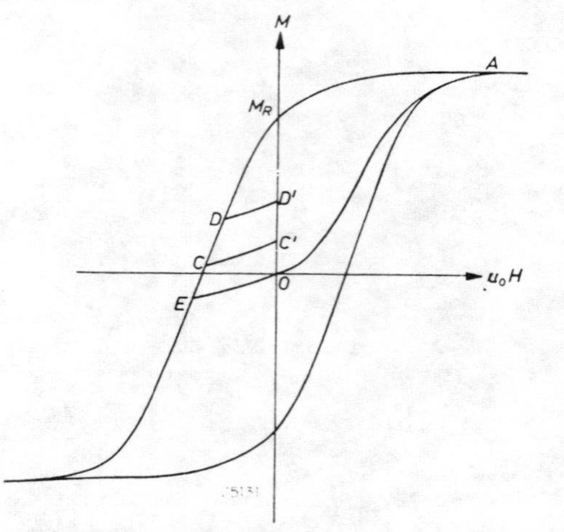

Fig. 15. Schematization of the d.c. recording method.

Firstly, in the absence of a signal the tape leaves the recording head in a demagnetized state. The fluctuations of the flux being less for a demagnetized than for a magnetized tape this means that for an a.c. biased tape the background noise is less than for a d.c. biased tape. Secondly, the linear range of the recording characteristic is more extended in the case of a.c. biasing, which means that the a.c. biasing method results in a greatly improved signal-to-noise ratio.

In principle it should also be possible to choose the direct current bias such that a point E in fig. 15 is reached under the influence of this biasing field and the origin after this field has faded. But the point E does not necessarily coincide with the point of inflexion C, and if not, this limits the recording range. Further a careful adjustment of this bias is necessary, and, owing to the inhomogeneity of the field in front of the recording head, the appropriate value is only reached for a small layer of the tape.

*) Continued from Philips Res. Rep. **8**, 148-157, 161-183, 1953.

Reprinted with permission from *Philips Res. Rep.*

Finally, even if the tape as a whole reaches the origin in this way, the state it has thus acquired is different from the virgin demagnetized state.

Owing to these reasons the improvement of the signal-to-noise ratio for direct-current recording, if possible at all, is very difficult to achieve.

2. Current explanations of the a.c. biasing method

An explanation of the a.c. biasing method was first given by Camras [21]), and independently along the same lines by Toomin and Wildfeuer [22]), the latter supporting it by an oscillographic study. Their explanation runs as follows: if the amplitude of the a.c. biasing field is such that a minor hysteresis loop is followed that has its reversal points on the steep branches of the major loop, then the application of an additional direct field will result in the shift of the minor loop along the steep branch. In fig. 16 a field H_1 will shift the loop $ABCD$ to the loop $A'B'C'D'$. Because the flanks of the major hysteresis loop are linear over a considerable length, the shift of the minor loops is proportional to the direct field applied.

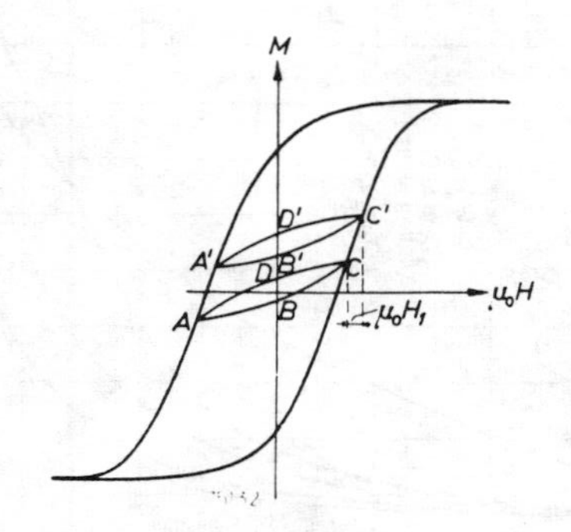

Fig. 16. Schematization of the a.c. recording method.

If now in the case of the loop $A'B'C'D'$ the a.c. field decreases, a path will be traced lying inside this loop and resulting in a magnetization about halfway between B' and D' when the a.c. field as well as the direct field have completely faded. This remanent magnetization too will be a linear function of the applied direct field.

In the above explanation it is assumed that stationary minor loops are followed. This, however, is only the case if a large number of cycli of the a.c. field are traversed while the direct field remains constant. In front of an actual head the fieldstrength varies rapidly with the distance to the gap, so that for a tape passing the head this condition is not fulfilled. To avoid this difficulty Holmes and Clark [23]) give an explanation in which they try to follow the magnetizing forces for successive parts of the tape in more detail. The magnetization curves, however, are insufficiently known to predict the ultimate magnetization of the tape with the desired accuracy. To come to conclusions a comparison is made with an AB-class push-pull amplifier, a method which yields the same results as those discussed above but has the disadvantage that the physical processes are less easily understood.

Both explanations are based upon properties of the magnetization curve without entering into the physical processes underlying these properties.

3. Relation between a.c. and ideal magnetization

Before discussing a physical model that explains the linearizing effect of an a.c. bias, we wish to draw attention to the close relation that exists between the method of a.c. biasing and the process of ideal or anhysteresic magnetization as proposed by Steinhaus and Gumlich [24]). In this method magnetization is accomplished by superimposing on the magnetic field an a.c. field of sufficiently large amplitude and then gradually decreasing this amplitude to zero. Thus a magnetization is brought about that is independent of the history of the magnetic material. Lying above the virgin curve, the ideal magnetization curve is always concave with respect to the H-axis. Further, if plotted against the internal field, the ideal magnetization curve rises perpendicular from the origin. If plotted against the external field the angle between the tangent in the origin and the M-axis is determined by the demagnetization factor.

The difference between the method of ideal magnetization and that in magnetic recording is that in the former the magnetic field remains constant during

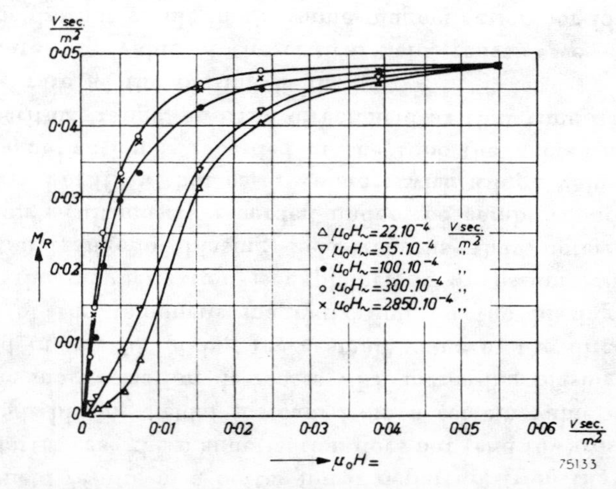

Fig. 17. Remanent magnetization as a function of direct field if an idealization process has been carried out starting from the indicated peak value of the idealizing field.

the decrease of the a.c. field, while in the latter it decreases at the same rate.

This may be illustrated by figs 17 and 18, taken from unpublished measurements of Dr. J. J. Went on magnetic tapes in homogeneous fields. The coercive force of the tape in question was $\mu_0 H_c = 75.10^{-4}$ Vsec/m². In both figures the remanent magnetization after the tape has been subjected to a superposition of a direct field and an a.c. field is plotted against the direct field for different values of the a.c. field. But, whereas in fig.17 the a.c. field was first removed and then the direct field, as in the case of ideal magnetization, in fig.18 both were reduced simultaneously and at the same rate, as is the case in magnetic recording. Experimentally this is achieved by drawing the tape out of a coil fed by superposed direct and alternating currents.

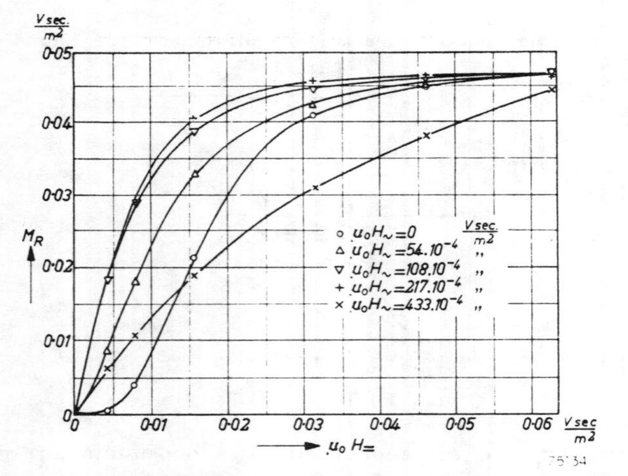

Fig.18. Remanent magnetization for the case where the direct field decreases at the same rate as the idealizing field, as a function of the starting value of the direct field.

It is seen from fig.17 that as the initial value of the a.c. field increases, the remanent magnetization increases also but approaches more and more a limiting value, the ideal curve. For a value of $\mu_0 H_\sim = 300.10^{-4}$ Vsec/m² the remanent curve already coincides with the ideal curve. In fig.19 the remanent magnetization is plotted against the idealizing fieldstrength, for a smal value of the direct field. It appears that the steepest part of this curve, at $\mu_0 H_c = 75.10^{-4}$ Vsec/m², coincides with the coercive force of the tape in question.

Comparison of figs 17 and 18 shows that in the latter the magnetization obtained is always smaller than in the former. This is to be expected, for in the latter case the direct field is smaller than in the former during all stages of the recording process.

It further appears that according to both methods an increase of the alternating field results in a better linear departure from the origin, but that in the latter case the magnetization does not tend towards a limiting value for high biasing fields. Instead, from a certain value of this field onwards, the magnetization decreases with increasing biasing field. This is illustrated in fig. 20, giving for a small value of the direct field the dependence of the remanent magnetization on the biasing field. The decrease of the magnetization is explained if we assume that there is only a certain range of biasing fieldstrengths that determines what magnetization is ultimately recorded under the combined action of a.c. and d.c. fields. For if, in that case, we start with a high value of the a.c. field the magnetization is recorded at the time the decreasing a.c. field passes this critical range, and the direct field at that time has decreased at the same rate as the biasing field.

To estimate the value of the critical fieldstrength we will compare the direct fields that result in the same magnetization for ideal and a.c.

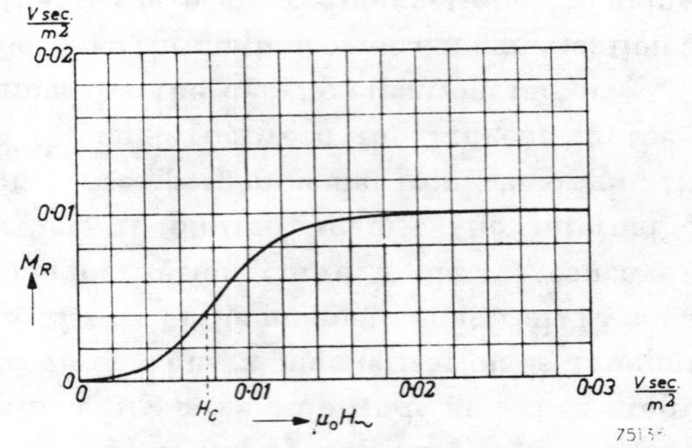

Fig.19. Remanent magnetization as a function of idealizing fieldstrength.

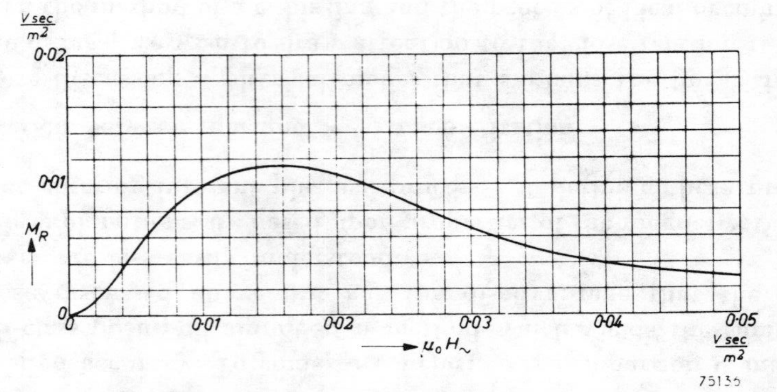

Fig.20. Remanent magnetization for the case where the direct field decreases with the idealizing field.

bias magnetization. If the a.c. field in the latter case is such that we are well over the maximum of fig.20 we may expect that these fields are related as the critical fieldstrength and the starting value of the a.c. field. Thus, by comparing the ideal curve of fig.17 with the curve for $\mu_0 H_{\sim} = 433.10^{-4}$ Vsec/m² of fig.18 we obtain for different magnetizations the following results.

$M_R \cdot 10^4$ Vsec/m²	$\mu_0 H_{id} \cdot 10^4$ from fig. 17 Vsec/m²	$\mu_0 H_m$ 10^4 from fig. 18 for $\mu_0 H_r = 433.10^{-4}$ Vsec/m²	$\mu_0 H_{crit} \cdot 10^4 = \frac{H_{id}}{H_m} .433$ Vsec/m²
50	5	32	67
100	10	69	63
150	15	114	57
200	22	164	58
250	30	225	58
300	41	293	61
350	60	392	66
400	85	514	72

Accordingly the critical fieldstrength is somewhat lower than the coercive force.

It should be remembered that there is in fact not one critical fieldstrength, but a small region that determines the recorded magnetization. This is evident also from fig.19. If there were one definite critical field it should be expected that the remanent magnetization be zero for idealizing fields below and have its full strength for fields above this critical value.

4. Magnetic model explaining the a.c. magnetization

We will now give a simple model that explains the ideal as well as the a.c. bias magnetization. Suppose that in a particle a number of magnetic states are possible. Each magnetic state corresponds to a minimum of the potential energy and is separated from other minima by potential barriers. For reasons of simplicity we will assume that the potential barriers are equally high, though this is not essential to our argument.

Application of a magnetic field will diminish the potential energy for those states where the direction of the magnetization is more in accordance with the direction of the field. Hence the potential barriers will decrease on the one side and increase on the other side of the potential minimum. For a certain value of the fieldstrength the first barrier will have disappeared and the magnetization will jump to a state with lower energy. Since all the barriers were supposed to be of equal height they will all be crossed at the same fieldstrength and therefore saturation will be obtained. Reversal of the field will effect saturation in the opposite direction. For an a.c. field of sufficient strength the magnetization will alternate between these two directions. If now the amplitude of the a.c. field is decreased gradually, there will be an equal chance for the magnetization to reach any of the potential minima. Thus a medium built up of a large number of independent particles will reach a state of zero magnetization.

If, however, a small direct field is added to the decreasing a.c. field there will be a preference for the direction of the direct field. In our case where all the potential barriers are of equal height the satured state will be reached provided only that the decrease of the a.c. field between two successive peak values is smaller than twice the strength of the direct field. Let a be the amplitude of the a.c. field at a certain moment, Δ the decrease of the a.c. field between two peaks, d the strength of the direct field and h the fieldstrength at which a potential barrier can just no longer be crossed; then the fieldstrength h will be reached for the first time when the a.c. and d.c. fields are of opposite direction, and this can just happen when $a - d = h$. When the a.c. field has altered its direction the amplitude of the next extreme will be $a - \Delta + d = h - \Delta + 2d$. This will surpass the fieldstrength h if $\Delta < 2d$, and accordingly the potential barriers on the other side of the minimum in question will be crossed. Therefore the only final state possible is that state where saturation in the direction of the applied field has been reached.

This is the explanation given by Steinhaus and Gumlich of the ideal magnetization. In effect, in a diagram of magnetization vs. internal field, the ideal curve will rise perpendicularly from the origin.

In reality a medium is not magnetized to saturation by an arbitrarily small direct field. This is because the magnetization of the particle gives rise to a demagnetizing field opposite in direction to the external field. The magnetization will increase to that value for which the external field is completely cancelled by the demagnetizing field. For only if the internal field is zero will no further increase of the magnetization take place during the idealizing process.

As the demagnetizing field is proportional to the magnetization and equals the direct field in strength, the residual magnetization will be proportional to the external field. The conditions to be fulfilled are that the a.c. field should be of sufficient strength for the magnetization to overcome the potential barriers for all the particles, and that there will be a sufficient number of stable positions of the magnetization.

Because the coercive force of a medium is that fieldstrength for which only half of the particles have reversed their magnetization, the first condition, viz. that all the barriers are overcome means that the idealizing field has to be well above coercivity, in accordance with fig.19. In fact it has to be above the highest coercive force met with in any of the particles.

If the number of stable magnetizations in one particle is limited, the latter condition will be fulfilled as well if the number of particles is sufficiently large.

In general we may explain the linearizing action of the a.c. field as the excitation of a magnetization which results in an internal field equal to the external field but of opposite direction.

If during the decrease of the alternating field the direct field changes also, as is the case in magnetic recording, the recorded magnetization is determined by that value of the direct field which existed at the moment that the amplitude of the alternating field was just sufficient to make the magnetization cross the potential barriers. In the case where the particles have different coercive forces, each particle will obtain the remanent magnetization that belongs to the value of the direct field existing when the a.c. field equalled its own coercive force.

5. Possible mechanisms of the magnetization process

In the above model the mechanism of the magnetization process was kept out of the discussion. It was only assumed that a number of magnetic states separated by potential barriers are possible and that these states are accompanied by a demagnetizing field proportional to the magnetization. For instance, a particle built up of two domains of different magnetization, separated by a Bloch wall that has a number of preferred positions, should match these requirements. We shall now see whether this can be brought in accordance with what is known about the magnetic particles.

From electronmicroscopic studies it appears that the particles γFe_2O_3, used in the magnetic tape studied above, are less than 0·1 μ in diameter. It depends upon the magnetic properties whether a Bloch wall can exist in these small particles, or whether it is energetically more favourable that the particle is magnetized as a whole.

The constants to be known are the Curie temperature T_c, the lattice constant a, the saturation magnetization M_s, and the anistropy constant K. The thickness δ of a 90° wall may then be estimated from

$$\delta = \left(\frac{kT_c}{a}\right)^{1/2} K^{-1/2},$$

and the wall energy per unit surface from

$$\sigma = \left(\frac{kT_c}{a}\right)^{1/2} K^{1/2} \text{ [25]}.$$

No value for the anisotropy constant of γFe_2O_3 being available, K can be estimated from the "approach to saturation" of the magnetization curve; because, if all irreversible processes have taken place, a further increase in magnetization can be accomplished only by rotation of the magnetization against the anisotropy forces. For practically independent particles with cubic symmetry the approach to saturation is given [26]) by

$$\frac{M}{M_s} = 1 - \frac{8}{105}\left(\frac{K}{M_s}\frac{1}{H}\right)^2 - \frac{192}{5005}\left(\frac{K}{M_s}\frac{1}{H}\right)^3$$

which satisfies approximately the experimental approach (fig. 21) by insertion of $\mu_0 K/M_s = 0{\cdot}08$ Vsec/m². With $M_s = 0{\cdot}44$ Vsec/m² this yields $K = 28000$ N/m² ($= 28.10^4$ ergs/cm³).

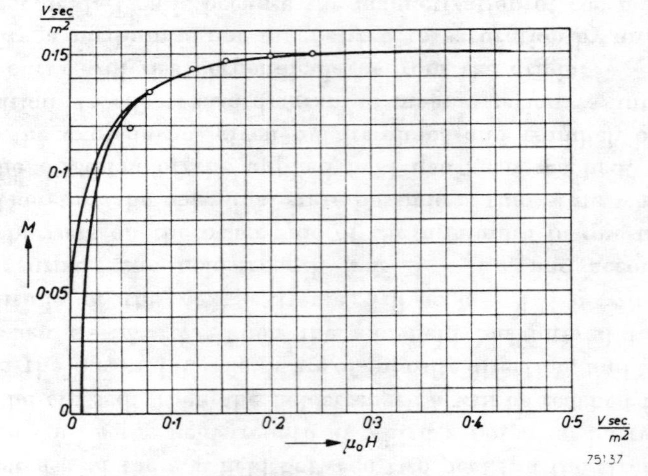

Fig. 21. Hysteresis loop with approach to saturation. The points are calculated from eq. (1).

With $T_c = 1000$ °K, $a = 3$ Å the thickness of a 90° wall is $\delta = 350$ Å, and the energy $\sigma \approx 1{\cdot}2.10^{-3}$ N/m ($= 1{\cdot}2$ erg/cm²).

If a spherical particle is magnetized to saturation the magnetostatic energy density is $M_s^2/(6\,\mu_0)$ N/m² per unit volume, and thus of a particle of radius r

$$E_m = \frac{2\pi}{9} r^3 \frac{M_s^2}{\mu_0} \text{ Nm}.$$

On the other hand the introduction of two perpendicular 90° walls in the particle will introduce a configuration in which the flux path is internally closed and the magnetostatic energy consequently very small. The energy then is that of the wall, $E_w \approx 2\pi r^2 \sigma$, and the critical diameter for which the energies are equal is $r_{cr} = 750$ Å. For particles with a diameter of less than 0·15 μ the saturated configuration has the lesser energy.

It seems therefore improbable that walls should occur in the magnetic powder studied above, unless positions of the Bloch wall in the particles are possible with an energy lower than the energy according to the formula for σ given above.

If, on the other hand, the particles are magnetized to saturation a change can only be obtained by rotation from one direction of easy magnetization to another. For a spherical particle with cubic lattice the coercive force is $2K/M_s$ if the direction of the applied magnetic field is opposite to the direction of magnetization. For a medium built up of a number of such particles oriented at random the average force was calculated by Néel [27]) as 0·64 K/M_s. For the magnetic powder of the recording tape studied this should give a coercive force of 640.10^{-4} Vsec/m² instead of the observed 75.10^{-4} Vsec/m². Therefore, if the particles are magnetized as a whole they are either not spherical, or not independent. In the latter case clusters of particles may be imagined in which the particles influence each other in such a way that the coercive force is less than in the case of separate particles. In this cluster a demagnetizing field proportional to the magnetization will occur.

IV. CALCULATION OF THE FIELDS IN AND AROUND THE TAPE

1. Basis of the calculation

When the recording process has taken place and the tape has left the recording head the recorded magnetization will give rise to an external field but also to a demagnetizing field in the tape. This field will be directed opposite to the recorded magnetization and thus result in a decrease of the latter. This decrease will be the more pronounced the stronger the demagnetizing field and the greater the permeability of the tape. When the permeability equals unity, the magnetization will not be affected by this or any other field provided the fieldstrength does not exceed the coercive force.

In this case of unit permeability, the equations for the response curve of the preceding section hold. Here it makes no difference if the recorded magnetization is directed in the direction of motion of the tape (longitudinal magnetization) or perpendicular to the plane of the tape (perpendicular magnetization), although the demagnetizing fields are quite different in the two cases, especially for long wavelengths.

This is evident from the fact that the apparent magnetic poles which give rise to the demagnetizing field are equal to $\mathrm{div}\boldsymbol{M}/\mu_0$. Thus for longitudinal magnetization the poles, and therefore also the demagnetizing field, tend towards zero for increasing wavelength. For perpendicular magnetization, on the contrary, there are always poles on the surface of the tape that give rise to a demagnetizing field, independent of the wavelength.

We shall now proceed to calculate the residual magnetization under the influence of the demagnetizing field of a tape in which the magnetic moment is not rigidly fixed or, in other words, that has a relative permeability greater than unity.

For the tape the Maxwell equations hold

$$\mathrm{curl}\ \boldsymbol{H} = 0\,, \tag{1}$$

$$\mathrm{div}\ \boldsymbol{B} = \mathrm{div}(\boldsymbol{M} + \mu_0 \boldsymbol{H}) = 0\,. \tag{2}$$

To solve the problem a relation between $\boldsymbol{M}$ and $\boldsymbol{H}$ must be known. For this we shall use

$$\boldsymbol{M} = \boldsymbol{M}_0 + (\mu - 1)\,\mu_0 \boldsymbol{H}. \tag{3}$$

For $\boldsymbol{M}_0 = 0$ this is the normal relation for linear isotropic media, where μ is the relative reversible permeability. Eq.(3) involves that after irreversible processes have taken place such that a remanent magnetization $\boldsymbol{M}_0$ is brought about, a magnetic field produces the same change in magnetization as before, independent of the magnitude of the remanent magnetization. This of course is only true in a limited range of remanent magnetization, but it is to be expected that it holds with sufficient accuracy in the range used in magnetic recording.

A proof for the independence of the permeability of the permanent magnetization $\boldsymbol{M}_0$ is found from the measurements of the selfinductance of a coil filled with tape. From these measurements it appears that the permeability does not change more than 2% after the tape has been subjected to direct fields, up to saturation strength.

That no irreversible changes occur may be seen from the fact that the recorded magnetization does not decrease noticeably after repeated playback. For during the passage of a head consisting of high-permeable metal the demagnetizing field is decreased considerably. If irreversible processes are taking place, a decrease of $\boldsymbol{M}_0$ is to be expected. Since the voltage induced in the reproducing head is a measure for $\boldsymbol{M}_0$ the result would be a decrease of the output voltage after repeated playback, an effect which is only observed in a slight degree after the first few times.

From eqs(1), (2) and (3) it follows that

$$\Delta \boldsymbol{H} = -\frac{1}{\mu\mu_0}\,\mathrm{grad\ div}\,\boldsymbol{M}_0. \tag{4}$$

M_0 is the magnetization recorded by the recording head. Thus, if the boundary conditions are known, $\boldsymbol{H}$ may be solved from this equation.

To make the boundary problem more easy to solve we shall suppose the tape to be of infinite length and width, the thickness of the tape being d. As we shall see the influence of the neighbourhood of a high-permeable reproducing head cannot be neglected. The head is introduced by the supposition that in a plane at a distance a from the tape the lines of force are perpendicular to this plane. This is equivalent with a head of infinite dimensions and of infinite permeability with its surface in this plane, the gap of which is so small that its influence on the field distribution may be neglected.

We shall use a coordinate system with its origin in the boundary plane of the tape nearest to the head, with the x-axis in the direction of motion and the y-axis normal to the plane of the tape (fig. 22). Thus the other boundary plane of the tape is $y = -d$, and the boundary plane of the head is $y = a$.

The recorded magnetization $\boldsymbol{M}$ be sinusoidal and a function of x only.

The direction of $\boldsymbol{M}$ is arbitrary in the xy-plane. In order to make the formulae less intricate we shall treat the longitudinal magnetization (directed along the x-axis) and the perpendicular (directed along y-axis) separately. The general case may be found from these two by superposition.

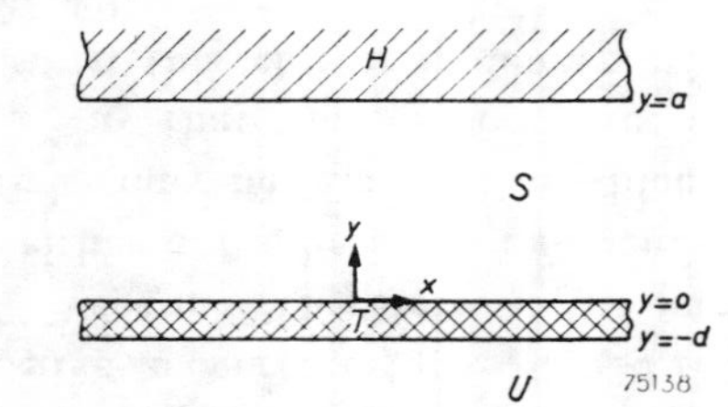

Fig.22. Schematic view of tape T and head H, the latter represented by the plane $y = a$ above which the permeability is infinite.

2. Longitudinal magnetization

The recorded magnetization being directed along the x-axis we may write for the tape

$$M_{0x} = M_0 \sin kx, \quad M_{0y} = 0; \quad (k = 2\pi/\lambda).$$

Thus eq.(4) becomes

$$\Delta H_x = \frac{M_0}{\mu\mu_0} k^2 \sin kx, \quad \Delta H_y = 0.$$

This is solved by

$$H_x = \left(A_1 e^{ky} + A_2 e^{-ky} - \frac{M_0}{\mu\mu_0}\right) \sin kx,$$

$$H_y = (-A_1 e^{ky} + A_2 e^{-ky}) \cos kx,$$

where A_1 and A_2 are arbitrary constants.

Outside the tape analogous solutions hold with $M_0 = 0$. Thus we have three sets of solutions each with two constants.

The six constants are determined by the boundary conditions

$$y = -\infty : H_x = H_y = 0, \qquad y = 0 : H_x \text{ and } B_y \text{ continuous},$$
$$y = -d \;: H_x \text{ and } B_y \text{ continuous}, \qquad y = a : H_x = 0.$$

This gives six equations for the determination of the constants. By solving for the constants and inserting the values we obtain the solution satisfying the differential equation and the boundary conditions

for $-\infty < y < -d$:

$$\begin{matrix}\mu_0 H_x \\ = \\ \mu_0 H_y\end{matrix} \quad \begin{matrix}-\sin kx \\ \\ +\cos kx\end{matrix} \times \frac{M_0 \tanh kd \{\tanh ka + (1/\mu)\tanh (kd/2)\} \exp \{k(y+d)\}}{1 + \tanh ka + (\mu \tanh ka + 1/\mu) \tanh kd}; \tag{5a}$$

for $-d < y < 0$:

$$\begin{matrix}\mu\mu_0 H_x + M_0 \sin kx \\ = \\ \mu\mu_0 H_y\end{matrix} \quad \begin{matrix}+\sin kx \\ \\ -\cos kx\end{matrix} \times \left\{ A \begin{matrix}\cosh k(y+d/2) \\ \sinh k(y+d/2)\end{matrix} - B \begin{matrix}\cosh ky \\ \sinh ky\end{matrix} \right\},$$

where

$$A = \frac{M_0 \{2 \cosh (kd/2) + (1/\mu) \sinh (kd/2)\}}{\cosh kd \{1 + \tanh ka + (\mu \tanh ka + 1/\mu) \tanh kd\}},$$

$$\text{and } B = \frac{M_0 (1 - \tanh ka)}{\cosh kd \{1 + \tanh ka + (\mu \tanh ka + 1/\mu) \tanh kd\}}; \tag{5b}$$

for $0 < y < a$:

$$\begin{matrix}\mu_0 H_x \\ = \\ \mu_0 H_y\end{matrix} \quad \begin{matrix}\sin kx \sinh k(a-y) \\ \\ \cos kx \cosh k(a-y)\end{matrix} \times \frac{-M_0 \tanh kd \{1 + (1/\mu) \tanh (kd/2)\}}{\cosh ka \{1 + \tanh ka + (\mu \tanh ka + 1/\mu) \tanh kd\}}. \tag{5c}$$

It is seen from eq.(5b) that in general the field, and thus the magnetization in the tape, is a function of the depth in the tape. If, however, the wavelength is large in comparison with the thickness of the tape, or $|ky| < kd \ll 1$, then in first approximation

$$\mu_0 H_x \approx -M_0 kd \frac{\tanh ka}{1+\tanh ka} \sin kx\,. \tag{6}$$

The field is directed against the magnetization and constant over the depth of the tape. To this approximation a demagnetization factor D can be introduced by the equation $D = -\mu_0 H_x/M_x \approx -\mu_0 H_x/M_0 \sin kx$, which with eq.(6) gives

$$D = kd \frac{\tanh ka}{1+\tanh ka}\,.$$

D is inversely proportional to the wavelength and depends, moreover, on the space a between head and tape, decreasing when a decreases. For tape and head in close contact ($a = 0$) the demagnetizing factor is zero, while for a tape free in space ($a = \infty$) $D = \pi d/\lambda$.

If the wavelength is not large in comparison with the thickness of the tape, $\boldsymbol{H}$ and $\boldsymbol{M}$ are different functions of the depth in the tape, and it is therefore senseless to introduce a demagnetizing factor.

For a tape in free space the demagnetizing field has a maximum and the magnetization a minimum in the centre of the tape. This is illustrated in fig. 23*a*, giving the computed induction $B_x = M_x + \mu_0 H_x = M_0 \sin kx + \mu\mu_0 H_x$ as a function of the depth in the tape for a tape with a permeability $\mu = 4$, and for different values of $d/\lambda = kd/2\pi$. Fig. 23*b* gives the same data for a tape in close contact with the head ($a = 0$).

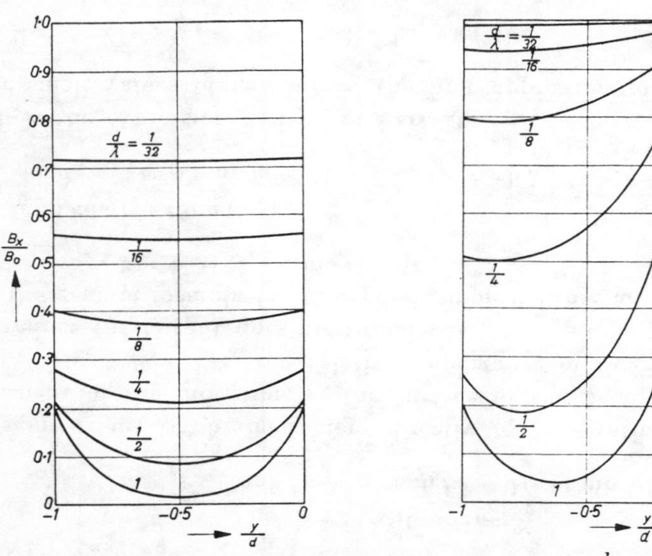
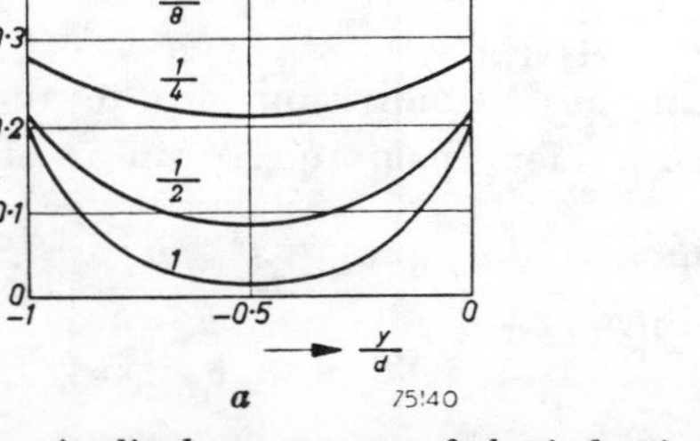

Fig. 23. Longitudinal component of the induction in a tape of permeability $\mu = 4$. *a*. tape free in space; *b*. tape on one side ($y = 0$) in contact with a head.

It is seen that by the presence of the head the demagnetizing field is decreased and thus the original magnetization partly restored.

It is most important to know the flux passing through the coil of the reproducing head due to the presence of the magnetized tape. This may be calculated by considering the plane $x = x_0$, where x_0 is the abscissa of the (infinitely short) gap. From div $\boldsymbol{B} = 0$ it follows that the balance of the flux traversing this plane is zero. Thus the flux in the head through this plane is found by calculating the flux in the tape and subtracting the flux closing in the space under the tape, U (fig. 22), and between tape and head, s. If the head is constructed properly, then the flux through the head is forced through the coil by the reluctance of the gap.

The amplitude Φ_1 of the flux through a width b of the tape is found from

$$\Phi_1 \sin kx_0 = b \int_{-d}^{0} B_x \mathrm{d}y = b \int_{-d}^{0} (\mu\mu_0 H_x + M_0 \sin kx_0)\, \mathrm{d}y =$$
$$= \Phi_0 \frac{\tanh kd}{kd} \frac{1 + (2/\mu)\tanh (kd/2) + \tanh ka}{1 + \tanh ka + (\mu \tanh ka + 1/\mu)\tanh kd} \sin kx_0\,.$$

Here $\Phi_0 = M_0 bd$ is the amplitude of the flux variations in the tape at long wavelengths. The dependence of Φ_1/Φ_0 on a/λ is represented in fig. 24.

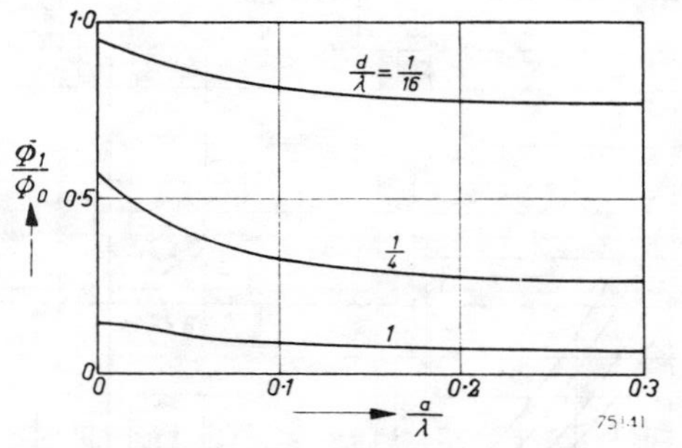

Fig. 24. Dependence of the flux in the tape on the separation between head and tape for a tape of permeability $\mu = 4$.

In the same way the flux Φ_2 below the tape is found to be

$$\Phi_2 = -\Phi_0 \frac{\tanh kd}{kd} \frac{\tanh ka + (1/\mu)\tanh (kd/2)}{1 + \tanh ka + (\mu \tanh kd + 1/\mu)\tanh kd},$$

and the flux Φ_3 closing between head and tape

$$\Phi_3 = -\Phi_0 \frac{\tanh kd}{kd} \frac{\{1 + (1/\mu)\tanh (kd/2)\}(1 - 1/\cosh ka)}{1 + \tanh ka + (\mu \tanh ka + 1/\mu)\tanh kd}.$$

Thus the flux Φ through the coil of the head is

$$\Phi = \Phi_1 - \Phi_2 - \Phi_3 =$$

$$= \Phi_0 \frac{\tanh kd}{kd} \frac{1 + (1/\mu)\tanh(kd/2)}{1 + \tanh ka + (\mu \tanh ka + 1/\mu)\tanh kd} \frac{1}{\cosh ka} . \quad (8)$$

The same result is obtained more directly by integrating the vertical component of the induction at the surface of the head from $x = 0$ to $x = x_0$. Then, however, the relative importance of the various factors that reduce the output does not stand out.

In the general formula (8) Φ depends upon d/λ, a/λ and μ in a rather complex way. For $\mu = 1$ we may write

$$\frac{\Phi}{\Phi_0} = \frac{\tanh kd}{kd} \frac{1 + \tanh(kd/2)}{1 + \tanh kd} \frac{1}{(1 + \tanh ka)\cosh ka} = \frac{1 - e^{-2\pi d/\lambda}}{2\pi d/\lambda} e^{-2\pi a/\lambda}, \quad (8a)$$

which is the expression derived in Part II.

For small values of ka and kd the first approximation of (8) gives

$$\Phi/\Phi_0 \approx 1 - ka - kd/2\mu . \quad (8b)$$

For larger values of kd, $\tanh(kd/2)$ and $\tanh kd$ are approximately 1 and thus

$$\frac{\Phi}{\Phi_0} \approx \frac{1}{kd} \frac{1}{1 + \mu \tanh ka} \frac{1}{\cosh ka} = \frac{1}{kd} \frac{2}{\mu + 1} \frac{e^{-ka}}{1 - e^{-2ka}(\mu - 1)/(\mu + 1)} . \quad (8c)$$

Since $\tanh 1{\cdot}5 = 0{\cdot}9$ this is a reasonable approximation from $d/\lambda = 0{\cdot}5$ onwards.

It is easily seen from the general formula (8) that an increase of kd and ka always results in a decrease of Φ/Φ_0. However, an increase of the permeability may sometimes enlarge the flux in the head as is seen from (8b) for small values of kd. For larger values of kd (eq.8c) the flux decreases with increasing μ.

In fig. 25*a* the variation of Φ/Φ_0 with d/λ is given for a constant ratio of the space a between head and tape to the thickness d of the tape, $a/d = \frac{1}{4}$, and some values of the relative permeability μ of the tape. Fig. 25*b* gives the variation with d/λ for a tape with $\mu = 4$ and some values of a/d.

In order to illustrate the causes of the flux loss in the case of a tape with $\mu = 4$ passing the head at a distance $a = 0{\cdot}1d$, fig. 26 represents as a function of d/λ the fraction of the flux in the tape closing beneath the tape (curve *a*) and the flux closing between tape and head (curve *b*). The remaining flux finds its way through the head (curve *c*).

From the approximation (8c) it is seen that, since Φ_0 is proportional to the tape thickness d, the flux Φ in the head is independent of tape thickness if the wavelength λ is comparable to d. This was observed already by Kornei [28]) and Herr, Murphy and Wetzel [29]). The reasons why only a surface layer contributes to the reproduced flux are twofold. On the one hand the demagnetizing field increases for deeper layers

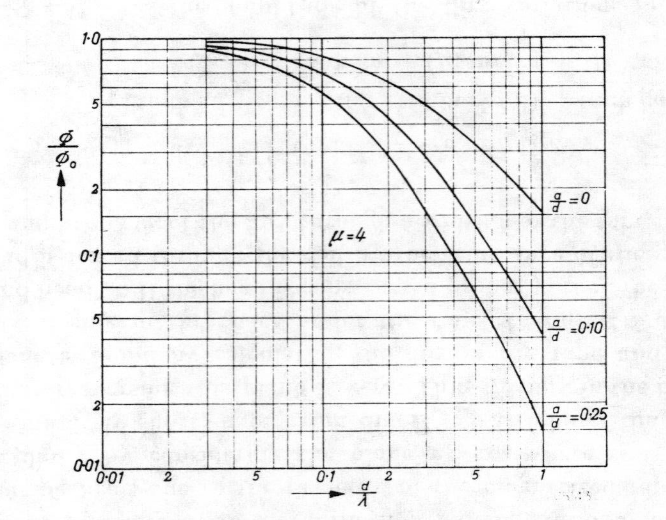

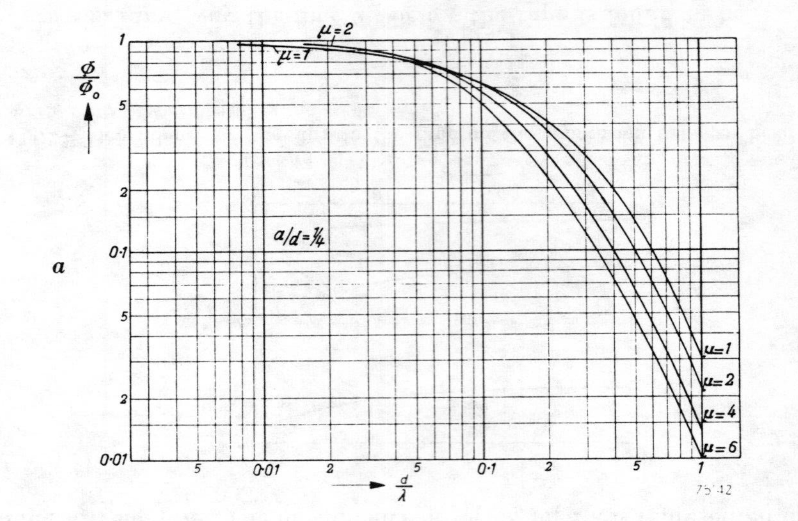

Fig. 25. Reproduced flux as a function of wavelength.
a. for some values of the permeability, and $a/d = \frac{1}{4}$,
b. For some values of the space between head and tape and $\mu = 4$.

so that for a tape in contact with the head the flux is decreased everywhere except on the contacting side, as is illustrated in fig. 22*b*; and on the other hand the deeper layers do not contribute to the flux in the head because of the rapid decrease of the reproduced flux with increasing distance.

Another fact is that the recorded magnetization will decrease with the distance from the recording head, while here the recorded magnetization was supposed to be constant over the depth of the tape.

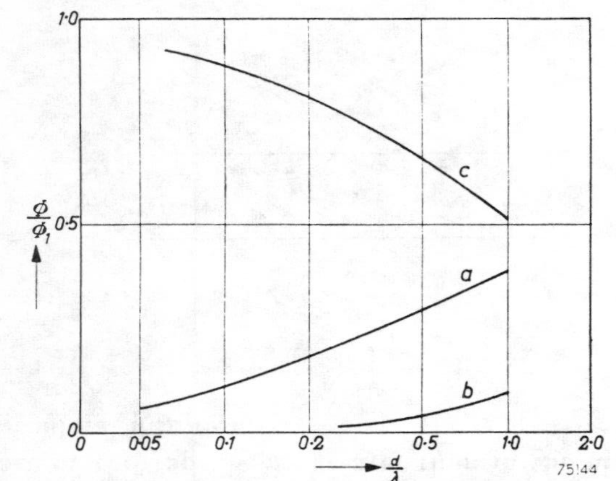

Fig. 26. Fraction of the flux in the tape passing through the space U (fig. 22) beneath the tape (a), through the space S between tape and head (b), and through the head H (c), as a function of d/λ.

3. Perpendicular magnetization

In the case of perpendicular magnetization the recorded magnetization may be written as

$$M_{0x} = 0, \quad M_{0y} = M_0 \sin kx,$$

and therefore eq. (4) becomes

$$\Delta H_x = 0, \quad \Delta H_y = 0\,.$$

The solution is obtained along the same lines as in the case of longitudinal magnetization in the preceding section. With the same boundary conditions it is found that

for $-\infty < y < -d$:

$$\begin{matrix} \mu_0 H_x \\ = \\ \mu_0 H_y \end{matrix} \quad \begin{matrix} \cos kx \\ \\ \sin kx \end{matrix} \times \frac{M_0 \tanh kd \{\tanh ka \tanh (kd/2) + 1/\mu\} \exp \{k(y+d)\}}{1 + \tanh ka + (\mu \tanh ka + 1/\mu) \tanh kd}; \qquad (9a)$$

for $-d < y < 0$:

$$\begin{matrix} \mu\mu_0 H_x \\ = \\ \mu\mu_0 H_y \end{matrix} \quad \begin{matrix} \cos kx \\ \\ \sin kx \end{matrix} \times \left\{ A \begin{matrix} \sinh k(y+d/2) \\ \\ \cosh k(y+d/2) \end{matrix} + B \begin{matrix} \sinh ky \\ \\ \cosh ky \end{matrix} \right\}, \qquad (9b)$$

where

$$A = \frac{-2M_0 \tanh ka \{\cosh (kd/2) + \mu \sinh (kd/2)\}}{\cosh kd \{1 + \tanh ka + (\mu \tanh ka + 1/\mu) \tanh kd\}},$$

$$B = \frac{-M_0(1 - \tanh ka)}{\cosh kd \{1 + \tanh ka + (\mu \tanh ka + 1/\mu) \tanh kd\}};$$

for $0 < y < a$:

$$\begin{matrix} \mu_0 H_x \\ = \\ \mu_0 H_y \end{matrix} \quad \begin{matrix} -\cos kx \sinh k(a-y) \\ \times \\ +\sin kx \cosh k(a-y) \end{matrix} \times \frac{M_0 \tanh kd \{1/\mu + \tanh(kd/2)\}}{\cosh ka \{1 + \tanh ka + (\mu \tanh ka + 1/\mu) \tanh kd\}}. \qquad (9c)$$

If the wavelength is large compared with the thickness of the tape ($kd \ll 1$) then in first approximation

$$\mu_0 H_y \approx -\frac{M_0}{\mu}\left(1 - \frac{1}{1 + \tanh ka}\,\frac{kd}{\mu}\right) \sin kx\,. \qquad (10)$$

Thus if a demagnetization factor is defined as $D = -\mu_0 H_y / M_y$ then it follows from eq. (10), in combination with $M_y = M_0 \sin kx + (\mu - 1)\mu_0 H_y$, that to a first approximation

$$D = 1 - kd\,\frac{1}{1 + \tanh ka}\,.$$

The demagnetization factor is unity for long wavelengths, independent of the question whether the tape is in contact with the head or not. This is easily seen in the case of a tape magnetized homogeneously in the perpendicular direction. For then, if $\operatorname{div} \boldsymbol{B} = \operatorname{div} (\boldsymbol{M} + \mu_0 \boldsymbol{H}) = 0$ is applied to a surface element of the tape it appears that $\mu_0 H_y = -M_y$, thus $D = 1$. It is further seen that in this case $M_y = M_0/\mu$. The magnetization has decreased by a factor μ under the influence of the demagnetizing field, and it is consequently to be expected that at low frequencies the output decreases at the same rate.

If the wavelength is not long compared with the thickness of the tape, then the field as well as the magnetization are functions of the depth in the tape. To obtain the flux through a reproducing head, with its gap in the plane $x = x_0$, at a distance a from the tape, we shall again calculate separately the flux through the cross-section of this plane with the tape, the flux closing beneath the tape, and between tape and head. As before, the result can also be obtained in a shorter way by integrating the vertical component of the induction at the surface of the head, in this case from $x = \lambda/4$ to $x = x_0$. The amplitude Φ_1^1 of the flux through the cross-section bd of the tape in the plane $x = x_0$ is found from

$$\Phi_1^1 \cos kx_0 = b \int_{-d}^{0} \mu\mu_0 H_x \mathrm{d}y,$$

which gives

$$\Phi_1^1 = \Phi_0 \frac{\tanh kd}{kd} \frac{(1-\tanh ka)\tanh(kd/2)}{1+\tanh ka + (\mu \tanh ka + 1/\mu)\tanh kd}.$$

For a free tape ($a = \infty$, and hence $\tanh ka = 1$) it is seen that $\Phi_1^1 = 0$, thus there is no resultant longitudinal flux component. The configuration of the magnetization in the tape is sketched in fig. 27*a*. The longitudinal components of the magnetization in the upper and the lower half of the tape are of opposite direction and cancel each other out owing to symmetry. If now the tape is brought in the vicinity of a reproducing head the symmetry is disturbed (fig. 27*b*), the longitudinal induction in the lower part of the tape being greater than in the upper part. Thus a residual longitudinal flux results.

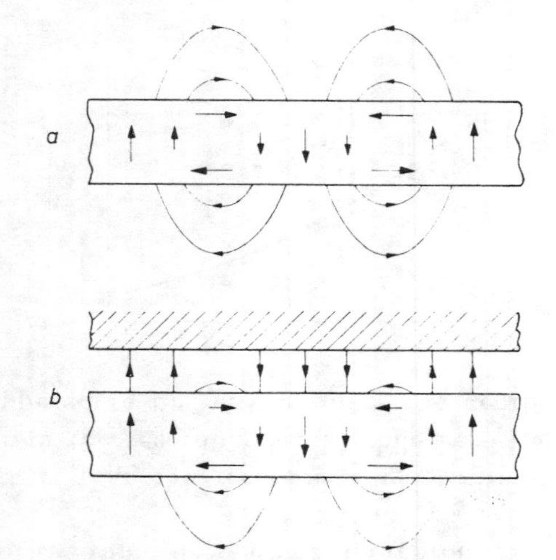

Fig. 27. Schematic illustration of the magnetization in the tape for perpendicular magnetization

a. free tape,
b. tape approached from one side by a reproduction head.

The amplitude of the flux closing beneath the tape is

$$\Phi_2^1 = \Phi_0 \frac{\tanh kd}{kd} \frac{\tanh ka \tanh(kd/2) + 1/\mu,}{1+\tanh ka + (\mu \tanh ka + 1/\mu)\tanh kd},$$

and of the flux closing between tape and head

$$\Phi_3^1 = \Phi_0 \frac{\tanh kd}{kd} \frac{\{1/\mu + \tanh(kd/2)\}(1 - 1/\cosh ka)}{1+\tanh ka + (\mu \tanh ka + 1/\mu)\tanh kd}.$$

It is seen from the formulae (9), and from fig. 27*b*, that the flux beneath the tape has the same direction as the residual flux in the tape; therefore the amplitude of the flux variations in the head, and thus in the coil, is $\Phi_p = \Phi_1^1 + \Phi_2^1 - \Phi_3^1$. Hence

$$\Phi_p = \Phi_0 \frac{\tanh kd}{kd} \frac{1/\mu + \tanh(kd/2)}{1+\tanh ka + (\mu \tanh ka + 1/\mu)\tanh kd} \frac{1}{\cosh ka}. \quad (11)$$

The recorded magnetization being a sine and the reproduced flux a cosine function there is a 90° phase difference between the two. In the case of the longitudinal magnetization no phase shift occurred.

In fig. 28*a* the values of Φ_p/Φ_0 computed from eq. (11) are plotted against d/λ for a constant ratio $a/d = \frac{1}{4}$ and several values of the permeability. In fig. 28*b* these curves are drawn for a constant value $\mu = 4$ of the permeability and some values of the parameter a/d.

For small values of ka and kd the first approximation of (11) gives

$$\Phi_p/\Phi_0 = (1/\mu)\{1 - ka + kd(\mu/2 - 1/\mu)\}. \quad (11)$$

In effect the output is inversely proportional to the permeability as a consequence of the demagnetization.

For shorter wavelengths, $kd/2 > 1$, Φ_p/Φ_0 is approximately equal to

$$\frac{\Phi_p}{\Phi_0} \approx kd \frac{1}{1+\mu \tanh ka} \frac{1}{\cosh ka} = \frac{1}{kd} \frac{2}{\mu+1} \frac{e^{-ka}}{1 - e^{-2ka}(\mu-1)/(\mu+1)}. \quad (11c)$$

This is identical with (8c), thus for short wavelengths longitudinal and perpendicular magnetization yield the same output.

4. Discussion

We have seen above that for of a tape of permeability other than unity the calculated output differs from that in the case of unit permeability, while the permeability enters the output formula in an intricate way. Moreover the output depends on whether the magnetization is longitudinal or perpendicular. We shall here first review the most important formulae for the two cases, longitudinal and perpendicular magnetization.

If the magnetization is given by $M_0 \sin kx$, directed along the x-axis and along the y-axis respectively, the flux through a reproducing head with its gap in the plane $x = x_0$ is

$$\frac{\tanh kd}{kd} \frac{1 + (1/\mu) \tanh (kd/2)}{N \cosh ka} \Phi_0 \sin kx_0 , \tag{8}$$

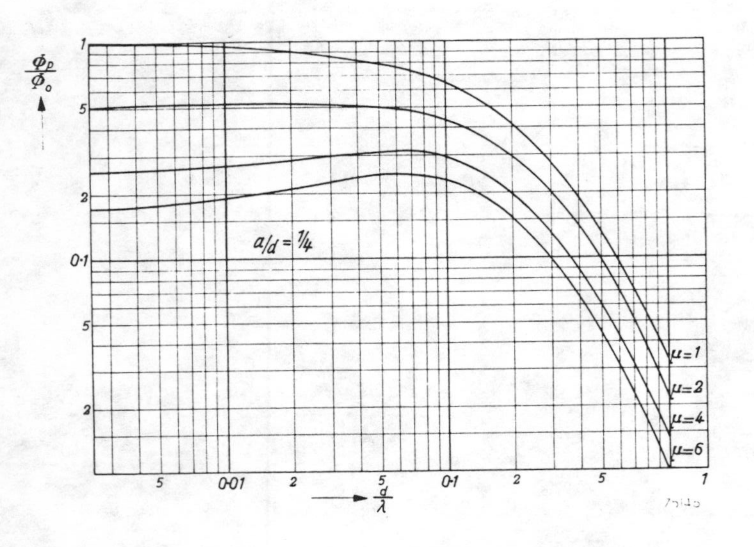

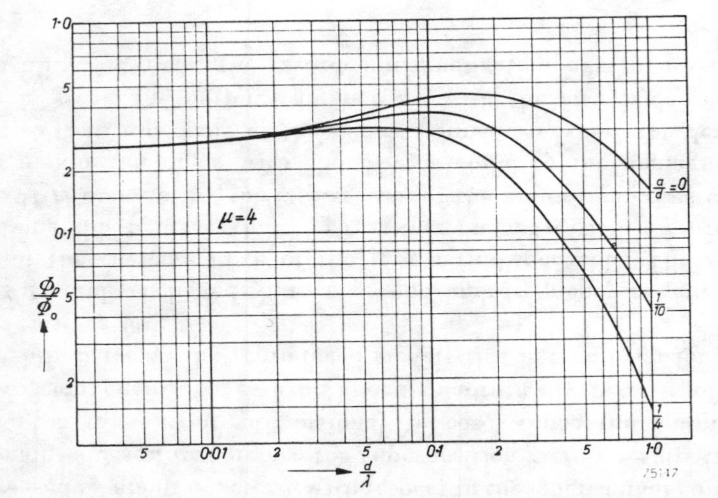

Fig. 28. Reproduced flux as a function of wavelength in the case of perpendicular magnetization.

a. for some values of the permeability, and $a/d = \frac{1}{4}$,
b. ,, ,, ,, ,, ,, space between head and tape, and $\mu = 4$.

and

$$\frac{\tanh kd}{kd} \frac{1/\mu + \tanh (kd/2)}{N \cosh ka} \Phi_0 \cos kx_0 \tag{11}$$

respectively, where

$$N = 1 + \tanh ka + (\mu \tanh ka + 1/\mu) \tanh kd \text{ and } \Phi_0 = M_0 bd .$$

If the wavelength is comparable with the thickness of the coating we may use in both cases as an approximation for these formulae

$$\Phi/\Phi_0 \approx \frac{1}{kd} \frac{2}{\mu + 1} \frac{e^{-ka}}{1 - e^{-2ka} (\mu - 1)/(\mu + 1)} . \qquad \text{(8c), (11c)}$$

For wavelengths much longer than the thickness of the coating the approximations

$$\Phi/\Phi_0 \approx 1 - ka - kd/2\mu \qquad \text{(8b) (longitudinal)}$$

and

$$\Phi_p/\Phi_0 \approx (1/\mu) \{ 1 - ka + kd (\mu/2 - 1/\mu) \} \qquad \text{(11b) (perpendicular)}$$

may be used.
In this approximation the demagnetizing field is given by

$$\mu_0 H_x = - kd \frac{\tanh ka}{1 + \tanh ka} M_0 \sin kx \tag{6}$$

and

$$\mu_0 H_y = - \frac{1}{\mu} \left(1 - \frac{kd}{\mu} \frac{1}{1 - \tanh ka} \right) M_0 \sin kx \tag{10}$$

respectively.

Comparison of eqs (8) and (11) shows that for $\mu = 1$ the amplitude factors are identical and can moreover be brought into the more simple form

$$\frac{1 - e^{-2\pi d/\lambda}}{2\pi d/\lambda} e^{-2\pi a/\lambda} . \tag{8a}$$

This is in conformity with the statement of section 4 in Part II where it was deduced that for a head of finite gap length the amplitude of the output variations is the same for longitudinal and perpendicular magnetization. It may be shown that this holds generally for any type of head that is symmetrical with respect to the y-plane provided that the field-strength vanishes at infinity.

In practice the magnetization will not be purely longitudinal or perpendicular. If the direction of magnetization makes an angle α with the positive x-axis, then the general solution for the output is obtained by a superposition of the solutions (8) and (11), where Φ_0 has to be replaced by $\Phi_0 \cos\alpha$ and $\Phi_0 \sin\alpha$ respectively. At short wavelengths, where the solutions for both cases are the same apart from the 90° phase shift, the superposition merely results in a phase shift of the reproduced signal over an angle a. At long wavelengths the output of the perpendicular component is decreased by about a factor μ with respect to the longitudinal component. For small values of a therefore the superposition merely results in the reduction of the output proportional to $\cos a$. Since the amplitudes of the two components have to be added quadratically because of the 90° phase shift between the reproduced signals this is true even for angles a rather close to 90°.

It is generally believed that for a good high-frequency response a high value of the coercive force of the tape is indispensable. The reasoning underlying this belief is that the resulting magnetization is reprensented in the M-H diagram by the intersection of the major hysteresis loop with a line making an angle with the M-axis equal to the demagnetization factor; so that, especially at the higher frequencies where the demagnetization is great, the output will increase with the coercive force.

In reality, however, the recorded magnetization has to be such that the major loop is never reached; for this would mean that the peaks of a recorded sine wave are cut off, which would result in a distortion of the signal. It is therefore obvious that in the formulae (8) and (11) the coercive force does not appear.

The real influence of the coercive force on the reproduction of the high frequencies has to be seen as follows. Firstly there is a general but not very strict relation between coercive force and permeability of magnetic materials, in the sense that a low coercive force is accompanied by a high permeability, and vice versa [30]). Since a high permeability reduces the high-frequency output this may also roughly be said of a low coercive force.

Secondly, a low coercive force in principle limits the level at which a good frequency response can be obtained without distortion at the high frequencies. But even in the case of the low coercive tape ($\mu_0 H_\sim = 75.10^{-4}$ Vsec/m^2) used in our experiments other sources of distortion come into play before this level is reached.

Dynamical Interpretation of Magnetic Recording Process

SHUN-ICHI IWASAKI AND TOSHIYUKI SUZUKI

***Abstract*—The demagnetization effect in magnetic recording must be evaluated, not as static self-demagnetization, but as a dynamic demagnetizing field at an instant when the head field is applied to the medium. From this fact it becomes necessary to obtain a self-consistent magnetization distribution in the medium. A method of calculation and its results are described. The relation between the longitudinal and the vector magnetization is clarified. The experimental results of the recording demagnetization in sinusoidal recording and the pulse width and the peak shift in digital recording are interpreted as the new phenomena that is related to the dynamical behavior of the demagnetizing field in the recording process.**

I. Introduction

THEORETICAL and experimental studies to obtain the magnetization distribution in the recording medium have been made by many researchers with the aim of attaining a higher storage density in magnetic recording. According to recent publications, these studies may be classified as follows.

Manuscript received March 4, 1968. Paper 3.1, presented at the 1968 INTERMAG Conference, Washington, D. C., April 3–5.

The authors are with the Research Institute of Electrical Communication, Tohoku University, Sendai, Japan.

A. *Theoretical*

1) In sinusoidal recording self-demagnetizing fields are calculated to determine the intensity of magnetization on the demagnetization curve [1].

2) In digital recording by means of the harmonic analysis, the self-demagnetization effect is taken into account for each harmonic component as in 1) [2]–[4].

3) The magnetization distribution in the medium is so determined that the maximum demagnetizing field produced by magnetic charges in the transition region may not exceed the coercivity of the medium [5]–[8].

4) In digital recording, regarding the transition regions between a positive and a negative magnetization as domain walls, the transition length may be determined by minimizing the energy involved [9].

B. *Experimental*

5) Remanent magnetization at each elemental volume in the medium is determined by tracing on a family of hysteresis curves of the medium according to the history which the elemental volume attains by passing through the head field [10]–[14].

Reprinted from *IEEE Trans. Magn.*, vol. MAG-4, pp. 269–276, Sept. 1968.

6) The magnetization distribution in the medium may be experimentally obtained with the aid of a scaled-up model of the record head and the medium [15].

In these methods the following difficulties are usually present. In the theoretical approach it is hardly possible to introduce the effect of the head field strength and to evaluate accurately the nonlinear effect of the medium. In the experimental approach, on the other hand, the difficulty lies, for example, in condition 5), in tracing the magnetization process including demagnetizing fields, and in condition 6), in judging the effect of the various parameters of the medium and in interpreting the results obtained. Therefore, the magnetization distributions obtained so far are only approximate ones and still far from satisfactory in explaining a number of phenomena in magnetic recording.

In our laboratory the phenomena in high-density recording have been studied for several years, and the conclusion reached is that it is necessary, in the theoretical approach, to consider the demagnetizing field together with the recording head field in determining the magnetization distribution [16]. By this fact it is not sufficient to evaluate the demagnetizing field only as a self-demagnetizing field, as is usually the case, which acts after the head field is removed, but it should be dynamically interpreted as a kind of reaction field against the head field in the recording process.

Focusing on this problem, a dynamical interpretation of the recording mechanism is given in the following sections. In the calculation a relatively thick medium and a short gap length are chosen. Such a condition is difficult to analyze, but significant in practice.

II. Demagnetization Mechanism in Recording Process

A. *Approach for Evaluating Effects of Demagnetizing Fields*

The recording field configuration of the magnetic head has been estimated in detail by theoretical calculation or experiment for the case where a recording medium does not exist in front of the head. In the theoretical determination of the magnetization distribution in the recording medium, the head field has been used to find an initial magnetization distribution on a given magnetization curve; then the distribution has been modified by the self-demagnetizing field after the head field is removed. This procedure has often been followed for the longitudinal magnetization.

In magnetic recording, however, the aim has been to store information as combinations of strong magnetic poles on the medium; because of the high intensity of the magnetization of the medium, it may not be correct to neglect the possible influence of the magnetization of the medium on the head field. The influence can be estimated by correctly evaluating the demagnetizing field which is produced by apparent magnetic charges due to the magnetization.

Because the magnetic medium possesses the nonlinear characteristics, the demagnetizing field must be evaluated at an instant when the head field is applied to the medium. Therefore, the sum of the demagnetizing field and the head field determine the initial magnetization distribution. When the head field is removed, it is sufficient to modify the magnetization distribution by the demagnetizing field alone (self-demagnetization). This results in the fact that the magnetization decreases along the demagnetization curve without appreciable change in the position of the maximum density of apparent magnetic charges. The magnetization distribution thus determined is quite different from the conventional one which is determined from self-demagnetization. This consideration of the effects of the demagnetizing field $\boldsymbol{H}_d$ leads to the conclusion that the effective internal field $\boldsymbol{H}_e$, which may act on the medium when the head field $\boldsymbol{H}_h$ is applied, must satisfy

$$\boldsymbol{H}_e = \boldsymbol{H}_h + \boldsymbol{H}_d. \tag{1}$$

Here $\boldsymbol{H}_d$ can be expressed by

$$\boldsymbol{H}_d(\boldsymbol{r}) = \int_{v'} \rho_m(\boldsymbol{r}') \frac{\boldsymbol{r} - \boldsymbol{r}'}{|\boldsymbol{r} - \boldsymbol{r}'|^3} dv' + \int_{s'} \sigma_m(\boldsymbol{r}') \frac{\boldsymbol{r} - \boldsymbol{r}'}{|\boldsymbol{r} - \boldsymbol{r}'|^3} ds'. \tag{2}$$

The volume density ρ_m and the surface density σ_m of apparent magnetic charges inside the medium and on its surfaces are given by

$$\rho_m(\boldsymbol{r}') = -\operatorname{div} \boldsymbol{M}(\boldsymbol{r}') \tag{3a}$$

and

$$\sigma_m(\boldsymbol{r}') = \boldsymbol{n} \cdot \boldsymbol{M}(\boldsymbol{r}') \tag{3b}$$

where dv' and ds' are the volume and the surface element at a point $\boldsymbol{r}'$, and $\boldsymbol{r}$ is the field point where $\boldsymbol{H}_d$ is to be evaluated, as shown in Fig. 1. $\boldsymbol{n}$ is the unit vector normal to the surface.

The magnetization distribution $\boldsymbol{M}$ in (3a) and (3b), which produces the demagnetizing field $\boldsymbol{H}_d$ in (2), must be a self-consistent magnetization that is directly determined by the sum of $\boldsymbol{H}_d$ and $\boldsymbol{H}_h$, as shown in Fig. 2(a). The magnetization distribution $\boldsymbol{M}$ determined in this way is called the self-consistent magnetization in the following.

The concept of the self-consistent magnetization distribution should naturally be applied to the magnetic head in determining its recording field. In this paper, however, the problem of the nonlinearity of the recording medium is studied exclusively, and the conventional expressions for $\boldsymbol{H}_h$ [17] are used in the analysis.

B. *Method of Calculating Self-Consistent Magnetization Distribution*

A self-consistent magnetization distribution is, in practice, computed with a digital computer by using the iteration method. By this method a certain stable value for the magnetization distribution $\boldsymbol{M}$, which forms a closed loop as shown in Fig. 2(a), is obtained.

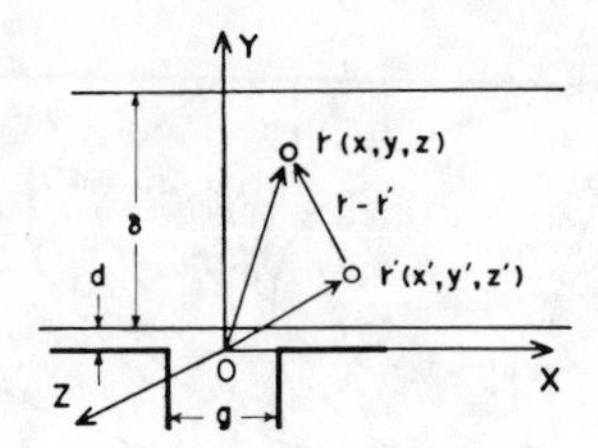

Fig. 1. Head tape coordinate system.

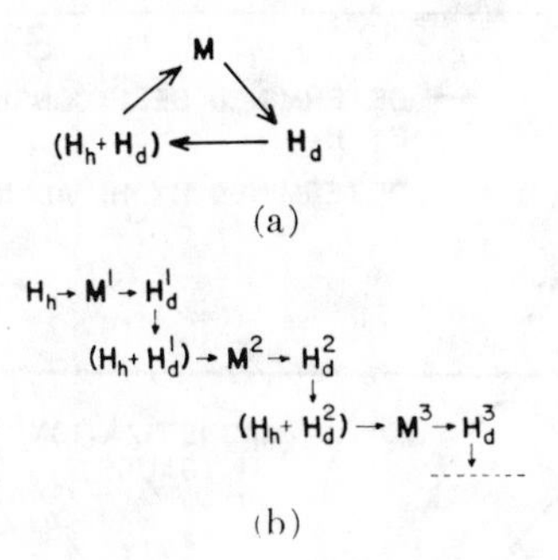

Fig. 2. (a) Self-consistent magnetization $\boldsymbol{M}$. (b) Iteration procedure.

For the calculation it is assumed that the recording medium is infinite in the z direction in Fig. 1, and the z components of magnetic fields and magnetization are zero.

Assuming square meshes of the size h in the medium, the coordinates (mh, nh) of the mesh point are represented by (m, n) on the new M–N coordinates, the origin of which is chosen as that of the X–Y coordinates in Fig. 1.

After integrating over z and setting $x = mh$, $y = nh$, $dxdy = h^2$, (2) may be expressed by the following summations:

$$H_{dx}(m,n) = \sum_{m'}\sum_{n'} \frac{2h\rho_m(m',n')(m-m')}{(m-m')^2+(n-n')^2} + \sum_{m'} \frac{2\sigma_m(m',n')(m-m')}{(m-m')^2+(n-n')^2} \quad (4a)$$

$$H_{dy}(m,n) = \sum_{m'}\sum_{n'} \frac{2h\rho_m(m',n')(n-n')}{(m-m')^2+(n-n')^2} + \sum_{m'} \frac{2\sigma_m(m',n')(n-n')}{(m-m')^2+(n-n')^2}. \quad (4b)$$

Substituting with a difference equation, (3) becomes

$$\rho_m(m',n') = -\left\{\frac{M_x(m'+1,n')-M_x(m'-1,n')}{2h} + \frac{M_y(m',n'+1)-M_y(m',n'-1)}{2h}\right\} \quad (5a)$$

and

$$\sigma_m(m',n') = \begin{cases} -M_y(m',n') & n' = d/h \\ M_y(m',n') & n' = (d+\delta)/h \end{cases} \quad (5b)$$

where d is the head-to-medium spacing and δ the thickness of the medium.

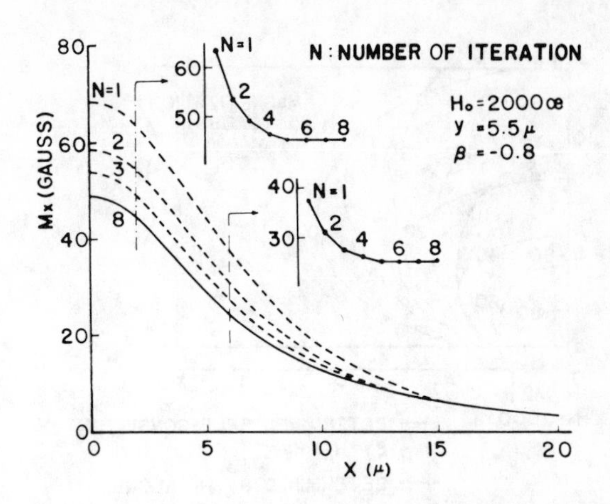

Fig. 3. Magnetization distribution with number of iteration as parameter. Solid line represents self-consistent magnetization distribution.

Finally, the equations for the head field are expressed in the M–N coordinates as

$$H_{hx}(m,n) = \frac{H_0}{\pi}\left\{\tan^{-1}\frac{m+g/2h}{n} - \tan^{-1}\frac{m-g/2h}{n}\right\} \quad (6a)$$

and

$$H_{hy}(m,n) = \frac{H_0}{2\pi}\ln\left\{\frac{(m-g/2h)^2+n^2}{(m+g/2h)^2+n^2}\right\} \quad (6b)$$

where g is the record gap length.

The self-consistent magnetization distribution can be found by the iterative modification of the magnetization distribution, which is determined by the head field alone, by summing the head field and the demagnetizing field, as shown in Fig. 2(b). The modification is performed by

$$M_{x,y}{}^n = M_{x,y} + \beta(M_{x,y} - M_{x,y}{}^{n-1}). \quad (7)$$

That is, the $(n-1)$th magnetization $M_{x,y}{}^{n-1}$ must be modified to a new nth magnetization $M_{x,y}{}^n$ by a magnetization $M_{x,y}$, which is directly obtained by summing the head field H_h and the demagnetizing field $H_d{}^{n-1}$ produced by $M_{x,y}{}^{n-1}$. β in (7) is a converging factor for the iteration and does not affect the final results, because in the limit, $M_{x,y}{}^n$ and $M_{x,y}{}^{n-1}$ converge to $M_{x,y}$, and the second term approaches zero.

In determining the absolute magnetization M by the absolute effective internal field, it is necessary to specify the magnetization curve for each mesh point according to the initial value of the magnetization at the point. For example, when a positive head field is suddenly applied to the ac-erased tape, a positive virgin curve can be specified for every point, because the initial value of the magnetization is equal to zero at the point. The following are approximated equations for the virgin curves of the oriented and the nonoriented γ-Fe_2O_3 tapes, respectively.

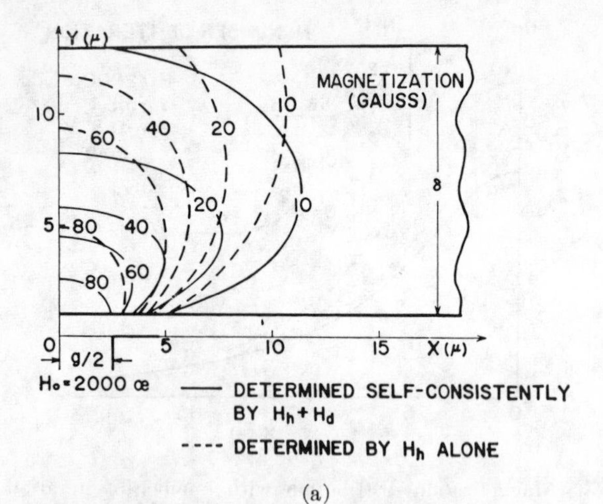

(a)

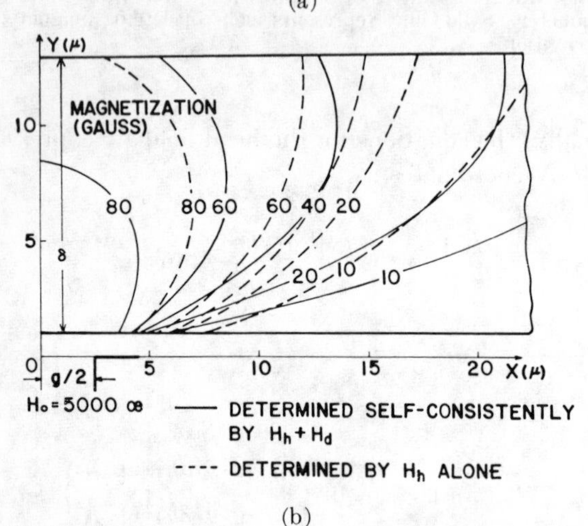

(b)

Fig. 4. Longitudinal magnetization distribution when head field H_0 is applied to ac-erased tape. (a) $H_0 = 2000$ Oe. (b) $H_0 = 5000$ Oe.

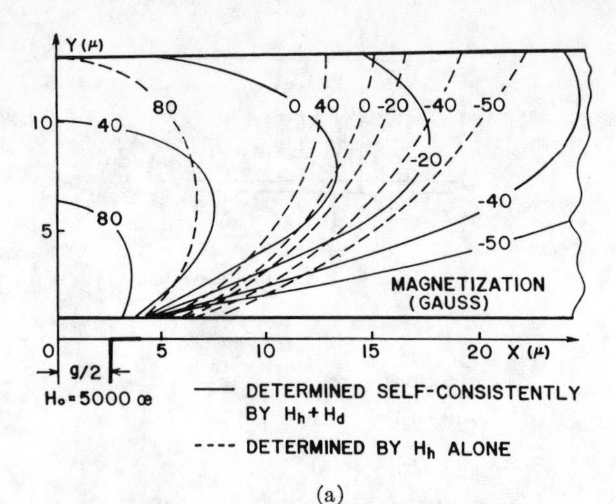

(a)

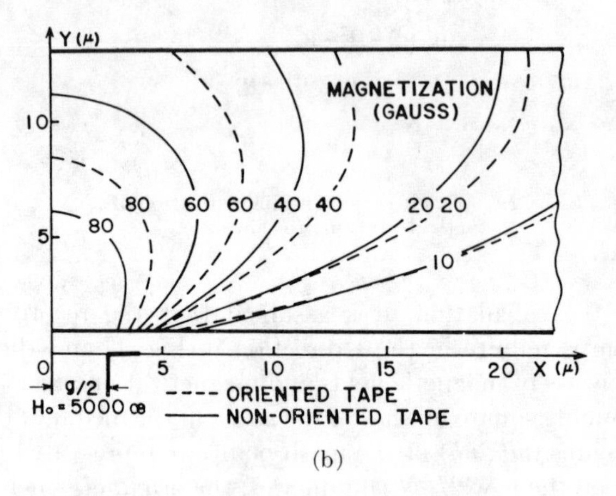

(b)

Fig. 5. (a) Longitudinal magnetization distribution when positive head field of $H_0 = 5000$ Oe is applied to negatively saturated tape. (b) Comparison of self-consistent magnetization distributions for oriented and nonoriented tapes.

For the oriented γ-Fe_2O_3 tape

$$M = 38.178 + 0.004\,H + 29.883 \tan^{-1}(0.0125\,H - 3.3125). \quad (8)$$

For the nonoriented γ-Fe_2O_3 tape

$$M = \frac{32.2625\,H^2}{10.5625 \times 10^4 + H^2} + \frac{54.0675\,H^3}{34.3281 \times 10^6 + H^3}. \quad (9)$$

Fig. 3 shows how the magnetization distribution converges. With a properly chosen β, the self-consistent magnetization can be obtained after several iterations.

The major errors due to the numerical approximation described above are as follows.

1) Error caused by replacing the differential equation by the difference equation. The error may depend upon the size of mesh h.

2) Error caused by replacing the integration by the summation. This error may also depend on h.

3) Error associated with the termination of an infinite series. The error may depend on the record gap length and the head field strength.

Evaluation of these errors by a model calculation has revealed that the first error is the principal one and that the total error may be kept within several percent if the size of mesh is made less than one tenth of the record gap length.

III. Magnetization Distribution in Recording Medium

A. Longitudinal Magnetization Distribution

Some results calculated by the method described in the previous section are presented. Fig. 4 shows the lines of equal intensity of the longitudinal magnetization in the medium at the instant when the head fields of $H_0 = 2000$ and 5000 Oe, respectively, in the gap are applied to an ac-erased γ-Fe_2O_3 tape (oriented). In this calculation a record gap length was taken to be $g = 5\ \mu$, a medium thickness $\delta = 12\ \mu$, a head-to-medium spacing $d = 1\ \mu$, and the y component in (3) was assumed to be zero. The approximated magnetization curve of (8) was also used. In the figure the solid lines represent the magnetization distribution that was self-consistently determined by summing the x components of the demagnetizing field and of the head field, while the broken lines show the distribution determined by the head field alone. It is seen from the figure that in the recorded region near the gap the de-

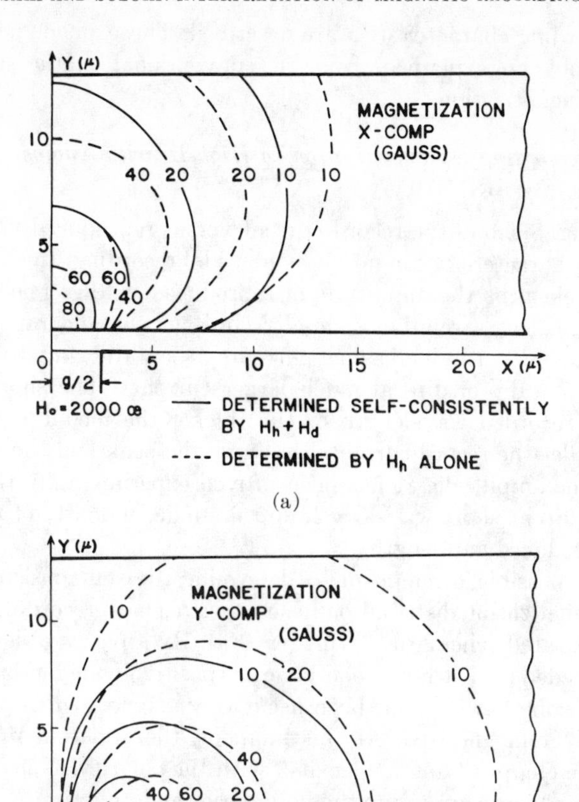

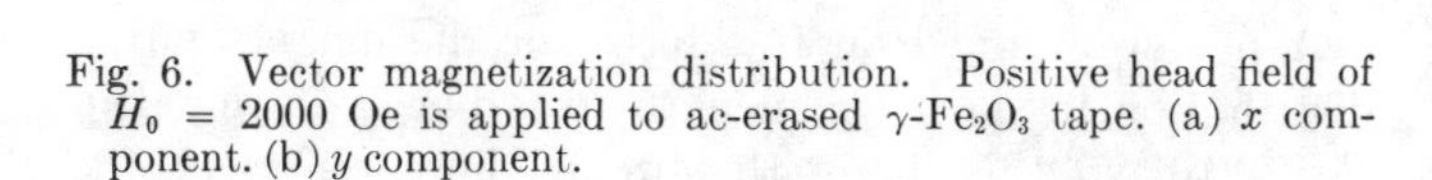

Fig. 6. Vector magnetization distribution. Positive head field of $H_0 = 2000$ Oe is applied to ac-erased γ-Fe_2O_3 tape. (a) x component. (b) y component.

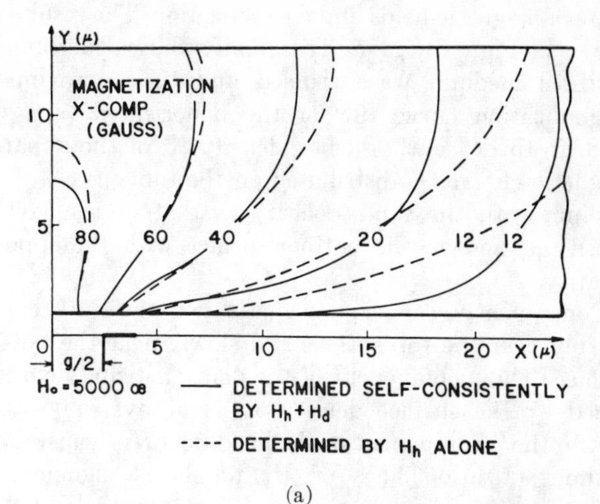

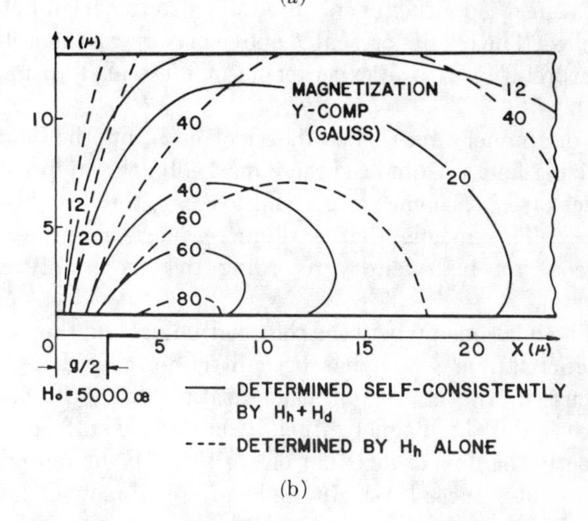

Fig. 7. Vector magnetization distribution. As in Fig. 6, except that $H_0 = 5000$ Oe.

magnetizing field H_{dx} acts so as to weaken the head field H_{hx}, and in the region far from the gap H_{dx} adds to H_{hx} with the same polarity, resulting in the fact that the boundary of the recorded region considerably shifts outward. The shift is, in fact, largely dependent on the head field strength H_0. It has not been able to obtain such results as presented here by the conventional analysis of self-demagnetization. The magnetization distribution shown in Fig. 4 may well represent, in principle, a distribution in RZ recording.

Fig. 5 shows how the magnetic characteristics of the medium and the recording field affect the self-consistent magnetization distribution. Illustrated in Fig. 5(a) are the distributions obtained at the instant when a positive head field is applied to the medium previously magnetized to a negative saturation. In the calculation an approximated major loop of the magnetization curve was used, which is almost equivalent to doubling the value in magnetization in (8). As a consequence, the acting demagnetizing field becomes stronger by almost twice as much. As a result, the penetration of the magnetization around the gap center line is significantly suppressed, and the negative equimagnetization lines shift outward, extending the recorded region to the x direction. From the results shown in the figure, it is possible to estimate the effect of doubling the saturation magnetization M_s of the medium with a constant H_c. That is, if the value of the remanence, 65 gauss, is added to each one of the equimagnetization lines in Fig. 5(a), a magnetization distribution corresponding to Fig. 4(b) can be obtained. It may be possible from these results to estimate the fundamental influence of the saturation magnetization M_s and the coercivity H_c of the medium.

In Fig. 5(b), the effect of particle orientation on the magnetization distribution is shown for γ-Fe_2O_3 tapes with constant M_s. The solid lines represent the self-consistent magnetization for the nonoriented tape, and the broken lines for the oriented tape. With the same packing density, the oriented tape always possesses a higher value of magnetization than the nonoriented tape for different values of H, except in the small field region. Therefore, a larger demagnetizing field arises for the oriented tape and shifts the equimagnetization lines outward. This is a very important fact in considering the genuine effects of the squareness ratio M_r/M_s.

B. Vector Magnetization Distribution

The vector self-consistent magnetization distributions were calculated by considering both the x and y components of magnetic fields and magnetization. The results are shown in Figs. 6 and 7. To simplify the calculation, an isotropic medium was assumed and the approximated magnetization curve (9) for the nonoriented tape was used. In this case, the surface density σ_m of the apparent magnetic charges is distributed on the top surface of the medium with the same polarity as that of the volume density ρ_m and on the bottom surface with the opposite polarity.

Consequently, each equimagnetization line of M_x curves sharply near the top surface and slowly near the bottom surface. The y component of the demagnetizing field produced by the surface density σ_m is always opposite in sign to the y component of the head field H_{hy}, causing the equimagnetization lines of M_y to shrink significantly. This effect is markedly enhanced if a stronger head field is applied. The results presented above may give a theoretical interpretation to the experiments of the scaled-up model [15].

For thinner tapes, since the effect of σ_m on the demagnetizing field is enhanced more markedly, the suppression effect on M_y becomes larger and localizes the distribution of σ_m. This means that a thinner tape is more advantageous for high-density recording than is usually considered.

The relationship between the longitudinal and the vector magnetization, which has been little considered, is now discussed. In the vector magnetization distribution, the direction of the magnetization M on the top surface gives directly the flow of flux that passes through the reproduce head and induces EMF. In the longitudinal magnetization distribution, on the other hand, the decrement (or increment) of the magnetization gives a source (or sink) of the flux. This is equivalent to assuming at the magnetization boundary a virtual distribution of magnetic charges that gives rise to lines of perpendicular magnetic forces, corresponding to the flow of flux in the vector magnetization. Therefore, in the longitudinal magnetization distribution, the region from which the lines of magnetic force go out of the medium is represented by the distribution of ρ_m, and the counterpart in the vector magnetization distribution is the distribution of σ_m on the top surface. Comparison between Fig. 5(b) (nonoriented tape) and Fig. 7(b) shows that these regions correspond closely to each other. This fact means that a quantitative evaluation is possible, with good accuracy, in analyzing the following phenomena in terms of the *self-consistent longitudinal magnetization*, even if the vector record field is applied to the medium. The following analyses are carried out on the basis of these considerations.

IV. New Phenomena in Magnetic Recording Characteristics

The calculated results of the self-consistent magnetization distributions in the preceding section afford a new and powerful insight into understanding of phenomena in high-density recording that suffer severe demagnetization. In this section a few phenomena recently discovered in the recording characteristics are described. These phenomena cannot be explained by the conventional static self-demagnetization.

A. *Recording Demagnetization in High-Density Analog Recording* [18], [19]

Fig. 8 shows the record current versus reproduced voltage characteristics in no-bias sinusoidal recording. In short wavelengths the amplitude of reproduced voltage reaches its peak at a relatively small record current and rapidly decreases with increasing currents; then the amplitude shows a dip or dips at much larger currents. The shorter the recorded wavelength or the thicker the medium, the smaller the record currents that give the peak and the dip in the amplitude. It has been proved experimentally that the dip at short wavelength appears independently of the reproduced gap length.

A possible mechanism to determine the self-consistent magnetization distribution in short wavelength recording is illustrated schematically in Fig. 9(a). By a positive signal preceded by a negative one, zone 2 is positively magnetized, and zone 1, which has been negatively magnetized, is partially remagnetized. At this moment, the apparent magnetic charges (volume density) with the polarity, as shown in the figure, appear on the boundaries and give rise to the demagnetizing fields H_{d1} and H_{d2} in zones 1 and 2, respectively. The polarity of H_{d1} is the same as H_h, and hence the effective field H_e in the medium is enhanced, and the boundary of zones 1 and 2 is moved to the dotted line. It should be noted that H_{d1} is a function of the recorded wavelength and becomes larger for shorter wavelengths.

A model obtained on this basis for the magnetization distribution in short wavelength recording is depicted in Fig. 9(b). In this figure the broken lines represent the boundaries of positively and negatively magnetized regions on which virtual flows of flux assumed in Section III-B are shown by the black arrows. The external flux that contributes to reproduction must seriously suffer the canceling effect of the positive and the negative virtual flows of flux in the model. This is in effect equivalent to supposing flux closures in the vector magnetization distribution.

Fig. 9(c) shows the relative loss in the output as a function of θ in Fig. 9(b) and represents the limiting case of the canceling effect. Since change in θ corresponds to change in the record current, Fig. 9(c) explains, in principle, the fact that a dip or dips appear in the reproduced output shown in Fig. 8.

As a result of the quantitative comparison between θ and the record current, it is understood that in short wavelengths the boundary of the self-consistent magnetization distribution is determined on the equimagnetization line of the applied field which is equal to about half the coercivity of the medium.

B. *Relation Between Pulse Width and Squareness Ratio of Medium in Digital Recording* [20]

It has been thought that increasing the squareness ratio of the medium is useful for improving the recording resolu-

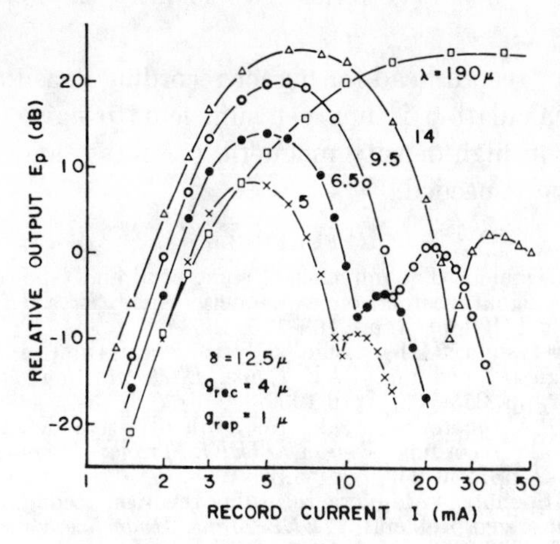

Fig. 8. Relative output versus record current in no-bias sinusoidal recording.

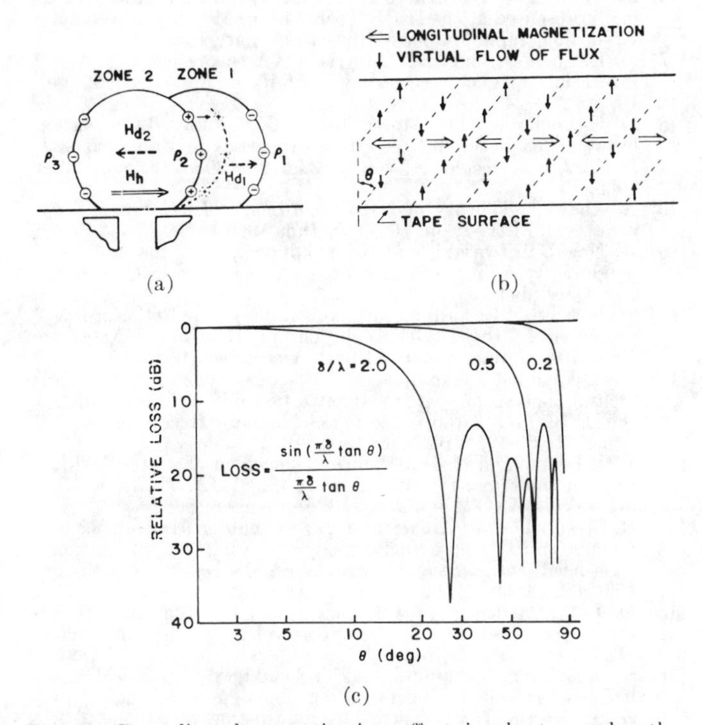

Fig. 9. Recording demagnetization effect in short wavelengths. (a) Schematic representation of recorded zones. (b) Schematic representation of virtual flows of flux. (c) Relative loss in reproducing process calculated from model.

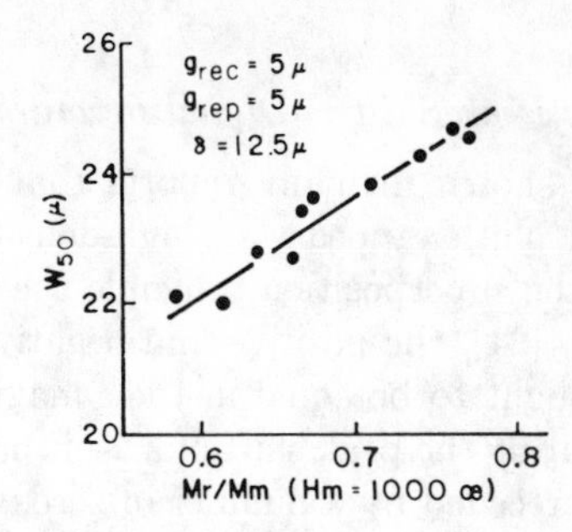

Fig. 10. Pulse width W_{50} versus squareness ratio.

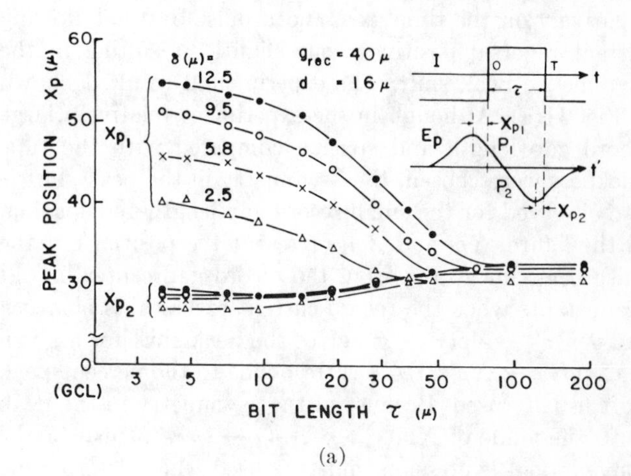

(a)

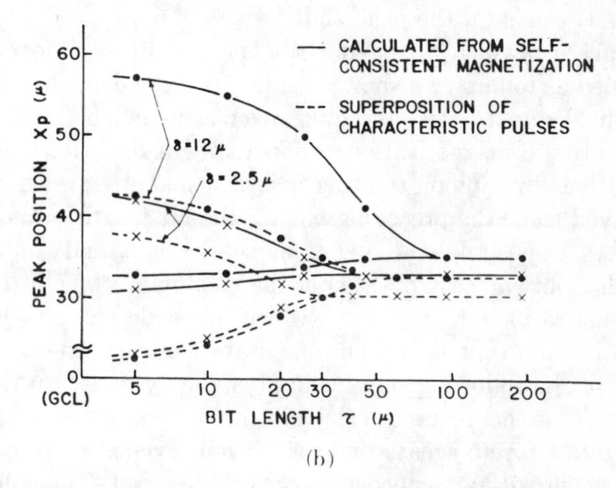

(b)

Fig. 11. Asymmetric peak shift of 2-bit pattern (...001100...) (a) Experimental. (b) Calculated.

tion, and the particle orientation has been performed. Increase in the squareness ratio is indeed accompanied by increase in the anhysteretic susceptibility and, consequently, is effective for improving the recording sensitivity in the ac-bias recording. However, especially in the saturation type recording, the result shown in Fig. 5(b) seems to predict that higher orientation might deteriorate the resolution. In order to ascertain this, the dependence of the pulse width W_{50} on the squareness ratio M_r/M_m of the medium was measured. The result is shown in Fig. 10. Samples used in the measurement were prepared from γ-Fe_2O_3 powder tapes of the same packing density by applying different strengths of orientation field so as to possess several values of M_r/M_m. Each sample has almost the same coercivity but different remanence according to the orientation. The tendency is clearly seen from the figure that the pulse width becomes wider as M_r/M_m increases.

Bate reported the relation between the peak shift and the orientation angle [21]. He showed experimentally that the smallest peak shift was obtained for the measuring angle of 65 degrees to the particle orientation direction, and the behavior of the peak shift was closely correlated to H_c/M_r.

These experimental tendencies seem to be understood by the dynamic behavior of the demagnetizing field described above. It is seen from Fig. 5(b) that the lines of equal intensity of the self-consistent magnetization for the oriented tape always shift more outward than the non-oriented tape. This fact clearly shows that the peak shift increases and the pulse becomes wider.

C. *Asymmetric Peak Shift in Digital Recording* [22]

It has been shown in many reports that positive and negative peak shifts appear in high-density recording. According to the superposition principle used in magnetic recording [23], [24], the positive and negative peak shifts have been thought to be equal in their magnitude. However, by measuring the peak shift of a 2-bit pattern (2 ones followed and preceded by a number of zeros) with the aid of phase reference pulses, it is observed that a pulse recorded by the preceding current reversal shifts largely and negatively on the time axis, and a pulse by the following current reversal positively but slightly, resulting in the asymmetric peak shift. The experimental result is shown in Fig. 11(a). Although in the experiment relatively large record gap length and spacing compared with the tape thickness were chosen, the asymmetry in the peak shift is also observed for the small record gap length and spacing. In the figure, Xp_1 and Xp_2 represent the positions of the pulse peaks measured from the record gap center line at the instants when the record current reverses its polarity, and $Xp(\tau) - Xp(\tau \rightarrow \infty)$ gives the peak shift for a given τ. $(Xp_1(\tau) - Xp_2(\tau))/2\tau$ corresponds to the percent peak shift usually used. Because of the asymmetry in the peak shift, the value of $Xp_1(\tau) - Xp_1(\tau \rightarrow \infty)$, instead of the conventional expression, must be used when determining the allowance for the peak shift.

Causes of such an asymmetric peak shift can be explained as follows. As shown in Fig. 5(a), the demagnetizing field due to the preceding reversal in magnetization extends the magnetization region to the x direction (the negative direction on the time axis); then another reversal delayed from the preceding one by the bit length τ magnetizes oppositely the region near the gap and shifts further outward the position of the maximum slope in the magnetization distribution (which may determine the position of a reproduced pulse). On the contrary, the position of the following pulse is approximately given by the position of the maximum slope in the self-consistent magnetization for an isolated magnetization reversal and does not move without another succeeding reversal. The solid lines shown in Fig. 11(b) are the calculated result and correspond to the measured result in Fig. 11(a). It is seen that the tendency of the measured asymmetry is explained satisfactorily. As is clearly shown, the broken lines in the figure, which are obtained by superposing measured characteristic pulses, do not explain the asymmetry at all.

V. Conclusion

In this paper, the basic but dynamic interpretation of the magnetic recording process and a few interesting phenomena that have recently been found are described. In the theory of magnetic recording, it has been a practical and fundamental problem to give an accurate quantitative evaluation to the nonlinearity of the medium. A clue to this problem is believed to be given by this paper. The present dynamical interpretation of the recording process is of great significance for improvements of the recording medium or record head for higher recording densities. The present calculation is not yet sufficient to cover all the problems in high-density magnetic recording, and further calculation is needed.

References

[1] E. D. Daniel, "The influence of some head and tape constants on the signal recorded on a magnetic tape," *Proc. IEE* (London), vol. 100, pt. 3, pp. 168–175, 1953.

[2] B. Kostyshyn, "A harmonic analysis of saturation recording in a magnetic medium," *IRE Trans. Electronic Computers*, vol. EC-11, pp. 253–263, April 1962.

[3] ——, "A theoretical model for a quantitative evaluation of magnetic recording system," *IEEE Trans. Magnetics*, vol. MAG-2, pp. 236–242, September 1966.

[4] D. E. Speliotis, "Magnetic recording theories: accomplishments and unsolved problems," *IEEE Trans. Magnetics*, vol. MAG-3, pp. 195–200, September 1967.

[5] C. W. Chapman, "Theoretical limit on digital magnetic recording density," *Proc. IEEE (Correspondence)*, vol. 51, pp. 394–395, February 1963.

[6] M. Nishikawa, "Magnetization mechanism in digital recording," presented at the IECE Tech. Group Meeting of Magnetic Recording, Japan, paper MR64-8, February 1965.

[7] D. E. Speliotis and J. R. Morrison, "A theoretical analysis of saturation magnetic recording," *IBM J. Res. Develop.*, vol. 10, pp. 233–243, May 1966.

[8] D. E. Speliotis, J. R. Morrison, and J. S. Judge, "Correlation between magnetic and recording properties in thin surfaces," *IEEE Trans. Magnetics*, vol. MAG-2, pp. 208–212, September 1966.

[9] A. Aharoni, "Theory of NRZ recording," *IEEE Trans. Magnetics*, vol. MAG-2, pp. 100–109, June 1966.

[10] K. Nagai, S. Iwasaki, and K. Yokoyama, "On the ac-bias of magnetic recording," *Electro-Tech. J. Japan*, vol. 3, pp. 131–135, December 1957.

[11] S. Iwasaki, "Recording mechanism in magnetic recording," presented at the IECE Tech. Group Meeting of Magnetic Recording, Japan, paper MR64-1, December 1964.

[12] K. Yokoyama, J. Kasama, and F. Higashiyama, "An experimental analysis of vector magnetization process," presented at the IECE Tech. Group Meeting of Magnetic Recording, Japan, paper MR65-19, September 1965.

[13] J. U. Lemke and R. J. McClure, "The effect of vector field history on the remanence of magnetic tape," *IEEE Trans. Magnetics*, vol. MAG-2, pp. 230–232, September 1966.

[14] R. Straubel, "Der Aufzeichnungsvorgang der Magnetspeichertechnik mit Wechselfeldvormagnetisierung in phänomenologischer Sicht," *Hochfrequenztech. Electroakustik*, vol. 75, pp. 153–162, October 1966.

[15] D. L. A. Tjaden, "A 5000:1 scale model of the magnetic recording process," *Philips Tech. Rev.*, vol. 25, pp. 319–329, 1963–1964.

[16] S. Iwasaki, Y. Nakamura, and T. Suzuki, "A theoretical calculation of recorded zone including demagnetizing fields," presented at the IECE Tech. Group Meeting of Magnetic Recording, Japan, paper MR67-15, September 1967.

[17] G. Schwantke, "Beitrag zur Darstellung des Spaltfeldes beim Magnetton und Anwendung auf den Wiedergabevorgang," *Acustica*, vol. 7, pp. 363–371, August 1957.

[18] S. Iwasaki and Y. Nakamura, "A study of magnetizing process in short wave-length recording," *J. ITE Japan*, vol. 18, pp. 638–646, October 1964.

[19] S. Iwasaki and Y. Suto, "Investigations on the recording demagnetization phenomena in magnetic recording process," *Sci. Rept. Res. Inst. Tohoku Univ.*, ser. B, vol. 18, no. 3/4, pp. 223–236, 1967.

[20] T. Suzuki and S. Iwasaki, "Effects of squareness ratio of medium on digital recording characteristics," presented at the Joint Conv. of Four Inst. of Elec. Engrs., Japan, Tohoku Chapter, paper 5B-14, 1967.

[21] G. Bate, "Thin metallic films for high-density digital recording," *IEEE Trans. Magnetics*, vol. MAG-1, pp. 193–205, September 1965.

[22] S. Iwasaki and T. Suzuki, "The effect of demagnetizing field on 2-bit pattern in digital recording process," presented at the IECE Tech. Group Meeting of Magnetic Recording, Japan, paper MR67-4, May 1967. Also in *Sci. Rept. Res. Inst. Tohoku Univ.*, ser. B, vol. 19, no. 3, 1968.

[23] A. S. Hoagland and G. C. Bacon, "High density digital magnetic recording techniques," *IRE Trans. Electronic Computers*, vol. EC-9, pp. 2–11, March 1960.

[24] M. Nishikawa, "A study in the reproducing process in saturation-type magnetic recording," *Elec. Commun. Lab. Tech. J.*, vol. 12, no. 6, pp. 717–794, 1963.

AN ANALYTICAL MODEL OF THE WRITE PROCESS IN DIGITAL MAGNETIC RECORDING

M. L. Williams and R. L. Comstock
IBM Corporation, San Jose, Ca. 95114

ABSTRACT

An analytical model of saturation NRZ recording is used to calculate transition lengths in media with hysteresis curves which are not square. The maximum gradient of transition magnetization is computed, including the field gradients of the major hysteresis curve and the spatial gradients of the recording-head field and transition demagnetization fields, by assuming the transition has an arctangent shape. The model is an approximation to, and is compared with, numerical self-consistent analyses.

INTRODUCTION

The principal problem remaining in the analysis of the digital magnetic recording process is the prediction of the length of the recorded magnetization transition under general conditions of head-medium spacing, head-gap length and medium magnetic properties. The recording process is highly nonlinear in contrast to the readback process which is linear and has been analyzed for a variety of recording head configurations.

The purpose of this paper is to present a simple analytical calculation of the write process in digital magnetic recording. The model used to represent the write process includes in an approximate way the effects of demagnetization resulting from the presence of the transition and the finite gradient of the head magnetic field function in the plane of the recording medium. The model is, therefore, an approximation to the more exact self-consistent write-process numerical analysis of Iwasaki and Suzuki[1] Curland and Speliotis[2] and Potter and Schmulian.[3]

WRITE PROCESS SLOPE CRITERION

The following criterion is used to establish an approximation for the length of the transition: the spatial slope of the transition at the origin is evaluated exactly by assuming that the transition shape is arctangent and the length of the transition is then evaluated. The process by which this analysis is carried out is shown schematically in Fig. 1.

ANALYTICAL MODEL OF THE WRITE PROCESS

For the first part of the analysis, it is assumed that an infinite pole length head which has a linearly varying magnetomotive force across its gap (Karlquist head) acts to reverse the magnetization of a previously saturated recording medium in its remanent state $+M_r$. Attention is focused on point I which, following remagnetization with susceptibility χ, will lie at the center of the final transition ($M = H_d = 0$ at $x = 0$). At point I the intermediate value of magnetization gradient is given by the following differential equation

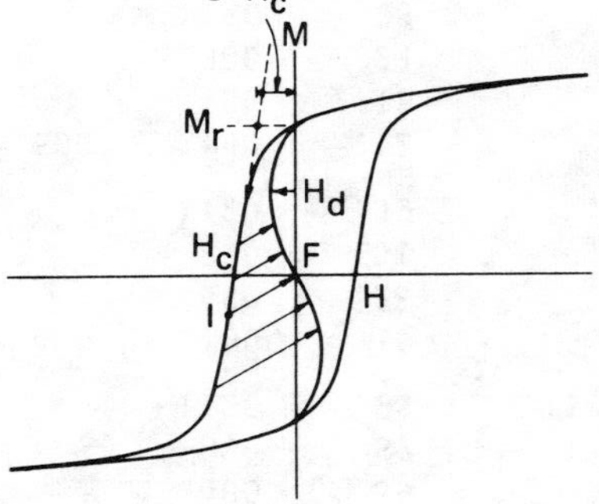

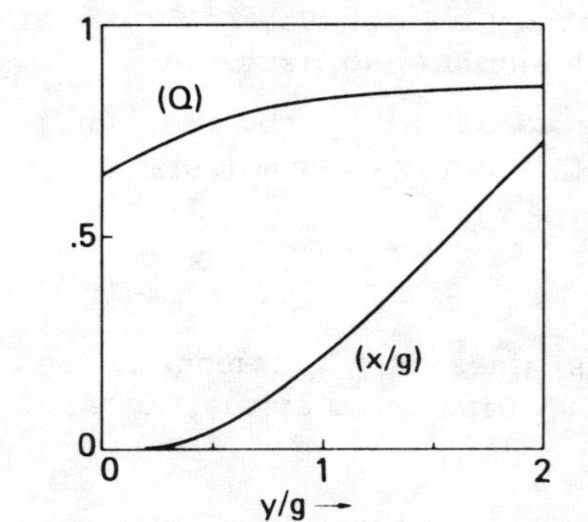

Fig. 1 Paths of magnetization in transition.

$$\frac{dM_I}{dx}(x=0) = \frac{dM_I}{dH_I}\left(\frac{dH_h}{dx} + \frac{dH_{dI}}{dx}\right). \quad (1)$$

The intermediate and final magnetization distributions are taken to be arctangent functions with transition parameters a_1 and a_2, respectively. For these functions[4] $M = (2M_r/\pi)\tan^{-1}(x/a)$ and $dM_I/dx = 2M_r/\pi a$. For simplicity the slope of the major hysteresis curve at point I is assumed to be unchanged from its value at $H = -H_c$, and $dM_I/dH_I = M_r/[H_c(1 - S^*)]$, where $H_I = -H_c/r$ and where $r = 1 - \chi(1 - S^*)H_c/M_r$. The parameter S^* varies between 0 and 1 and is a measure of hysteresis curve squareness at H_c. The Karlquist head field function at a plane spaced a distance y from the head-pole faces is given by $H_h = (H_g/\pi)[\tan^{-1}(x/y) - \tan^{-1}((x+g)/y)]$, and the gap field (gap g) is assumed to be adjusted such that, at the effective head-medium spacing y, dH_h/dx is a maximum, which is given by QH_I/y, where the curve $Q(y/g)$ is shown in Fig. 2. Also shown is the distance x/g from the trailing edge of the gap where the maximum occurs. The demagnetizing field gradient dH_{dI}/dx at point I for small medium thickness δ is $dH_{dI}/dx = -4M_r\delta/a_1^2$. Experiments on various types of recording media have shown that χ is given approximately by $M_r/4H_c$ and this value is used in subsequent calculations.

Fig. 2 Maximum head field gradient and distance from gap for Karlquist head.

Reprinted with permission from *17th Annu. AIP Conf. Proc.*, Part I, no. 5, 1971, pp. 738-742.

The four derivatives when combined in Eq. (1) result in the following solution for a_1,

$$a_1/r = y(1-S^*)/(\pi Q) + \{[y(1-S^*)/(\pi Q)]^2 + (2M_r\delta/H_c)(2y/Qr)\}^{1/2} . \quad (2)$$

For the second part of the analysis, the final magnetization gradient at the center of the transition is related to the initial magnetization gradient calculated in the first part of the analysis. The magnetization slopes are given by

$$dM_f/dx - dM_I/dx = \chi(dH_{df}/dx - dH_I/dx), \quad (3)$$

$dH_I/dx = (H_c(1-S^*)/M_r)\,dM_I/dx$ and $dH_{df}/dx = (2\pi\delta/a_2)\,dM_f/dx$ is derived assuming the medium is thin enough to satisfy $\delta \ll a_2$ so that the slope of the final trajectory dM_f/dH_{df} near $x = 0$ is given by $a_2/2\pi\delta$. By substituting the magnetization gradient expressions for arctangent transitions and eliminating the field gradient expressions in Eq. (3), the following result is obtained for the final transition length

$$a_2 = a_1/2r + [(a_1/2r)^2 + 2\pi\chi\delta a_1/r]^{1/2} , \quad (4)$$

and a_2 is greater than a_1, i.e., the transition broadens as it moves away from the head gap.

To include the shunting effect of the infinitely permeable Karlquist recording head spaced a distance y from the medium, on the intermediate transition length, the demagnetization field gradient is reduced according to $dH_{dI}/dx = 4M_r\delta[a_1^{-2} - (a_1+2y)^{-2}]$. This result is obtained by using the method of images following Miyata and Hartel.[5] An expression for the intermediate transition length including the shunting effect is

$$a_1/r = B + [B^2 + 4M_r\delta y V/(H_c Qr)]^{1/2} , \quad (5)$$

where $V = 4y(a_1+y)/(a_1+2y)^2$ and $B = (1-S^*)y/(\pi Q)$.

In high magnetization media with low recording-head spacing, the transition recorded by the process described above may not be in equilibrium; that is, the demagnetization energy savings resulting from increasing $\underline{a}$ may exceed the work required to do that spreading. Equilibrium is established when $\partial U/\partial a = -\partial W/\partial a$, where

$$U = -\frac{1}{2}\int_{-\infty}^{\infty} \bar{H}_d \cdot \bar{M}\,dx \quad \text{and} \quad W = \int_{-\infty}^{\infty}\int_{M_r}^{M} (\bar{H}\cdot d\bar{M})\,dx .$$

These expressions have been evaluated for a quadratic transition which has

$$M = \mathrm{sgn}(x)(M_r x/\pi a)(2 - x/\pi a) \quad \text{for } |x| < \pi a$$
$$= \mathrm{sgn}(x)\,M_r \quad \text{for } |x| > \pi a ,$$

and a hysteresis curve with $M = M_r(H + H_c)(S^*H + H_c)^{-1}$ in the second quadrant. The result is

$$a_d = 2M_r\delta F(S^*)/H_c ,$$

where $F(S^*) \approx 3.3 - 2.3\,S^*$. This value of a_d is to be used in place of a_2 from Eq. (3) whenever $a_d > a_2$.

The transition lengths calculated by the present model for a recording system using a Karlquist head are compared in Table I to those calculated by Kostyshyn (harmonic analysis),[6] Middleton,[7] and Curland and Speliotis (self-consistent iterative analysis)[2] for a typical set of magnetic parameters. The head-medium spacing d is 0.75 micron, the head gap is 1 micron, $S = S^* = 0.8$, and the medium thickness δ is 0.125 micron.

Table I Transition length parameters (a)

H_c Oe	M_r G	Harmonic	Middleton	Iterative	Williams/Comstock
		microns			
200	160	.24	.21	.86	.84
	480	.39	.61	1.49	1.48
	800	.57	1.01	2.27	1.96
	1120	.82	1.42	2.74	2.37
400	160	.15	.11	.55	.57
	480	.23	.30	.87	1.02
	800	.31	.51	1.19	1.35
	1120	.39	.71	1.49	1.65
600	160	.12	.08	.40	.48
	480	.18	.21	.66	.84
	800	.23	.34	.87	1.10
	1120	.28	.47	1.16	1.30

Readback calculations for a recording system with identical record and readback heads are given in Table II for the following set of magnetic parameters: $H_c = 600$ Oe, $M_r = 800$ G, $S = S^* = 0.8$. The effective spacing y for writing was chosen in parallel with the readback analysis of Middleton as $y = [d(d + \delta)]^{1/2}$, the pulse width readback from an isolated arctangent transition is given by $PW_{50} = [g^2 + 4(d + a)(d + a + \delta)]^{1/2}$.

Table II Readback pulse widths (PW_{50}) in microns

Spacing microns	Medium thickness microns 0.1	0.2	0.4
0.5	2.88	3.72	5.17
0.75	3.63	4.59	6.19
1.0	4.35	5.41	7.14
1.25	5.06	6.20	8.04

REFERENCES

1. S. I. Iwasaki and T. Suzuki, IEEE Trans. Magn. 4, 269 (1968).
2. N. Curland and D. E. Speliotis, IEEE Trans. Magn. 6, 640 (1970).
3. R. I. Potter and R. J. Schmulian, IEEE Trans. Magn., to be published.
4. R. I. Potter, J. Appl. Phys. 41, 1647 (1970).
5. J. J. Miyata and R. R Hartel, IRE Trans. Electron. Comput. 8, 159 (1959).
6. B. Kostyshyn, IEEE Trans. Magn. 2, 236 (1966).
7. B. K. Middleton, IEEE Trans. Magn. 2, 225 (1966).

An isotropic particulate medium with additive Hilbert and Fourier field components

James U. Lemke

Spin Physics Incorporated, San Diego, California 92121-1199

Interference effects between planar and normal components of magnetization in an isotropic tape are used to separate the contribution of each and address the question of transition lengths for so-called vertical recording. For low level record signals (−4 dB relative to optimum band edge current) and high coercivity isotropic or anisotropic media, the dominant magnetization at short wavelength is found to be normal to the plane (vertical), and the transition length is nonexistent. The isotropic medium supports twice the normal component of the anisotropic medium. Constructive interference of the two components causes apparent departure from the Wallace equations although each component is exactly described by the Wallace model. Increasing current causes a log/linear attenuation which may be caused by factors other than transition length. Using the isotropic tape, densities of 10 000 transitions/mm (250 000 f.c.i.) are seen.

PACS numbers: 75.60. − d, 85.70. − w

INTRODUCTION

Recent interest in so-called perpendicular magnetic recording, wherein the medium has a preferred magnetic axis normal to the plane of recording, has led to some conjecture concerning the magnitude of the transition length[1,2,3,4], a_t, for such normal magnetization reversals. The generally accepted term, $(a_t + d)$, where d is the spacing between medium and head, is conspicuous by the extended log/linear character of its output/frequency with conventional media. The transition length arises from an assumed $\tan^{-1}$ transition between opposing planar magnetizations due to demagnetization effects; planar magnetic anisotropy is assumed.

Lacking a suitable magnetic medium which will support only normal magnetizations, it is nonetheless possible to explore this question through an isotropic medium which will equally retain both normal and planar magnetizations.

The consideration of both vectors has been generally shunned in the literature since, it is argued, the practice of orienting the magnetic particles in the plane favors that component. The more important tacit reason is that the addition of the normal component inordinately complicates the already formidable mathematical task. The early work by Wallace[5] considered both components and assumed an isotropic medium. However, utilization of the full theory has not been easy since the low density recordings of the past have required that the magnetization through the entire magnetic layer be considered with its attendant phase shifts and general non-linear interaction problems. Recent very high density recordings which penetrate only on the order of a micron into the medium allow a simplification of the recording model to almost trivial dimensions with very good experimental confirmation. The planar magnetization process gives rise to fields which can be analyzed through Fourier transforms whereas the normal magnetization yields Hilbert transforms. These transforms are related through a phase rotation of $-\pi/2$ and simplify the consideration of complex waveforms. In this paper it will be necessary to consider only single frequencies to examine the spectral response, i.e. cos and -sin functions will suffice.

TECHNIQUE

The separation of the normal, M_{ry}, and planar, M_{rx}, magnetizations can be achieved through a study of the widely observed interference effects[6] which occur at long record gap length and low magnetization. All conventional tapes examined to date have exhibited the effect since none has complete anisotropy. Figures 1 and 2 show interferences with 300 Oersted instrumentation tape and 500 Oersted C_rO_2 tape.

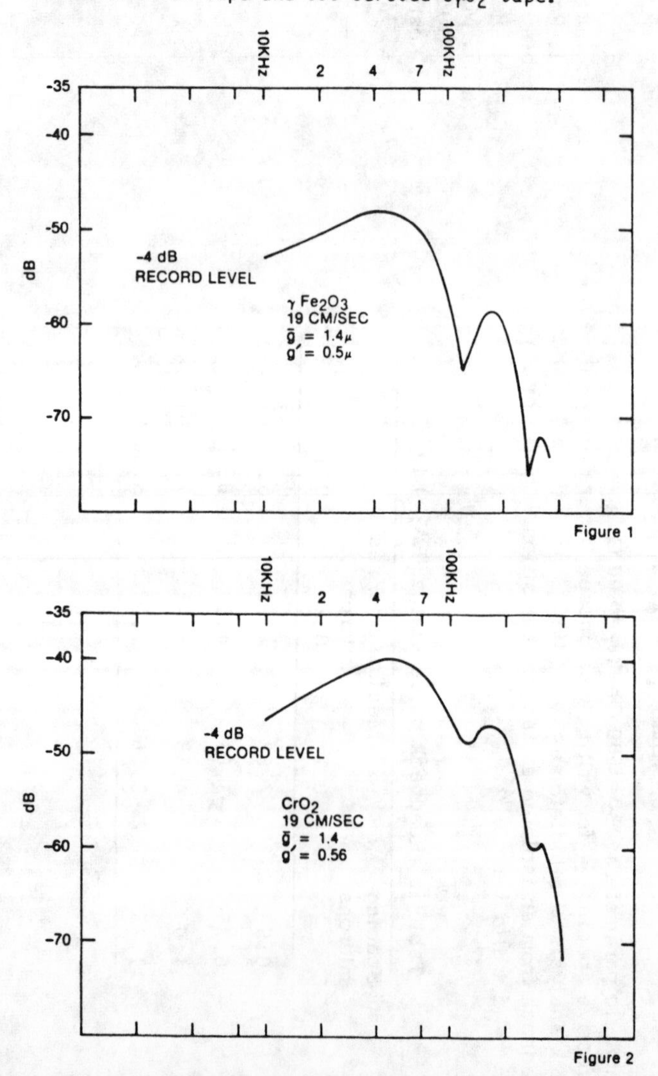

Figure 1

Figure 2

Interference effects can be observed from record gaps of less than 1μ and may be inferred from gaps as small as 0.15μ. Since recording is done with a low amplitude field (comparable to H_c), the two gap edges comprise separate recording sites. Contours of constant H_x emanate from and return to gap edges (see Figure 3). Consequently, M_{rx} is established at the leading gap edge and, for low fields, is unreversed by subsequent transit of the trailing gap edge. H_y, however, reverses from the leading head pole to the trailing pole and M_{ry} tends to be detmined largely by the field at the latter. The Karlqvist equations(7) define the fields accurately down to about g/10 above the poles.

$$H_x(x,y) = \frac{H_o}{\pi}\left[\tan^{-1}\left(\frac{x + g/2}{y}\right) - \tan^{-1}\left(\frac{x - g/2}{y}\right)\right] \quad (1)$$

$$H_y(x,y) = \frac{H_o}{2\pi}\ln\left[\frac{(x + g/2)^2 + y^2}{(x - g/2)^2 + y^2}\right] \quad (2)$$

$$H_o = \frac{4\pi N i n}{10g} \quad (3)$$

where H_o is the deep gap field, N is the number of turns, i the current, and η the head efficiency. Typically, heads are designed for an efficiency of at least 50% even for the smallest gap.

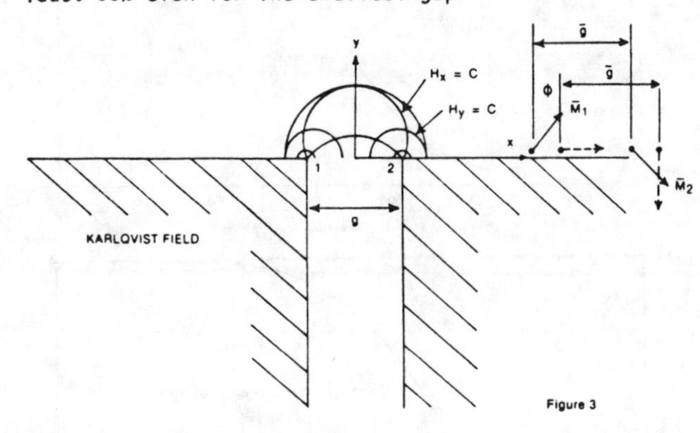

Figure 3

Figure 4(a) shows how constructive interference, (+), occurs for an effective gap distance ḡ; Figure 4(b) shows destructive interference, (-). During recording, a magnetization which can be represented by a cos function will be generated at the leading gap edge (x = 0), and a -sin function of magnetization will be recorded near the trailing edge at ḡ. Upon playback at wavelengths where the -sin function is in phase with the cos function at x = 0, (+) interference results. At wavelengths where they are out of phase, (-) interference results. Multiple (+) and (-) interferences will be produced at the following wavelengths:

$$\bar{g} = (4n + 1)\frac{\lambda(+)}{4} \quad (4)$$

$$\bar{g} = (4n + 3)\frac{\lambda(-)}{4} \quad (5)$$

$$n = 0,1,2,...$$

The locations of λ(+) and λ(-) allow an accurate determination of ḡ which is very close to the actual record gap length. The same value of ḡ is obtained for a wide range of tape coercivities due to the high field gradient of H_y.

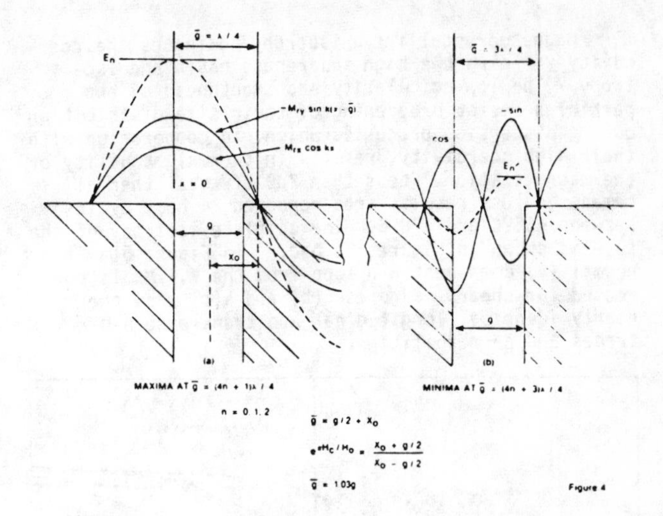

Figure 4

From equation (2) and $H_o \approx H_c$, ḡ can be estimated.

$$\bar{g} = g/2 + X_o \quad (6)$$

$$e^{\pi H_c/H_o} \approx \frac{X_o + g/2}{X_o - g/2} \quad (7)$$

$$\bar{g} \approx 1.03g \quad (8)$$

M_{rx} and M_{ry} can be determined from the maxima and minima voltages, E_n, of the response curve.

$$|M_{ry}| + |M_{rx}| = E_n \quad (9)$$

$$|M_{ry}| - |M_{rx}| = E'_n \quad (10)$$

Normalizing to M_{ry},

$$M_{rx} = \beta M_{ry} = \beta \quad (11)$$

$$\beta = \frac{E_n - E'_n}{2} \quad (12)$$

By fitting the Wallace equations to the M_{rx} and M_{ry} components and comparing with the composite data, it is possible to measure the residual loss to be accounted for by (a_t + d) and determine the amount to be attributed to each component.

TAPE

The isotropic tape utilizes unoriented cobalt doped γFe_2O_3 particles of approximately 0.2μ length, acicularity of 2 to 3, and a very small size distribution. Typical magnetic properties are:

H_c = 790 Oersteds
B_s = 1,850 Gauss
B_r = 1,575 Gauss
Sq. = .85
S.F.D. = .43

Magnetocrystalline anisotropy dominates the coercivity yielding the high squareness ratio and isotropy. The low acicularity and smoothness of the particles resist broadening of their size distribution during dispersion processes which, in cooperation with their high coercivity, results in thermal stability of the magnetization. Less than 2dB of total thermal remanence loss results from repeated 24 hour cycles between +55°C and -20°C. The complete isotropy of the tape is shown in Figures 5a,b,c. In Figure 5(a) the normal (y) component B-H loop from the V.S.M. is corrected for shear. Figures 5(b) and 5(c) show the nearly identical longitudinal and transverse B-H traces and permeabilities.

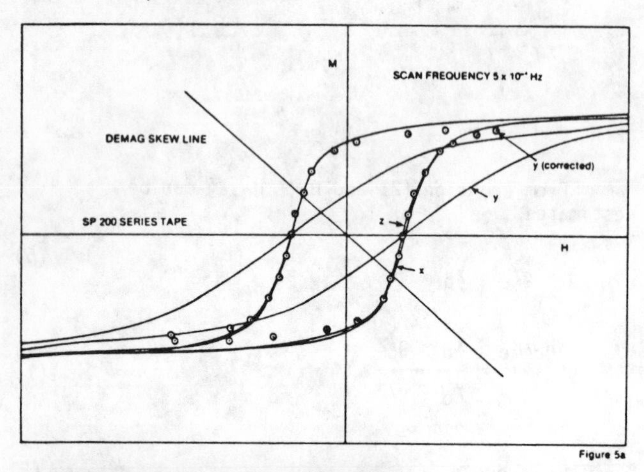

Figure 5a

LONGITUDINAL

Figure 5b

SP 243 TAPE
H_C = 786 Oersteds
H_M = 3000 Oersteds
Φ = .85
SFD = .43

TRANSVERSE

Figure 5c

Figure 6 shows the d.c. remanence and Figure 7 the 180° rotational remanence. The rotational remanence was obtained by applying a field in the plane of the tape at 90° to the direction of tape travel, rotating the sample 180° in the plane, removing the field, and measuring the remanent vectors. This simulates the field history of a short wavelength test volume traversing the gap region where the y demagnetization is not a factor.

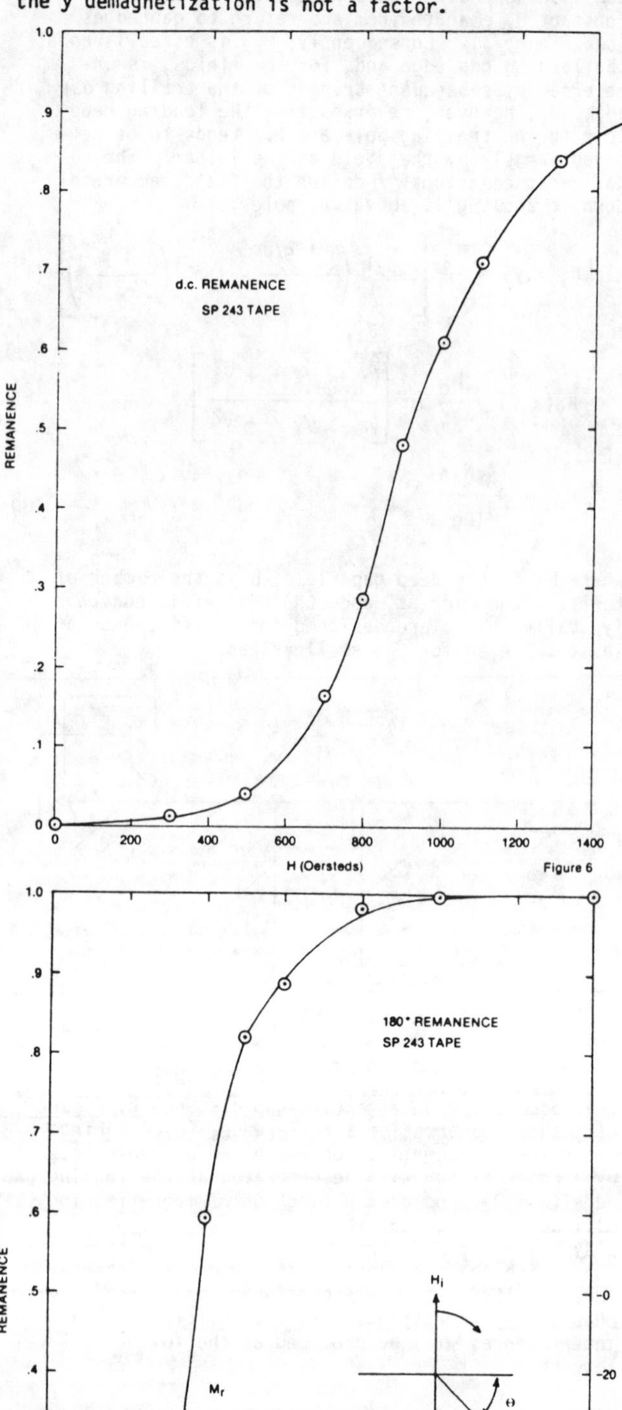

Figure 6

Figure 7

By comparison, a highly anisotropic video tape of H_c = 620 Oersteds is shown in Figures 8 and 9 for d.c. remanence and 180° rotational remanence, respectively. It is interesting to note that the rotational remanence of the isotropic tape inceases monotonically whereas the anisotropic tape peaks and then reduces to about 50%.

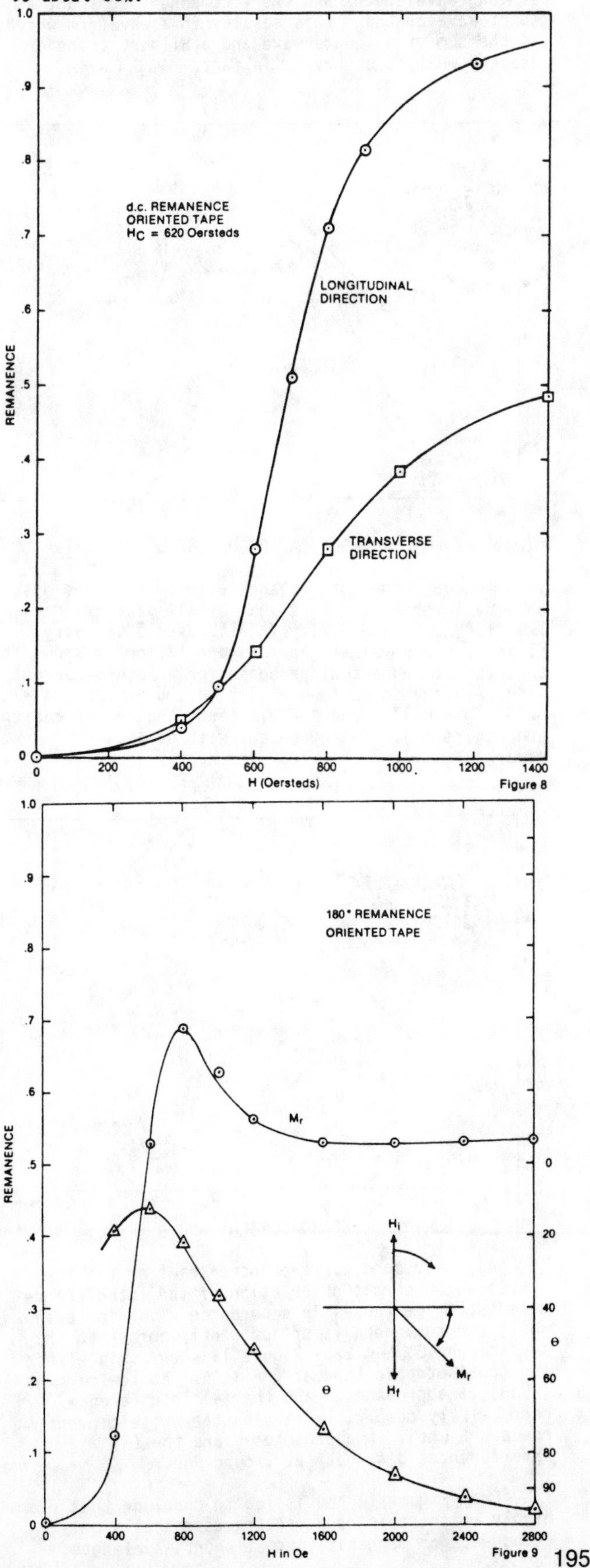

Figure 8

Figure 9

Figure 10 shows the rotational $|M_{rx}|$ and $|M_{ry}|$ for the two tapes. Two results are striking: at levels of H_0 typical for normal recording both tapes become magnetized predominantly normal to the surface, and the remanence of the isotropic tape exceeds that of the anisotropic tape by about a factor of two. This result puts in question the common practice of modelling with only the M_{rx} component.

In addition to its relatively larger normal component, the isotropic tape has a high B_r due to the compactness of its small particles which yields up to a total of 9dB output level advantage at short wavelengths.

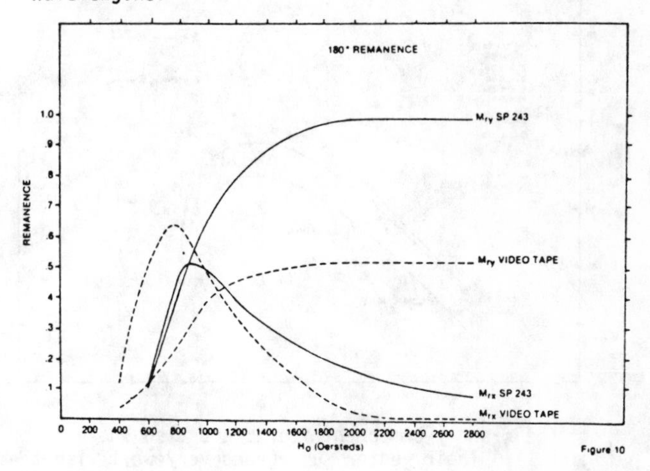

Figure 10

MODELLING

The playback voltage from a magnetic medium uniformly magnetized by a sinusoidal signal is, in dB,

$$e_0 \propto \log \left[M_r k\delta \cos kx \left(\frac{1 - e^{-k\delta}}{k\delta} \right) \cdot \left(\frac{\sin \pi g'/\lambda}{\pi g'/\lambda} \right) \left(e^{-kd} \right) \right] \quad (13)$$

where $k = 2\pi/\lambda$, δ = recording depth, g' = the playback gap length, and d = the spacing between the head and the medium. The term, e^{-kd}, is often adapted to $e^{-k(a_t + d)}$ to account for the anomalous log/linear losses attributed to the transition length. For recording zones separated by $\bar{g}$ with phase relationship of $-\pi/2$,

$$e_0 \propto \log \left[\left(M_{rx} \cos kx - M_{ry} \sin k(x - \bar{g}) \right) k\delta \cdot \left(\frac{1 - e^{-k\delta}}{k\delta} \right) \left(\frac{\sin \pi g'/\lambda}{\pi g'/\lambda} \right) \left(e^{-kd} \right) \right] \quad (14)$$

The spectral response is,

$$e_0 \propto \log \left[(\beta^2 + 2\beta \sin k\bar{g} + 1)^{\frac{1}{2}} (1 - e^{-k\delta}) \cdot \left(\frac{\sin \pi g'/\lambda}{\pi g'/\lambda} \right) e^{-kd} \right] \quad (15)$$

Equation (15) has broad applicability to a variety of heads and tapes if the recording field is sufficiently low ($\geq$ 4dB below the current for peak output at $\lambda \approx 1\mu$). For example, Figure 11 shows the plot of equation (15) and actual data for the subject isotropic tape where $\beta = 0.5$ was determined from E_n and E'_n, $\delta = 1.35\mu$ to account for the maximum output, $\bar{g} = 2.04\mu$ from E'_o, and $g' = 0.5\mu$. The physical gap length of the record head was measured to be 1.98μ. Due to the high surface finish of the tape, $d \approx 0$.

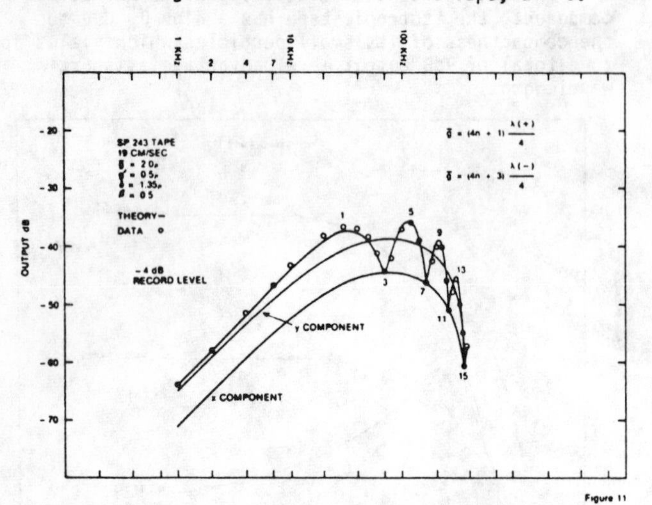

Figure 11

The x and y component fields are also shown in Figure 11. Their vector sum agrees very well with the measured response. It is interesting to note that the short wavelength data are fully accounted for without recourse to an $(a_t + d)$ term. The implication of this fact is that no appreciable demagnetization factor exists at short wavelength for either the x or the y components of magnetization in high coercivity isotropic tapes.

LONG WAVELENGTHS

Demagnetization plays a dominant role at long wavelengths resulting in essentially planar magnetization. When the medium is in contact with the head poles, imaging dictates the use of the demagnetization factor at the midpoint of a normal charge distribution with effective length 2δ.

$$M'(x,\lambda) = \Big[M'_{rx} \cos kx - M'_{ry}(1 - e^{-k\delta}) \sin k(x - \bar{g}) \Big] \tag{16}$$

$$M'(\lambda) = \Big[\beta'^2 + (1 - e^{-k\delta})^2 + 2\beta'(1 - e^{-k\delta}) \sin k\bar{g} \Big]^{\frac{1}{2}}, \tag{17}$$

where β' is the ratio of $|M'_{rx}|/|M'_{ry}|$ for 180° rotational remanence about the z axis starting at $x = 0$. The crossover from planar to normal magnetization occurs when $\delta/\lambda \approx 0.2$. Figure 12 shows the effect of the y demagnetization at long wavelength when $\beta = 0.5$ at short wavelength. Although the data fit is excellent, assumption of an unphysically large δ is required. The determination of β as a function of wavelength and medium anisotropy properties is not a simple task. Neglecting the long wavelength demagnetization and using a value of β appropriate for short wavelengths results in only about 1 dB of error near the crossover wavelength and very good agreement elsewhere.

It is clear from consideration of Figures 11 and 12 that the latter more correctly represents the true situation. If the y component dominates at long wavelengths, equalized square wave response would appear as a Hilbert transform rather than as a square wave. Figure 12 shows the x component to be dominant at long wavelengths and the y component dominant at short wavelengths. This results in a waveform which is the sum of a square wave and a Hilbert transform - exactly what is observed, in fact.

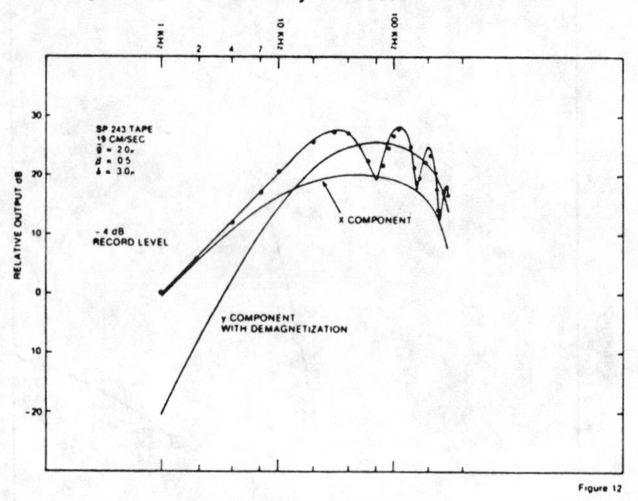

Figure 12

SHORT GAP RECORD HEAD

The gap edge interference effect is caused not only by large record gaps, but by all gaps tried: 1.9, 1.4, 0.8, and 0.3μ. In all cases $\bar{g}$ was very close to the physical gap. Figure 13(lower) shows the theoretical and actual response for a record current -4dB relative to maximum at 300 KHz, $\bar{g} = 0.36\mu$, $\delta = 0.76\mu$, $g' = 0.37\mu$, and $\beta = 0.2$ (from the record current and Figure 10). The agreement with data is remarkable - less than 1dB over nearly 9 octaves.

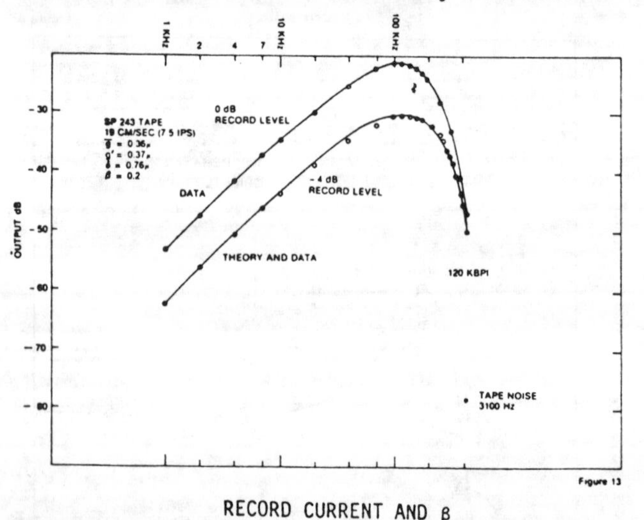

Figure 13

RECORD CURRENT AND β

As the record current increases from the low levels which permit observation of the interference effects, β decreases in accordance with Figure 10 until the remanence is predominantly normal to the surface and $E'_n \longrightarrow E_n$. Figure 14 shows this effect for the isotropic tape at $\bar{g} = 1.4\mu$. At low current (-9dB), β approaches 1 and the (-) interference becomes very deep. Increasing the write current reduces β until it approaches 0 and the (-) interferences disappear at around 0dB.

From Figure 10 one is led to conclude that even highly anisotropic tapes record essentially only y components of magnetization at short wavelength.

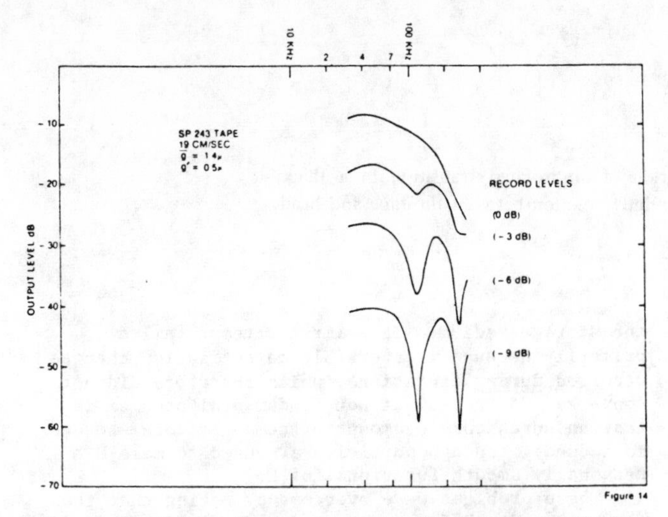

Figure 14

Replotting the curves of Figure 13 on a log/linear scale is shown in Figure 15. When the gap loss term is subtracted from the -4dB curve, the response is essentially flat except for the interference, I, caused by $\beta = 0.2$.

$$I = 20 \log \left[\beta^2 + 2\beta \sin k\bar{g} + 1\right]^{\frac{1}{2}} \quad (18)$$

which accounts for 1.7dB. Subtracting that term yields zero anomalous loss.

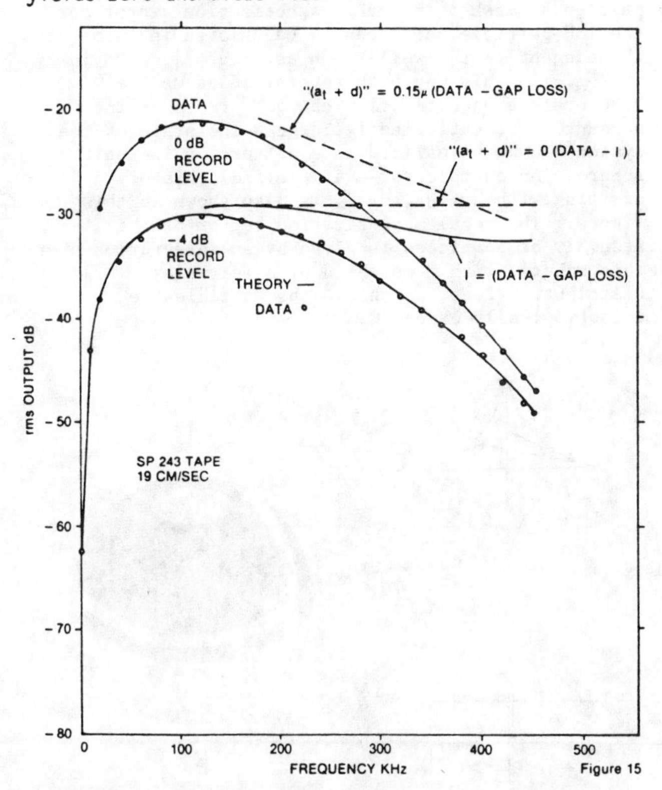

Figure 15

The 0dB record current response curve of Figure 13(upper) is shown on the log/linear plot of Figure 14. When the gap loss is subtracted from it, the familiar linear plot of "effective" $(a_t + d)$ results. However, this attenuation would appear not to be due to a transition length, since it is absent at low levels, but, rather, is caused by a remagnetization of the remanence in the high gradient H_y head field. The log/linear character results from the nearly linear remagnetization curve at fields comparable to H_G and the ln(x) nature of the H_y field. The analysis of this effect is beyond the scope of this paper and will be reported separately.

SHORT WAVELENGTH RECORDING

A short record gap creates a high field gradient in H_y which is desirable for short wavelength recording. Such gaps, when coupled with an isotropic medium, yield so-called perpendicular, or "vertical," recording and show no transition length. Figure 16 shows response out to a wavelength of 0.2μ wavelength for an effective density of 10,000 transitions/mm (250,000 f.c.i.).

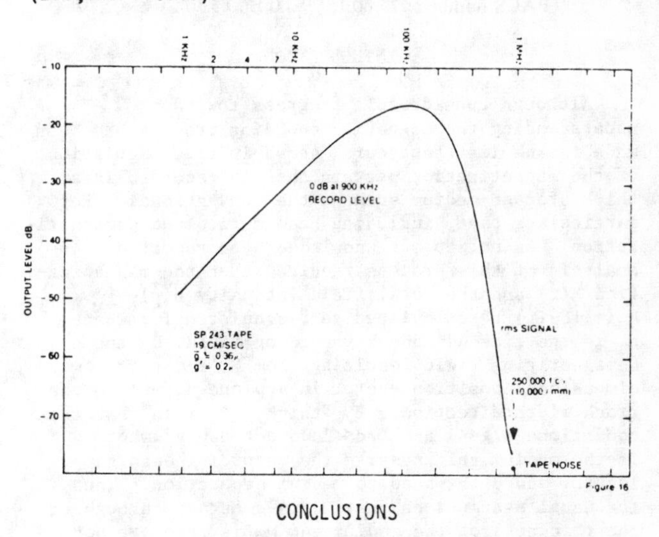

Figure 16

CONCLUSIONS

1. Separation of the x and y components of magnetization permits the examination of each for transition length effects at low recording levels - none is seen in isotropic media.
2. High coercivity anisotropic and isotropic media record short wavelength magnetizations which are predominantly normal to the surface, and M_{ry} (isotropic) ≈ $2M_{ry}$ (anisotropic) for the two tapes considered.
3. Small record gaps are preferred for short wavelength recording.
4. Without considering long wavelength demagnetization, very good data fit is possible through the consideration of both x and y magnetizations with δ as the only free parameter.

ACKNOWLEDGEMENTS

The author is indebted to three of his associates who painstakingly and accurately measured the data contained herein: Fred Jeffers for the V.S.M. data, and Chuck Wright and Dan Geyer for the tape spectral data.

REFERENCES

(1) M.L. Williams and R.L. Comstock, American Institute of Physics Conference Proceedings on Magnetism and Magnetic Materials, No. 5, rpt. 1, 738-742, 1971.
(2) H.N. Bertram and R. Niedermeyer, IEEE Trans. Magn., Vol. Mag-14, 5, 743-745, 1978.
(3) E. Koester and D. Pfefferkorn, IEEE Trans. Magn., Vol. Mag-16, 1, 1980.
(4) B.K. Middleton and T. Brown, The Radio and Electronic Engineer, Vol. 50, 9, 467-473, 1980.
(5) R.L. Wallace, Jr., Bell System Technical Journal, 1145-1173, 1951.
(6) J.G. Woodward and M. Pradervand, Journal of the Audio Engineering Society, Vol. 9, 4, 254-300, 1961.
(7) O. Karlqvist, K. Tekn, Hg. Handl., 86, 1, 1954.

A self-consistent calculation of the transition zone in thick particulate recording media

Irene B. Ortenburger and Robert I. Potter

IBM Research Laboratory, San Jose, California 95193

We describe a fully two dimensional self-consistent calculation of an isolated transition in a thick particulate magnetic recording tape. The calculation includes relative motion between the tape and head, as well as a vector magnetic hysteresis model.

PACS numbers: 75.60.Jp, 75.60.Ej, 85.70. – w, 02.70. + d

INTRODUCTION

Although considerable progress toward understanding the magnetic recording process has been made in the last ten years, an ab initio calculation of the magnetization pattern that is recorded in a thick storage medium such as the conventional γ-Fe_2O_3 particulate tape, including head motion and demagnetization, has not to our knowledge been reported. An analysis of this process requires that the magnetization $\vec{M}(\vec{r})$ and the total field intensity $\vec{H}(\vec{r})=\vec{H}_a(\vec{r})+\vec{H}_d(\vec{r})$ be calculated self-consistently, where $\vec{H}_a$ is the time-dependent recording field, $\vec{H}_d$ the demagnetizing field resulting from $\vec{M}$, and $\vec{r}$ the two dimensional position vector in a plane normal to the track width direction. By "thick," we mean that the conditions $\delta/g<<1$ and $\delta/d<<1$ do not apply, where δ is the medium thickness, g the recording head gap length, and d the head to medium separation. Thus the usual assumptions that M_x is constant throughout the thickness of the medium and M_y is zero are not valid.

Several models of the recorded magnetization pattern for thick media have been proposed. For example, Chang and Perez [1] have extended the isolated transition arctangent and sinusoidal models by tilting the transition zone. Middleton [2] has postulated a semi-circular magnetization pattern for which $\nabla\cdot\vec{M}$ is zero everywhere except on the tape surface. A self-consistent calculation would shed light on these models, as well as be useful for investigating new types of recording heads and media.

Two dimensional self-consistent calculations have been performed for a stationary head by Iwasaki and Suzuki [3] and Suzuki [4], and with head-medium motion but for a relatively thin Fe_3O_4 medium by Ortenburger, Cole and Potter [5], referred to below as OCP. The OCP hysteresis model consisted of independent scalar hysteresis models for the x and y directions and is not extendable to thick media for which M_y is not always small compared to M_s.

In this paper we report a two dimensional calculation for a 4 μm thick magnetically isotropic particulate storage medium using an unrestricted vector hysteresis model. The notation, coordinate system, and numerical techniques are those of OCP unless otherwise noted.

THE VECTOR HYSTERESIS MODEL

Our first attempt at a thick medium calculation used the hysteresis model described by Hartman, Potter and Ortenburger [6] but with particle interactions suppressed. This model is similar to that of Suzuki [4] in that it consists of an assembly of noninteracting Stoner-Wohlfarth particles [7] at each mesh point of the storage medium, but differs in that it has the particles distributed in three dimensions instead of two. Thus the squareness ratio is $M_r/M_s\simeq0.5$. Note that although these particles are noninteracting within the context of the hysteresis model, interactions are included in a mean field sense when iterating over the mesh that represents the storage medium. This first attempt failed primarily because irreversible magnetization changes occurred during iterations, which therefore did not converge. A second but not fundamental problem is that an unreasonable amount of computer core storage is needed if enough particles are used to make $\vec{M}$ a reasonably smooth functional of $\vec{H}$.

These problems were overcome by noting that the magnetic state S due to a particular field history for a system of N Stoner-Wohlfarth particles is given by $\{\psi_i\}$, i=1,2,...N, where

$$\psi_i = (\vec{m}_i \cdot \hat{p}_i)/m_i = \pm1.$$

Here $\vec{m}_i$ is the magnetization vector for the ith particle when the field intensity is reduced to zero and $\hat{p}_i$ the unit vector along the semi-major axis. Thus the history at a given mesh point can be stored as N binary bits. Given ψ_i and the total field intensity $\vec{h}$ (in reduced units of $2K/M_s$) at a particular mesh point, the magnetization vector for the ith particle there can be calculated in terms of the tangent to the well-known astroid [7] shown in Fig. 1. In this way both reversible as well as irreversible magnetization changes are taken into account. For sufficiently large N the state of the system can be visualized as a pattern on the unit sphere. An example for a simple field history starting with $\vec{h}=-\hat{y}h_o$, $h_o>1$, is also shown in this figure. The total magnetization vector of the assembly of particles is given by an integration over this sphere, with ψ weighted by a suitable distribution function should the particles be anisotropically oriented.

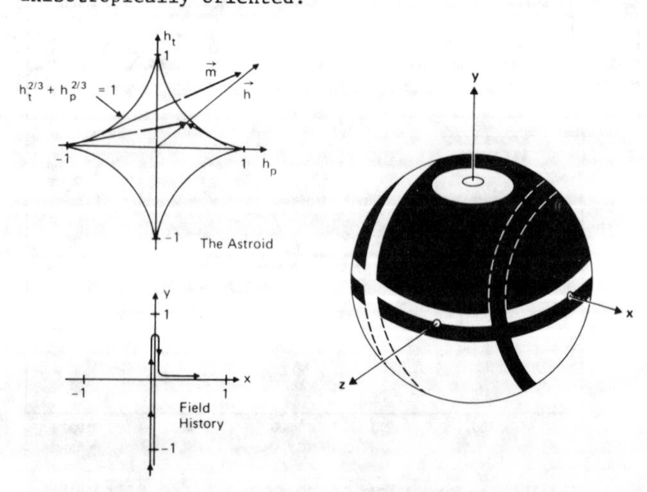

Fig. 1. Upper left: The astroid from which the magnetization of a uniaxial particle can be calculated given the field $\vec{h}$ and one of two possible states ψ of the particle. See ref. 7.
Right: The unit sphere on which the state S of an assembly of particles is recorded. Dark areas correspond to ψ=1 or $\vec{m}$ nearly parallel $\vec{p}$.
Lower left: Field history giving the pattern shown.

Reprinted with permission from *J. Appl. Phys.*, vol. 50, no. 3, pp. 2393–2395, Mar. 1979.

As a practical matter N must be finite. Therefore the discontinuities in the macroscopic $\vec{M}$ versus $\vec{H}$ curves are not entirely eliminated, but the point is that for workable N the storage requirements are modest and that by freezing S during iterations convergence can be obtained.

The calculation described below was done using 99 particles distributed approximately uniformly over one fourth of the sphere. This gives 33 different switching fields for $\vec{h}$ in the $\hat{x}$ or $\hat{y}$ directions.

OTHER DETAILS

In view of the exploratory nature of the calculation, many refinements that could have been included were not. The familiar Karlquist expressions were used for the applied field and readback voltage calculations. Image terms [5] were included but the head was not removed and replaced between recording and readback. The record current was not optimized. Rather, it was chosen sufficiently large to switch the bottom layer while not exceeding the deep gap field of ~2500 Oe obtainable from ferrite heads.

The parameters of the calculation are M_s=160 emu/cc, M_r=80 emu/cc, H_c=300 Oe, d/g=0.5 and δ/g=2. If g=2 μm then the medium thickness is δ=4 μm and the separation is d=1 μm. We assume these specific values for the sake of clarity. The medium is magnetically isotropic.

The numerical techniques have been described by OCP. Here, the mesh consists of 31 points in the x or longitudinal direction and 5 points in the y or vertical direction with a separation Δx=Δy=1 μm between points. The top and bottom rows of points are on the surface. An analytic continuation of the magnetization beyond the ends of the tape segment considered reduces end effects but it still is necessary to discard the three outer columns of points. Thus, the mesh shown in the figures below has 25 points in the x direction.

RESULTS AND DISCUSSION

The magnetization pattern for an isolated transition is shown as an arrow plot in Fig. 2. The recording current was reversed when the head was located relative to the medium at x=16 μm as indicated in this figure. Also shown is the corresponding demagnetizing field $\vec{H}_d$. The peak horizontal component of $\vec{H}_d$ in the midplane of the medium is 122 Oe at x=15 μm. The vertical component of $\vec{H}_d$ at the surface reaches 220 Oe. It is clear from both $\vec{M}$ and $\vec{H}_d$ that $\nabla\cdot\vec{M}$ is appreciably different from zero in the interior of the medium.

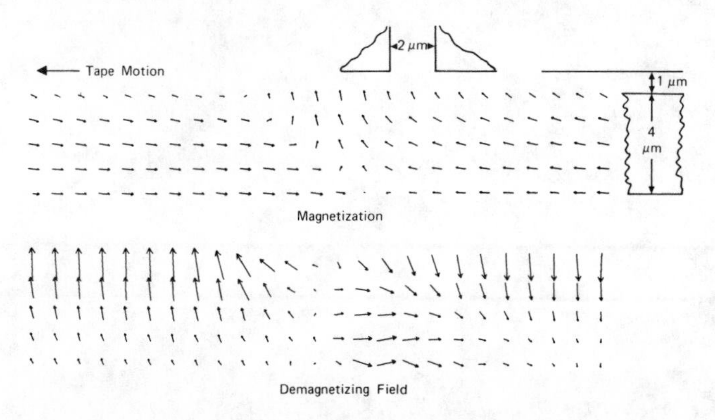

Fig. 2. The isolated transition. Head is shown at position where current was reversed.

The magnetization components for the five layers are plotted in Figs. 3 and 4 as a function of x. Here it is evident that the vertical component of $\vec{M}$ approaches the remanent magnetization.

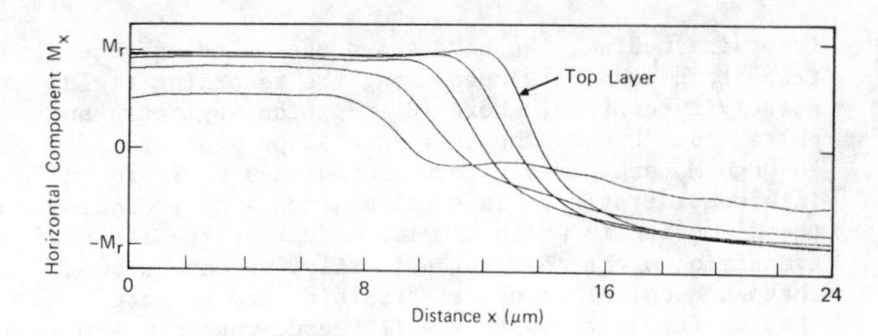

Fig. 3. Horizontal magnetization components for the five layers as a function of x.

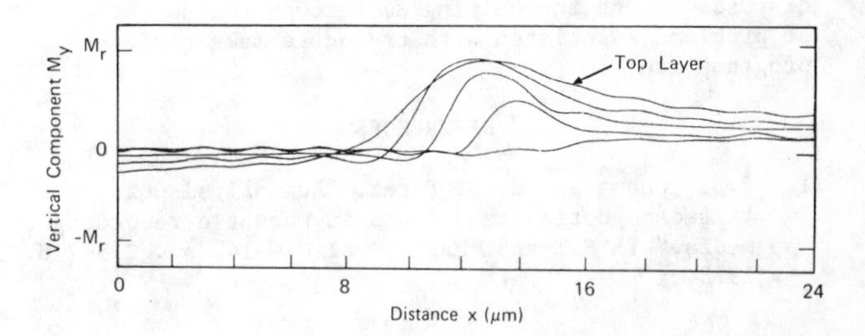

Fig. 4. Vertical magnetization components.

Shown in Fig. 5 are the M_x and M_y contributions to the total readback pulse as computed by reciprocity. Here it is evident that the large y component of magnetization does not cause pulse asymmetry. Rather, when its contribution is summed with the relatively sick contribution from the x component the result is a reasonably symmetric and considerably sharper total pulse than the x component would produce alone. The pulse width is 7.1 μm. The total readback pulse is decomposed into the contributions from each of the five layers in Fig. 6, where it is seen that the second layer provides the largest contribution.

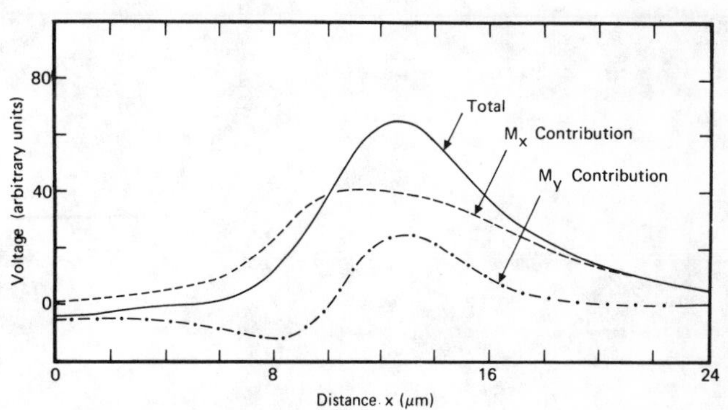

Fig. 5. The readback pulse.

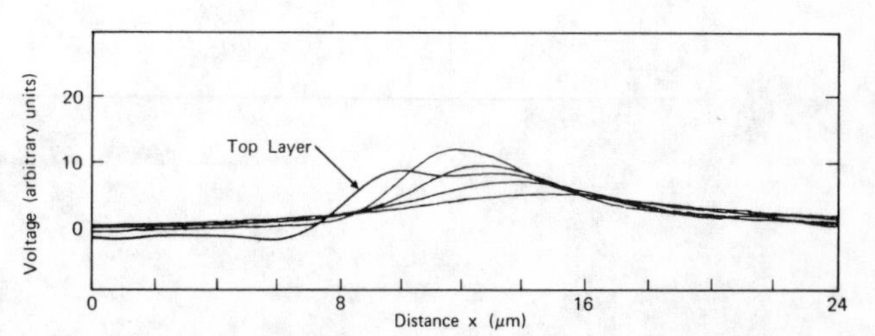

Fig. 6. Contributions to the readback pulse from the various layers.

In summary, we have performed what we believe to be the first two dimensional self-consistent calculation with head motion for a thick particulate recording medium. We have shown that although the transition zone is sloped along the recording field coercivity contour, where $|\vec{H}_a|=H_c$, the asymmetry and pulse width broadening that this slope produces through M_x are largely compensated when M_y is taken into consideration. This calculation also indicates the direction in which simpler models of the isolated transition might go. But primarily, we have shown that such calculations are feasible, and we expect that in future years they will become routine.

ACKNOWLEDGMENT

We wish to thank W. E. Rudge for interesting discussions and for helping us through the quagmire of problems associated with trying to make computer programs run.

REFERENCES

1. P. T. Chang and H. S. Perez, "An analysis of tilted magnetic transitions in magnetic recording media," IEEE Trans. Magn., vol MAG-14, pp. 213-218, 1978.
2. B. K. Middleton, "The replay signal from a tape with magnetization components parallel and normal to its plane," IEEE Trans. Magn., vol MAG-11, pp. 1170-1172, 1975.
3. S. Iwasaki and T. Suzuki, "Dynamical interpretation of the magnetic recording process," IEEE Trans. Magn. vol MAG-4, pp. 269-276, 1968.
4. K. Suzuki, "Theoretical study of vector magnetization distribution using rotational magnetization model," IEEE Trans. Magn. vol MAG-12, pp. 224-229, 1976.
5. I. B. Ortenburger, R. W. Cole and R. I. Potter, "Improvements to a self-consistent model for the magnetic recording properties of non-particulate media," IEEE Trans. Magn. vol MAG-13, pp. 1278-1283, 1977.
6. K. Hartman, R. I. Potter and I. B. Ortenburger, "A vector model for magnetic hysteresis based on interacting dipoles," IEEE Trans. Magn. vol MAG-14, pp. 223-227, 1978.
7. See, for instance, M. Prutton, Thin Ferromagnetic Films. London: Butterworths, 1964, pp. 54-62. The formalism for uniaxial particles and films is the same.

SELF-CONSISTENT COMPUTER CALCULATIONS FOR PERPENDICULAR MAGNETIC RECORDING

Robert I. Potter and Irene A. Beardsley

Abstract - A comparison of perpendicular and longitudinal recording indicates that conventional heads are suitable for recording on high coercivity media having perpendicular magnetic anisotropy. At equal bit shift and head to disk separation perpendicular recording gives a factor of 2.5 increase in the linear density attained with media in use today.

INTRODUCTION AND SUMMARY

Perpendicular magnetic recording has been studied in the past by Hoagland [1], Fan [2], Westmijze [3] and others. It has recently received a resurgence of interest due to the work of Iwasaki [4], Iwasaki, Nakamura and Ouchi [5], Iwasaki and Ouchi [6], and Iwasaki and Nakamura [7], who have shown that films with suitable coercivity and perpendicular magnetic anisotropy can be recorded upon at a linear density of up to 4000 magnetization reversals/millimeter (mr/mm). These results were obtained with a "single pole" or probe type of recording head that is magnetized by a ferrite core solenoid located on the opposite side of the thin, flexible storage medium substrate. Reading is accomplished by a separate conventional head because single pole heads reported to date do not exhibit sufficient read resolution. This single pole recording head structure is more suitable for tape or thin flexible disks than for rigid disk recording.

Lemke [8] has shown that an equally high linear density can be achieved by reducing the head to medium separation and head gap length and recording on a magnetically isotropic medium. (Most media have in-plane anisotropy.) Although part of this increase is a consequence of geometric scaling, a sizeable part of it is because in an isotropic medium the recording process generates a substantial perpendicular component of magnetization [9]. This is advantageous because in perpendicular recording the demagnetizing field aids rather than opposes the recording process. That such a magnetization component exists is not surprising, because the recording field near the edge of a conventional ferrite or thin film head points predominantly in the perpendicular direction.

It is the purpose of this paper to study the perpendicular recording process using our self-consistent vector field recording model [9], and to address what would happen if a more practical, conventionally shaped head of the type in use today were used with a perpendicularly oriented medium of the type described by Iwasaki et al. The ground rules for a perpendicular versus longitudinal recording comparison should be to use the same head to medium separation (we choose d=0.25 μm) and the same head saturation magnetization or deep gap field, but to allow whatever changes in head design and medium parameters improve the results. The work presented here follows this approach.

Our conclusion is that existing heads work well in perpendicular recording. They are capable of recording on a perpendicularly oriented medium having a higher coercivity than its longitudinal recording counterpart. Perpendicular recording with ferrite or thin film heads gives higher linear density, lower bit shift and is better suited to the use of equalization techniques, even when both media have the same coercivity. Moreover, the absence of strong demagnetization at high density leaves the recording head fringe field as the dominant and controlling factor in the recording process, and opens the possibility for further advances through suitable head design.

THE RECORDING MODEL AND DETAILS OF THE CALCULATIONS

The self-consistent vector field recording model is an outgrowth of the Potter-Schmulian model [10] and has been previously described [9]. As before, the basic idea is to solve simultaneously and as a function of head-to-medium location two relationships. The first is the total field intensity, which depends on the magnetization throughout the medium as well as the applied recording field, and the second is the functional hysteresis relationship between the total field and the magnetization.

The only recent change in the recording model is to partition the storage medium cross section into N squares, assume the magnetization vector is constant within each square, and calculate the demagnetizing field at each center due to the magnetization discontinuities on the four edges of all N squares. This allows a matrix formulation of the demagnetizing field calculation that is similar to the way [11] that recording heads are analyzed, and permits us to compute the field using a fast Fourier transform in place of numerical integration and a finite difference approximation to div $\underline{M}$.

The vector hysteresis model [9,12] affixed to the center of each square is based on an assembly of particles or crystallites with uniaxial magnetic anisotropy. The angular distribution of the easy axes of these particles determines the magnetic anisotropy and hysteresis loop shape of the film that they represent. Some care needs to be taken in choosing the distribution function such that for a tractable number of particles the magnetization does not exhibit numerically troublesome Barkhausen-like discontinuities caused by several particles switching almost simultaneously. In this work we placed all particles in the xy plane and used the following distribution functions, expressed as the angular increment $\Delta\phi$ between particles and starting with $\phi=\phi max$:

$$\Delta\phi \sim \frac{1}{\cos^2\phi}\,\frac{\pi}{2N}, \qquad \phi_{max}=0.88\,\frac{\pi}{2N}, \qquad N=200$$

$$\Delta\phi \sim \frac{1}{\cos^{14}\phi}\,\frac{\pi}{2N}, \qquad \phi_{max}=0.38\,\frac{\pi}{2N}, \qquad N=800$$

The ratio of the two orthogonal remanent magnetizations given by these distributions is two and five, respectively, as shown in Fig. 1. An iterative solution for the direction in which the magnetization of each particle points is avoided by deriving an expression [12] for the switching astroid tangency condition [13].

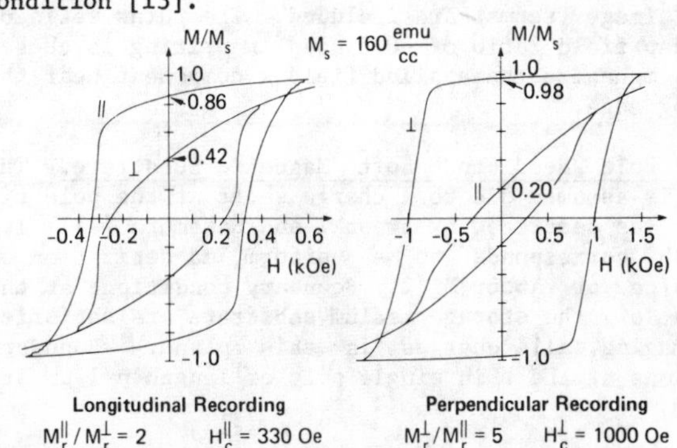

Fig. 1. Major hysteresis loops.

Manuscript received March 8, 1980.
The authors are with the IBM Research Laboratory, San Jose, California 95193.

Reprinted from *IEEE Trans. Magn.*, vol. MAG-16, pp. 967-972, Sept. 1980.

The two independent material parameters in addition to the anisotropy are the saturation magnetization M_s and the coercivity H_c. The latter is set through a suitable choice of particle acicularity ratio or anisotropy constant K. We could but did not include a distribution of switching fields 2K/M, where M is the particle magnetization, thus irreversible magnetization changes occur only when the field intensity is between K/M and 2K/M. We used 2K/M=600 and 1492 Oe for the longitudinal and perpendicular calculations, respectively. These values and the angular distribution functions described above give the coercivities indicated in Fig. 1.

The hysteresis model produces major loops, minor loops and responds to an applied vector field in the same overall way that bulk samples do, and for the 5:1 remanence ratio the computed loops agree closely with the unsheared Co-Cr loops published by Iwasaki and Ouchi [6]. It remains, however, an assumption on our part that the physics of magnetic hysteresis in continuous films such as Co-Cr is adequately represented by a noninteracting assembly of uniaxial crystallites, and in view of this the model is best regarded as phenomenological.

In this work we consider four types of recording heads and treat them as follows.

Ferrite Head. The Karlqvist expressions are used for the fringe field with Hg=3.142 kOe, corresponding to a record current of 0.25 A when the gap length g=1 μm. The boundary conditions at the head are handled within the spirit of Karlqvist's approximation by imaging the storage medium magnetic charge -div $\underline{M}$ in the air bearing surface.

Thin Film Head. Analytic expressions for the fringe field are obtained by differentiating the approximate scalar potential suggested by Potter [14]. The x and y components are (cgs units):

$$H_x = \frac{I_o}{\pi g}\left(\tan^{-1}\frac{x+g/2}{y} - \tan^{-1}\frac{x-g/2}{y}\right) - \frac{I_o}{2\pi}\left\{\frac{y}{x^2+y^2} - \frac{p+g/2}{2(x^2+y^2)}\left[\frac{x^2-y^2}{x^2+y^2}\left(\tan^{-1}\frac{x+p+g/2}{y} - \tan^{-1}\frac{x-p-g/2}{y} - \pi\right) + \frac{xy}{x^2+y^2}\ln\frac{(x+p+g/2)^2+y^2}{(x-p-g/2)^2+y^2}\right]\right\}$$

$$H_y = \frac{I_o}{2\pi g}\ln\frac{(x+g/2)^2+y^2}{(x-g/2)^2+y^2} - \frac{I_o}{2\pi}\left\{\frac{x}{x^2+y^2} + \frac{p+g/2}{4(x^2+y^2)}\left[\frac{4xy}{x^2+y^2}\left(\tan^{-1}\frac{x+p+g/2}{y} - \tan^{-1}\frac{x-p-g/2}{y} - \pi\right) - \frac{x^2-y^2}{x^2+y^2}\ln\frac{(x+p+g/2)^2+y^2}{(x-p-g/2)^2+y^2}\right]\right\}$$

The calculation described below with g=1 μm and pole tip length p=0.5 μm has Hg=3.142 kOe and no image terms. The calculation in the appendix with g=0.75 and p=2 has Hg=4.195 kOe and Karlqvist-like image terms as a first approximation to the true Green's function.

Double Gap Head. The applied field is obtained by superimposing displaced Karlqvist expressions as discussed [11] in the analysis of the magnetoresistive head. Image terms are included. The turns ratio or deep gap field ratio of -0.2 used in writing is chosen to de-emphasize the applied field x component near the trailing edge.

Single Pole Head and Soft Magnetic Substrate. The field is assumed due to a charge sheet at the pole tip surface as described by Iwasaki and Nakamura [7]. Its strength corresponds to a uniform magnetization of 400 emu/cc or about $M_s/2$. Boundary conditions at the surface of the storage medium substrate are satisfied by imaging all charges in this plane. Boundary conditions at the thin single pole of length p=1 μm are ignored.

In all cases the head to medium separation is 0.25 μm and removal and replacement of the head between recording and reading is omitted. The 8 μm long segment of the 0.5 μm thick storage medium is broken up into 64 volume elements of 0.25×0.25 μm square cross section. One calculation was repeated with four times as many volume elements in order to demonstrate that elements 0.25 μm on a side are sufficiently small. The difference in the results is negligible. In spite of an analytic extension of the magnetization beyond this segment the outer three columns of elements are unreliable and are discarded. With respect to the medium, head motion is from left to right and the magnetization is iterated to convergence every 0.25 μm or whenever the record current Io is changed by at most 0.25 Io. The convergence criterion is on the field intensity and is 12 Oe.

In order to establish a common basis for comparison and to separate the recording merits of the various heads from their reading characteristics, we computed the distance derivative of the flux (times two) through the plane normal to the longitudinal direction extending from y=d to infinity as indicated in Fig. 2. This quantity can also be regarded as the readback signal obtained with an infinite permeability, infinite pole tip length head in the limit as gap length approaches zero. The flux-through-the-plane viewpoint, however, avoids certain unnecessary conceptual difficulties associated with satisfying boundary conditions, and is shown as a dotted curve along with the readback signal that is obtained from the same head that did the recording. Cubic spline function interpolation is used in generating the readback plots.

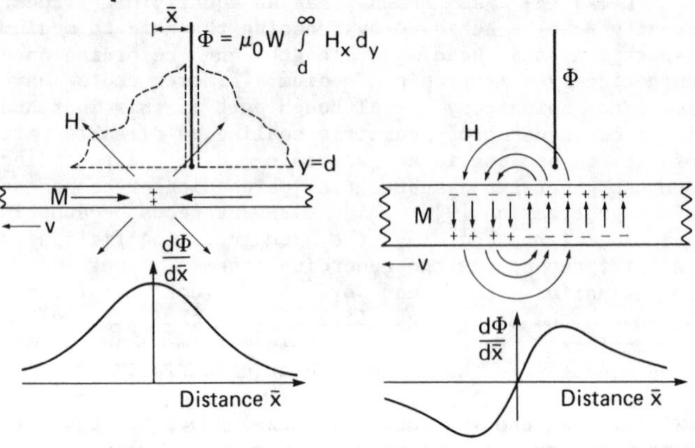

Fig. 2. Schematic waveforms and definition of $d\Phi/d\bar{x}$.

As an example of the recording calculations consider the more familiar longitudinal case shown in Fig. 3, which serves as the reference calculation for perpendicular recording. The applied fringe field components are shown at the top, the magnetization pattern that exists just before and after the current is reversed come next, then the stabilized isolated transition, and at the bottom is the readback pulse and the related quantity $d\Phi/d\bar{x}$ defined in Fig. 2. A similar calculation differing only in that the medium thickness is 1 μm appears in Fig. 4. Note that the transition in Fig. 3 differs considerably from the top two layers of Fig. 4.

RESULTS FOR PERPENDICULAR RECORDING

Shown in Fig. 5 is what happens when a high H_c medium with perpendicular anisotropy is substituted for the longitudinal one used in Fig. 3. The two readback waveforms are compared in Fig. 6. Because in perpendicular recording it is desirable to detect the point of maximum slope, the quantity analogous to the pulse width P_{50} in longitudinal recording is the distance S_{50} between the points at which the slope drops to 50% of its maximum value. The waveform for the opposite polarity transition (shown later in Fig. 17) is virtually the negative of the one in Fig. 5 even though the initial magnetization condition is <u>not</u> the negative. This indicates that overwriting old data will not be a problem.

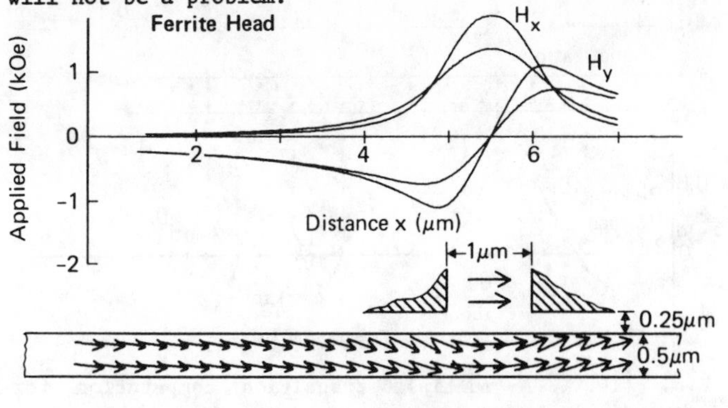

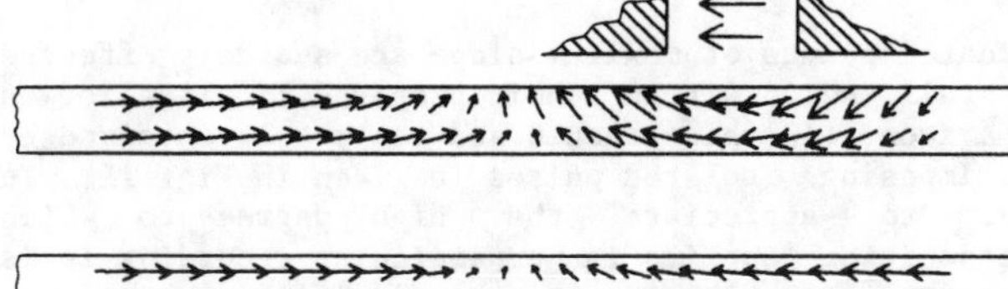

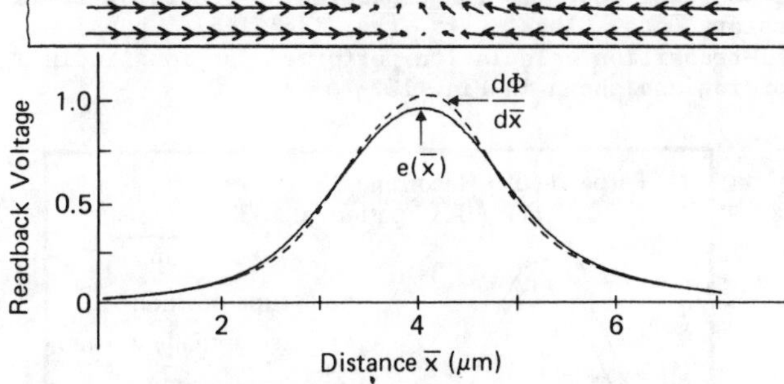

Fig. 3. Longitudinal recording.

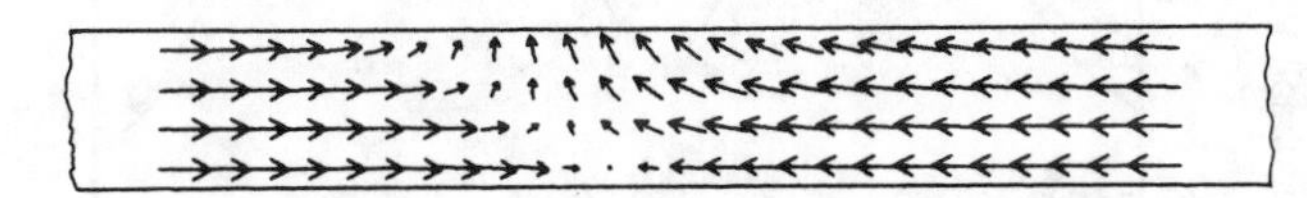

Fig. 4. As above, but with 1 μm medium thickness.

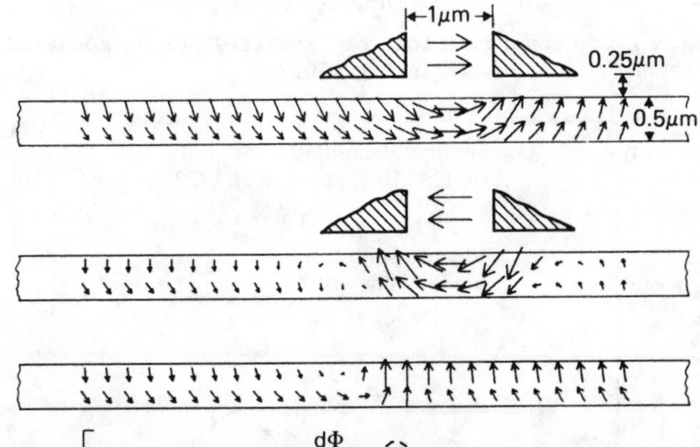

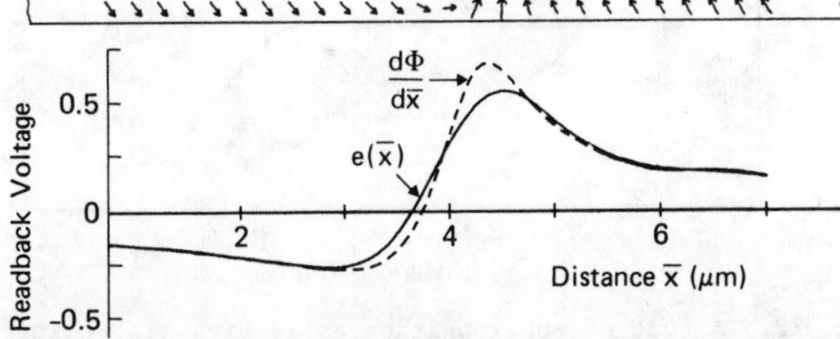

Fig. 5. As in Fig. 3, but with medium with perpendicular anisotropy.

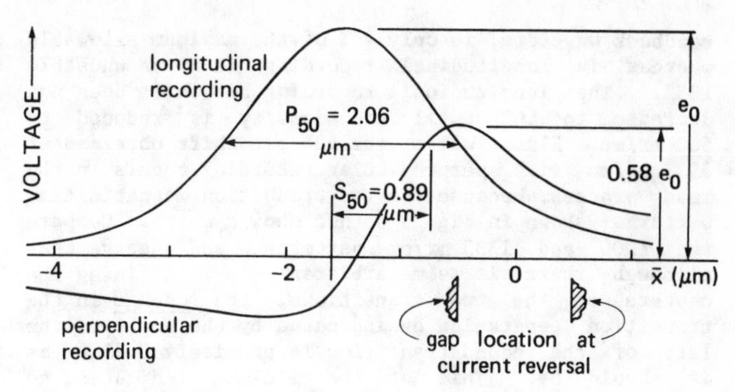

Fig. 6. Comparison of the g=1 μm waveforms in Figs. 3 and 5. See text for definition of S_{50}.

A comparison of the demagnetizing fields that exist during and after an isolated transition is recorded is given in Fig. 7. The main point is that prior to reversing the current the demagnetizing field in the trailing edge region points in the same direction that the applied field will point after the current is reversed. Thus this demagnetizing field is not one that has to be overcome in order to switch the medium, as is the case in longitudinal recording. The demagnetizing field is small in the transition region and approaches $4\pi M_y$ elsewhere. It tends to demagnetize only the wings of the transition and leaves the central portion essentially untouched, with the net result that the transition is sharper. In longitudinal recording the demagnetizing field is small far from the transition and is maximum close to the transition's center. The effect is a broadening of the transition.

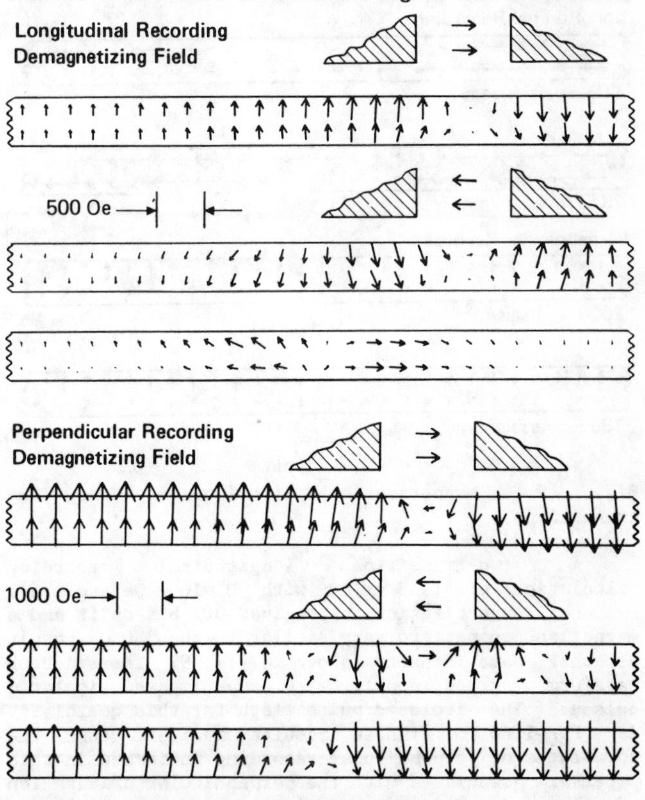

Fig. 7. Demagnetizing field comparison.

Computations for two transitions at various densities are summarized in Fig. 8. Additional calculations at intermediate densities would indicate a broad peak around 800 mr/mm in the perpendicular recording amplitude curve. The amplitude at 1333 mr/mm (33,858 mr/inch) is barely changed from the isolated pulse value. The bit shift at this density, as computed from the points of maximum slope on the readback waveform, is only 40% of the maximum allowable whereas in longitudinal recording it is an unusable 133%. The longitudinal recording bit shift does not decrease to 40% until the density is reduced to 500 mr/mm. Almost all of the 40% bit shift obtained at 1333 mr/mm with perpendicular recording occurs in the read process, because the two-transition magnetization patterns shown in Fig. 9 do not show a shift. Compare the 1000 and 1333 mr/mm patterns, and observe that although there is some arbitrariness in defining the centers of the two transitions, the change in the transition separation as indicated by the shift to the left of the second transition is precisely 0.25 μm as it should be. This at the minimum indicates no increase in bit shift in going from 1000 to 1333 mr/mm.

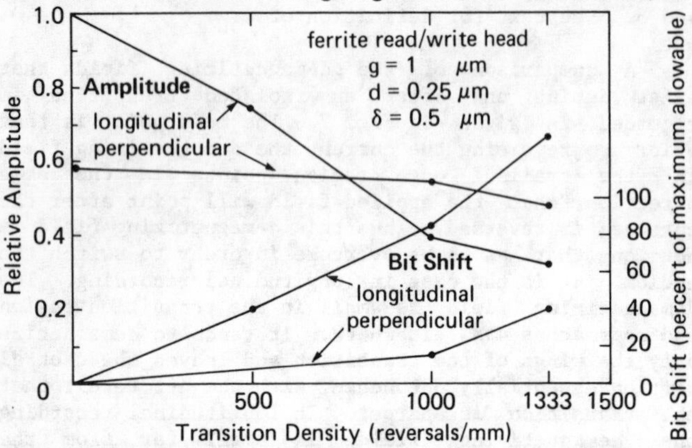

Fig. 8. Amplitude and bit shift for two transitions. Maximum allowable shift is one cell length.

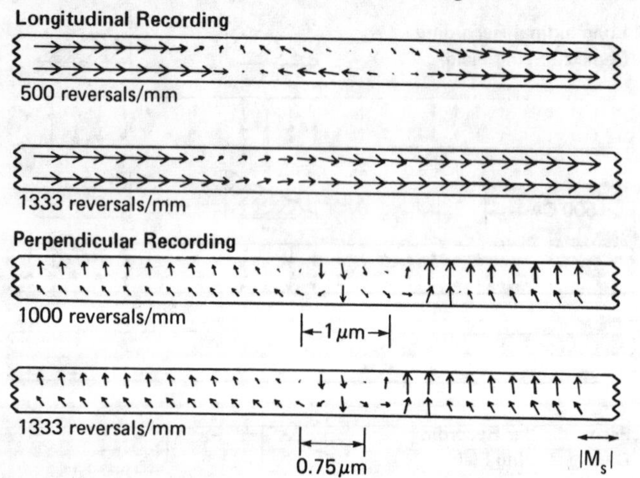

Fig. 9. Magnetization distributions for two transitions.

A two-transition longitudinal recording calculation at 1333 mr/mm with H_c=1000 Oe and a 5:1 remanent magnetization ratio gives 80% bit shift and a magnetization pattern very similar to the 330 Oe result at that same density shown in Fig. 9. The readback waveform agreed poorly with superimposed isolated pulses. The isolated pulse width for this coercivity is P_{50}=1.66 μm. These results indicate that the advantages of perpendicular recording indicated in this paper are associated with the perpendicular orientation of the easy axis of magnetization and not simply the use of different coercive force values.

A multiple transition calculation was performed as indicated in Fig. 10. Due to the close proximity of the outer transitions to the ends of the track segment, the readback waveform baseline shift is sensitive to the way in which the magnetization is extended for purposes of the readback voltage calculation, but the indicated points of maximum slope are scarcely affected and fall well within their proper 0.75 μm cells. A comparison of this result and the result obtained by superimposing isolated pulses is given in Fig. 11. In order to appreciate the high degree to which superposition applies in perpendicular recording it is necessary to look at the identical, hopeless, multi-transition calculation performed for longitudinal recording and shown in Fig. 12.

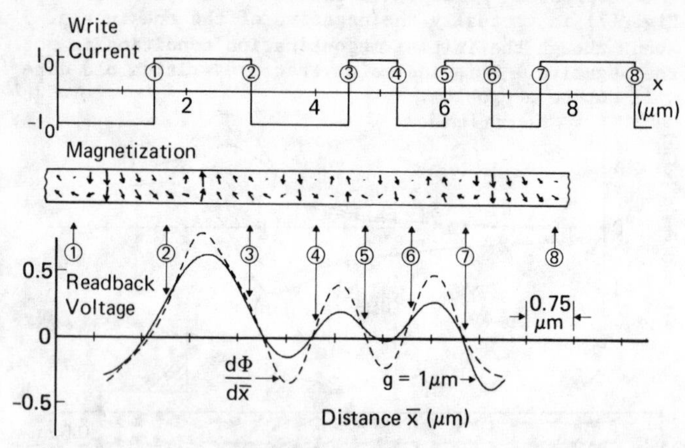

Fig. 10. A multiple transition computation for perpendicular recording.

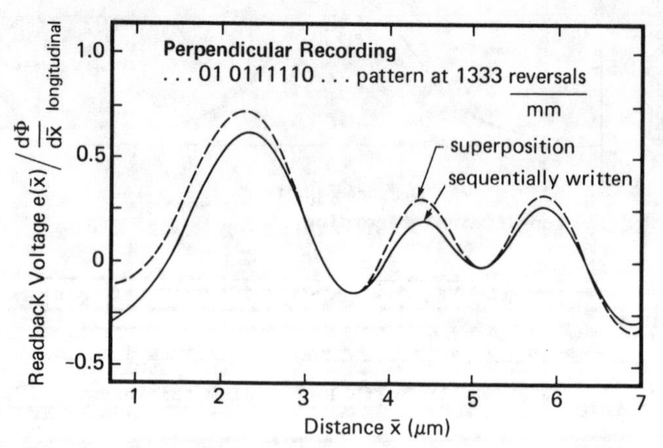

Fig. 11. Superposition of isolated pulses compared with the result shown in Fig. 10.

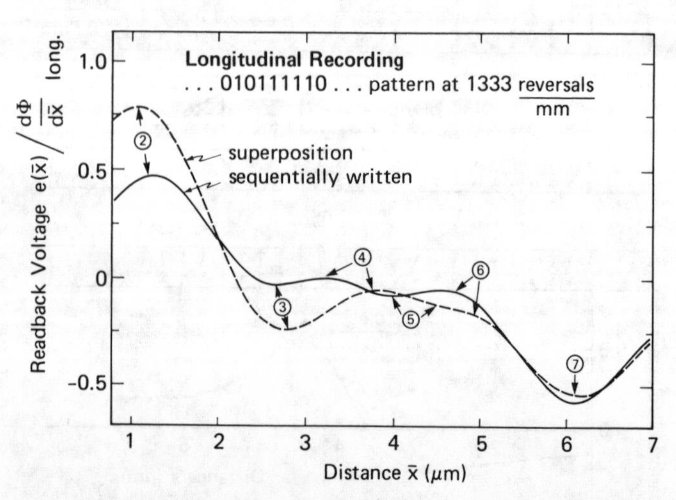

Fig. 12. A test of superposition as in Fig. 11 but for longitudinal recording. Circled numbers correspond to current reversals in Fig. 10.

Additional calculations using various recording head geometries are shown in Figs. 13, 14, 15 and 16. The single pole and soft magnetic substrate geometry described by Iwasaki, Nakamura and Ouchi [5] produces a pleasingly antisymmetrical magnetization pattern and $d\Phi/d\bar{x}$ waveform, but the pulse width obtained by reading with the single pole head is no better than what can be obtained in longitudinal recording. The overall quality of the transitions recorded by four different heads are compared in Fig. 17.

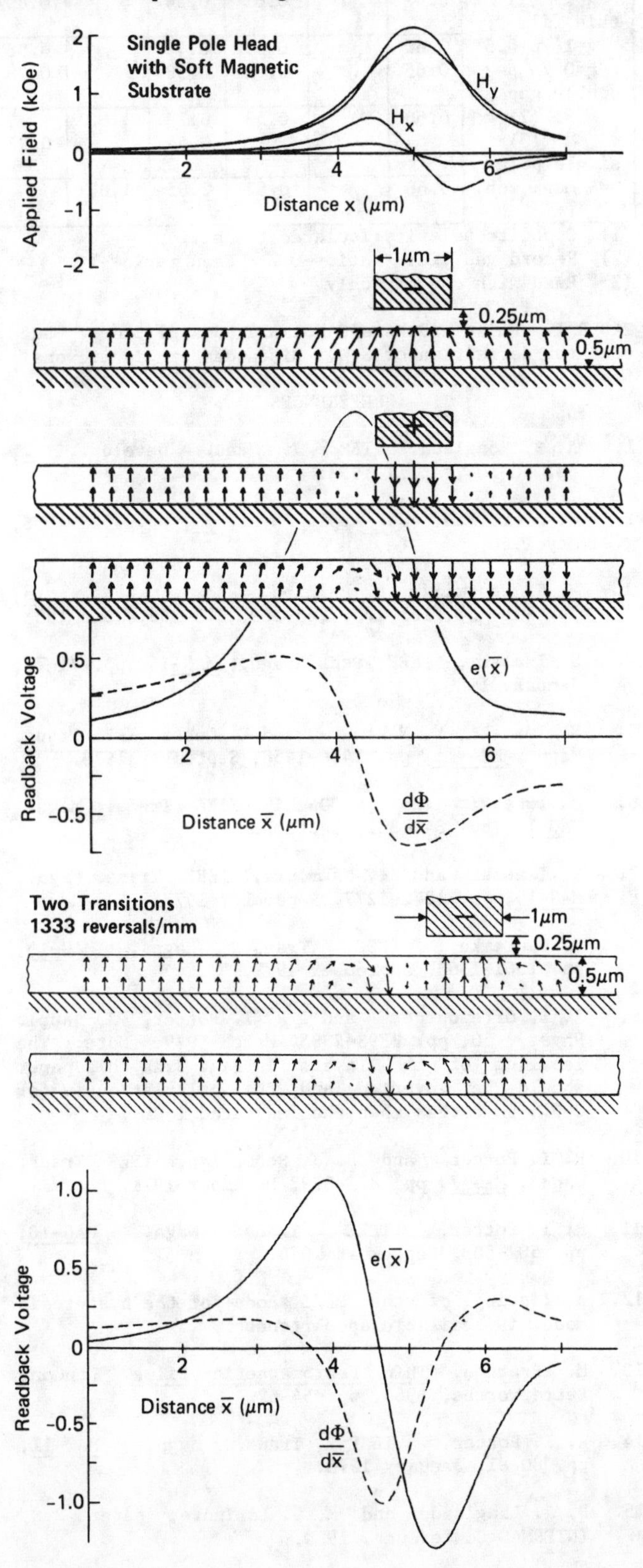

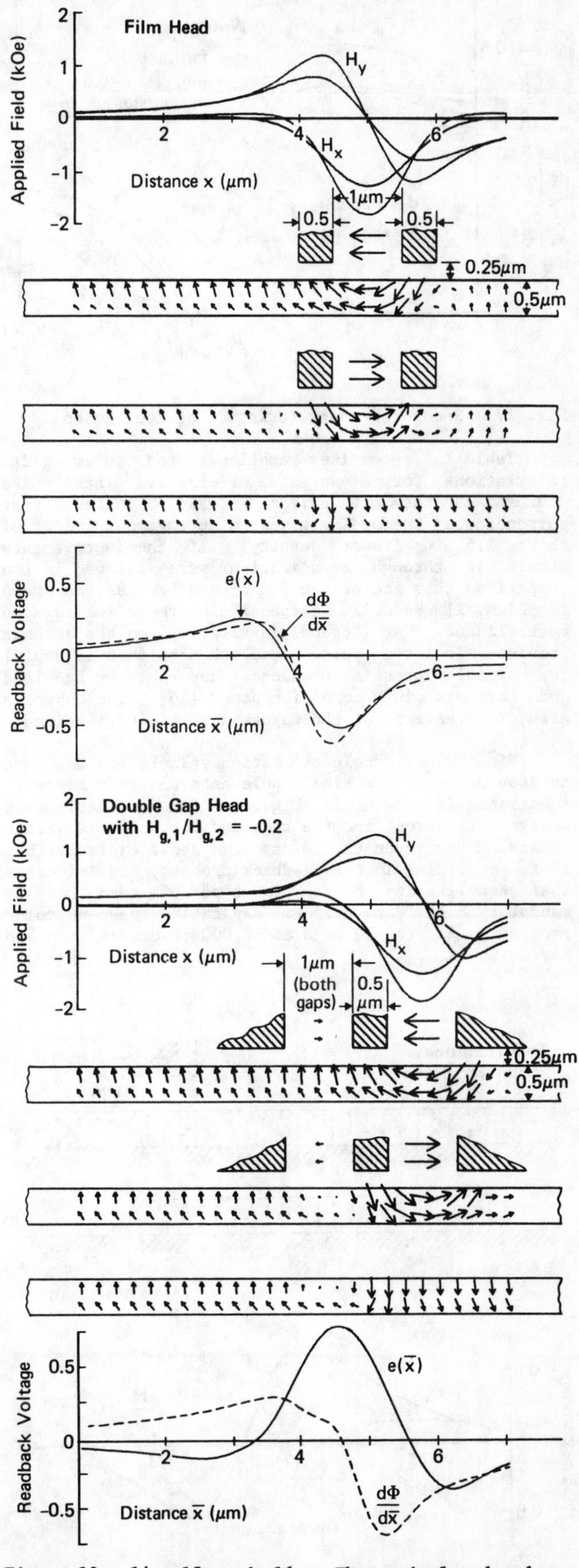

Figs. 13, 14, 15 and 16. Three isolated and one two-transition calculations as indicated. Several possibilities for reading are available with the double gap head as noted in Table I.

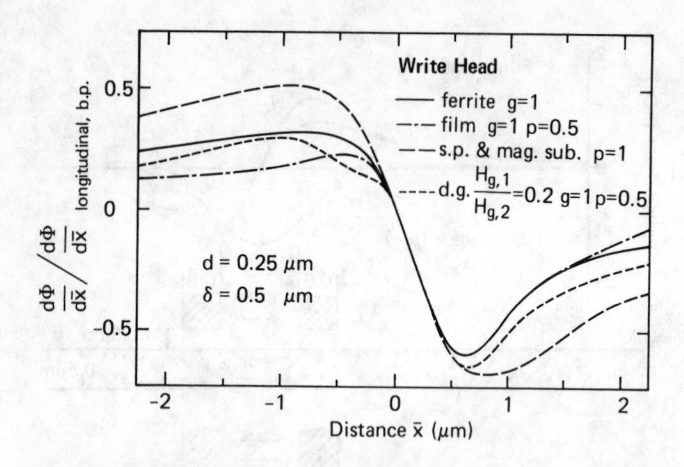

Fig. 17. Recording characteristics of four heads.

Table I summarizes amplitude and pulse width calculations for a variety of cases. These pulse widths and the bit shifts in Fig. 8 indicate that perpendicular recording gives an improvement factor of about 2.5 in linear density. If the improvements attainable through equalization were allowed in the comparison this factor would increase because, as shown in Figs. 11 and 12, perpendicular recording is also more linear. An appendix available from the authors provides isolated pulse readback data that is useful for signal processing studies as described by Langland and Larimore in a companion paper [15]. The appendix also gives several of the magnetization distributions.

Finally, a double transition calculation recorded at 2666 mr/mm with a single pole head and soft magnetic substrate is shown in Fig. 18. The magnetization pattern is strong and the recorded bit shift is within limits. But even the $d\Phi/d\bar{x}$ curve shows excessive bit shift occurring in the readback process, and indicates that equalization or a new type of read head is mandatory if at 0.25 μm separation perpendicular recording is to be used at 67,000 reversals/inch and beyond.

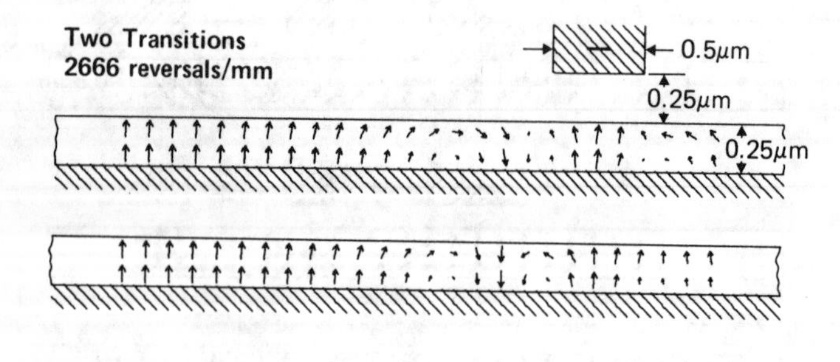

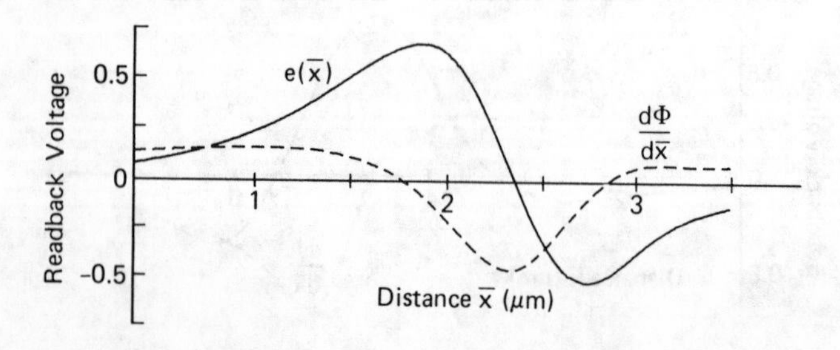

Fig. 18. Two transitions at 67,716 reversals/inch.

	$d\Phi/d\bar{x}$			Read with Write		
LONGITUDINAL ferrite	Amp.	P_{50}	S_{50}	Amp.	P_{50}	S_{50}
g=0.5	1.07	1.79		1.05	1.84	
g=1	1.00	1.90		0.94	2.06	
g=1 δ =1	1.46	2.64		1.41	2.73	
PERPENDICULAR ferrite						
g=0.75	0.72		0.62	0.62		0.74
g=1	0.67		0.63	0.55		0.90
g=1 (1)	0.67		0.61	0.54		0.89
thin film						
g=1 p=0.5	0.60		0.59	0.49		0.88
g=0.75 p=2	0.65		0.62	0.58		0.78
double gap						
(2)	0.66		0.53	0.78	1.32	
(3)				0.54		0.97
single pole & mag. sub.	0.68		0.62	2.05	1.87	

(1) Opposite polarity isolated transition
(2) Record gap field ratio=-0.2 Read ratio=-1
(3) Read with one gap only

TABLE I. Pulse Widths and Normalized Amplitudes. See text for definition of S_{50}. Distances are in microns.

REFERENCES

1. A. S. Hoagland, IBM J. Res. Develop., 2, pp. 90-104, April 1958.

2. G. J. Y. Fan, J. Appl. Phys., 31, pp. 402S-403S, May 1960.

3. W. K. Westmijze, Philips Res. Rep., 8, pp. 148-157, 161-183, 245-269, 343-366, June 1953.

4. S. Iwasaki, IEEE Trans. Magn., MAG-16, pp. 71-76, January 1980.

5. S. Iwasaki, Y. Nakamura, and K. Ouchi, IEEE Trans. Magn., MAG-15, pp. 1456-1458, September 1979.

6. S. Iwasaki and K. Ouchi, IEEE Trans. Magn., MAG-14, pp. 849-851, September 1978.

7. S. Iwasaki and Y. Nakamura, IEEE Trans. Magn., MAG-13, pp. 1272-1277, September 1977.

8. J. U. Lemke, IEEE Trans. Magn., MAG-15, pp. 1561-1563, September 1979.

9. I. B. Ortenburger and R. I. Potter, J. Appl. Phys., 50, pp. 2393-2395, March 1979. (Note: The labeling of the layers in Fig. 3 in this paper should be inverted and the calculation is for d=0.5 μm rather than the indicated 1 μm.)

10. R. I. Potter and R. J. Schmulian, IEEE Trans. Magn., MAG-7, pp. 873-879, December 1971.

11. R. I. Potter, IEEE Trans. Magn., MAG-10, pp. 502-508, September 1974.

12. A listing of the PL/I code for the hysteresis model is available upon request.

13. M. Prutton, Thin Ferromagnetic Films. London: Butterworths, 1964, pp. 54-62.

14. R. I. Potter, IEEE Trans. Magn., MAG-11, pp. 80-81, January 1975.

15. B. J. Langland and M. G. Larimore, paper 2.8, INTERMAG Conference, 1980.

PERPENDICULAR RECORDING

B. K. Middleton[+] and C. D. Wright[+]

Summary

A simple theoretical study of the write process in perpendicular recording is presented. The width of the transitions recorded in media having various properties are predicted and compared with those predicted for for planar recording. Results suggest narrower transitions in perpendicular recording and the available experimental evidence is shown to support this view. It is also shown that the sum of transition width a and head-to-tape separation on replay d, i.e. (a+d), is an important parameter of recording performance and that available values indicate areas worthy of consideration for further study in perpendicular recording.

1. Introduction

Present day recording systems are often thought to utilise magnetisation components oriented essentially parallel to the planes of the recording media and to achieve high densities of recorded information highly coercive, thin, well oriented magnetic coatings are used. However indications from experimental work (refs. 1-7), from theoretical predictions (refs. 6-10), and from large scale modelling (ref. 11) suggest that increased linear recording densities may be attainable by recording with components of magnetisation directed normal to the plane of the recording media. This recording mode is termed 'perpendicular recording'.

In the main part of this paper, the process by which an isolated magnetisation reversal, or transition, is written in a medium with purely perpendicular magnetisation components is studied. Particular attention is paid to the width over which magnetisation transitions occur as this plays a crucial role in determining the limits to the packing densities attainable. As part of this study the suitablility of different head structures for record and replay are considered and comparisons are made with predictions and experimental results for conventional planar recording.

S.I. units are used throughout the paper with both magnetic field H and magnetisation M being measured in amperes per metre and being related to the magnetic induction by the relationship $B=\mu_o(H+M)$, where μ_o is the permeability of free space.

2. The record process: theory

2.1 Head structures

Conventional ring and single pole heads, similar to those used by Iwasaki (ref. 12), are considered and shown diagrammatically in figs. 1(c) and 1(d). The field components produced by these heads, in directions parallel and normal to the planes of the recording media, i.e. in the x and y

[+] Department of Electrical and Electronic Engineering, Manchester Polytechnic, Chester Street, Manchester M1 5GD, England.

Reprinted with permission from *4th Int. IERE Conf. Proc.*, no. 54, 1982, pp. 181-192.

directions defined in fig. 1(a), are also shown in figs. 1(c) and 1(d). The general similarity between some of these components is obvious. The similar field distributions may differ in fine detail although they are to a first approximation given by well known formulae. For example the x and y components of field produced by a conventional ring head are given by, for $g < y$, (ref. 13)

$$H_{rx}(x,y) = \left(\frac{H_g}{\pi}\right)\left[\arctan\left(\frac{g+x}{y}\right) + \arctan\left(\frac{g-x}{y}\right)\right] \quad (1a)$$

$$H_{ry}(x,y) = \left(\frac{-H_g}{2\pi}\right)\ln\left[\frac{y^2 + (g+x)^2}{y^2 + (g-x)^2}\right] \quad (1b)$$

where H_g is the x component of head field along the head face i.e. along $-g < x < g$, $y=0$. The x and y components of field produced by the single pole head are given by expressions similar to (1a) and (1b) so that (refs. 1,8)

$$H_{spx}(x,y) = -H_{ry}(x,y) \quad (2a)$$

$$H_{spy}(x,y) = H_{rx}(x,y) \quad (2b)$$

where H_g, for the single pole head, should be interpreted as the y component of field just under the pole face. For simplicity only the small gap (and pole) approximations to (1a) and (1b) will be used in the calculations in the next section. The formulae are

$$H_{rx}(x,y) = \frac{H_g 2g}{\pi} \frac{y}{x^2+y^2} \quad (3a)$$

$$H_{ry}(x,y) = \frac{-H_g 2g}{\pi} \frac{x}{x^2+y^2} \quad (3b)$$

Our calculations have shown that the results obtained using these convenient approximations differ little from those obtained using the more complete formulae (1a) and (1b).

2.2 Record theory

The analytical methods used in this work are developments of those used by Maller and Middleton (ref. 14) and Williams and Comstock (ref. 15) when investigating planar recording.

It is assumed that the magnetisation variation produced by a single reversal of current in the coil of the record head, see fig 1(b)(11), can be reasonably approximated by

$$M_y = \frac{2}{\pi} M_o \arctan \frac{x_o}{a_p} \quad (4)$$

where a_p is the transition width parameter for perpendicular recording and M_o represents the levels between which the magnetisation is switched. M_o may differ from the remanent magnetisation M_r and can be determined by considering the self-demagnetising fields $\underline{H}_d(\underline{r})$ at positions $\underline{r}$ in the medium containing the distribution (4). These can be calculated using the general formula (ref.16)

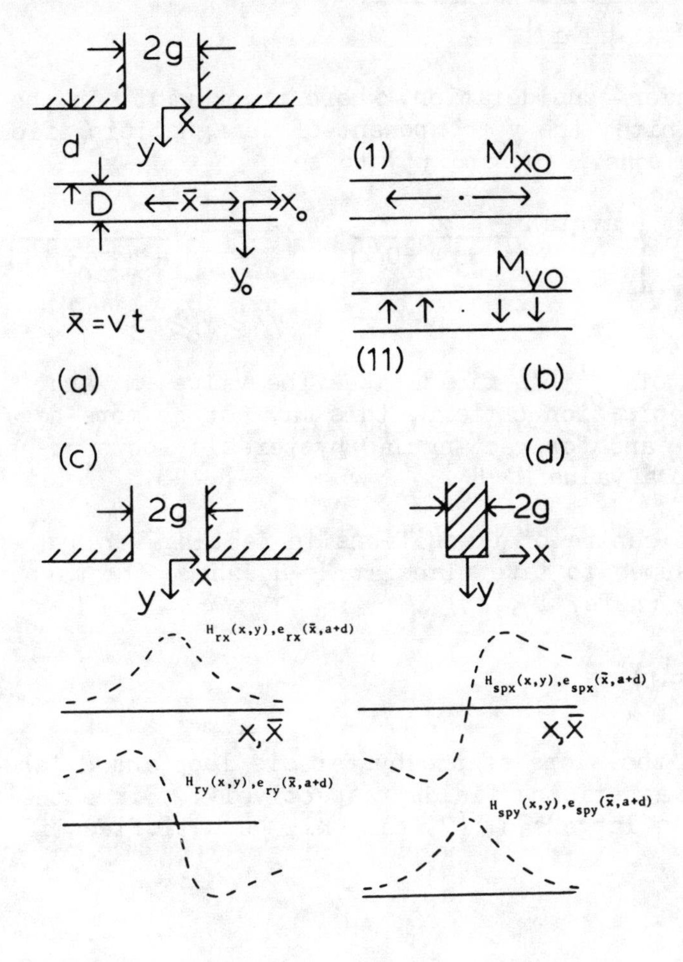

2g
d
D
x̄
x_o
y_o
x̄ = vt
(a)
(1)
M_{xo}
M_{yo}
(11)
(b)
(c)
(d)
2g
x
y
$H_{rx}(x,y), e_{rx}(\bar{x}, a+d)$
$x, \bar{x}$
$H_{ry}(x,y), e_{ry}(\bar{x}, a+d)$
$H_{spx}(x,y), e_{spx}(\bar{x}, a+d)$
$H_{spy}(x,y), e_{spy}(\bar{x}, a+d)$

Fig 1

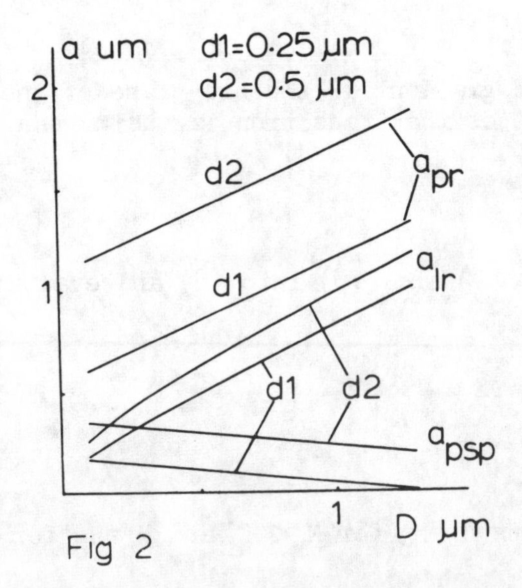

a um
d1=0.25 µm
d2=0.5 µm
2
1
d2
d1
a_{pr}
a_{lr}
a_{psp}
D µm

Fig 2

$$\underline{H}_d(\underline{r}) = \frac{-1}{4\pi} \int_{V'} \frac{\nabla . M(\underline{r}') \, (\underline{r}-\underline{r}') \, dV'}{|\underline{r}-\underline{r}'|^3} \tag{5}$$

For the case under consideration, where track width in the z_o direction is assumed infinite, the y_o component of demagnetising field can be determined from eqns. (4) and (5) to be

$$H_{dy}(x_o,y_o) = \frac{-M_o}{\pi} \left\{ \arctan \frac{x_o}{a_p + |y_o - D/2|} + \arctan \frac{x_o}{a_p + |y_o + D/2|} \right. \qquad D/2 \geq y_o \geq -D/2 \tag{6}$$

At large values of x_o this field takes the value $-M_o$ and, clearly, for any stable magnetisation pattern, this may not be more negative than $-H_c$. Therefore $M_o \leq H_c$ and for rectangular hysteresis loops possessing $M_r > H_c$ M_o is limited to a value $M_o = H_c$.

Recording at the centre of transitions in rectangular hysteresis loop materials is assumed to take place at $H=-H_c$ where the magnetisation gradient is given by (refs. 14, 15)

$$\frac{dM}{dx} = \frac{dM}{dH}\left(\frac{dH_h}{dx} + \frac{dH_d}{dx}\right) \tag{7}$$

where (dM/dH) is the slope of the hysteresis loop and H_h and H_d are the head and self-demagnetising fields respectively. Since the slope of the hysteresis loop is large eqn. (7) can only be satisfied if

$$\frac{dH_h}{dx} = \frac{-dH_d}{dx} \tag{8}$$

Taking firstly the case of recording with a ring head it is assumed that the head field amplitude is adjusted to give maximum head field gradient at $H_h=-H_c$. The latter can easily be shown, using eqn. (3b), to be given by

$$\frac{dH_{ry}(x,y)}{dx} = \frac{H_c}{2\sqrt{3}} \cdot \frac{1}{y} \tag{9}$$

The demagnetising field gradient in the mid-plane of the recording medium, $y_o=0$, and at the centre of the transition is, using eqn. (6)

$$\frac{dH_{dy}(0,0)}{dx_o} = \frac{-2M_o}{\pi} \frac{1}{a_p+D/2} \tag{10}$$

Substitution from eqns. (9) and (10) into (8) and evaluation leads to

$$a_{pr} = \frac{4\sqrt{3}}{\pi} \frac{M_o}{H_c} y - \frac{D}{2} \tag{11a}$$

$$= \frac{4\sqrt{3}}{\pi} \frac{M_o}{H_c} \left(d+\frac{D}{2}\right) - \frac{D}{2} \tag{11b}$$

subject to the conditions $M_o=M_r$ for $M_r < H_c$ and $M_o=H_c$ for $M_r > H_c$.

The single pole head, by contrast, produces a much higher field gradient of

$$\frac{dH_{spy}(x,y)}{dx} = \frac{\sqrt{3}H_c}{2} \frac{1}{y} \tag{12}$$

Substitution from eqns. (12) and (10) into eqn. (8) and evaluation gives the corresponding transition width as

$$a_{psp} = \frac{4}{\pi\sqrt{3}} \frac{M_o}{H_c} y - \frac{D}{2} \tag{13a}$$

$$= \frac{4}{\pi\sqrt{3}} \frac{M_o}{H_c}\left(d+\frac{D}{2}\right) - \frac{D}{2} \tag{13b}$$

subject to $M_o=M_r$ for $M_r<H_c$ and $M_o=H_c$ for $M_r>H_c$.

Comparison of eqns. (13) and (11) shows clearly that the single pole head produces much narrower transition widths than the ring head, as would be expected from the relative magnitudes of their head field gradients.

2.3 Planar recording

In planar (or longitudinal) recording the usual assumption of the form of the magnetisation, see fig. 1(b)(1), is

$$M_{xo}(x_o) = \frac{2}{\pi} M_r \arctan \frac{x_o}{a_1} \tag{14}$$

where a_1 is the transition width parameter for longitudinal recording. The above distribution gives rise to a demagnetising field (refs. 16, 17) given by

$$H_{dx}(x_o,y_o) = \frac{-M_r}{\pi}\left\{\arctan \frac{a_1+D/2-y_o}{x_o} + \arctan \frac{a_1+D/2+y_o}{x_o} - 2\arctan \frac{a_1}{x_o}\right\}$$

$$D/2 > y_o > -D/2 \tag{15}$$

The corresponding demagnetising field gradient at the centre plane of the medium and at the centre of the transition can be shown to be

$$\frac{dH_{dx}}{dx_o} = \frac{-M_r D}{\pi} \frac{1}{a_1(a_1+D/2)} \tag{16}$$

Using the optimised field gradient for a ring head, which is

$$\frac{dH_{rx}}{dx} = \frac{\sqrt{3}}{2} \frac{H_c}{y} \tag{17}$$

leads, using eqns. (16) and (17) in eqn. (8), to

$$a_{lr} = -\frac{D}{4} + \sqrt{\frac{D^2}{16} + \frac{2M_r(d+D/2)D}{\sqrt{3}\,H_c\,\pi}} \tag{18}$$

2.4 Predictions of the theory

Fig. 2 shows transition widths predicted using eqns. (11b), (13b), and (18) plotted as a function of medium thickness for different head to medium separations. (M_r/H_c) was taken to have a value of 4, which is reasonable for current particulate media and for CoCr films (ref. 2), while the demagnetisation process limits (M_o/H_c) in the perpendicular recording case

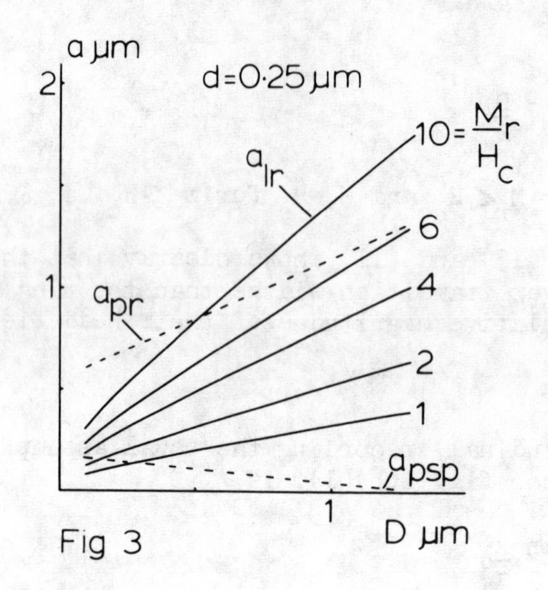
a µm
2
1
d=0·25 µm
10 = Mr/Hc
a lr
6
4
2
1
a pr
a psp
1
D µm
Fig 3

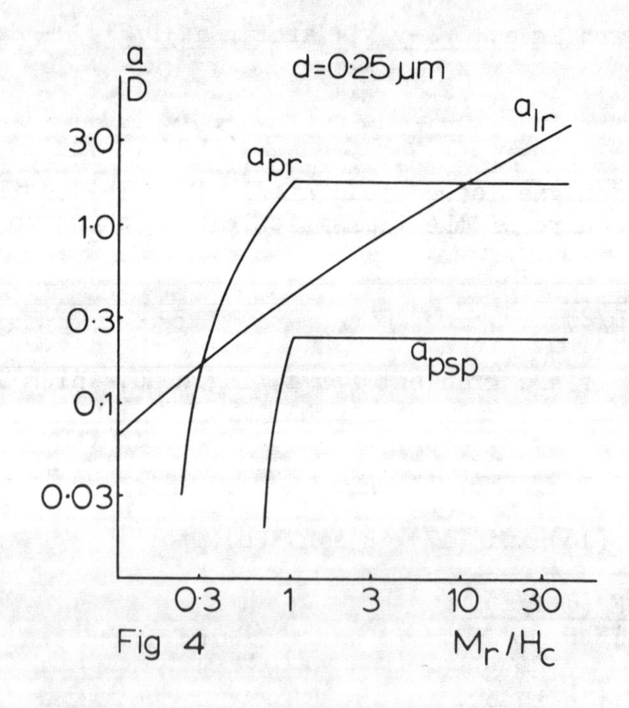
a/D
d=0·25 µm
a lr
3·0
a pr
1·0
0·3
a psp
0·1
0·03
0·3
1
3
10
30
Mr/Hc
Fig 4

to a value of unity. Clearly perpendicular recording with a single pole head results in narrowest transitions while longitudinal and perpendicular recording with ring heads give rise to wider transitions. With regard to perpendicular recording the narrower transitions are obtained when using a single pole head and are due to the high field gradients produced by that head while the improvement of perpendicular over planar recording is due to the different demagnetising field gradients in the two modes of recording. The importance of head to tape separation in perpendicular recording is also worthy of note.

Some comments about the role of self-demagnetising fields seem appropriate. In planar recording the demagnetising field gradients are proportional to M_r and large; see eqn. (15). This results in fairly wide recorded transitions. In perpendicular recording the demagnetising field gradients are proportional to M_o, see eqn. (10), which is kept small by self-demagnetising effects. This helps to keep the recorded transitions narrow. Therefore demagnetisation effects are much less harmful in perpendicular than in planar recording.

Fig. 3 shows transition widths plotted as functions of medium thickness, for d=0·25μm and for a range of values of (M_r/H_c). It is seen that in planar recording transition widths increase for higher values of (M_r/H_c) but that in perpendicular recording they do not vary since M_o is limited to a value $H_c < M_r$. The role of M_r is emphasised in fig. 4 which shows the various values of normalised transition width plotted as functions of (M_r/H_c). In planar recording transition widths rise continuously with increasing (M_r/H_c) while in perpendicular recording they reach a maximum value determined by M_o being limited to H_c. The message from fig. 4 is again that perpendicular recording with a single pole head gives rise to narrowest transitions widths. In view of the strong dependence of transition widths on head to medium separation in perpendicular recording it is to be expected that reductions in same would lead to further improvements in favour of this type of recording.

3. The replay process: theory

3.1 Reciprocity and output pulse shapes

Replayed voltage waveforms $e_x(\bar{x})$ and $e_y(\bar{x})$, respectively from planar and perpendicular components of magnetisation, are most easily predicted using the well known 'reciprocity' formulae

$$e_x(\bar{x}) = -\mu_o v \int_{-\infty}^{+\infty}\int_{d}^{d+D}\int_{-\infty}^{+\infty} \frac{dM_x(x-\bar{x})}{d\bar{x}} \frac{H_x(x,y)}{i}\, dxdydz \qquad (19a)$$

$$e_y(\bar{x}) = -\mu_o v \int_{-\infty}^{+\infty}\int_{d}^{d+D}\int_{-\infty}^{+\infty} \frac{dM_y(x-\bar{x})}{d\bar{x}} \frac{H_y(x,y)}{i}\, dxdydz \qquad (19b)$$

where $H_x(x,y)/i$ and $H_y(x,y)/i$ are the x and y components of head field per unit current in their windings. The form of the head field distributions was discussed in section 2.1 and they are shown diagrammatically in figs. 1(c) and 1(d). In the case of replay from very thin film media, $D \ll d$, supporting wide recorded tracks and transitions of the shape defined by eqns. (4) and (14) it can easily be shown that the above outputs become

$$e_x(\bar{x}) \propto H_x(\bar{x}, a_l + d) \qquad (20a)$$

$$e_y(\bar{x}) \propto H_y(\bar{x}, a_p + d) \qquad (20b)$$

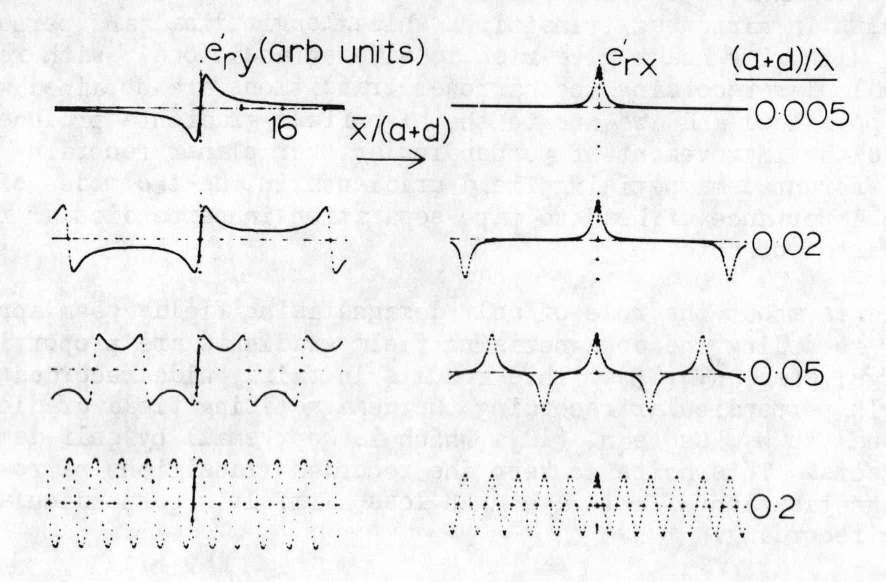

e'ry(arb units)
16
x̄/(a+d)
e'rx
(a+d)/λ
0·005
0·02
0·05
0·2

Fig 5

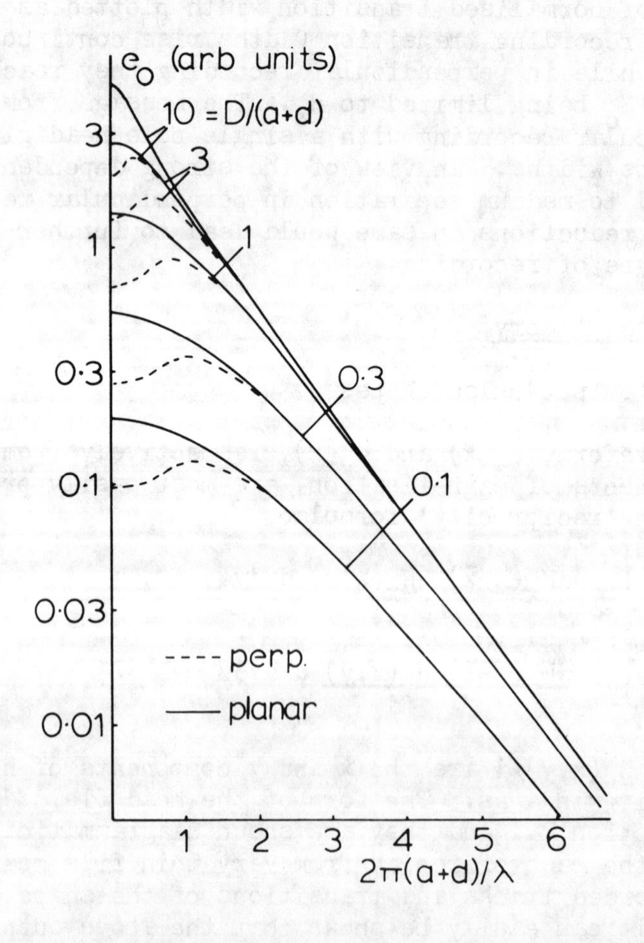

eo (arb units)
10 = D/(a+d)
3
3
1
1
0·3
0·3
0·1
0·1
0·03
0·01
--- perp.
— planar
1
2
3
4
5
6
2π(a+d)/λ

Fig 6

The corresponding pulse shapes for conventional ring and single pole heads are labelled in figs. 1(c) and 1(d) using an obvious notation. Hence it is easy to see that

$$e_{rx}(\bar{x},a_1+d) \propto e_{spy}(\bar{x},a_p+d) \quad (21a)$$

$$e_{ry}(\bar{x},a_p+d) \propto e_{spx}(\bar{x},a_1+d) \quad (21b)$$

and therefore any advantage on the replay side accruing simply from re-orienting the magnetisation from planar to perpendicular could just as easily be obtained by changing the type of the replay head. Consequently the merits of perpendicular recording are not to be judged on the particular symmetry or shape of the output pulses arising from particular heads but on the nature of the distributions recorded into the media, particularly the transition widths, and how these effect outputs.

3.2 Pulse crowding

The replayed pulse shapes derived for thin film recording media using eqns, (3a), (3b), (20a) and (20b) are shown in fig. 5. Isolated pulse and pulse crowded waveforms are shown for planar and perpendicular magnetisation components: the linear superposition of isolated pulse shapes separated by distances $\lambda/2$ being used for the prediction of crowded waveforms. It is clear that perpendicular recording gives rise to pulses with a much longer tail and that the effects of pulse crowding are evident at much lower packing densities than in planar recording. However, in the former case, the overlapping is constructive whereas, in the latter, it is destructive. Consequently perpendicular outputs rise with increasing but low packing densities before beginning to fall off at high densities. The variation of output amplitude with reduced packing density is shown in fig. 6 and compared with that for planar recording. These results were obtained using eqns. (3a), (3b), (4), (14), (19a), and (19b) and carrying out the necessary integrations and summations of pulse shapes. It can be seen that although the output voltage amplitudes of the two modes of recording differ at low packing densities they coincide at high densities. In both cases the output amplitudes vary, at short wavelengths, as

$$e_o \propto e^{-2\pi(a+d)/\lambda} \quad (22)$$

Therefore it would appear that (a+d) is a major factor in determining output amplitudes at high packing densities and so is a parameter giving a measure of planar and perpendicular recording performance for comparison purposes; smaller values of (a+d) giving lower rates of fall of output being desirable.

4. Experimental results

4.1 Pulse shapes and pulse crowding

Output voltage pulses generated in conventional and single pole heads by thin films of CoCr and by composite films of CoCr having backing layers of NiFe alloy have been shown (refs. 1, 3, 4) to have the odd and even symmetries, respectively, shown in figs. 1 and 5. In addition the general shapes of the pulse crowding curves shown in fig. 6 have also been observed (refs. 1, 3, 4, 7). Further the pulse crowding curve obtained by replaying from a composite film of CoCr and NiFe using a single pole head has been fitted to theoretical predictions by Iwasaki et al (ref. 6) assuming the values $a_p=0$ and $d=0{\cdot}2$ µm and using pulse superposition. These results generate a measure of confidence in the initial assumptions for the magnetisation distributions, the nature of the replay process, and the validity of pulse superposition.

In particulate media the predictions of Potter and Beardsley (ref. 9) and the large scale modelling of Monson et al (ref. 11) suggest replayed pulses which have contributions from odd and even symmetry pulses. Such circumstances are not covered by this work.

With regard to pulse superposition there is considerable evidence for its validity in conventional recording (ref. 18) and the work of Iwasaki et al (ref. 6) is evidence for its validity in perpendicular recording. The computations of Potter and Beardsley (ref. 9) add further support in its favour.

4.2 (a+d) values

By fitting theoretical predictions of output voltage variations with packing density to observed results values of (a+d) have been found by a number of workers. The most common method adopted is, essentially, to plot ln (e_o) against ($2/\lambda$) and from the slope of the graph determine the appropriate value of (a+d) using eqn. (22). A wide range of values of (a_1+d), for conventional recording, have been quoted in the literature (refs. 18-21); the lowest being 0·44 µm (ref. 21). For perpendicular recording the only value yet to appear in print is that, already quoted, by Iwasaki et al (ref. 6) of d=0·2 µm on the assumption that a_p=0. However the latter theory could be repeated without the assumption on the value of a_p=0 and the result would become (a_p+d)=0·2 µm. In examining these numbers it is informative to try to separate out the contributions of a and d to their sums. In perpendicular recording it might be expected in the case (a_p+d)=0·2 µm that the major contribution comes from d and that a_p is very small as originally predicted by Iwasaki et al. For the numbers quoted by Koster and Pfefferkorn (ref. 21), for conventional recording, the value of a_1 would be considerably larger. Using the suggested value of d=0.27 µm leads to a value of a_1=0·17 µm. The relative magnitudes of these numbers is in line with the theoretical predictions of section 2.

To achieve high packing densities it is obviously necessary to make (a+d) extremely small. In doing this it is essential that d is minimised by ensuring smooth head and tape surfaces; this also ensured that a is small since it is a function of d. The authors are aware that values of (a_1+d), for conventional recording, of less than 0·44 µm have been obtained using smooth head and particulate tape surfaces when care was taken to maintain good head-tape contact. However some results also obtained using particulate tapes and narrow gap record heads have shown effective separations of 0·05 µm and the ability to sustain extremely high digital data packing densities (ref. 22). In this work perpendicular components of magnetisation were thought to play a significant role. Therefore these additional results continue to support the potential for perpendicular recording.

5. Discussions and conclusions

A theory previously employed for predicting recorded transition widths in planar recording has been modified to apply for perpendicular recording. The predictions are clear in that perpendicular recording with a single pole head can result in very narrow transition widths, considerably narrower than those in conventional planar recording. The use of different types of replay head has been shown to have little or no bearing on the relative merits of perpendicular and planar recording but it has been shown that it is the transition width in conjunction with replay head to tape separation, namely (a+d), which is of importance in determining achievable packing densities.

An examination of (a+d) values observed in planar and perpendicular recording has been made. Iwasaki et al (ref. 6) have reported a value, interpreted here as $(a_p+d)=0{\cdot}2\ \mu m$, while the corresponding values for conventional recording on particulate media are generally somewhat larger. However some other measurements, also on particulate media, have shown that extremely high density recording is possible as a result of very small effective head to tape separations when perpendicular components of magnetisation seem to have had a role to play. It seems, therefore, that recording with perpendicular components of magnetisation does offer potential for recording at densities in excess of those obtained in conventional recording and that these may be achieved either by recording onto thin media with a high perpendicular anisotropy or by recording onto suitably prepared particulate media.

Acknowledgment

Thanks are due to the Science and Engineering Research Council for the provision of support by way of Research Grant GR/B/28453.

References

1. Iwasaki,S and Nakamura, Y., 'An analysis for the magnetisation mode for high density magnetic recording', IEEE Trans. on Magnetics MAG-13, 1272-1277, Sept., 1977.
2. Iwasaki, S. and Ouchi, K. K., 'CoCr recording films with perpendicular magnetic anisotropy', IEEE Trans. on Magnetics MAG-14, 849-851, Sept., 1978.
3. Iwasaki, S., Nakamura, Y. and Ouchi, K., 'Perpendicular magnetic record- with a composite anisotropy film', IEEE Trans. on Magnetics MAG-15, 1456-1458, Sept., 1979.
4. Iwasaki, S., 'Perpendicular magnetic recording', IEEE Trans. on Magnetics MAG-16, 71-76, Jan., 1980.
5. Iwasaki, S., Ouchi, K. and Honda, N., 'Studies of perpendicular magnetisation mode in CoCr sputtered films', IEEE Trans. on Magnetics MAG-16, 1111-1113, Sept., 1980.
6. Iwasaki, S., Nakamura, Y. and Muraoka, H., 'Wavelength response of perpendicular magnetic recording' IEEE Trans. on Magnetics MAG-17, 2535-2537, Nov., 1981.
7. Yamamori, K., Nishikawa, R., Asano, R. and Fujiwara, T., 'Perpendicular magnetic recording performance of double layer media', IEEE Trans. on Magnetics MAG-17, 2538-2540, Nov., 1981.
8. Satake, S., Hayakawa, K., Hokkyo, J. and Simamura, T., 'Field theory of twin pole head and a computer simulation model of perpendicular recording', Paper of IECEJ Tech. Group Meeting on Magnetic Recording MR 77-26, 33-42, Nov., 1977.
9. Potter, R. I. and Beardsley, I. A., 'Self consistent computer calculations for perpendicular magnetic recording', IEEE Trans. on Magnetics MAG-16, 967-972, Sept., 1980.
10. Johnson, R. A. and Chi, C. S., 'Fundamental properties of perpendicular magnetic recording' IEEE Trans. on Magnetics MAG-17, 2544-2546, Nov., 1981.
11. Monson, J. E., Fung, R. and Hoagland, A. S., 'Large scale model studies

of vertical recording', IEEE Trans. on Magnetics MAG-17, 2541-2543, Nov., 1981.

12. Iwasaki, S. and Nakamura, Y., 'The magnetics field distribution of a perpendicular recording head', IEEE Trans. on Magnetics MAG-14, 436-438, Sept., 1978.
13. Karlqvist, O., 'Calculation of the magnetic field in the ferromagnetic layer of a magnetic drum', Trans. Roy. Inst. Tech. Stockholm 86, 3-27, 1954.
14. Maller, V. A. J. and Middleton, B. K., 'A simplified model of the writing process is saturation magnetic recording', The Radio and Electronic Engineer 44, 5, 281-285, May 1974.
15. Williams, M. L. and Comstock, R. L., 'An analytical model of the write process in digital magnetic recording', Proc. A.I.P. Conf. on Magnetism and Magnetic Materials, 1971.
16. Potter, R. I., 'An analysis of saturation magnetic recording based on arctangent magnetisation transitions', J. Appl. Phys. 41, 4, 1647-1651, March 1970.
17. Middleton, B. K., PhD Thesis, University of Salford, Oct. 1970.
18. Middleton, B. K. and Wisely, P. L., 'Pulse superposition and high density recording', IEEE Trans. on Magnetics MAG-14, 5, 1043-1050, Sept., 1978.
19. Mallinson, J. C., 'A unified view of high density digital recording', IEEE Trans. on Magnetics MAG-11, 5, 1166-1169, Sept., 1975.
20. Bertram, H. N. and Niedermeyer, R., 'The effect of demagnetisation fields on recording spectra', IEEE Trans. on Magnetics MAG-14, 743-745, Sept., 1978.
21. Koster, E. and Pfefferkorn, D., 'The effect of remanence and coercivity on short wavelength recording', IEEE Trans. on Magnetics MAG-16, 56-58, Jan., 1980.
22. Lemke, J. U., 'Ultra-high density recording with new heads and tapes', IEEE Trans. on Magnetics MAG-15, 1561-1563, Nov., 1979.

OBSERVATION OF RECORDED MAGNETIZATION PATTERNS
BY ELECTRON HOLOGRAPHY

K. Yoshida, T. Okuwaki, N. Osakabe, H. Tanabe
Y. Horiuchi, T.Matsuda, K.Shinagawa
A. Tonomura and H. Fujiwara.

Abstract-Magnetization patterns recorded on Co thin films prepared by oblique incidence vacuum deposition were directly observed by electron holography. Complicated ellipse-like interference fringes were observed along the sawtooth-like walls in the transition regions. Stray magnetic flux was observed in empty space beyond the edge of the film.
The following results were obtained from analysis of these interference images.

The smaller the product of the remanence and the film thickness, and the higher the coercivity, the narrower is the transition length that determines the limit of the recording density of a medium.

It was confirmed that longitudinal magnetic recording of a density of up to 170 kBPI is attainable using Co film 30 nm thick and with coercivity of 112 kA/m.

It was quantitatively proven that the intensity of the recorded magnetization in a medium is equal to its remanence when the magnetized area is distinguishable from the transition region.

INTRODUCTION

Recent increases in recording density for magnetic recording have been remarkable, and in the near future it is expected that it will become feasible to record with a wave length narrower than 1 µm. In order to attain such a high recording density, it is desirable to clarify the micromagnetic structure of the transition region which determines the limit of the magnetic recording medium.

Recently, Tonomura, et al. [1], developed an electron holography technique that enabled them to observe the microscopic distribution of magnetic flux and to measure its density. Using this technique, we have observed longitudinally recorded magnetization patterns on Co films prepared by oblique incidence vacuum deposition. This was done in order to investigate the relation between the behavior of magnetization in the transition region and the magnetic properties of the recording media, and also to search for the upper limit to the density of longitudinal recording.

PRINCIPLE BEHIND OBSERVATION MAGNETIC FLUX BY ELECTRON HOLOGRAPH

The principle underlying the observation of magnetic flux by electron holography is as follows.

Consider an electron beam passing through a magnetic thin film and focused at one point as shown in Fig.1. The phase difference ($\Delta\Phi(P_1,P_2)$) between two points (P_1 and P_2) below the specimen due to the vector potential ($\vec{A}$) from the magnetic specimen can be given [2] as

$$\Delta\Phi(P_1,P_2)=-\frac{e}{\hbar}\int\vec{A}d\vec{r}=-\frac{e}{\hbar}\int\vec{B}d\vec{S} \qquad (1)$$

Here, $\vec{A}$ is a vector potential, while $\vec{B}$ is magnetic induction. e is a charge of electron and $\hbar=h/2\pi$, where h is Planck's constant. The two integrals are performed along two electron trajectories($\vec{r}$), and the surface($\vec{S}$) enclosed by the trajectories. If magnetic field does not leak from the specimen, these equations lead to the following results: (1) the phase difference is equal to zero, if P_1 and P_2 lie along a magnetic line of force in the specimen; (2) the phase difference is just one wave length (2π) when two trajectories contain a magnetic flux of $h/e\{=4.1\times10^{-15}$ Wb}. Therefore, if a contour map of the electron wave front can be obtained, magnetic lines of force in the specimen can be directly observed as contour lines. And the intensity of magnetic field can be quantitatively measured, because neighboring contour lines contain a constant flux of h/e.

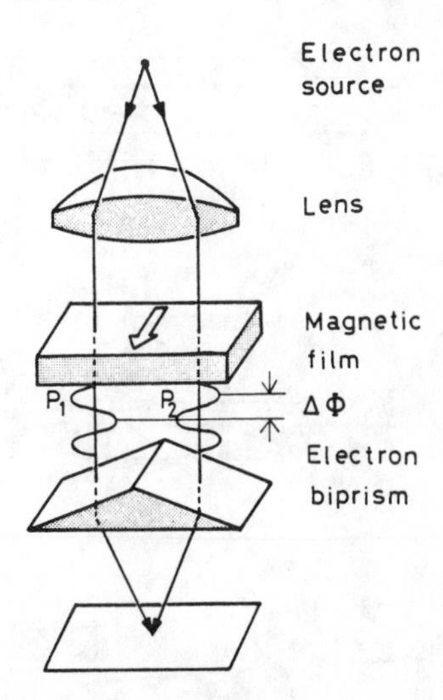

Fig. 1. Phase difference ($\Delta\Phi$)between the two parts of an electron beam transmitted through a magnetic thin film. $\Delta\Phi$ is not zero when the specimen has the magnetization direction indicated by the arrow in the figure, though the optical path lengths are equal to each other.

Manuscript received February 28, 1983

The authors are with the Central Research Laboratory, Hitachi Ltd., Kokubunji, Tokyo, Japan

Reprinted from *IEEE Trans. Magn.*, vol. MAG-19, pp. 1600-1604, Sept. 1983.

Observation of the contour map was performed in two stages, formation of an electron hologram and optical reconstruction. The electron hologram was formed in a field-emission electron microscope by means of a well collimated electron beam source. A schematic diagram of our setup is shown in Fig.2. An object was illuminated with an electron beam and its image was formed through the objective lense. A reference beam was projected on the image plane by an electron biprism, forming an off-axis image hologram. The interference pattern was then magnified and recorded on a photographic plate, forming a hologram.

Reconstruction was performed in the optical system shown in Fig.3. A collimated laser beam from a He-Ne laser was split into two beams by beam splitter A. One beam illuminated the electron hologram so as to construct the image, which was again focused by lenes E and F on the observation plane. The other beam (comparison beam) from the splitter was superimposed on the observation plane to form the interference image. Phase contour map for an electron beam transmitted through the specimen was obtained using the comparison beam that was parallel to the object beam.

The advantage of this holographic technique is that it makes it possible to obtain phase-amplified interference images having a factor of n(integer) [3]. Thus it becomes possible to observe details by means of this phase-amplification, even if phase differences are less than 2π.

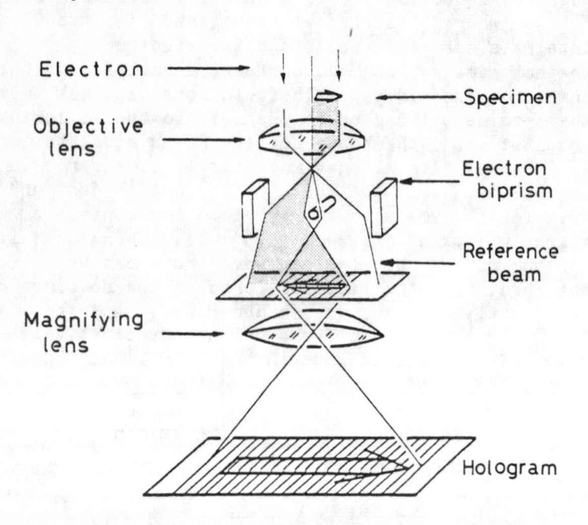

Fig. 2. Schematic diagram of electron hologram formation [1].

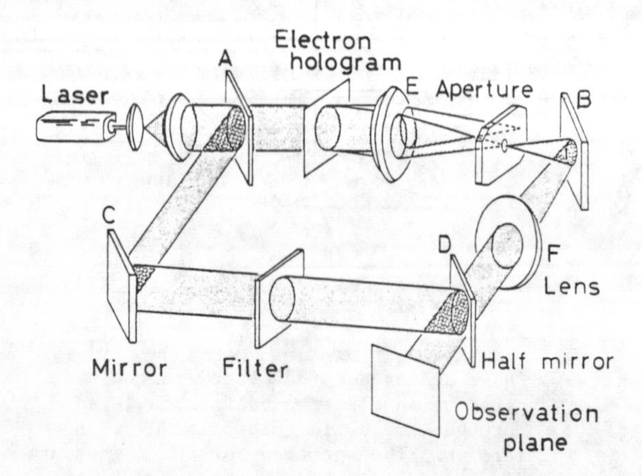

Fig. 3. Optical reconstruction system for interference microscopy.

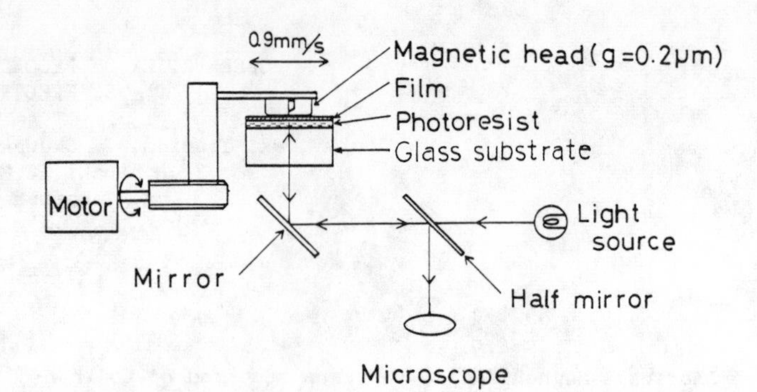

Fig. 4. Method of magnetic recording.

EXPERIMENTAL PROCEDURE

Sample Preparation

Co thin films were prepared by oblique incidence vacuum deposition. 20 mm-square Corning 7059 plates or 30 mm-diameter optical flats were used as substrates. These substrates were precoated with photoresist about 0.5 µm in thickness, to allow removal of the Co films after performance of magnetic recording.

Co was evaporated while admitting a small amount of O_2, under the following experimental conditions: an incident angle of 70°, deposition rate of 15 nm/min, and substrate temperature of 100°C. Co film thickness ranged from 30 nm to 50 nm.

Saturation induction and film coercivity were varied by controlling the amount of O_2 admitted into the bell jar. Saturation induction ranged from 0.5 to 1.1 T, coercivity from 30 to 120 kA/m, and the squarness ratio was over 0.9.

Film magnetic properties and thickness were respectively measured by VSM(vibrating sample magnetometr) and a surface roughness tester.

Magnetic Recording Method

Magnetic recording was performed on the samples using a tripad-type head of Mn-Zn ferrite having an 0.2 µm gap length and 300 µm track width, that was supported by gimbal spring. Head-to-medium relative speed was 0.9 mm/s, and rectangular recording currents were used.

Head-to-medium separation was maintained at less than 0.1 µm. This was done by observing the separation by optical interference (Fig.4).

RESULTS AND DISCUSSIONS

Observation of Recorded Magnetization Patterns

Recording was performed on a Co film of 45 nm thickness, and a pattern having a bit length of 5 µm was formed. The coercivity of the film was 27 kA/m and its saturation induction was 1.1T. Figure 5-a shows the recorded magnetization pattern observed by Lorentz microscopy, and Figure 5-b shows an interference micrograph of the same area as Fig. 5-a produced by means of electron holography. The horizontal axis of the pictures is the driven direction of the magnetic head, and the vertical is the direction of the track width.

Well known sawtooth-like walls [4-6] are observed in the transition region in the Lorentz image(Fig.5-a). However, no information could be obtained about the magnetic flux distribution either in the magnetized area, or in empty space. In the transition region, on the other hand, complicated ellipse-like interference fringes are obtained in the electron holographic reconstructed image [7], as shown in Fig.5-b.

When compared with the Lorentz microscopic image, it can be seen that the abrupt turning of the interference fringes occurs at the magnetic walls. Moreover, arc-like interference fringes are observed in the magnetized area and in empty space.

The interference fringes observed in empty space directly represent the magnetic lines of the stray field at the track edge. However, the interference fringes in the film do not directly represent the magnetization distribution, because they include both information regarding the magnetization in the film and the surface stray field that exists above and below the film. In order to truly know the magnetization distribution in the film, it is necessary to presume a certain magnetization distribution in the film and calculate the interference fringes--then to compare the result with the actual interference micrograph.

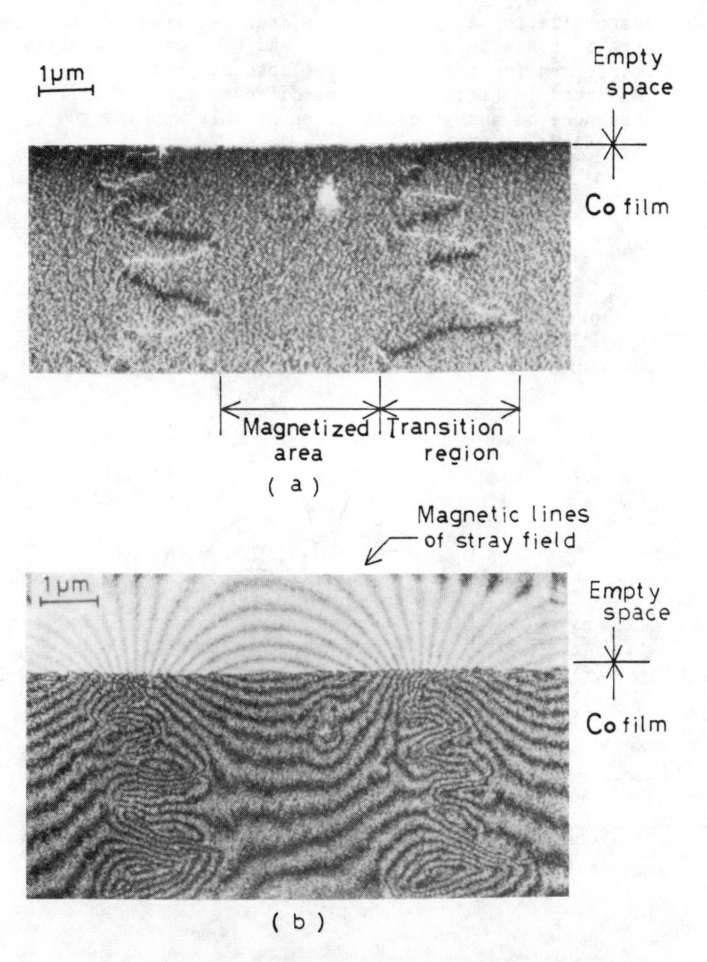

Fig. 5. Recorded magnetization pattern on a Co film (film thickness=45nm, coercivity= 27kA/m, saturation induction=1.1T). A bit length is 5μm. (a) Lorentz micrograph. (b) Interference micrograph.

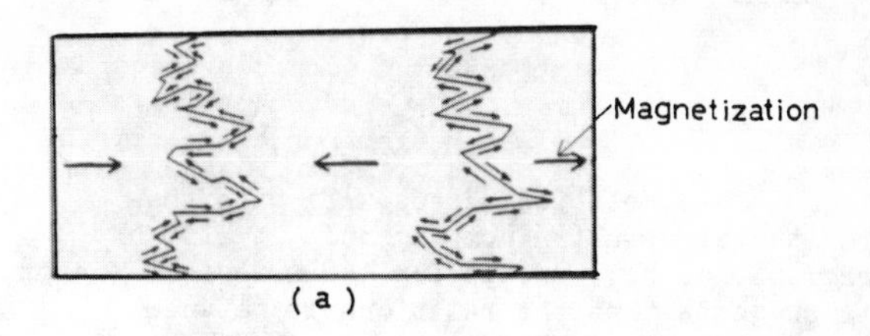

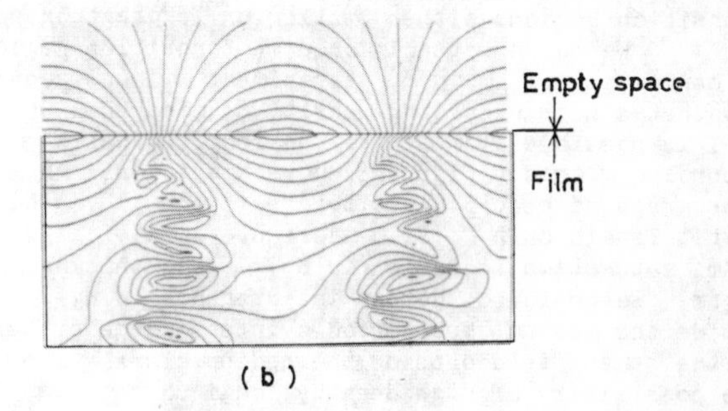

Fig. 6. Calculated interference image.
(a) Presumed magnetization distribution.
(b) Calculated interference image using the model of (a).

Figure 6-a shows the magnetization distribution model for a sawtooth-like domain proposed by Minnaja and Nobile [8]. In the vicinity of a magnetic wall the direction of the magnetization begins to tilt around an axis parallel to the magnetic wall, and a 180° magnetic wall is formed. Figure 6-b shows the interference image calculated using the model. In comparison with Fig. 6-b and Fig. 5-b, the calculated interference image coincides rather well with the actual interference image. Therefore, the magnetization distribution in the transition region is supposed to be very much like the one shown in Fig. 6-a. Strictly speaking, though, the appearances of both interference images are different, especially in the regions near the tips of the sawtooth-like walls, where a larger number of closed fringes are observed in the actual image than in the calculated one.

Relation between Transition Length and Film Properties

The attainable recording density of a medium is determined by the transition length recorded on that medium. As to the relationship between transition length and film properties, some results have been reported where Lorentz microscopy was used [4][5][9]. However, in those experiments the transition lengths were more than 1 μm because films with low coercivities were used. These results do not represent sufficient information about high density recording. Therefore, we examined the relationship using high coercive films, where the transition lengths were less than 1 μm under the conditions of in-contact NRZ recording mentioned above.

The results, shown in Fig. 7, indicate that the transition lengths (l) can be expressed by the following equation, in terms of the coercivity (H_c), remanence (M_r), and thickness (δ) of films in SI-units.

$$l = \frac{\delta \cdot Mr}{Hc} \qquad (2)$$

This relation agrees well with the experimental results from Curland, et al., and Yamagishi, et al., except for the numerical factor. This suggests that the relationship between transition length and film properties is the same for transition regions either smaller or larger than 1μm.

We could actually observe transition regions as narrow as 0.2 μm in our experiment. Thus we can say that a high recording density of more than 100 kBPI is possible by means of the longitudinal recording method. Figure 8 shows the interference micrograph of recording pattern written with an 0.15 μm bit length on a Co film where coercivity is 112 kA/m, saturation induction is 0.7 T, and thickness is 30 nm. We could not detect interference fringes inside the medium, but periodic interference patterns in the stray field outside the medium clearly show the possibility of high density longitudinal recording at an 0.15 μm bit length.

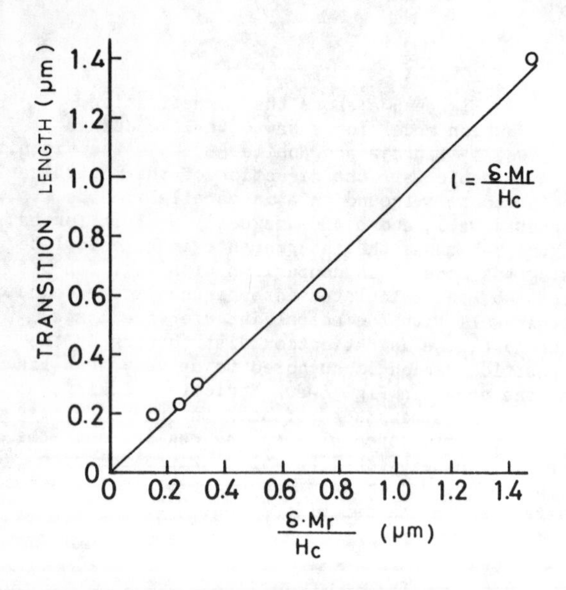

Fig. 7. Transition length as a function of $\delta M_r/H_c$.

Fig. 8. Interference micrograph of a 0.15μm bit length recorded on a Co film (thickness= 30nm, coercivity=112kA/m and saturation induction=0.7T).

Quantitative Analysis of Recorded Magnetization in Medium

One of the outstanding features of electron holography is that it allows direct observation of magnetic lines of force in empty space. The interference fringes in empty space shown in Fig. 5-b directly represent the magnetic lines of force straying from the track edge. The intensity of the recorded magnetization in the film itself can easily be determined from the number of the interference fringes at the track edges. This analysis is performed by calculating the relation between the recorded magnetization in a film and number of interference fringes.

At first, for simplicity, we calculated the number of interference fringes for an ideal case in which the transition length is zoro, (see Fig. 9). The number of interference fringes which issue from the track edge between points P_0 and P_1 is proportional to the phase difference, $\Delta\Phi(P_0,P_1)$, between them. $\Delta\phi$ is given by Eq.(1), where the vector potential, $\vec{A}$, at point P can be expressed as

$$\vec{A}(P) = -\frac{\mu_o}{4\pi}\int d\vec{v}_q \frac{[\vec{M}(Q)\times\vec{R}\ ]}{R^3_{pq}} \qquad (3)$$

Here, μ_o is permeability of vacuum, $\vec{M}(Q)$ is magnetization at point Q in a specimen, and R_{pq} is the distance between points P and Q. The integral is performed for the specimen. Especially in the case as shown in Fig.9, the phase difference, $\Delta\Phi(P_0,P_1)$ can here be analytically given by this approximate equation:

$$\Delta\Phi(P_0,P_1) = n\ \mu_o\ M_{rec} l_{bit}\ \delta\ \frac{e}{h} \qquad (4)$$

where M_{rec} is recorded magnetization on a magnetic thin film, δ is film thickness, l_{bit} is bit length and n is phase-amplification. The number of interference fringes (N) is

$$N = [\frac{\Delta\ \Phi}{2\ \pi}] \qquad (5)$$

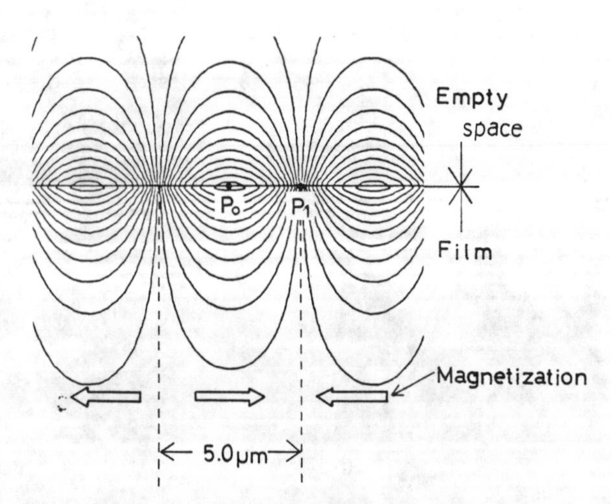

Fig. 9. Model of a magnetization distribution and the interference fringes calculated using the model. The transition length is assumed to be zero. The bit length is 5μm and $\mu_o\delta M_{rec}$ is 0.05μmT.

Here,[ ] means the integer part of the operand. Therefore, we can immediately obtain the recorded magnetization in a medium, if N, δ and l_{bit} are given.

Figure 10 shows the relation between the number of the interference fringes and the bit length, where the dots are for the experimental results and the solid line represents the calculation. The numbers were normalized to phase-amplification by a factor of 1, and reduced to a case in which $\mu_\circ M_r \delta$ is equal to 0.05 μmT, where M_r is remanence. Therefore, if the intensity of recorded magnetization is equal to the remanence of a film, the number of the interference fringes lies on the solid line. In Fig.10 every experimental point lies on the solid line, so we can conclude that saturation recording was performed for every recorded bit length, even at the very short bit length of 0.15 μm.

Although transition length is assumed to be idealy zero in the above discussion, the result is applicable to the case of actual magnetization configuration in the transition region. Figure 11 shows the calculated fringe pattern for a film of $\mu_\circ M_{rec} \delta$ = 0.05 μmT, assuming sawtooth-like transitions as shown in the same figure. Almost the same number of interference fringes can be seen coming out of the track edge as in Fig.9.

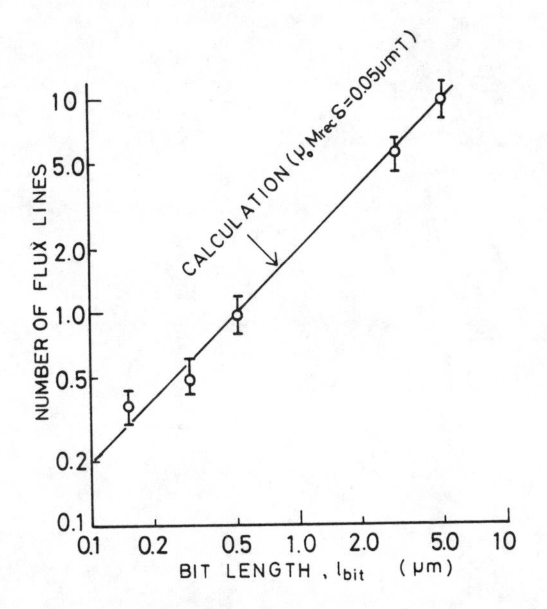

Fig. 10. Dependence of the number of stray field flux lines on bit length.

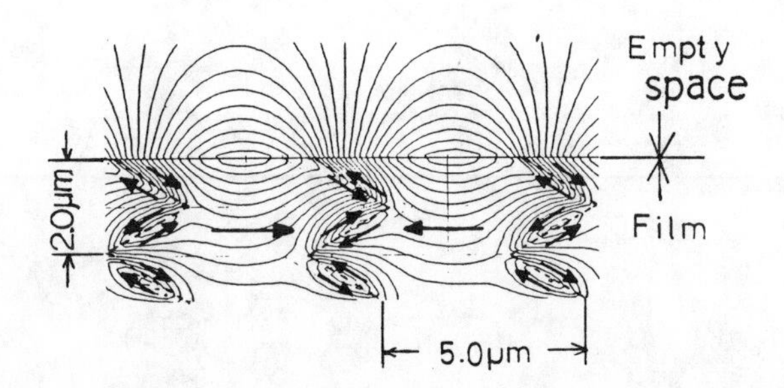

Fig. 11. The calculated interference fringes using sawtooth-like magnetization distribution. The bit length is 5μm and $\mu_\circ M_{rec} \delta$ is 0.05 μmT. The number of the stray field flux lines is the same as in Fig.9, though the magnetization distribution is different.

SUMMARY

The following facts were confirmed by observing recorded magnetization patterns using electron holography.

a) The smaller the product of the film thickness and the remanence, and the higher the coercivity, the narrower is the transition length of Co thin film.
b) The intensity of the recorded magnetization on Co thin film was measured by counting the magnetic lines of force coming out of the track edge. It was concluded that the intensity of the recorded magnetization is almost equal to the remanence of the film up to a recording density of 170 kBPI.
c) Ultra-high density longitudinal recording of 170 kBPI was attained using a 30 nm Co film prepared by oblique incidence vacuum deposition. The coercivity was 112kA/m and the saturation induction was 0.7T.
d) It was confirmed that such a magnetization configuration as had been proposed by Minnaja and Nobile for the sawtooth-like transition region gives almost the same fringe pattern as was observed, although some modification should be made in the configuration at the tip regions of the sawtooth in order for the calculated pattern to precisely fit the observed one.

ACKNOWLEDGEMENT

The authors would like to express sincere thanks to Dr.T.Doi at Hitachi Maxell Ltd. for promoting the present research and also for his encouragement. And we are indebted to Dr. M. Kudo for his support and encouragement and to Mr. J. Endo who kindly reconstructed skillful interference micrographs.

REFERENCES

[1] A.Tonomura, et al., Phys. Rev. Lett.44, 1430(1980).
[2] Y.Aharanov and D.Bohm, Phys. Rev.115, 485 (1959).
[3] J.Endo, T.Matsuda and A.Tonomura, Jpn. J.Appl. Phys. 18, 2291 (1979).
[4] N.Curland and D.Speliotis, J. Appl. Phys.41, 1099 (1970).
[5] D.D.Dresser and J.H.Judy, IEEE Trans. Mag.MAG-10, 674 (1974).
[6] T.Chen, IEEE Trans. Mag.MAG-17, 1181(1981).
[7] N.Osakabe, et al., Appl.Phys.Lett. 42, 746 (1983).
[8] N.Minnaja and M.Nobile, AIP Conf. Proc. No.5, 1001(1972).
[9] F.Yamagishi and S.Iwasaki, IECE Jpn., Tech. Rep., MR 75-29 (1975) in Japanese.

Part VI
Readback

The Reproduction of Magnetically Recorded Signals

R. L. WALLACE, JR.

(*Manuscript Received July 9, 1951*)

For certain speech studies at the Bell Telephone Laboratories, it has been necessary to design some rather specialized magnetic recording equipment. In connection with this work, it has been found experimentally and theoretically that introducing a spacing of d inches between the reproducing head and the recording medium decreases the reproduced voltage by $54.6\ (d/\lambda)$ decibels when the recorded wavelength is λ inches. For short wavelengths this loss is many decibels even when the effective spacing is only a few ten-thousandths of an inch. On this basis it is argued that imperfect magnetic contact between reproducing head and recording medium may account for much of the high-frequency loss which is experimentally observed.

Introduction

WITHIN the last few years there has been increasing use of magnetic recording in various telephone research applications (examples are various versions of the sound spectrograph used in studies of speech and noise). Some of these uses[1] have required a reproducing head spaced slightly out of contact with the recording medium. Experimental studies were made to determine the effect of such spacing and the results were found to be expressible in an unexpectedly simple form. The general equation derived is believed to be fundamental to the recording problem and to account for much of the high-frequency loss that is found in both in- and out-of-contact systems.

This paper discusses results of the experimental study and presents for comparison some theoretical calculations based on an idealized model.

Measurements of Spacing Loss

In order to measure the effect of spacing between the reproducing head and the medium, an experiment was set up as indicated in Fig. 1. The recording medium used was a 0.0003 inch plating of cobalt-nickel alloy[2] on the flat surface of a brass disc approximately 13 inches in diameter by $\frac{1}{4}$ inch thick.

This disc was made with considerable care to insure that the recording surface was as nearly plane and smooth as possible and that it would turn reasonably true in its bearings. Speeds of 25 and 78 rpm were provided.

[1] R. C. Mathes, A. C. Norwine, and K. H. Davis, "Cathode-Ray Sound Spectroscope," *Jl. Acous. Soc. Am.*, *21*, 527 (1949).
[2] Plating was done by the Brush Development Company.

The ring-type record-reproduce head shown in Fig. 1 was lapped slightly to obtain a reasonably good fit with the surface of the disc.

A single-frequency recording was made with the head in contact with the disc using a-c. bias in the usual way. Then the open circuit reproduced signal level was measured, first with the head in contact, and then after introducing paper shims of various thickness between the reproducing head and the medium. Thus the effect of spacing was measured at a particular frequency and recording speed. The signal was then erased and the process was repeated for other recorded frequencies and for several record-reproduce speeds. Measurements were also made for cases in which the recording and reproducing speeds were different. Considerable care was required to keep the disc and head sufficiently clean so that reproducible results could be obtained.

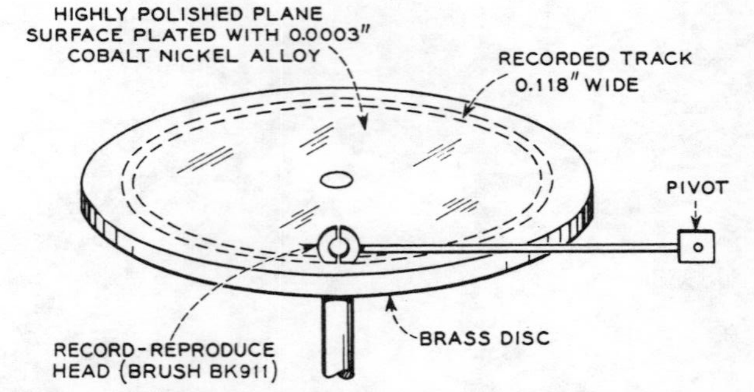

Fig. 1—Mechanical arrangement of recording set up. The one head served for recording, playback, and erase.

Figure 2 shows typical response curves measured at 21 in./sec. with the reproducing head in contact and with 0.004 inch spacing. The difference between these two curves will be called the spacing loss corresponding to this spacing and speed. From these data and more of the same sort it is found that, within experimental error, spacing loss can be very simply related to spacing and the recorded wavelength, λ, by the empirical equation,

$$\text{Spacing loss} = 55(d/\lambda)\ \text{decibels} \tag{1}$$

where spacing loss is the number of decibels by which the reproduced level is decreased when a spacing of d inches is introduced between the reproducing head and a magnetic medium on which a signal of wavelength λ inches has been recorded.

The fact that this expression fits the experimental data reasonably well is indicated in Fig. 3 where spacing loss data measured at a number of different speeds, frequencies, and spacings are plotted against d/λ.

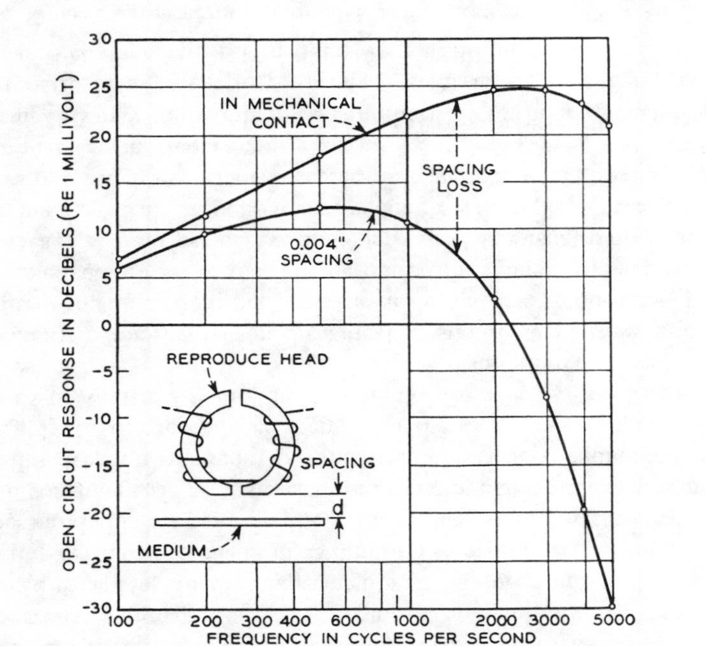

Fig. 2—Response curves taken at 21 in./sec. Recordings were made with head in contact and were played back first with head in contact and then with a spacing of 4 mils between head and disc.

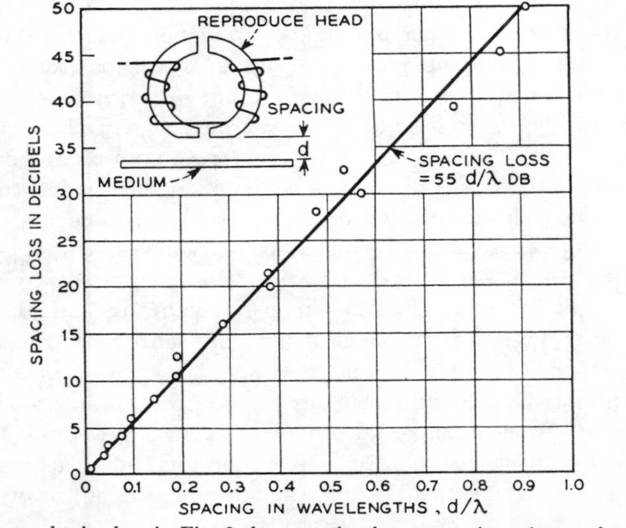

Fig. 3—Data obtained as in Fig. 2 show spacing loss approximately equal to $55(d/\lambda)$ decibels.

Implications of the Experiment

In this section it will be assumed that equation (1) holds true in all cases where the spacing d is sufficiently small and the recorded track is sufficiently wide so that end effects are negligible. If this is true, as it seems experimentally to be, then it is indeed surprising how great can be the effect of even a very small spacing when the recorded wavelength is small. For example, take the case of a 7500 cps signal recorded at 7.5 in./sec. in which case the wavelength is 0.001 inch. A particle of dust which separated the tape from the reproducing head by one-thousandth of an inch would decrease the reproduced level by 55 db. A spacing of 0.0001 inch would produce a quite noticeable 5.5 db effect and even at 0.00001 inch spacing the 0.55 db loss would be measurable in a carefully controlled experiment.

In view of the magnitudes involved, it seems probable that this spacing loss may play a significant role even in cases where the reproducing head is supposed to be in contact with the medium. For example, it has been known for some time that chattering of the tape on the reproducing head or changes in the degree of contact due to imperfect smoothness of the tape can result in amplitude modulation of the reproduced signal and thereby give rise to "modulation noise" or "noise behind the signal."

With the aid of equation (1) it is possible to estimate the magnitude of the noise provided some assumption is made about the wave form of the modulation. To take a simple case, suppose that the roughness of the tape were such as to sinusoidally modulate the spacing by a very small amount and at a low frequency. The reproduced signal would then be modulated and would contain a sideband on each side of the center frequency. The energy in these two sidebands constitutes the modulation noise in this case. If it is required that this noise be 40 db down on the signal, then one can calculate the maximum permissible excursion of the tape away from the reproducing head. This turns out to be $1.1(10)^{-5}$ cm. or about one-sixth of the wavelength of the red cadmium line! Of course, the one mil wavelength assumed in this example is about as short as is often used and the effect becomes less severe as the wavelength is increased. This is one of the reasons that speeds greater than 7.5 in./sec. are used for highest quality reproduction.

One can also make some rough qualitative inferences about the effect of the thickness of the recording medium on the shape of the response curve. As can be seen from equation (1) or from Fig. 2, low frequencies can be reproduced with very little loss in amplitude in spite of considerable spacing between the reproducing head and the medium while high frequencies (i.e. short wavelengths) may be appreciably attenuated by even 0.0001 inch spacing between the head and the medium. With this in mind it is easy to see that at high frequencies only a thin layer of the medium nearest the reproducing head will contribute to the reproduced signal. In this case (short λ) increasing the thickness of the medium beyond a certain amount can have no effect on the reproduced level simply because the added part of the tape is too far from the head to make its effect felt. Consider the effect of increasing the thickness of the medium from 0.3 mil to 0.6 mil when the wavelength is one mil. Since the spacing loss for 0.3 mil spacing at $\lambda = 1$ mil is 16.5 db, the signal contributed by the lower half of a medium 0.6 mils thick cannot be less than 16.5 db lower than that contributed by the upper half and hence the increase in thickness can do no more than to raise the reproduced level by 1.2 db.

At a lower frequency for which $\lambda = 100$ mil, however, the corresponding spacing loss is only 0.165 db and in this case the two halves of the tape can contribute almost equally with the result that doubling the thickness of the medium can almost double the reproduced signal voltage.

Qualitatively, then, one might expect that increasing the thickness of the recording medium, other things being equal, would increase the response to low frequencies and leave the high frequency response relatively unaltered. This is in agreement with data published by Kornei.[3]

The estimates of magnitudes just given rest on assumptions which cannot be proved except by further experiments. It has been implicitly assumed, for example, that the medium is uniformly magnetized throughout its thickness and this may not be the case. It does seem perfectly safe, however, to conclude that at a wavelength of one mil that part of the medium which lies deeper than about 0.3 mil from the surface cannot contribute appreciably to the reproduced signal. Furthermore, as the wavelength is decreased beyond this point the thickness of the effective part of the tape decreases in inverse proportion to λ with the result that the available flux also decreases. For this reason the "ideal" response curve cannot continue indefinitely to rise at 6 db per octave as it does at low frequencies. In fact, when the effective part of the tape becomes thin enough, the available flux will decrease at 6 db per octave and just cancel the usual 6 db per octave rise, giving an "ideal" response curve which rises 6 db per octave at low frequencies but which eventually becomes flat, neither rising nor falling with further increase in frequency.

Spacing loss may contribute in still another way to the frequency response characteristic of a magnetic recording system in which the reproducing head makes contact with the medium. It is well known to those who work

[3] Otto Kornei, "Frequency Response of Magnetic Recording," *Electronics*, p. 124, August, 1947.

with magnetic structures such as are used in transformers and the like that intimate mechanical contact between two parts of a magnetic circuit does not imply intimate magnetic contact. In fact, even when great care is taken in fitting such parts together, measurements invariably show an effective air gap between them and the effective width of this gap usually amounts to appreciably more than one mil. One reason for this is that the permeability of soft materials such as are used in the cores of transformers and reproducing heads is very sensitive to strain. Even the light cold working which a surface receives in being ground flat is sufficient to impair very seriously the permeability of a thin surface layer.

In view of this it is to be expected that the magnetic contact between reproducing head and medium is less than perfect. If cold working during the fabrication of the head or due to abrasion by the recording medium should result in an effective air space between head and medium amounting to as much as one mil, the effect on frequency response would be pronounced indeed. At a recording speed of 7.5 in./sec. this amount of spacing would cause a loss of 7.3 db at 1000 cps, 14.6 db at 2000 cps, 21.9 db at 3000 cps, 29.2 db at 4000 cps, etc.

It seems certain that in a practical recording system some loss of this sort must occur. The problem of determining the magnitude of the loss or in other words the amount of the effective spacing in a practical case is, however, a difficult one. So far, no direct experimental method for its determination has been found.

Theoretical Calculations for an Idealized Case

In the preceding section an experimentally determined spacing loss function has been discussed. It was shown that as the reproducing head is moved away from the recording medium the reproduced signal level decreases. This means that the magnetic flux through the head decreases. If the distribution of magnetization in the recording medium were known, it should be possible to compute the flux through the head and thereby to derive the spacing loss function on a theoretical basis. Unfortunately it seems almost impossible to do this calculation in an exact way because very little is known about the magnetization pattern in the medium and because the geometry of the usual ring type head makes the boundary value problem an exceedingly difficult one to solve.

It is possible, however, to obtain a solution for an idealized case which bears at least some resemblance to the practical situation and this solution will be presented. The results must, of course, be viewed with due skepticism until they can be proved experimentally or else recalculated on the basis of better initial assumptions. It is hoped, however, that in some measure they may serve as a guide to a better understanding of the magnetic reproducing process.

The Idealized Recording Medium

The problem will be reduced to two dimensions by assuming an infinitely wide and infinitely long tape of finite thickness δ. A rectangular coordinate system will be chosen in such a way that the central plane midway between the upper and lower surfaces of the recording medium lies in the x–y plane. It will be assumed that the medium is sinusoidally magnetized in such a way that in the medium the intensity of magnetization is given by

$$I_x = I_m \sin (2\pi x/\lambda)$$

$$I_y = I_z = 0. \tag{2}$$

Equations (2) say that the recording is purely longitudinal. In a practical case, of course, the recorded signal is neither purely longitudinal nor purely perpendicular but rather contains components of both sorts. In Appendix I it is shown that the frequency response does not depend on the relative amounts of these two components and hence that the computed results are equally valid whether the recorded signal is purely longitudinal, purely perpendicular, or a mixture of the two.

Appendix II contains calculations for the case of a round wire sinusoidally magnetized along its axis, and for a plated wire. These results, though much different in mathematical form, are shown to be very similar to the results for a flat medium.

The Idealized Reproducing Head

Figure 4 shows a semi-practical version of the sort of idealized reproducing head which will be treated.

It consists of a bar of core material with a single turn of exceedingly fine wire around it. This head is imagined to be spaced d inches above the surface of the recording medium. If the dimensions of the bar are made large enough, the amount of flux through it will obviously be as great as could be made to pass through any sort of head which makes contact with only one side of the tape and so the open circuit reproduced voltage per turn is as high as can be obtained with any practical head.

Suppose a very narrow gap is introduced in this head where the single turn coil was and that the magnetic circuit is completed by a ring of core material as shown in Fig. 5.

If the permeability of the head is very high and the gap very small then the flux which passed through the single turn coil of Fig. 4 will now pass

through the ring of Fig. 5 and can be made to thread through a coil of many turns wound on the ring. In so far as this is true, calculations based on this bar type head are applicable to ring type heads.

If the bar of Fig. 4 is now allowed to become infinite in length, width, and thickness, the flux density in it can be computed and the flux per unit width can be evaluated. This calculation is outlined in Appendix I. If the tape moves past the head with a velocity v in the x direction, the reproduced voltage should be proportional to the rate of change of flux. In the appendix this is shown to be

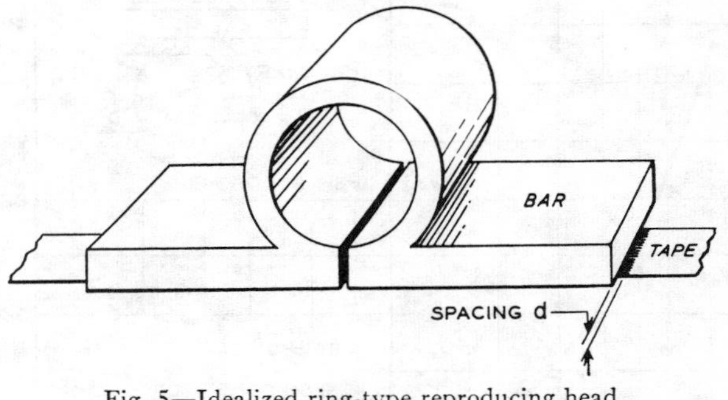

Fig. 4—Idealized bar-type reproducing head.

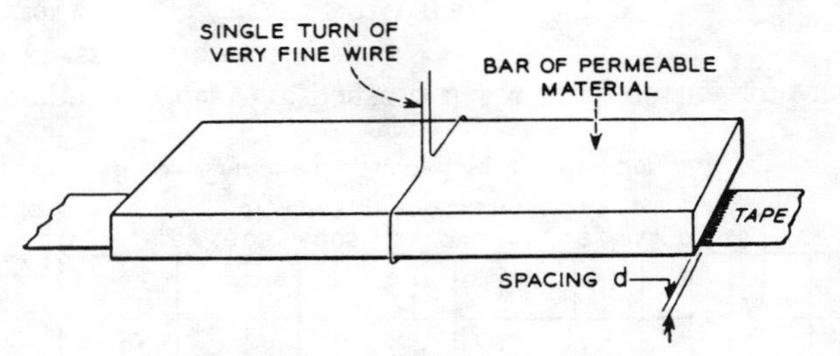

Fig. 5—Idealized ring-type reproducing head.

$$\frac{d\phi_x}{dt} = -\frac{\mu}{\mu + 1} 4\pi W v I_m (1 - e^{-2\pi\delta/\lambda}) e^{-2\pi d/\lambda} \cos (\omega t) \tag{3}$$

where $\frac{d\phi_x}{dt}$ is the rate of change of flux in W cm. width of the reproducing head measured in Maxwells per sec.,

μ is the permeability of the reproducing head,

W is the width in cm. of the reproducing head (and of the recorded track in a practical case),

v is the velocity in cm./sec. with which the recording medium passes the *reproducing* head.

I_m is the peak value of the sinusoidal intensity of magnetization in the recording medium measured in gauss,

δ is the thickness of the recording medium measured in the same units as λ,

λ is the recorded wavelength measured in any convenient units,

d is the effective spacing between the reproducing head and the surface of the recording medium measured in the same units as λ, and

ω is 2π times the reproduced frequency in cycles per sec.

Note that equation (3) applies to a ring type head with no back gap. If the head has a back gap then not all the available flux will thread through the ring. Some of it will return to the medium through the scanning gap and hence will not contribute to the reproduced voltage. This does not affect the shape of the frequency response curve but does contribute a constant multiplying factor (less than unity) to the right hand side of equation (3). The value of this factor depends on the reluctances of the gaps and of the magnetic parts of the reproducing head. If the reluctance of the magnetic parts is negligible and the reluctance of the back gap is equal to the reluctance of the front gap then the available flux will divide equally in the two gaps and the factor will be one-half. This factor will not be considered further in this paper because it does not contribute to the shape of the response curve but only to the absolute magnitude of the reproduced voltage. It could be interpreted as reducing the effective number of turns on the reproducing head to a value somewhat lower than the actual number of turns.

Spacing Loss

The term $e^{-2\pi d/\lambda}$ tells how the reproduced voltage depends on spacing. In order to compare this computed effect with the experimentally observed one it is necessary to put it in decibel form by computing twenty times the $\log_{10}$ of $e^{-2\pi d/\lambda}$. This gives

$$\text{Spacing Loss} = 54.6\ (d/\lambda) \text{ decibels}.$$

This agrees very well indeed with the experimentally determined equation (1) in which the constant is 55 instead of the computed 54.6. The computed spacing loss function is plotted in Fig. 6.

Thickness Loss

The effect of the thickness of the recording medium shows up in the term $(1 - e^{-2\pi\delta/\lambda})$. At low frequencies for which the wavelength is much greater than the thickness of the medium this reduces to $2\pi\delta/\lambda$. In this case the reproduced voltage is proportional to the thickness of the medium and to frequency. This is the familiar six db per octave characteristic.

At high frequencies, however, when $\lambda \ll \delta$ the term reduces to unity

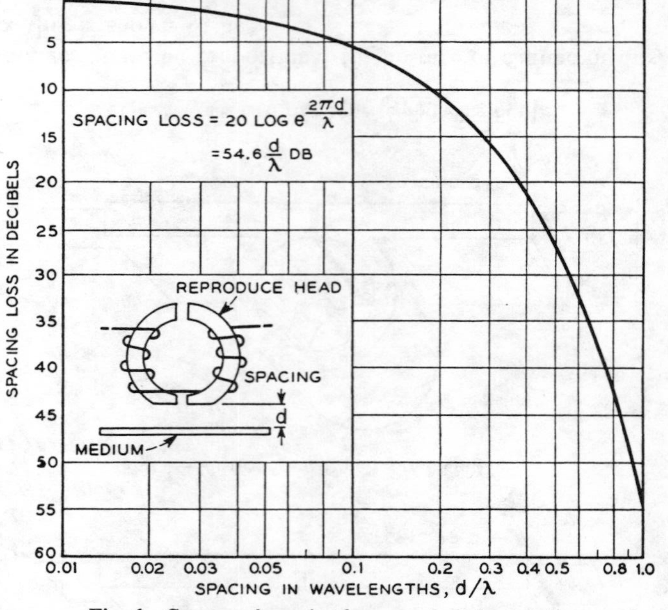

Fig. 6—Computed spacing loss as a function of d/λ.

and the computed "ideal" response is flat with frequency and independent of the thickness of the medium.

If the term $(1 - e^{-2\pi\delta/\lambda})$ is rewritten as

$$(2\pi\delta/\lambda)\left[\frac{1 - e^{-2\pi\delta/\lambda}}{2\pi\delta/\lambda}\right]$$

then the part in parenthesis accounts for a 6 db per octave characteristic and the part in brackets accounts for a loss *with respect to this 6 db per octave characteristic*. This loss, which will be called Thickness Loss[4], is given by

[4] It seems somewhat awkward to speak of "Thickness Loss" when nothing is actually lost by making the medium thick. The only excuse for this way of splitting the terms is that it makes for ease in comparing measured and computed curves.

$$\text{Thickness Loss} = 20 \log_{10} \frac{2\pi\delta/\lambda}{1 - e^{-2\pi\delta/\lambda}} \text{ db} \tag{5}$$

where λ is the recorded wavelength and δ is the thickness of the recording medium. This function is plotted in Fig. 7.

Comparison with Experiment

The most elementary consideration of the magnetic recording process indicates that when the recording signal current is held constant the open circuit reproduced voltage should be a function of frequency, increasing by 6 db for each octave increase in frequency. Experimental response curves tend to show this 6 db per octave characteristic when the recorded wave-

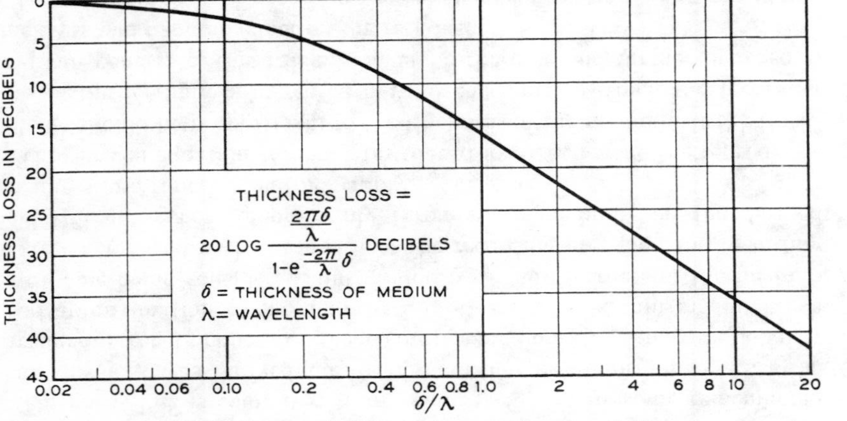

Fig. 7—Computed thickness loss as a function of δ/λ.

length is moderately long and the frequency moderately low. This makes it possible to draw a 6 db per octave line on the measured response characteristic in such a way as to coincide with the low-frequency part of the measured response characteristic. As the frequency is increased the measured curve tends to fall more and more below the 6 db per octave line. This is because several kinds of loss come into play as the wavelength decreases or as the frequency increases. Among these losses are:

1. Self demagnetization,
2. Eddy current and other losses in the recording and reproducing heads, and
3. Gap loss due to the finite scanning slit in the reproducing head.

The work presented in the first sections of this paper indicates that the

following two kinds of loss should be added to this list:

4. Spacing loss due to imperfect magnetic contact between the reproducing head and the recording medium, and
5. Thickness loss.

Of these five losses three can be evaluated quantitatively either by direct measurement or by calculation from theory. The remaining two are self-demagnetization and spacing loss.

In this section the known losses will be evaluated for a particular recording system. This leads to a response curve which can be compared with the measured curve. The difference between the two curves should be due to self-demagnetization and to spacing loss provided the above list of losses is complete.

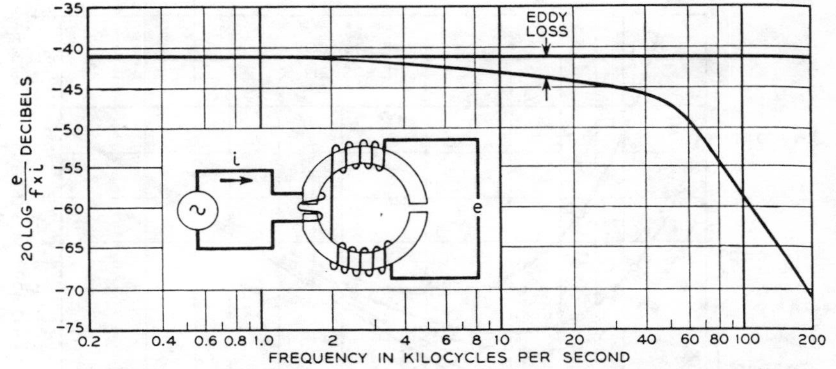

Fig. 8—Measured eddy current loss as a function of frequency.

The recording system used is the one shown in Fig. 1 with the speed set at 15.5 in./sec. for both recording and reproducing. A constant signal current of 0.1 ma was used for recording with the 55 kc bias adjusted to give maximum open circuit reproduced voltage.

Eddy current losses were measured as indicated in Fig. 8 by sending a measured constant current i through a small auxiliary winding around the pole tip and measuring the open circuit voltage developed across the normal winding of the head. Any departure of this measured voltage from a 6 db per octave increase with increasing frequency is due to losses in the head which will be loosely called eddy current losses. Other kinds of loss may enter into this measurement (as, for example, loss due to the self-capacitance of the winding) but in the frequency range of interest, eddy losses predominate.

By a completely different sort of measurement,[5] J. R. Anderson has arrived at a similar value for eddy current loss in this type of head and has shown that approximately the same loss occurs in both the recording and the reproducing process. For this reason it seems proper to assume that eddy currents account for just twice the loss measured by the method of Fig. 8.

The loss due to the finite gap in the reproducing head is computed from the well known relation.[6]

$$\text{Gap loss} = 20 \log_{10} \frac{\pi g/\lambda}{\sin (\pi g/\lambda)}$$

where g is the effective gap width in inches and λ is the recorded wavelength in inches.

Thickness loss is computed from equation (5). It must be remembered that this loss was derived on the assumption of uniform magnetization throughout the thickness of the recording medium. This may be a fairly good approximation to the actual state of affairs for a thin medium such as the one being considered, but obviously if the thickness of the medium is large compared with the width of the recording gap then the recording field will not penetrate uniformly through the medium and the derived thickness loss function will not apply.

The derived equation (3) indicates that at low frequencies the reproduced voltage should be proportional to the thickness of the medium. If the thickness of the medium is increased beyond the limit to which the recording field can penetrate, this will no longer be the case and further increase in thickness will have no effect on the response.

Data presented by Kornei[3] on the cobalt-nickel plating being considered here shows that the low-frequency response is approximately proportional to the thickness of the medium for values of thickness between 0.075 mil and 0.5 mil. This may be taken as an indication of approximately uniform penetration through these thicknesses and hence tends to indicate that the derived thickness loss function should be applicable in the case of the 0.3 mil plating being considered here.

The effects of these losses are shown in Fig. 9 along with measured frequency response data. Consider first the experimentally measured response data shown as circles falling near the lowest curve. Some of the measured points have been omitted to avoid crowding but enough remain to show the trend. At low frequencies these points fall along a line of approximately 6 db per octave.

A straight 6 db per octave line labeled 1 has been drawn through these points and extended as shown in the figure. This line is the base from which the various losses must be subtracted. Curve 2 shows the effect of subtracting the computed thickness loss. When eddy losses and gap loss are

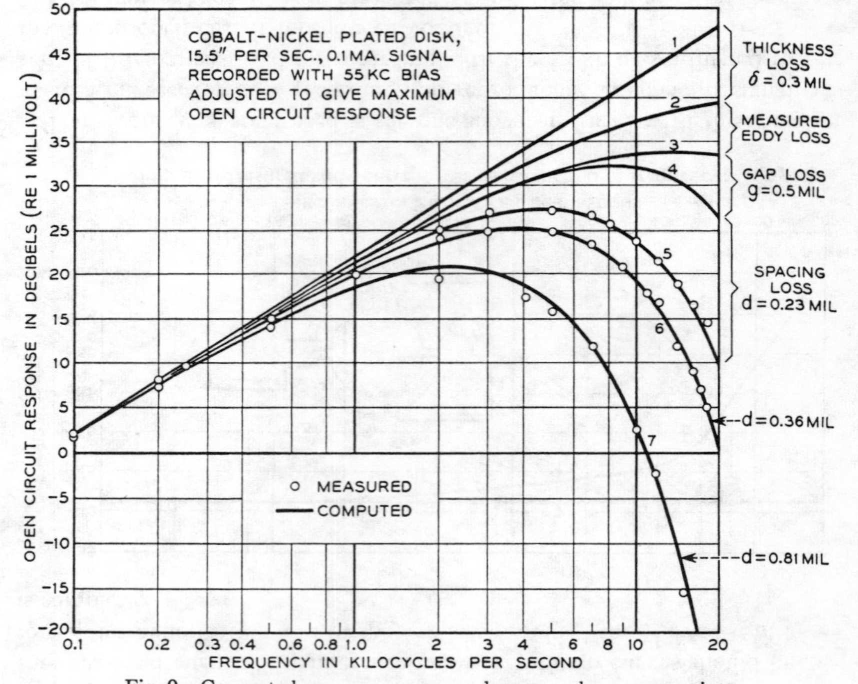

Fig. 9—Computed response curves and measured response points.

also taken into account, curve 4 is obtained. The difference between this curve and the lowest measured response points is presumably due either to self-demagnetization, to spacing loss, or perhaps to both.

There is one clue which may be of help in deciding how much of this loss should be attributed to self-demagnetization and how much to spacing loss. This clue comes from the fact that the form of the spacing loss function is known. Any part of the loss which is due to spacing must follow the equation

$$\text{Spacing Loss} = 54.6\ (d/\lambda)\ \text{db}$$

[5] In unpublished work, J. R. Anderson of the Bell Telephone Laboratories has made use of the fact that eddy losses depend on frequency while all other magnetic recording losses depend on wavelength. By recording a single frequency and playing back at various speeds he determined the loss on playback. By recording various frequencies with recording speed adjusted to give constant recorded wavelength and using a single playback speed he evaluates the eddy loss in the recording process.

[6] S. J. Begun, "Magnetic Recording," p. 84, Murray Hill Books, Inc., New York.

whereas there is reason to believe that the effects of self-demagnetization cannot possibly account for more than something like ten or fifteen db loss and hence could not follow the equation given above.

In view of this it seems reasonable to try as a first guess the assumption that all the unexplained loss is due to spacing.

If this assumption properly accounts for the shape of the measured response curve there will be at least some reason to suppose it may be correct; particularly so if the required amount of effective spacing seems reasonable.

The lowest solid curve, No. 7 of Fig. 9, has been computed on this basis. That is, a spacing loss corresponding to 0.81 mil effective spacing has been subtracted from curve 4. It is seen that this computed response curve fits reasonably well with the measured points. Furthermore, 0.81 mil effective spacing corresponds to quite reasonably good magnetic contact.

If this interpretation of the measured data is correct then it is obvious that the high-frequency response could be improved a great deal if more intimate magnetic contact between the reproducing head and the recording medium could be achieved. To this end an attempt was made to lap the surface of the head in such a way as to remove material very gently and slowly. After lapping, the response was appreciably improved as indicated by the set of measured points around curve 6. This curve was computed assuming an effective spacing of 0.36 mil. Note that the computed curve now fits the measured points very well indeed.

After still more lapping,[7] the measured response points around curve 5 were obtained. In this case it is necessary to assume only 0.23 mil effective spacing in order to account for the measured curve. Further lapping failed to give further improvement in response but a defect in the head which may account for this has since been found and it is believed that with great care one might actually measure something very close to curve 4.

To summarize, this is what seems to have been found. It is possible to compute a response curve taking into account gap loss, eddy current losses, and thickness loss. If this curve is compared with the final measured response curve it is found that the measured curve gives less high-frequency response than was computed. The difference between the two curves is just the right sort of function of frequency and of just the right magnitude to be accounted for by an effective spacing of 0.00023 inch between the reproducing head and the recording medium. It seems probable that the effective spacing could not have been much smaller than this value and therefore it may be correct to assume that practically all the unexplained

[7] After each lapping it was found that smaller values of bias current sufficed to give maximum reproduced voltage. This is presumably because the improved magnetic contact made the bias current more effective.

high-frequency loss is due to spacing. This would imply that, under the conditions of this experiment, self-demagnetization has a negligible effect on frequency response.

In any case it seems clear that the intimacy of magnetic contact between the reproducing head and the medium can have a very pronounced effect on high-frequency response. The condition of the surface of the reproducing head (and of the tape) may have more effect on high-frequency response than any other single factor.

In a very fine piece of pioneering work Lübeck[8] found empirically that a term of the form $e^{-\lambda_1/\lambda}$ was needed to account for the shape of measured response curves. Guckenburg[9] has recently written more on this subject. Both authors have assumed that this term has to do with self-demagnetization and that λ_1 is determined by the magnetic properties of the recording medium. The experiment just discussed and the theory presented in this paper suggest, on the other hand, that λ_1 is not a function of the magnetic properties of the recording medium but rather is determined by the intimacy of magnetic contact between reproducing head and medium. If this is the case then Lubeck's λ_1 is related to the d of this paper through the equation $\lambda_1 = 2\pi d$.

Guckenburg reports $\lambda_1 = 100\mu$ for the best available medium. This corresponds to $d = 0.625$ mil and yields a response curve a little better than the poorest measured curve of Fig. 9. The best measured curve of Fig. 9. corresponds to $\lambda_1 = 37\mu$.

ACKNOWLEDGEMENTS

The author wishes to express his appreciation for the encouragement and guidance of Mr. R. K. Potter, Mr. J. C. Steinberg, and Mr. W. E. Kock.

APPENDIX I

THE FIELD DUE TO A FLAT SINUSOIDALLY MAGNETIZED MEDIUM

The Field in Free Space

It is convenient to begin by evaluating the field inside and outside the recording medium when the medium is in free space. By making use of the

[8] H. Lübeck, "Magnetische Schallaufzeichnung mit Filmen und Ringkopfen," *Akustiche Zeit.*, 2, 273 (1937).

[9] W. Guckenburg, "Die Wechselbeziehungen zwichen Magnettonband und Ringkipf bie der Wiedergabe," *Funk und Ton*, 4, 24 (1950).

method if images, this solution can be used to find the fields which exist when the medium is under an idealized reproducing head of permeability μ. Also, the free space solution may be of use in evaluating the effect of self demagnetization since the demagnetizing field is computed.

Let the recording medium be an infinite plane sheet of thickness δ and choose rectangular coordinates so that the central plane of the medium lies in the x–y plane as shown in Fig. 10.

Let the permeability of the recording medium be unity and let the intensity of magnetization inside it be given by

$$\begin{aligned} I_x &= I_m \sin(2\pi x/\lambda) \\ I_y &= I_z = 0 \end{aligned} \tag{6}$$

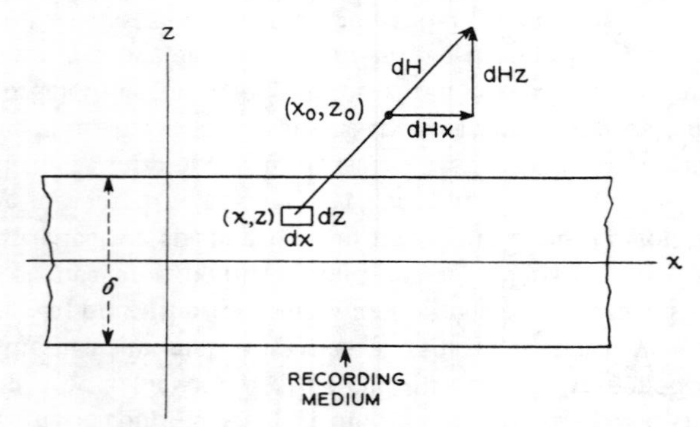

Fig. 10—Coordinate system for flat tape calculations.

This is equivalent to a volume density of "magnetic charge" given by

$$\begin{aligned} \rho &= -\operatorname{div} I \\ &= -\frac{dI_x}{dx} \\ &= -(2\pi I_m/\lambda)\cos(2\pi x/\lambda) \end{aligned} \tag{7}$$

The problem then is to compute the field at a point (x_0, z_0) due to this charge. Consider the field at (x_0, z_0) due to the element $dx\, dz$ at (x, z). This element amounts to an infinitely long line of uniform charge density. The field due to such a distribution is directed perpendicular to the line and has a magnitude equal to twice the linear charge density divided by the distance from the point to the line.

In the present case this leads to

$$\begin{aligned} dH_x &= -(4\pi I_m/\lambda)\frac{(x_0 - x)}{(x_0 - x)^2 + (z_0 - z)^2}\cos(2\pi x/\lambda)\, dx\, dz \\ dH_z &= -(4\pi I_m/\lambda)\frac{(z_0 - z)}{(x_0 - x)^2 + (z_0 - z)^2}\cos(2\pi x/\lambda)\, dx\, dz \end{aligned} \tag{8}$$

The total field at (x_0, z_0) is obtained by integrating with respect to x over the range $-\infty$ to $+\infty$ and with respect to z over the range $-\delta/2$ to $+\delta/2$. In carrying out the integration over x it is convenient to make the substitution

$$\begin{aligned} (x_0 - x)/(z_0 - z) &= p \\ dx &= -(z_0 - z)\, dp \end{aligned} \tag{9}$$

Neglecting terms which obviously integrate to zero, this gives

$$\begin{aligned} H_x &= (4\pi I_m/\lambda)\sin(2\pi x_0/\lambda)\int_{-\delta/2}^{\delta/2}\left[\int_{\infty}^{-\infty}\frac{p\sin[2\pi(z_0 - z)p/\lambda]}{1 + p^2}\, dp\right] dz \\ H_z &= (4\pi I_m/\lambda)\cos(2\pi x_0/\lambda)\int_{-\delta/2}^{\delta/2}\left[\int_{\infty}^{-\infty}\frac{\cos[2\pi(z_0 - z)p/\lambda]}{1 + p^2}\, dp\right] dz \\ & z_0 \geq z \end{aligned} \tag{10}$$

The integrals in brackets can be found in tables.[10] Carrying out the integration gives

$$\begin{aligned} H_x &= -(4\pi^2 I_m/\lambda)\sin(2\pi x_0/\lambda)\int_{-\delta/2}^{\delta/2} e^{-2\pi(z_0 - z)/\lambda}\, dz \\ H_z &= -(4\pi^2 I_m/\lambda)\cos(2\pi x_0/\lambda)\int_{-\delta/2}^{\delta/2} e^{-2\pi(z_0 - z)/\lambda}\, dz \\ & z_0 \geq z \end{aligned} \tag{11}$$

which integrate to

$$\begin{aligned} H_x &= -2\pi I_m \sin(2\pi x_0/\lambda)e^{-2\pi z_0/\lambda}[e^{\pi\delta/\lambda} - e^{-\pi\delta/\lambda}] \\ H_z &= -2\pi I_m \cos(2\pi x_0/\lambda)e^{-2\pi z_0/\lambda}[e^{\pi\delta/\lambda} - e^{-\pi\delta/\lambda}] \\ & z_0 \geq \delta/2 \end{aligned} \tag{12}$$

[10] D. Bierens de Haan, "Nouvelles Tables D'Integrales Définies," p. 223, Leide, Engels, 1867.

Below the recording medium, that is for $z_0 \leq -\delta/2$,

$$\begin{aligned} H_x &= -2\pi I_m \sin(2\pi x_0/\lambda) e^{+2\pi z_0/\lambda} [e^{\pi\delta/\lambda} - e^{-\pi\delta/\lambda}] \\ H_z &= 2\pi I_m \cos(2\pi x_0/\lambda) e^{+2\pi z_0/\lambda} [e^{\pi\delta/\lambda} - e^{-\pi\delta/\lambda}] \\ & z_0 \leq -\delta/2 \end{aligned} \tag{13}$$

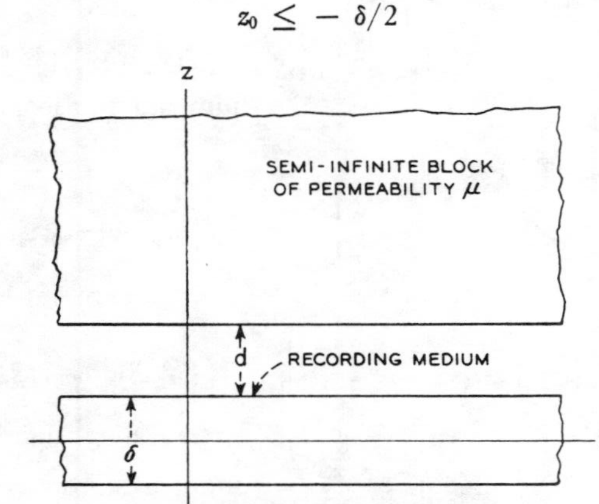

Fig. 11—Flat tape under idealized reproducing head.

Inside the recording medium,

$$\begin{aligned} H_x &= -(4\pi^2 I_m/\lambda) \sin(2\pi x_0/\lambda) \\ &\cdot \left[\int_{-\delta/2}^{z_0} e^{-2\pi(z_0-z)/\lambda}\, dz + \int_{z_0}^{\delta/2} e^{+2\pi(z_0-z)/\lambda}\, dz\right] \\ H_z &= -(4\pi^2 I_m/\lambda) \cos(2\pi x_0/\lambda) \\ &\cdot \left[\int_{-\delta/2}^{z_0} e^{-2\pi(z_0-z)/\lambda}\, dz - \int_{z_0}^{\delta/2} e^{+2\pi(z_0-z)/\lambda}\, dz\right] \end{aligned} \tag{14}$$

which integrate to

$$\begin{aligned} H_x &= -2\pi I_m \sin(2\pi x_0/\lambda)[2 - e^{-\pi\delta/\lambda}(e^{-2\pi z_0/\lambda} + e^{2\pi z_0/\lambda})] \\ H_z &= 2\pi I_m \cos(2\pi x_0/\lambda) e^{-\pi\delta/\lambda}(e^{-2\pi z_0/\lambda} - e^{2\pi z_0/\lambda}) \\ & \delta/2 \geq z_0 \geq -\delta/2 \end{aligned} \tag{15}$$

The Fields in and Under the Reproducing Head

The idealized reproducing head amounts simply to a semi-infinite block of high permeability material with a flat face spaced a distance d above the surface of the recording medium as shown in Fig. 11.

The problem of most interest is that of finding the x component of magnetic induction, B_x, at any point (x_0, z_0) in the idealized head and integrating this with respect to z_0 to determine the total flux passing through unit width (in the y direction) of a plane $x = x_0$. This plane will then be allowed to move with a velocity v by putting $x_0 = vt$ and the time rate of change of flux will be computed. Except for the effects of eddy currents, self demagnetization, gap loss, etc. (which are treated separately) this rate of change of flux should be proportional to the open circuit reproduced voltage. This is the only result of which direct use will be made but for the sake of completeness all the field components will be evaluated not only in the idealized head but also at all other points.

This problem is completely analogous to the problem of a point charge in front of a semi-infinite dielectric treated by Abraham and Becker[11] and can be solved by use of the method of images.

The Field Inside the High Permeability Head

By analogy with the treatment of Abraham and Becker, the value of B in the high permeability head is computed as though this head filled all space and as though the recording medium were polarized to a value $2\mu/(\mu + 1)$ times the actual value of polarization present. This gives directly from equations (12),

$$\begin{aligned} B_x &= -[2\mu/(\mu + 1)]2\pi I_m \sin(2\pi x_0/\lambda) e^{-2\pi z_0/\lambda}(e^{\pi\delta/\lambda} - e^{-\pi\delta/\lambda}) \\ B_z &= -[2\mu/(\mu + 1)]2\pi I_m \cos(2\pi x_0/\lambda) e^{-2\pi z_0/\lambda}(e^{\pi\delta/\lambda} - e^{-\pi\delta/\lambda}) \\ & z_0 \geq d + \delta/2 \end{aligned} \tag{16}$$

The Field Below the Reproducing Head

Again by analogy with the treatment of Abraham and Becker, the field outside the idealized head is computed as though no head were present. The field is that due to the actual magnetized medium plus the field due to an image of the medium (centered about $z = 2d + \delta$). The intensity of magnetization of the image medium is $-(\mu - 1)/(\mu + 1)$ times the intensity of magnetization of the actual medium.

The field due to the image medium is computed from equations (13) after suitable modification. The required modifications are:

1. Multiply the right hand sides by $-(\mu - 1)/(\mu + 1)$ to take account of the magnitude and sign of the image magnetization as just discussed, and
2. Replace z_0 by $z_0 - (2d + \delta)$ to take account of the position of the image.

[11] M. Abraham and R. Becker, *The Classical Theory of Electricity and Magnetism*. p. 77, Blackie and Son Limited, London, 1937.

This gives the field due to the image plane as

$$H_{xi} = 2\pi I_m \frac{\mu - 1}{\mu + 1} \sin (2\pi x_0/\lambda) e^{2\pi(z_0 - 2d - \delta)/\lambda} (e^{\pi\delta/\lambda} - e^{-\pi\delta/\lambda})$$

$$H_{zi} = -2\pi I_m \frac{\mu - 1}{\mu + 1} \cos (2\pi x_0/\lambda) e^{2\pi(z_0 - 2d - \delta)/\lambda} (e^{\pi\delta/\lambda} - e^{-\pi\delta/\lambda}) \tag{17}$$

$$z_0 \leq d + \delta/2$$

To this must be added the field due to the real medium which is given by equations (12) when $\delta/2 \leq z_0 \leq d + \delta/2$, by equations (15) when $-\delta/2 \leq z_0 \leq \delta/2$, and by equations (13) when $z_0 \leq -\delta/2$.

Performing this addition gives the following results:

Between the head and the recording medium,

$$H_x = -2\pi I_m \sin (2\pi x_0/\lambda) e^{-2\pi z_0/\lambda} (e^{\pi\delta/\lambda} - e^{-\pi\delta/\lambda}) \cdot \left[1 - \frac{\mu - 1}{\mu + 1} e^{-2\pi(2d + \delta - 2z_0)/\lambda}\right]$$

$$H_z = -2\pi I_m \cos (2\pi x_0/\lambda) e^{-2\pi z_0/\lambda} (e^{\pi\delta/\lambda} - e^{-\pi\delta/\lambda}) \cdot \left[1 + \frac{\mu - 1}{\mu + 1} e^{-2\pi(2d + \delta - 2z_0)/\lambda}\right] \tag{18}$$

$$d + \delta/2 \geq z_0 \geq \delta/2$$

Inside the recording medium,

$$H_x = -2\pi I_m \sin (2\pi x_0/\lambda) \cdot \left[2 - e^{-\pi\delta/\lambda}(e^{2\pi z_0/\lambda} + e^{-2\pi z_0/\lambda}) - \frac{\mu - 1}{\mu + 1} e^{-2\pi(2d + \delta - z_0)/\lambda} (e^{\pi\delta/\lambda} - e^{-\pi\delta/\lambda})\right]$$

$$H_z = -2\pi I_m \cos (2\pi x_0/\lambda) \cdot \left[e^{-\pi\delta/\lambda}(e^{2\pi z_0/\lambda} - e^{-2\pi z_0/\lambda}) + \frac{\mu - 1}{\mu + 1} e^{-2\pi(2d + \delta - z_0)/\lambda} (e^{\pi\delta/\lambda} - e^{-\pi\delta/\lambda})\right] \tag{19}$$

$$\delta/2 \geq z_0 \geq \delta/2$$

Below the recording medium,

$$H_x = -2\pi I_m \sin (2\pi x_0/\lambda) e^{2\pi z_0/\lambda} (e^{\pi\delta/\lambda} - e^{-\pi\delta/\lambda}) \cdot \left[1 - \frac{\mu - 1}{\mu + 1} e^{-2\pi(2d + \delta)/\lambda}\right]$$

$$H_z = 2\pi I_m \cos (2\pi x_0/\lambda) e^{2\pi z_0/\lambda} (e^{\pi\delta/\lambda} - e^{-\pi\delta/\lambda}) \cdot \left[1 - \frac{\mu - 1}{\mu + 1} e^{-2\pi(2d + \delta)/\lambda}\right] \tag{20}$$

$$z_0 \leq -\delta/2$$

The Flux per Unit-Width in the Idealized Reproducing Head

The desired flux per unit width is computed from

$$\phi_x = \int_{d+\delta/2}^{\infty} B_x \, dz \tag{21}$$

where B_x is given by equation (16). Performing the indicated integration gives

$$\phi_x = -\frac{2\mu}{\mu + 1} 2\pi\delta I_m \sin (2\pi x_0/\lambda) \left[\frac{1 - e^{-2\pi\delta/\lambda}}{2\pi\delta/\lambda}\right] e^{-2\pi d/\lambda} \tag{22}$$

If the reproducing head moves past the recording medium with a velocity v so that $x_0 = vt$,

$$\frac{d\phi_x}{dt} = -\frac{\mu}{\mu + 1} 4\pi v I_m (1 - e^{-2\pi\delta/\lambda}) e^{-2\pi d/\lambda} \cos (\omega t) \tag{23}$$

where ω is 2π times the reproduced frequency. This is the result for unit width of the reproducing head. For a width of W cm.,

$$\frac{d\phi_x}{dt} = -\frac{\mu}{\mu + 1} 4\pi W v I_m (1 - e^{-2\pi\delta/\lambda}) e^{-2\pi d/\lambda} \cos (\omega t) \tag{24}$$

The Case of Perpendicular Magnetization

Equation (23) was derived for the case of pure longitudinal magnetization as defined by equations (6). It will now be shown that this same result is obtained for $d\phi_x/dt$ if the magnetization is purely perpendicular, that is if

$$I_z = -I_m \cos (2\pi x/\lambda)$$

$$I_x = I_y = 0 \tag{25}$$

In this case the divergence of I is zero except at the surface of the tape and this magnetization is equivalent to a surface distribution of magnetic charge on the top and bottom surfaces of the tape. The magnitude of this charge density is just equal to I_z so that on the top surface of the tape there is a surface density of charge given by

$$\sigma = -I_m \cos (2\pi x/\lambda) \qquad \text{at } z = \delta/2 \tag{26}$$

and on the bottom surface of the tape there is a surface density of charge given by

$$\sigma = I_m \cos (2\pi x/\lambda) \qquad \text{at } z = -\delta/2 \tag{27}$$

Since the permeability of the recording medium is assumed to be unity, this problem reduces to that of finding $d\phi_x/dt$ due to two infinitely thin tapes of the sort to which equation (23) applies. One of these tapes is at $z = \delta/2$ and the other at $z = -\delta/2$.

The problem then is to rewrite equation (23) for a very thin tape and in terms of surface density of charge. As δ approaches zero, equation (23) reduces to

$$\frac{d\phi_x}{dt} = -\frac{\mu}{\mu + 1} 4\pi v I_m (2\pi\delta/\lambda) e^{-2\pi d/\lambda} \cos(\omega t) \tag{28}$$

From equation (7), the volume density of charge in this tape is

$$\rho = -(2\pi I_m/\lambda) \cos(2\pi x/\lambda)$$

But as δ approaches zero, the longitudinally magnetized tape to which equation (28) applies becomes equivalent to a surface distribution of magnetic charge of surface density equal to $\delta\rho$. This amounts, for the thin longitudinally magnetized tape, to a surface charge density of

$$\sigma_1 = -(2\pi\delta/\lambda) \cos(2\pi x/\lambda) \tag{29}$$

But the charge density on the top side of the perpendicularly magnetized tape is given by equation (26). Comparing these two values shows that the surface charge density in the thin longitudinally magnetized tape is just $2\pi\delta/\lambda$ times as great as the surface charge density on top of the perpendicularly magnetized medium. This means that $d\phi_x/dt$ due to the top side of the perpendicularly magnetized tape can be obtained by dividing the right hand side of equation (28) by $2\pi\delta/\lambda$. This gives

$$\frac{d\phi_x}{dt} = -\frac{\mu}{\mu + 1} 4\pi v I_m e^{-2\pi d/\lambda} \cos(\omega t) \tag{30}$$

due to the top side of the tape.

The contribution from the bottom side is obtained from equation (30) by replacing d by $d + \delta$ (since the bottom side is spaced $d + \delta$ from the reproducing head) and changing the sign. Adding these two contributions gives for the total

$$\frac{d\phi_x}{dt} = -\frac{\mu}{\mu + 1} 4\pi v I_m (1 - e^{-2\pi\delta/\lambda}) e^{-2\pi d/\lambda} \cos(\omega t) \tag{31}$$

This is the same as equation (23) and so the desired result has been established.

Note from equations (6) and (24) that in order to get the same result for the perpendicular and longitudinal cases it was necessary to assume a 90-degree phase difference between I_x and I_z. The usual type of recording head lays down a pattern of magnetization which is neither purely perpendicular nor purely longitudinal but the two components are always in phase. This means that the two contributions to $d\phi_x/dt$ add as vectors at 90 degrees. If the intensity of magnetization in the recording medium is held constant while the relative values of perpendicular and longitudinal components are changed, the only effect on the reproduced signal is a change of phase.

APPENDIX II

THE FIELD DUE TO A ROUND WIRE

In Appendix I the field due to a sinusoidally magnetized flat medium such as a tape has been calculated and the rate of change of flux in an

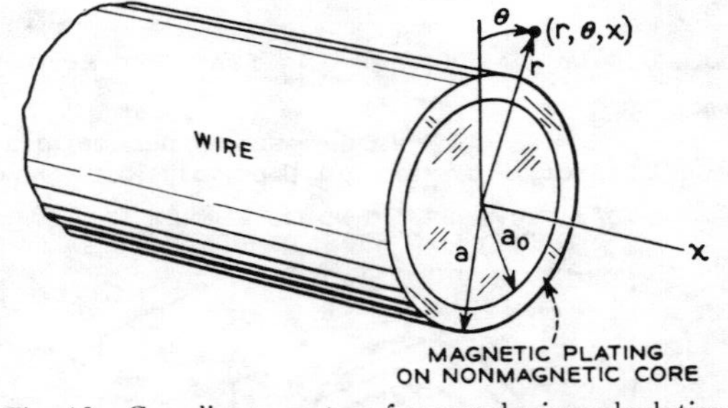

Fig. 12—Coordinate system for round wire calculations.

idealized reproducing head has been evaluated. The analogous calculations for a round wire have also been carried through and it is the purpose of this section to present some of the results. The derivation of these results seems too tedious and long to be presented here.

The Recording Medium

Let the recording medium be a wire, the axis of which lies along the x axis as shown in Fig. 12. Let the radius of the wire be a. To take account of plated wires as well as solid magnetic ones, let the wire have a nonmagnetic core of radius a_0. Let the cylindrical shell between a_0 and a be magnetized sinusoidally in the x direction so that

$$I_x = I_m \operatorname{Sin}(2\pi x/\lambda)$$
$$I_r = I_\theta = 0 \tag{32}$$

By putting $a_0 = 0$ in the expressions which follow it will be possible to obtain the result for a solid magnetic wire.

The Field in Free Space

If no reproducing head is present to disturb the field distribution, the computed field components at a point (x_0, r_0) are

$$\begin{aligned} H_x &= -4\pi I_m \text{ Sin } (2\pi x_0/\lambda)K_0(2\pi r_0/\lambda)[(2\pi a/\lambda)I_1(2\pi a/\lambda) \\ &\qquad - (2\pi a_0/\lambda)I_1(2\pi a_0/\lambda)] \\ H_r &= -4\pi I_m \text{ Cos } (2\pi x_0/\lambda)K_1(2\pi r_0/\lambda)[(2\pi a/\lambda)I_1(2\pi a/\lambda) \\ &\qquad - (2\pi a_0/\lambda)I_1(2\pi a_0/\lambda)] \\ & r_0 \geq a \end{aligned} \tag{33}$$

A discussion and tabulation of the I and K functions can be found in Watson's "Theory of Bessel Functions."[12]

The field due to a solid magnetic wire is obtained by setting $a_0 = 0$ in equations (33). This gives

$$\begin{aligned} H_x &= -4\pi I_m \text{ Sin } (2\pi x_0/\lambda)(2\pi a/\lambda)K_0(2\pi r_0/\lambda)I_1(2\pi a/\lambda) \\ H_r &= -4\pi I_m \text{ Cos } (2\pi x_0/\lambda)(2\pi a/\lambda)K_1(2\pi r_0/\lambda)I_1(2\pi a/\lambda) \\ & r_0 \geq a \end{aligned} \tag{34}$$

The Rate of Change of Flux in an Idealized Head

It has not been possible to carry out the calculations for an idealized head which is a satisfactory approximation to the grooved ring-type head often used in wire recording. The results presented below will apply only to reproducing heads which completely surround the wire. In this case the idealized head is an infinitely large block of core material of permeability μ pierced by a cylindrical hole of radius R in which the wire is centered as shown in Fig. 13. At any point (x_0, r_0) in the permeable medium the components of flux density can be shown to be

$$\begin{aligned} B_x &= \alpha H_x \\ B_r &= \alpha H_r \\ r_0 &\geq R \end{aligned} \tag{35}$$

where

$$\alpha = \frac{\mu}{(\mu - 1)(2\pi R/\lambda)I_0(2\pi R/\lambda)K_1(2\pi R/\lambda) + 1} \tag{36}$$

and H_x and H_r are given by equation (33).

[12] G. N. Watson, "A Treatise on the Theory of Bessel Functions," p. 79, 361, 698, Cambridge Univ. Press, 1922.

The total flux through a plane $x = x_0$ in the permeable medium is obtained by integrating

$$\phi_x = \int_R^\infty B_x(2\pi r)\, dr \tag{37}$$

This gives

$$\phi_x = -2\lambda^2 \alpha I_m \sin (2\pi x_0/\lambda)(2\pi R/\lambda)K_1(2\pi R/\lambda)[(2\pi a/\lambda)I_1(2\pi a/\lambda) - (2\pi a_0/\lambda)I_1(2\pi a_0/\lambda)] \tag{38}$$

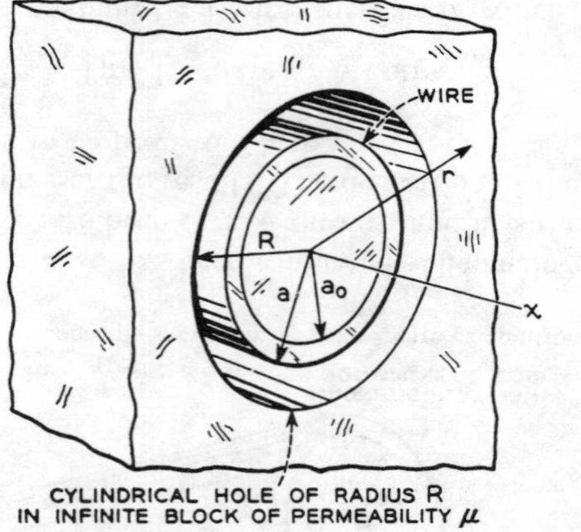

Fig. 13—Round wire surrounded by idealized reproducing head consisting of an infinite block of core material of permeability μ.

If the plane $x = x_0$ moves with a velocity v with respect to the wire so that $x_0 = vt$, then

$$\frac{d\phi_x}{dt} = -4\pi\lambda\alpha v I_m \cos (\omega t)(2\pi R/\lambda)K_1(2\pi R/\lambda)[(2\pi a/\lambda)I_1(2\pi a/\lambda) - (2\pi a_0/\lambda)I_1(2\pi a_0/\lambda)] \tag{39}$$

where $\omega = 2\pi f$ and f is the reproduced frequency.

Special Cases

Equation (39) can be used to compute the response of a simple reproducing head consisting of a single turn of very fine[13] wire as shown in Fig. 14.

In this case $\mu = 1$ and equation (36) shows that $\alpha = 1$. Furthermore if the wire is solid so that $a_0 = 0$, equation (39) reduces to

[13] Unless the diameter of the wire is small compared to the recorded wavelength there will be additional loss not accounted for by 39.

$$\frac{d\phi_x}{dt} = -4\pi\lambda v I_m \cos(\omega t)(2\pi R/\lambda)(2\pi a/\lambda)K_1(2\pi R/\lambda)I_1(2\pi a/\lambda) \quad (40)$$

As λ approaches infinity, $K_1(2\pi R/\lambda)$ approaches $\lambda/2\pi R$ and $I_1(2\pi a/\lambda)$ approaches $\pi a/\lambda$ so that, for very long wavelengths, equation (40) reduces to

$$\frac{d\phi_x}{dt} = -4\pi I_m v(2\pi/\lambda)(\pi a^2)\cos(\omega t) \quad (41)$$

This relation (which could have been derived in a much simpler manner) should be useful for the experimental determination of the intensity of magnetization, I_m.

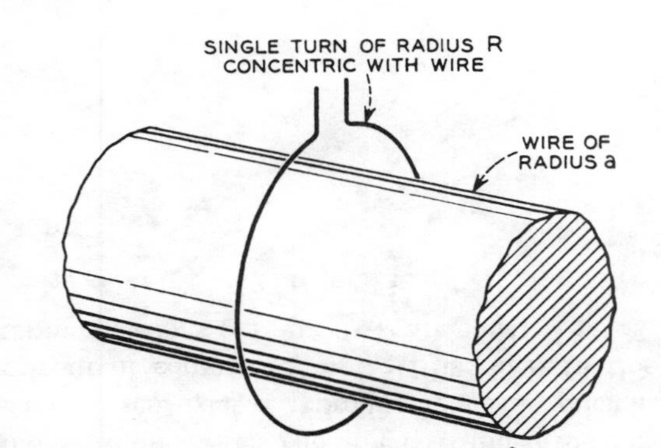

Fig. 14—Elementary reproducing head consisting of a single turn of wire.

Another case of some interest corresponds to a high permeability reproducing head which surrounds the wire. In this case μ is great enough so that equation (36) reduces to

$$\alpha = \frac{1}{(2\pi R/\lambda)I_0(2\pi R/\lambda)K_1(2\pi R/\lambda)} \quad (42)$$

If it is assumed, in addition, that the wire is solid so that $a_0 = 0$, then equations (42) and (39) give

$$\frac{d\phi_x}{dt} = -4\pi\lambda v I_m \cos(\omega t)(2\pi a/\lambda)I_1(2\pi a/\lambda)/I_0(2\pi R/\lambda) \quad (43)$$

Comparison between Round Wire and Flat Medium Response

It is interesting to compare equation (43) with equation (24) to see how the response characteristic of a round wire compares with that of a tape. Assuming $\mu \gg 1$, the appropriate equation for the flat medium is

$$\frac{d\phi_x}{dt} = -4\pi W v I_m \cos(\omega t)(1 - e^{-(2\pi\delta/\lambda)})e^{-2\pi d/\lambda} \quad (44)$$

To compare equations (43) and (44), consider first the limiting cases of very long and very short wavelength. As λ approaches infinity they reduce to

$$\frac{d\phi_x}{dt} = -\pi a^2(8\pi^2 v/\lambda)I_m \cos(\omega t) \quad (45)$$

for the wire and

$$\frac{d\phi_x}{dt} = -\delta W(8\pi^2 v/\lambda)I_m \cos(\omega t) \quad (46)$$

for the tape.

These two expressions are identical provided the cross section area of the wire, (πa^2), is the same as that of the recorded track on the tape, (δW).

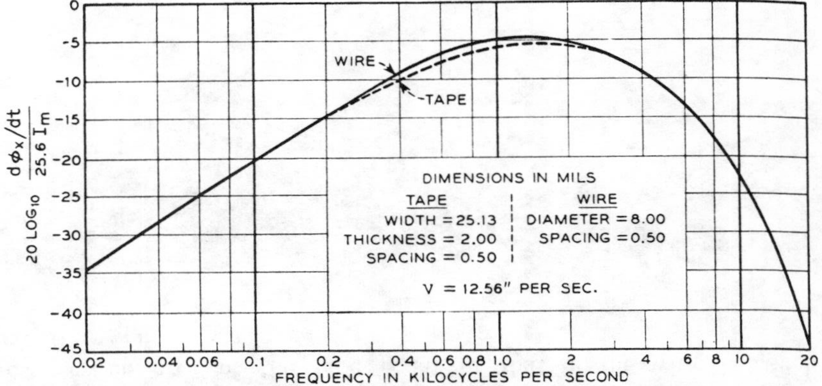

Fig. 15—Computed responses for wire and tape showing that the responses are very similar provided the dimensions of the wire and tape are suitably related.

As λ approaches zero, the two expressions reduce to

$$\frac{d\phi_x}{dt} = -4\pi v(2\pi a)\sqrt{R/a}\,e^{-2\pi(R-a)/\lambda}I_m \cos(\omega t) \quad (47)$$

for the wire, and

$$\frac{d\phi_x}{dt} = -4\pi v(W)e^{-2\pi d/\lambda}I_m \cos(\omega t) \quad (48)$$

for the tape.

Suppose that the reproducing head makes reasonably good contact with the wire so that $\sqrt{R/a} \doteq 1$. In this case equations (47) and (48) are identical provided the circumference of the wire, $(2\pi a)$, is the same as the width of the recorded track on the tape and provided also the effective spacing between reproducing head and medium is the same in the two cases, $(d = R - a)$. In both cases only a thin surface layer of the recording medium is effective in producing high frequency response. For this reason the high-frequency response is independent of the "thickness" of the medium and is directly proportional to the "width" of the track provided $2\pi a$ is interpreted as the width of track on a wire.

The comparisons which have just been made indicate that if the dimensions of a wire and of a tape are suitably related, the two media should give identical response at very high and very low frequencies provided they are equally magnetized. The dimensional requirements are

$$\begin{aligned} \pi a^2 &= \delta W, \\ 2\pi a &= W, \text{ and} \\ R - a &= d \end{aligned} \tag{49}$$

In order to show how the computed responses compare at intermediate frequencies, numerical calculations have been made for a special case in which equations (49) are satisfied. The case chosen is that of a wire 8 mils in diameter moving at a velocity of 12.56 in./sec. past a reproducing head which is effectively one half mil out of contact with the wire ($R - a = 0.5(10)^{-3}$ in.). By equations (49) the corresponding flat medium is a tape which is 2 mils thick and 25.13 mils wide. The tape is assumed to be moving with a velocity of 12.56 in./sec. past a reproducing head which is also effectively one half mil out of contact ($d = 0.5(10)^{-3}$ in.). In this case the numerical constants in equations (43) and (44) are equal. That is,

$$8\pi^2 av = 4\pi W v = 25.6 \text{ cm.}^2/\text{sec.}$$

and the quantity to be computed and compared for the two cases is

$$20 \log_{10} \frac{d\phi_x/dt}{25.6 I_m}$$

The computed curves are shown in Fig. 15 from which it can be seen that they coincide at low and high frequencies as planned and that furthermore they differ by no more than 1.5 db in the middle range of frequencies.

As has been pointed out, equation (43) applies only to the unusual case in which the head completely surrounds the wire. The similarity of the two curves of Fig. 15, however, suggests a way of computing approximately the response to be expected when the wire head makes contact with only a part of the circumference of the wire. It suggests that the computation be carried out as though the wire were a flat medium of suitably chosen dimensions. In order to make the high frequency end come out right one would expect that W in equation (44) should be given a value equal to the length of the arc of contact between the wire and the head. To make the low frequency end come out right, δ must be given a value which makes the cross section area of the tape equal to that of the wire, i.e. such that $\delta W = \pi a^2$.

Dependence of Reproducing Gap Null on Head Geometry

DENNIS A. LINDHOLM

Abstract—The spatial frequency response of reproduce heads with infinite depth, but finite pole length, is determined to very good accuracy by superposition of conformal map solutions for simpler geometries. This approach yields closed form approximations of the frequency response which are accurate to better than 7% for any head length-to-gap length ratio and any spatial frequency. In particular, the effect of this ratio on the location of the first gap null in the frequency response spectrum is explored for narrow pole length heads. The results lead to design considerations for extending the useful frequency range of thin film reproduce heads.

INTRODUCTION

With the advent of batch fabricated, thin film integrated recording head construction, the dimension of a pole tip may approach the length of the gap, resulting in a total head length which is only a few times larger than the gap length.

The reproducing flux response of such a head will consist of a series of oscillations associated with the total length dimension b and another series associated with the gap length g. The ratio b/g has a strong influence on fluctuations about zero output when measured against reciprocal wavelength or spatial frequency $1/\lambda$.

Even assuming infinite depth, infinite track width, and infinite permeability, recent computations [1]–[4] of the frequency response of finite pole-length reproduce heads are not very useful since they do not result in accurate formulae in closed forms. In principle, any two-dimensional infinite permeability head can be examined by means of conformal mapping techniques. However, the mapping function for an infinite depth head with arbitrary b/g is sufficiently complicated [5] that it is unlikely an exact closed form solution for the frequency response can be found.

In this work, a number of simpler conformal map solutions are combined to obtain useful approximate formulae valid for any ratio of $b/g \geqslant 1$.

FREQUENCY RESPONSE DEFINED

The track width, head depth, and permeability are all taken to be infinite, but the gap g and head length b may be finite.

Manuscript received February 24, 1975; revised July 21, 1975. Paper 16.7 was presented at the 1975 INTERMAG Conference, London, England, April 14–17.

The author is with the Ampex Corporation, Redwood City, Calif. 94063.

The pole length is $p = (b - g)/2$. The general geometry is shown in Fig. 4(d). The usual head coordinate system is chosen, in which the y axis is the vertical axis of symmetry and the x axis coincides with the top surface of the head.

It is well known from the reciprocity theorem that (in the frequency domain) the playback response of a magnetic recording system consists of a product of frequency or wavelength dependent terms, each of which is associated with some component or mechanism in the system. In particular, the reproduce head is described by a characteristic function [4] which gives the harmonic response [6] or frequency response of the head.

For the cases considered here, the frequency response function of the head S is the Fourier transform of the negative gradient of the magnetic scalar potential U taken along the x axis. Since the potential is an odd function of x, the gradient is even and S is written

$$S = -\int_{-\infty}^{\infty} (\partial U/\partial x)_{y=0} \cos(2\pi x/\lambda)\, dx \qquad (1)$$

so that S depends on the wavelength λ or the spatial frequency $1/\lambda$. In (1), S implicitly depends on the geometry. Therefore, in order to explicitly display the geometrical parameters g and b, the notation $S(g, b; 1/\lambda)$ is employed in what follows.

The function in (1) is always defined for the head in contact with the media ($y = 0$). For out-of-contact playback ($y > 0$), the response is multiplied by the usual spacing loss factor $\exp(-2\pi y/\lambda)$.

Special cases of the general geometry are given in the next section wherein the potential U at a point $z = x + iy$ in the complex z plane is represented by

$$U = -(1/2\pi)\, \mathrm{Re}\, \{i \ln (w - 1)\} \qquad (2)$$

The complex variable w in (2) relates the z plane to the transformed w plane by means of a conformal mapping function of the form

$$z = z(w, h). \qquad (3)$$

In (3), h is a scale factor which causes $z = -h$ to map into $w = 0$.

Boundary conditions are chosen without lack of generality so that $U = 0.5$ on the left pole and -0.5 on the right one. Potentials at an arbitrary point z may be difficult to obtain, since the back transformation in (3) to obtain w from z usually has to be performed numerically before (2) can be used.

Reprinted from *IEEE Trans. Magn.*, vol. MAG-11, pp. 1692–1696, Nov. 1975.

SPECIAL CASES

1. *Gapped, Infinite Pole Head*

In this geometry $g \neq 0$ but $b = \infty$. The mapping function given by Westmijze [7] is

$$z = (ih/2\pi)\ [2\sqrt{w} + \ln(\sqrt{w} - 1) - \ln(\sqrt{w} + 1)],$$

$$h = g/2. \tag{4}$$

The magnetic potential from (2) and (4) across the top of the left pole is plotted in the inset of Fig. 1.

The frequency response function has been calculated over a large frequency range in rather coarse steps by Westmijze and in fine detail up to and slightly beyond the first gap null by Wang [8]. Neither author suggests a closed form solution for low frequencies, but Westmijze gives a two-term formula for high frequencies.

Simple closed form approximations were found by curve fitting to the tables of Westmijze and Wang. The frequency response depends only on the ratio g/λ and is denoted by

$$S(g, \infty; 1/\lambda) = S_1(g/\lambda) \tag{5a}$$

where

$$S_1(f) = \begin{cases} \text{sinc}\,(1.11f), & f < 0.5 \\ 0.326f^{-2/3}\sin\pi(f + 1/6) + 0.0552f^{-4/3} \\ \quad \cdot \sin\pi(f - 1/6), & f \geq 0.5 \end{cases} \tag{5b}$$

with

$$\text{sinc}\,(f) = \sin(\pi f)/(\pi f). \tag{6}$$

Equations (5b) and (6) are function definitions with f being the "dummy" independent variable.

The "gap loss" function $S_1(g/\lambda)$ is characterized by a series of zeroes in the frequency response, i.e., "gap nulls." The first gap null occurs at $g/\lambda_1 = 0.8795$.

Excerpts of the Westmijze/Wang tables are plotted as discrete points along with (5) in Fig. 1. The agreement is better than 1%. The first coefficient for $f = g/\lambda \geq 0.5$ in (5b) is correctly given by Westmijze, but his second published coefficient is much too high, resulting in poor agreement with his own tabulated values. The value given in (5b) for this coefficient reproduces his table with negligible deviation. Equation (5), then, is a convenient representation in closed algebraic forms of the exact solution for this geometry.

Fan [9] gives an expression for the low frequency end of the response function based on Fourier series analysis. He obtains the first gap null correctly but the amplitude of S is in poor agreement with the tables in the range $0.8 < g/\lambda < 0.95$. Possibly this discrepancy arises from the inadequacy of a Fourier series for the magnetic field near the gap edge, where the field is discontinuous.

Karlqvist [10] also obtained the potential distribution for this case by a conformal map. However, in order to obtain a simple analytic expression for the fringe field, $H(x, y) = -\nabla U(x, y)$, he took the potential $U(x, 0)$ to vary linearly across the gap. Actually, the potential at the surface departs as much as 17% from linearity. The Karlqvist approximation results in a fair approximation for the fringe field, provided $y > g/2$, but it leads to a head frequency response varying as $\text{sinc}\,(g/\lambda) = (\pi g/\lambda)^{-1} \sin(\pi g/\lambda)$. This function predicts the first gap null at $g/\lambda_1 = 1$, which is too high by 14% as compared with the exact solution or its closed form equivalent, (5).

2. *Parallel Plate Head*

For this case, the pole is infinitely thin ($p = 0$) so that the gap length and total head length are identical ($g = b$). Equipotentials for this geometry have been plotted by Booth [11]. The mapping function is

$$z = (ih/\pi)\ [w + \ln(w - 1)], \qquad h = g/2 = b/2 \tag{7}$$

with respect to the potential in (2). The magnetic potential across the top left pole of this head as computed from (2) and (7) appears in the inset of Fig. 2.

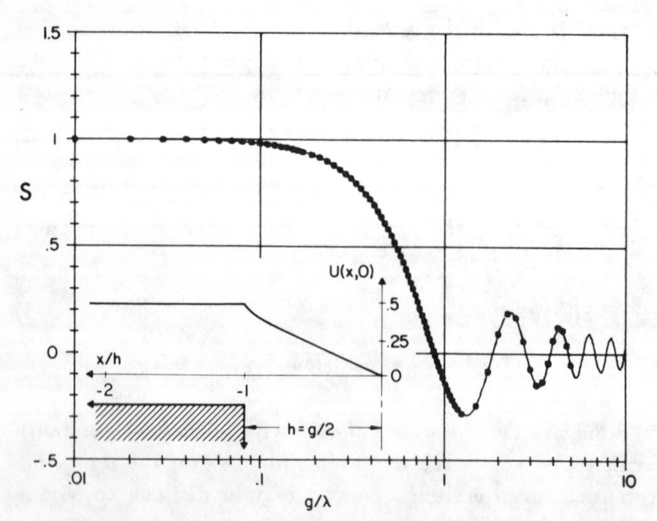

Fig. 1. Case 1 (gapped, infinite pole head) frequency response S (points from [7] and [8]) and magnetic potential distribution U.

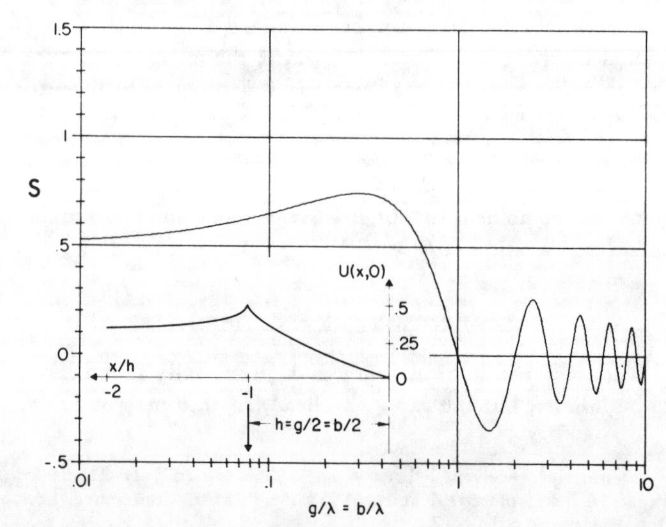

Fig. 2. Case 2 (parallel plate head) frequency response S and magnetic potential distribution U.

It is instructive to compare the potential drop *inside* the gap of this head, Fig. 2, with that of the previous head, Fig. 1. The parallel plate head has a sharper corner at $x = -g/2$ and hence, the potential drops a bit faster than in the previous case, yet the general shapes of the two curves inside the gap are quite close. In fact, a numerical comparison shows that the maximum percent difference between the potentials of Figs. 1 and 2 inside the gap, $|x| \leqslant g/2$, is less than 7%.

Hence, as the pole length shrinks from infinity to zero, keeping the same fixed gap length, the potential in the gap changes by less than 7% between these two extremes. This indicates that the potential drop inside the gap is relatively insensitive to the length of the pole beyond it.

The frequency response function has not been previously published. It is given exactly in terms of the gamma function Γ and depends only on $g/\lambda = b/\lambda$.

$$S(g, g; 1/\lambda) = S_2(g/\lambda) \tag{8a}$$

where [12]

$$S_2(f) = (1/2)\,\mathrm{sinc}\,(f)\,(e/f)^f\,\Gamma(f+1) \tag{8b}$$

with $e = 2.718 \cdots$. The frequency response function (8) is plotted in Fig. 2. The first gap null occurs at $g/\lambda = b/\lambda = 1$ exactly.

3. *Gapless, Finite Pole Head*

In this instance, the gap length is zero ($g = 0$), but the pole length p and the head length b are finite, with $p = b$. Duinker and Geurst [6] give the mapping function for the potential in (2) as

$$z = (2ih/\pi)\,[\sqrt{w}\sqrt{w-1} + \ln(\sqrt{w} + \sqrt{w-1})],$$
$$h = b/2. \tag{9}$$

The magnetic potential across the top left pole for this geometry, from (2) and (9), is plotted in the inset of Fig. 3.

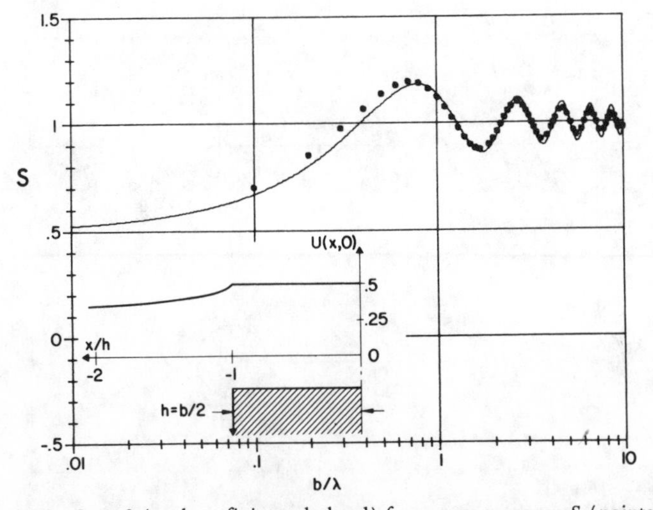

Fig. 3. Case 3 (gapless, finite pole head) frequency response S (points from [6]) and magnetic potential distribution U.

It is interesting to compare the potential drop *beyond* the pole of the parallel plate head, Fig. 2, with that of this head, Fig. 3. The sharper corner on the parallel plate head at the pole edge again results in a stronger decrease of the potential. Yet, numerical computations show that the maximum percent difference between the potentials of Figs. 2 and 3 beyond the pole, $|x| \geqslant b/2$, is again less than 7%.

The interpretation here is that for a fixed total length b the pole length p has very little effect on the potential *beyond* the pole edge, since in going from $p = 0$ to $p = b$ the potential distribution is changed by less than 7%.

The frequency response function, $S(0, b; 1/\lambda)$ has been computed numerically and is given in tabular form by Duinker and Geurst; Westmijze [7] gives an asymptotic formula. No closed form solutions which cover the entire range of spatial frequencies have been published.

$S(0, b; 1/\lambda)$ exhibits a series of low frequency oscillations above the base line, often referred to as "head bumps." Excerpts of the Duinker and Geurst table are plotted as discrete points in Fig. 3. The curve for S is an approximation by superposition to be discussed in the next section.

SUPERPOSITION

The special cases studied have revealed that the potential drop *inside* the gap, $|x| \leqslant g/2$, and *beyond* the pole, $|x| \geqslant b/2$, are both very insensitive to the pole length. Of course, the x-distances must be scaled to the characteristic length h appropriate to each region. To a very good approximation, then, the more general head, Fig. 4(d), can be formed from superpositions of the special cases studied independently above. It is perhaps conceptually easier to visualize the superposition by Method I in Fig. 4.

The gap potential is expressed by a Case 1 head in (a); the potential of a Case 3 head (b) is added to it. After (b), the entire left side of the head is too high by a magnetic potential of 0.5 and the right too low by -0.5. This is corrected by subtracting out the potential of the excess Case 1 head in (c). Finally, in (d) by Method I, the left pole remains at a magnetic potential of exactly 0.5 and the right at exactly -0.5. Moreover, it has the gap potential of a gapped, infinite pole head (Case 1) and the potential beyond the pole of a gapless finite pole head (Case 3). However, since no closed form solutions exist for the frequency response of a

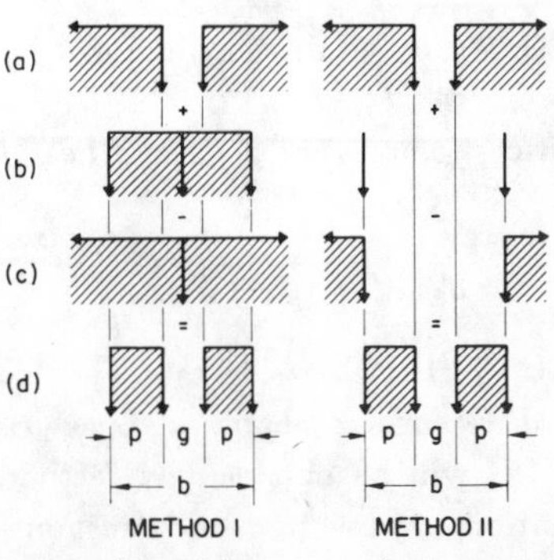

Fig. 4. Superposition of special cases to obtain general head geometry.

Case 3 head, it is more convenient to perform the superposition by Method II, Fig. 4.

Starting with the gap potential in (a), a parallel plate head having the desired head length is added in (b). After step (b) the potential beyond the pole, $|x| \geq b/2$, has the character of a Case 2 head, but the potential is too high by 0.5 for $x \leq -b/2$ and too low by -0.5 for $x \geq b/2$. For $|x| \leq b/2$, there is a mixture of gap- and constant-potentials. By subtracting the potential of a Case 1 head with $g = b$ in (c), the desired result is obtained. In Method II, the gap potential is a compromise between Case 1 and Case 2 heads, the potentials on the poles are not quite constant, and the potential beyond the pole of Case 2 type.

The comparison of the two methods of superposition is illustrated in Fig. 5 for the case $b/g = 2$. The maximum difference in potential anywhere along the x axis is less than 7%.

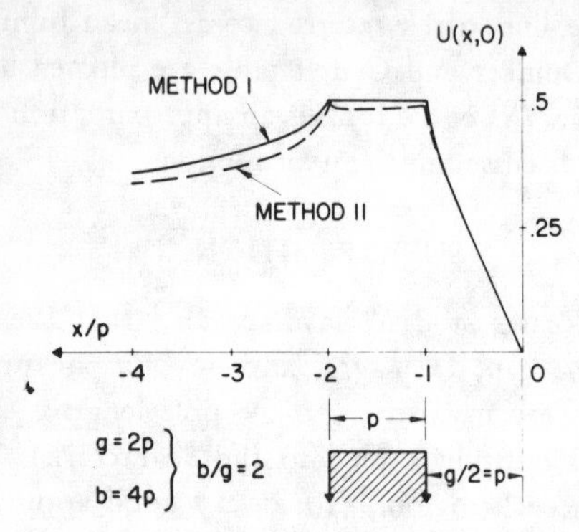

Fig. 5. Potentials by two method of superposition for $b/g = 2$.

It follows that for any value of the ratio b/g that superposition of independent solutions by either method will produce a potential accurate to better than 7% as compared with the extremely complicated double-cornered conformal map of the Fig. 4(d) geometry [5].

In the frequency domain, the frequency response function by superposition using Method II is given by

$$S(g, b; 1/\lambda) = S(g, \infty; 1/\lambda) + S(b, b; 1/\lambda) - S(b, \infty; 1/\lambda) \qquad (10)$$

which gives in closed form from (5) and (8)

$$S(g, b; 1/\lambda) = S_1(g/\lambda) + S_2(b/\lambda) - S_1(b/\lambda). \qquad (11)$$

Note in (11) that exact results are obtained in the cases $b = g$ and $b = \infty$.

In the specific geometry of a Case 3 head, $g = 0$, $b \neq 0$, we find

$$S(0, b; 1/\lambda) = 1 + S_2(b/\lambda) - S_1(b/\lambda). \qquad (12)$$

Equation (12) is plotted alongside tabulated data in Fig. 3. The maximum difference is about 6.2% at $b/\lambda = 0.3$. The same result as (11) will be obtained by Method I using (12) as an approximation for the frequency response of a Case 3 head.

RESULTS

The development of (11) by means of closed form functions allows both "head bump" and "gap loss" effects to be computed easily for any value of $b \geq g$. In particular, the location of the first gap null can be studied as a function of the ratio of head length-to-gap length.

Equation (11) can be normalized to the gap length ($g \neq 0$) so that

$$S(g, b; 1/\lambda) = S(1, b/g; g/\lambda). \qquad (13)$$

The first gap null occurs at $g/\lambda = g/\lambda_1 > 0$, where S first passes through zero:

$$S(1, b/g; g/\lambda_1) = 0. \qquad (14)$$

For a given b/g, g/λ_1 is varied until the first nonzero root of (14) is found.

The results are shown in Fig. 6. As b/g is increased from unity to infinity, the maximum value of g/λ_1 occurs when $g/\lambda_1 = 1$ and $b/g = 1$ (parallel plate head); the minimum lies near $g/\lambda_1 = 0.77$, where $b/g = 2.25$. Characteristically g/λ_1 oscillates about the line $g/\lambda_1 = 0.8795$ with decreasing amplitude as b/g approaches infinity (gapped, infinite pole head).

There can be abrupt and even discontinuous jumps from a low value of g/λ_1 to a much higher one with only a small increase in b/g as shown in Fig. 6. This can be explained with the aid of Fig. 7 where the frequency response spectrum for $b/g = 4.9$ is plotted.

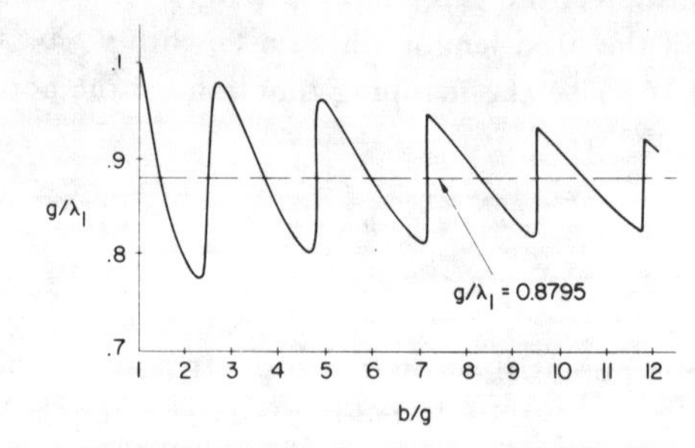

Fig. 6. Location of first gap null as function of length-to-gap ratio.

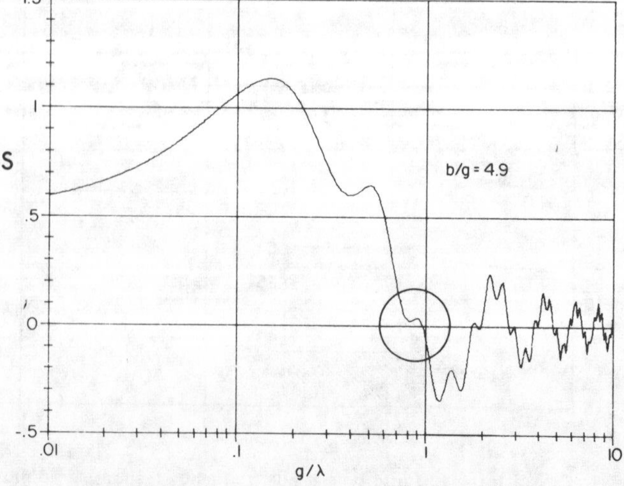

Fig. 7. Frequency response spectrum for $b/g = 4.9$.

At low values of b/g the series of ripples in the head bump spectrum associated with the head length b interfere with the zero crossings of the gap loss spectrum associated with the gap length g. The circled portion of Fig. 7 illustrates the point.

For $b/g < 4.9$ the head bump ripple would be further to the right and lower, shifting the zero crossing to a lower value of g/λ_1. When b/g approaches 4.9, the ripple shifts to the left and upward (as shown), resulting in a discontinuous jump of g/λ_1 to a higher value.

To avoid erratic high frequency response from narrow pole tip heads, it is advisable to choose b/g well removed from a region of a jump discontinuity in Fig. 6. As shown in Fig. 7, narrow pole tip heads afford a degree of enhancement at high frequencies below the first gap null. By controlling the ratio b/g as well as the head depth (limited to an infinite value here) a variety of frequency responses are possible [12].

CONCLUSION

The solution of the parallel plate head problem led heuristically to the discovery that the potentials in the gap and beyond the pole are nearly independent of pole length. The superposition utilized here works by using this result and maintaining each pole at (approximately) constant potential. The method used in this paper does not appear to be related to a more rigorous technique [13] applicable to a multiply connected region (e.g., multiple head structures).

REFERENCES

[1] M. J. Stavn, "Reading characteristics of a finite length pole piece recording head," *IEEE Trans. Mag.*, Vol. MAG-9, 698 (1973).

[2] J. C. van Lier, "Simple algorithm to compute magnetic field problems, in particular to integrated magnetic heads," *IERE Conference Proc.*, No. 26, 285 (1973).

[3] J. C. Mallinson, "On recording head field theory," *IEEE Trans. Mag.*, Vol. MAG-10, 773 (1974).

[4] Y. Ichiyama, "Effects of head shape on reproduction characteristics in high density magnetic recording," Paper MR-73-32, Tech. Group Meeting for Mag. Recording, Inst. Elec. and Commun. Engineers, Tokyo, Japan (1974).

[5] I. Elabd, "Study of the field around magnetic heads of finite length," *IEEE Trans. Audio*, Vol. AU-11, 21 (1963).

[6] S. Duinker and J. A. Geurst, "Long wavelength response of magnetic reproducing heads with rounded outer edges," *Phillips Res. Rep.*, Vol. 19, 1 (1964).

[7] W. K. Westmijze, "Studies in Magnetic Recording," *Phillips Res. Rep.*, Vol. 8, 161 (1953).

[8] H. S. C. Wang, "Gap loss function and determination of certain critical parameters in magnetic data recording instruments and storage systems," *Rev. Sci. Instrum.*, Vol. 37, 1124 (1966).

[9] G. J. Fan, "Study of the playback process of a magnetic ring head," *IBM J. Res. Developm.*, Vol. 5, 321 (1961).

[10] O. Karlqvist, "Calculation of the magnetic field in the ferromagnetic layer of a magnetic drum," *Trans. Roy. Inst. Technol.* (Stockholm), No. 86, 1 (1954).

[11] A. D. Booth, "On two problems in potential theory and their application to the design of magnetic recording heads for digital computers," *Brit. J. Appl. Phys.*, Vol. 3, 307 (1952).

[12] D. A. Lindholm, to be published.

[13] S. G. Mikhlin, *Integral Equations*, Pergamon Press (1957), Ch. VI.

Spectral Response from Perpendicular Media with Gapped Head and Underlayer

DAN S. BLOOMBERG, MEMBER, IEEE

Abstract—A general spectral formulation is presented, in the spirit of the Karlquist approximation, for the field from a gapped head with a permeable underlayer. For an infinite permeability underlayer, of particular interest in perpendicular recording, the fields and readback response are derived for sine-wave and square-wave magnetization, as well as for an isolated transition. The sine-wave response is given in a simple form, allowing interpretation in terms of a reduced pole density from the medium at the underlayer. The square-wave response is given as a Fourier series of sine-wave amplitudes. Approximate analytical functions are constructed for the response to an isolated transition.

I. Introduction

IT has been demonstrated both experimentally [1] and theoretically [2] that media with perpendicular anisotropy can be written and read with gapped heads, which have poor perpendicular field gradients. In this paper we consider a recording geometry with improved readback resolution; namely, the use of a highly permeable underlayer with a gapped head. This geometry gives sharp field gradients $\partial H_y/\partial x$ at the gap center, and acts to stabilize magnetostatically the recorded perpendicular transitions. However this configuration is not useful for *in-plane* recording for two reasons: the underlayer acts to impede writing by reducing the in-plane fields, and it reduces readback by acting as a keeper to shunt the flux from recorded transitions.

In the first investigation of this readback geometry (1954), Karlquist [3] used a Schwartz-Christoffel (S-C) transformation to find the fields from an ideal gapped head with an infinitely permeable underlayer. For a head-to-underlayer distance h, gap-length G, and deep-gap field H_g, the perpendicular field far from the gap is $H_y \simeq \pm H_g(G/2h)$. When h and G are comparable, most of the change in field occurs within the internal $-G/2 < x < G/2$. Karlquist also derived approximate expressions for the fields when the underlayer has finite permeability and thickness. These equations are not easily amenable to computation of the spectral (sine-wave and square-wave) response of the head.

Lopez *et al.* [4] numerically inverted the S-C transformation to find the perpendicular field, and then used reciprocity and superposition to compute the readback for a series of equally spaced transitions. This "square-wave" response is appropriate for perpendicular recording because the transition length is small due to magnetostatic coupling between adjacent bits.

Yamamori *et al.* [5] claimed that the presence of an infinite permeability underlayer in the perpendicular mode eliminates the magnetic poles at the interface of medium with underlayer, and hence the medium "appears" to be infinitely thick, having only poles at the top surface. This result is only true at zero separation, where the head field H_y varies inversely with medium thickness t, resulting in an $H_y t$ product that is independent of t. For any finite separation, it will be shown that readback has a thickness dependence, which is more pronounced for longer wavelength. Instead of using reciprocity, Yamamori *et al.* calculate the response using a method of images due to Wallace [6], which assumes a gapless head. Consequently their results are inaccurate in the short wavelength limit.

Toda *et al.* [7] recently reported measurements on double-layer perpendicular media, using all four combinations of probe and gapped heads for writing and reading. The best amplitude and resolution were obtained using a probe head to write and a gapped head for readback.

The content of this paper is as follows. First the vector potential for an ordinary gapped head is derived as a Fourier integral. Then the two-dimensional time-dependent diffusion equation is solved in the static (nonconducting) limit for the vector potential A_z, by specification of boundary conditions. The result is a generalization of Karlquist's results to an underlayer (or, in fact, a series of underlayers) with arbitrary permeability. The fields are derived as a spatial Fourier integral, so the sine-wave response (by reciprocity) is simply contained within the integrand. Reciprocity is used to find the sine-wave response for a perpendicularly magnetized medium (with finite thickness) and an infinite permeability underlayer. The effect of the underlayer is given geometric interpretations in terms of both poles at the medium-underlayer interface and image currents. The square-wave response is then derived as a superposition of sine-wave amplitudes, rather than a superposition of amplitudes from isolated transitions (which themselves must be found either by integration over all wave vectors of the sine-wave response or by inversion of a conformal transformation). The output from an isolated transition and two analytical approximations to the waveform are given in the Appendix.

These considerations only apply to readback. It is assumed that the underlayer is sufficiently thin that it is saturated at a low write current, enabling the write process to be adequately

Manuscript received August 24, 1982; revised February 2, 1983.

The author was with the Xerox Palo Alto Research Center, Advanced Development Laboratory, El Segundo, CA. He is now with Xerox Palo Alto Research Center, Mechanical Engineering Sciences, 141 Webber Ave., N. Tarrytown, NY 10591.

Reprinted from *IEEE Trans. Magn.*, vol. MAG-19, pp. 1493–1502, July 1983.

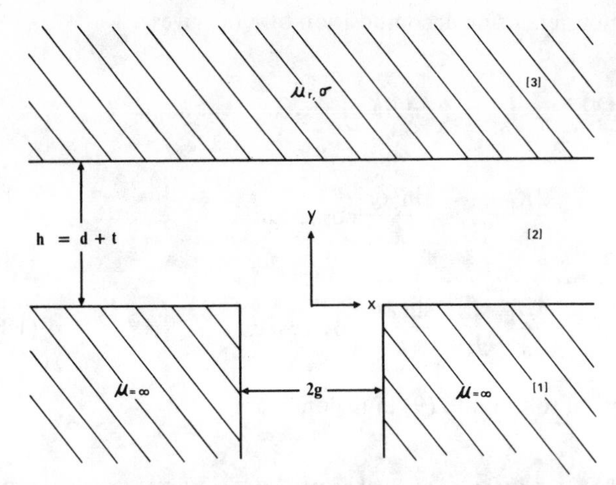

Fig. 1. Recording geometry for an ideal gapped head (region 1) with a permeable, conductive underlayer (region 3). Gap-length is $G = 2g$. Medium (not shown) occupies thickness t in region 2 adjacent to the underlayer. Head-medium spacing is d.

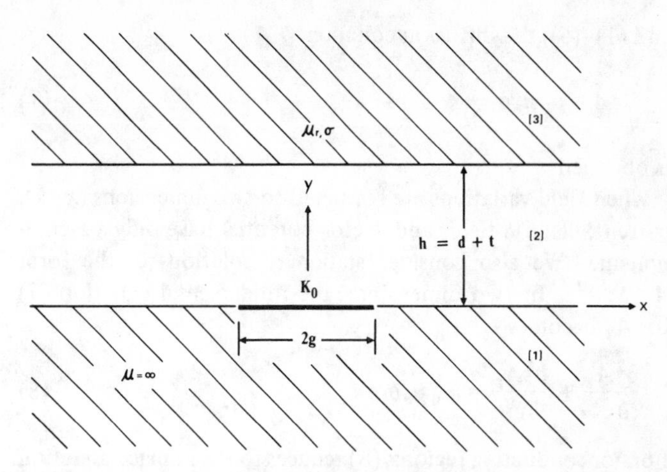

Fig. 2. Recording geometry with current density K_0 (amp/m) across the gap.

described by ignoring its presence. However, the underlayer may not be saturated during writing by a *finite pole-length* gapped head. This interesting situation will not be discussed here.

II. Recording Geometry

Fig. 1 shows the recording geometry of the gapped head with underlayer. The infinite permeability pole-pieces are at two magnetic equipotentials, differing by $2gH_g$, where H_g is the deep-gap field (at $y = -\infty$) and g is half the gap-length. Region 3 is the underlayer, taken to have conductivity σ, relative permeability μ_r, and infinite thickness. The recording medium (not shown) is in region 2.

The problem is formulated in terms of current sources J_z and the vector potential component A_z. The Karlquist approximation for the field from a gapped head is generated by a uniform current sheet at $y = 0$ between the pole-pieces. The fields in region 2 are calculated from both this current source and its image currents in the underlayer and the infinite permeability region $y < 0$.

For the purposes of specifying boundary conditions between regions 1 and 2, region 1 is taken to have infinite permeability in the gap as well as the pole-pieces. These boundary conditions are only enforced on those contributions to A_z arising from *induced* currents in regions 1 and 3; the *driving* current at $y = 0$ automatically satisfies the equipotential requirements for $|x| > g$ at $y = 0$. Because the image currents are much farther from region 2 than the driving current at $y = 0$, the error introduced by simplifying the boundary at $y = 0$ is small compared to the initial error in the Karlquist approximation.

The model recording geometry described above is given in Fig. 2, with the understanding that the boundary condition at $y = 0$ ($H_x = 0$) does not apply to the part of A_z arising directly from the current K_0.

For perpendicular recording, we are most interested in an underlayer with infinite permeability. If the underlayer is sufficiently thin (<1 μm), it will be saturated at low write current and will be unimportant in the writing process. This condition is essential; otherwise, the large perpendicular field will erase transitions previously recorded at the gap and final writing will occur from the outside corner of the (wide) trailing pole face!

For the purpose of analyzing readback, the underlayer is considered to be part of the writing "head." The alternative view, that as part of the medium it effectively doubles the medium thickness, leads to incorrect results because it neglects the infinite series of image poles due to multiple reflections between the head and underlayer.

III. Fundamental Equations

In the MKS system, the appropriate Maxwell equations for $\vec{B}$ and $\vec{E}$ are

$$\vec{\nabla} \times \vec{B} = \mu_0 \mu_r \vec{J} \tag{1}$$

and

$$\vec{\nabla} \times \vec{E} = -\frac{\partial \vec{B}}{\partial t}, \tag{2}$$

where $\mu_0 = 4\pi \times 10^{-7}$ is the permeability of free space and μ_r is the relative permeability.

The displacement current has been neglected in (1) because the signal propagation time across the head structure is much less than the inverse of useful data-rate frequencies. The fields are related to the potentials $\vec{A}$ and Φ by

$$\vec{B} = \vec{\nabla} \times \vec{A}, \quad (\vec{\nabla} \cdot \vec{A} = 0) \tag{3}$$

and

$$\vec{E} = -\frac{\partial \vec{A}}{\partial t} - \vec{\nabla}\Phi. \tag{4}$$

Because the only field sources are currents, $\Phi = 0$ and

$$\vec{J} = \sigma \vec{E} = -\sigma \frac{\partial \vec{A}}{\partial t}. \tag{5}$$

The conductivity σ has dimensions $(\Omega\text{-m})^{-1}$. Using the vector identify

$$\vec{\nabla} \times (\vec{\nabla} \times \vec{A}) = \vec{\nabla}(\vec{\nabla} \cdot \vec{A}) - \nabla^2 \vec{A} \tag{6}$$

and (1)-(5), the diffusion equation

$$\nabla^2 \vec{A} = \mu_0 \mu_r \sigma \frac{\partial \vec{A}}{\partial t} \tag{7}$$

is obtained.

When field variations are restricted to two dimensions (x-y), current, electric field, and vector potential have only a z-component. We also consider stationary solutions of the form $A \sim e^{i\omega t}$. In two dimensions, the fundamental equation (7) for A_z becomes

$$\frac{\partial^2 A_z}{\partial x^2} + \frac{\partial^2 A_z}{\partial y^2} = i\mu_0 \mu_r \sigma \omega A_z. \tag{8}$$

For nonconducting regions, (8) reduces to the Laplace equation

$$\nabla^2 A_z = 0. \tag{9}$$

Equation (8) can be written

$$\frac{\partial^2 A_z}{\partial x^2} + \frac{\partial^2 A_z}{\partial y^2} = \alpha^2 A_z \tag{10}$$

with

$$\alpha^2 = i\mu_0 \mu_r \sigma\omega = 2i/\delta^2. \tag{11}$$

By this definition,

$$\alpha = (1+i)/\delta \tag{12}$$

where the skin depth δ is given by

$$\delta = \sqrt{\frac{2}{\mu_0 \mu_r \sigma \omega}}. \tag{13}$$

The fields H_x and H_y, obtained from (3) and the relation

$$\vec{B} = \mu_0 \mu_r \vec{H} \tag{14}$$

are

$$H_x = \frac{1}{\mu_0 \mu_r} \frac{\partial A_z}{\partial y} \tag{15a}$$

$$H_y = -\frac{1}{\mu_0 \mu_r} \frac{\partial A_z}{\partial x}. \tag{15b}$$

IV. Potential Arising from the Current Sheet Alone

The method in the following two sections is based on Fourier cosine analysis and is similar to that given by Stoll [8]. Referring to Fig. 2, it is first necessary to find the relation between the gap-field H_g at $y = 0_+$ and the current density K_0 at $y = 0$.

The field H arising adjacent to the current sheet at $y = 0$ with current density K_0 (amp/m) is found by integrating

$$\vec{\nabla} \times \vec{H} = \vec{K}_0 \delta(y) \tag{16}$$

to give the Karlquist boundary condition for fields in region 2:

$$H_x(x, 0) = \frac{1}{\mu_0} \frac{\partial A_2^0}{\partial y}\bigg|_{y=0} = -\frac{K(x)}{2} = \begin{cases} -K_0/2, & |x| < g \\ 0, & |x| > g. \end{cases} \tag{17}$$

The Fourier cosine decomposition of $K(x)$ gives

$$\begin{aligned} K(x) &= \frac{1}{\pi} \int_0^\infty \cos kx\, dk \int_{-\infty}^{\infty} K(x') \cos kx'\, dx' \\ &= \frac{2K_0}{\pi} \int_0^\infty \frac{\sin kg}{k} \cos kx\, dk \\ &= \frac{4H_g}{\pi} \int_0^\infty \frac{\sin kg}{k} \cos kx\, dk. \end{aligned} \tag{18}$$

A general solution to (9) in region 2 is

$$A_2^0 = \int_0^\infty C(k) \cos kx\, e^{-ky}\, dk. \tag{19}$$

The coefficient function $C(k)$ is found, by substitution of (18) and (19) into the boundary condition (17), to be

$$C(k) = \mu_0 \frac{2H_g}{\pi} \frac{\sin kg}{k^2}. \tag{20}$$

Hence the potential due to the head (without underlayer) is given by

$$A_2^0 = \mu_0 \frac{2H_g}{\pi} \int_0^\infty \frac{\sin kg}{k^2} \cos kx\, e^{-ky}\, dk. \tag{21}$$

Using (15), the familiar Karlquist equations for the fields result:

$$H_x = -\frac{H_g}{\pi}\left[\tan^{-1}\frac{g+x}{y} + \tan^{-1}\frac{g-x}{y}\right] \tag{22a}$$

$$H_y = \frac{H_g}{2\pi} \ln\left[\frac{y^2 + (g+x)^2}{y^2 + (g-x)^2}\right]. \tag{22b}$$

V. General Solution With Underlayer

Solutions to (10) in region 3, which are spatially periodic in x, are of the form

$$\cos kx\, e^{\pm \gamma y} \tag{23}$$

where, by substitution,

$$\gamma = \sqrt{k^2 + \alpha^2} = \sqrt{k^2 + i\frac{2}{\delta^2}}. \tag{24}$$

At the frequencies and wavelengths of interest in magnetic recording $\delta > 1/k$. (The situation is a little more complicated, in that when μ_r is large, H_y in region 2 becomes insensitive to the (small) H_y in region 3, even if the contribution there from eddy currents is substantial.) Thus the underlayer can be considered to be nonconducting, and only the static solutions with $\gamma = k$ need to be found.

Because the form of the potential A_2 in region 2 is known, we can immediately write down the general solution in regions 2 and 3:

$$A_2 = A_2^0 + A_2^1 + A_2^2$$

$$= \mu_0 \frac{2H_g}{\pi} \int_0^\infty \frac{\sin kg}{k^2} \cos kx \cdot [e^{-ky} + C_1 e^{-ky} + C_2 e^{-k(h-y)}]\, dk \tag{25}$$

$$A_3 = \mu_0 \frac{2H_g}{\pi} \int_0^\infty \frac{\sin kg}{k^2} \cos kx\, C_3 e^{-k(y-h)}\, dk. \tag{26}$$

The coefficient (functions) C_1 and C_2 give the solutions which decay into region 2 from the $y = 0$ and $y = h$ interfaces, respectively. There are three boundary conditions:

$$\frac{\partial}{\partial y} (A_2^1 + A_2^2)\Big|_{y=0} = 0 \tag{27}$$

$$A_2(y = h) = A_3(y = h) \tag{28}$$

$$\frac{\partial A_2}{\partial y}\Big|_{y=h} = \frac{1}{\mu_r} \frac{\partial A_3}{\partial y}\Big|_{y=h}. \tag{29}$$

Equation (27) expresses the requirement, described in Section II, that H_x from the image sources is zero at $y = 0$. Equation (29) gives the continuity of H_x at the interface between regions 2 and 3.

Substituting (25) and (26) into (27), (28), and (29), the coefficients are found to be

$$C_1 = C_2 e^{-kh} \tag{30}$$

$$C_2 = \frac{(\mu_r - 1) e^{-kh}}{\mu_r(1 - e^{-2kh}) + (1 + e^{-2kh})} \tag{31}$$

and

$$C_3 = \frac{2\mu_r e^{-kh}}{\mu_r(1 - e^{-2kh}) + (1 + e^{-2kh})}. \tag{32}$$

The potentials are then

$$A_2 = \mu_0 \frac{2H_g}{\pi} \int_0^\infty \frac{\sin kg}{k^2} \cos kx \cdot \left[e^{-ky} + \frac{(\mu_r - 1) e^{-kh} \cosh ky}{\mu_r \sinh kh + \cosh kh}\right] dk \tag{33}$$

$$A_3 = \mu_0 \frac{2H_g}{\pi} \int_0^\infty \frac{\sin kg}{k^2} \cos kx \frac{\mu_r e^{-k(y-h)}}{\mu_r \sinh kh + \cosh kh}\, dk. \tag{34}$$

From (15) and (33), the fields in region 2 are

$$H_x = \frac{2H_g}{\pi} \int_0^\infty \frac{\sin kg}{k} \cos kx \cdot \left[-e^{-ky} + \frac{(\mu_r - 1) e^{-kh} \sinh ky}{\mu_r \sinh kh + \cosh kh}\right] dk \tag{35a}$$

$$H_y = \frac{2H_g}{\pi} \int_0^\infty \frac{\sin kg}{k} \sin kx \cdot \left[e^{-ky} + \frac{(\mu_r - 1) e^{-kh} \cosh ky}{\mu_r \sinh kh + \cosh kh}\right] dk. \tag{35b}$$

The first term in (35a) and (35b) is produced directly by the current sheet. The second term, which vanishes when $\mu_r = 1$, represents an infinite set of image currents reflected in regions 1 and 3. All image currents are reflected across the boundaries in the same direction ($\pm z$). By reciprocity, the integrands (excluding the sin kx and cos kx term) give the sine-wave response of the head in readback. The (sin kg)/k term is the gap-loss function.

For an infinite permeability underlayer, the fields simplify further:

$$H_x = \frac{2H_g}{\pi} \int_0^\infty \frac{\sin kg}{k} \cos kx \cdot \left(-e^{-ky} + \frac{e^{-kh} \sinh ky}{\sinh kh}\right) dk \tag{36a}$$

$$H_y = \frac{2H_g}{\pi} \int_0^\infty \frac{\sin kg}{k} \sin kx \cdot \left(e^{-ky} + \frac{e^{-kh} \cosh ky}{\sinh kh}\right) dk. \tag{36b}$$

The image terms decrease H_x in (36a) whereas they increase H_y in (36b), and have a geometric interpretation. For example, in (36b), expansion of the denominator in

$$(e^{-k(2h-y)} + e^{-k(2h+y)})/(1 - e^{-2kh}) \tag{37}$$

gives two infinite sets of image currents, in the underlayer and head, respectively. The currents in each series are of the same polarity and are separated by $2h$.

Representative in-plane and perpendicular fields are given in Figs. 3 and 4, where the fields are calculated at $y = G/2$ and the underlayer distance $h = G$. (In this and following graphs, all distances are normalized to the full gap $G = 2g$.) Note that the perpendicular field with underlayer in Fig. 4 reaches an asymptotic value for $|x| > G$ of $H_y = \pm H_g(G/2h)$, as is expected from a simple magnetic circuit argument. The slope of H_y at $x = 0$ is greater with the underlayer, suggesting a better frequency response.

The method used here can easily be extended to multiple underlayers. From each interface there are two solutions which decay exponentially away from the interface. Hence, two unknown functions $C_i(k)$ are introduced for each interface. There are also two boundary conditions, requiring continuity of both A_z and H_x.

At the $y = h$ interface, $H_x = 0$ and the integral for H_y can

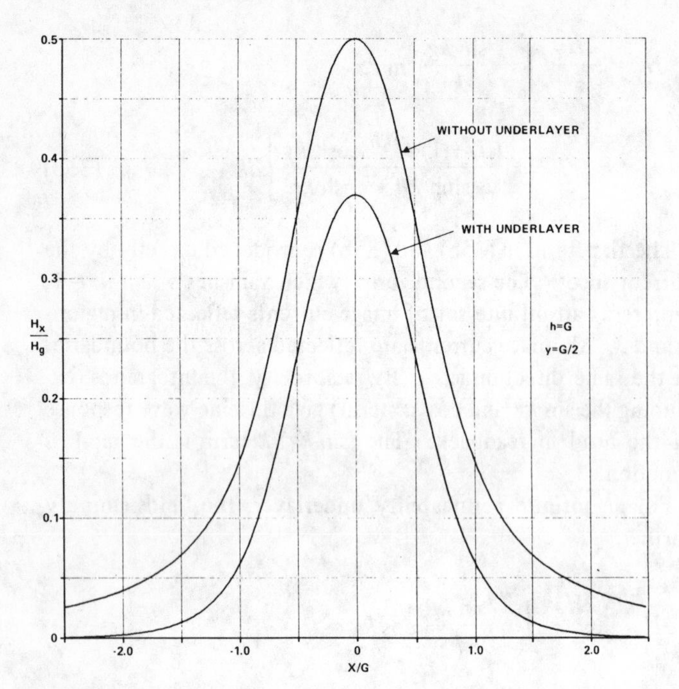

Fig. 3. In-plane field from a gapped head, with and without underlayer. Head-underlayer distance $h = G = 2g$. Field is evaluated at $y = G/2$.

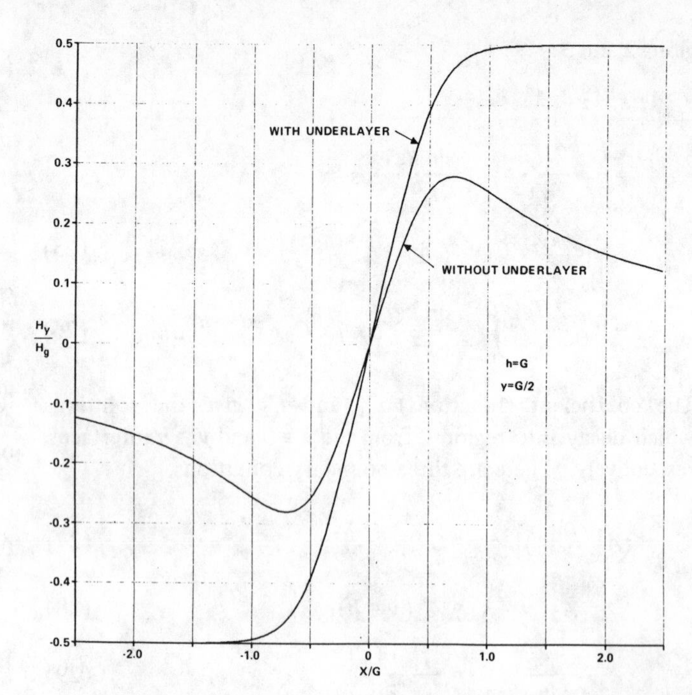

Fig. 4. Perpendicular field from a gapped head, with and without underlayer. Head-underlayer distance $h = G$. Field is evaluated at $y = G/2$.

be performed using [9, (4.114.1)]:

$$H_y(y=h) = \frac{2H_g}{\pi}\int_0^{\infty} \frac{\sin kg}{k} \sin kx\, e^{-kh}(1+\coth kh)\, dk$$

$$= \frac{2H_g}{\pi} \tanh^{-1}\left[\tanh\frac{\pi g}{2h}\tanh\frac{\pi x}{2h}\right]$$

$$= \frac{H_g}{\pi}\ln\left[\frac{\cosh(\pi(g+x)/2h)}{\cosh(\pi(g-x)/2h)}\right]. \quad (38)$$

This special case was derived by Karlquist [3, eq. (23)].

VI. Spectral Response

Assume that a perpendicularly magnetized medium of thickness $t < h$ fills the space adjacent to the underlayer. The head-medium distance d is then given by $d = h - t$. The magnetization M_y is assumed to be independent of y. By reciprocity, as an isolated step transition passes the gap, the induced voltage in the head is proportional to the head field (36b), averaged over the medium thickness. Because perpendicular transitions are expected to be sharp, even at high density, the square-wave response can be found by readback superposition from an infinite string of such isolated transitions. However because the integrand in (36b) gives the sine-wave response, a more direct method is to Fourier decompose the square-wave magnetization and use superposition on the sine-wave components.

A. Sine-Wave Response

The reciprocity convolution integral [10] gives the output voltage when the head is centered at x as

$$\epsilon(x) = wvN\int_{-\infty}^{\infty} dx' \int_d^{d+t} dy' \bar{H}_y(x-x', y') \cdot \frac{\partial}{\partial x'}(\mu_0 M_y(x')) \quad (39)$$

with

w track width (m)
v head velocity (m/s)
N number of read turns

and

$\mu_0 M_y$ magnetization (Wb/m²).

$\bar{H}_y$ is the head field (per ampere) and is found from (36) by using $H_g = \eta I/2g$:

$$\bar{H}_y = \frac{\eta}{\pi g}\int_0^{\infty}\frac{\sin kg}{k}\sin kx \cdot \left(e^{-ky} + \frac{e^{-kh}\cosh ky}{\sinh kh}\right) dk. \quad (40)$$

The readback efficiency η should be close to 1.

Carrying out the integration over y', with $h = d + t$,

$$\epsilon(x) = \frac{wvN\eta\mu_0}{\pi g}\int_{-\infty}^{\infty} dx' \int_0^{\infty} dk' \frac{\sin k'g}{k'^2} e^{-k'd} \cdot \left[(1-e^{-k't}) + \left(1 - \frac{\sinh k'd}{\sinh k'(d+t)}\right) e^{-k't}\right] \cdot \sin k'(x-x') \frac{\partial M_y(x')}{\partial x'}. \quad (41)$$

Using a sinusoidal magnetization

$$M_y(x') = M_r \cos kx' \tag{42}$$

and integrating first over dx' and then over dk',

$$\epsilon(x) = \epsilon(k) \cos kx \tag{43}$$

where the sine-wave response $\epsilon(k)$ is

$$\frac{\epsilon(k)}{R} = \frac{\sin kg}{kg} e^{-kd} \left[(1 - e^{-kt}) + \left(1 - \frac{\sinh kd}{\sinh k(d+t)}\right) e^{-kt} \right]$$

$$= \frac{\sin kg}{kg} \frac{\sinh kt}{\sinh k(d+t)} \tag{44}$$

with

$$R = wvN\eta\mu_0 M_r.$$

Consider the intermediate expression in (44). The first two factors are the gap-loss and conventional spacing-loss functions. The factor

$$1 - e^{-kt} \tag{45}$$

gives the thickness loss for the Karlquist head (without underlayer). It has a simple interpretation in terms of the magnetic charge at near and far surfaces of the medium; namely, the induced voltage from a sinusoidal surface charge is proportional to e^{-ky}, where y is the distance from pole-face to surface charge. The charge on the back surface at $y = d + t$ has the same magnitude as (but opposite polarity to) the charge on the near surface at $y = d$. For thin media (or long wavelength), the response is proportional to kt. The second factor

$$\left(1 - \frac{\sinh kd}{\sinh k(d+t)}\right) e^{-kt} \tag{46}$$

gives the added response due to image currents in head and underlayer. For long wavelengths it is proportional to $t/(d + t)$, a finite dc response. Adding the thickness factors (45) and (46), one obtains

$$1 - \frac{\sinh kd}{\sinh k(d+t)} e^{-kt} \tag{47}$$

as the thickness loss factor with underlayer. There is an effective fractional charge at the underlayer

$$\rho = \frac{\sinh kd}{\sinh k(d+t)} \tag{48}$$

which is zero only for zero spacing ($d = 0$). For long wavelengths it has an asymptotic value

$$\rho \to \frac{d}{d+t} \qquad (\lambda \to \infty). \tag{49}$$

The thickness dependence of the spectral response for a gapped head with underlayer (47) and without (45) are shown in Figs. 5 and 6, for six different wavelengths, with $d/G = 0.5$. The output at all wavelengths is greater for the underlayer medium.

The wavelength dependence of the sine-wave response (from (44)), up to the first gap null $\lambda = G$, is shown in Fig. 7. The spacing and thickness are both one-half the gap-length. It should be noted that for a head with underlayer, the spacing loss is no longer a simple exponential that decays at 54.6 dB/wavelength. Instead, the spacing loss and thickness loss are combined. At long wavelength, the combined loss is $t/(d + t)$,

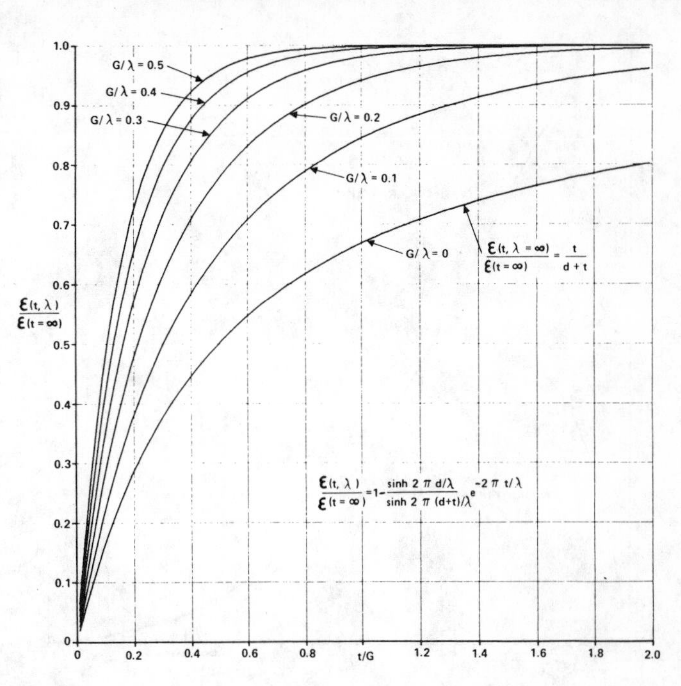

Fig. 5. Medium thickness dependence of readback from gapped head *with* underlayer. Medium has thickness t, at a separation d from the head. Distance from head to underlayer is $d + t$. Graph gives output from perpendicular sine-wave magnetization as a function of t, normalized to $t = \infty$, for six wave-lengths. $d = 0.5G$.

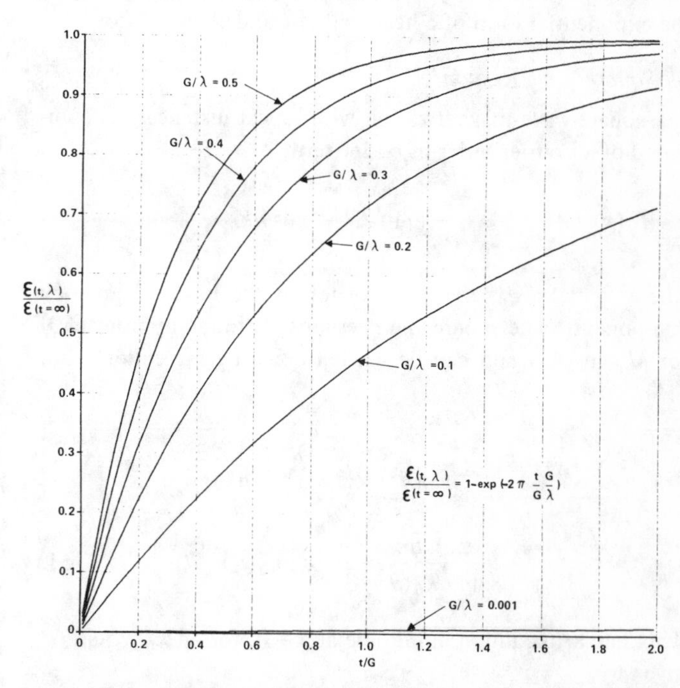

Fig. 6. Medium thickness dependence of readback from gapped head *without* underlayer. Medium thickness is t, at a separation d from the head. Graph gives output from perpendicular sine-wave magnetization as a function of t, normalized to $t = \infty$, for six wavelengths. $d = 0.5G$.

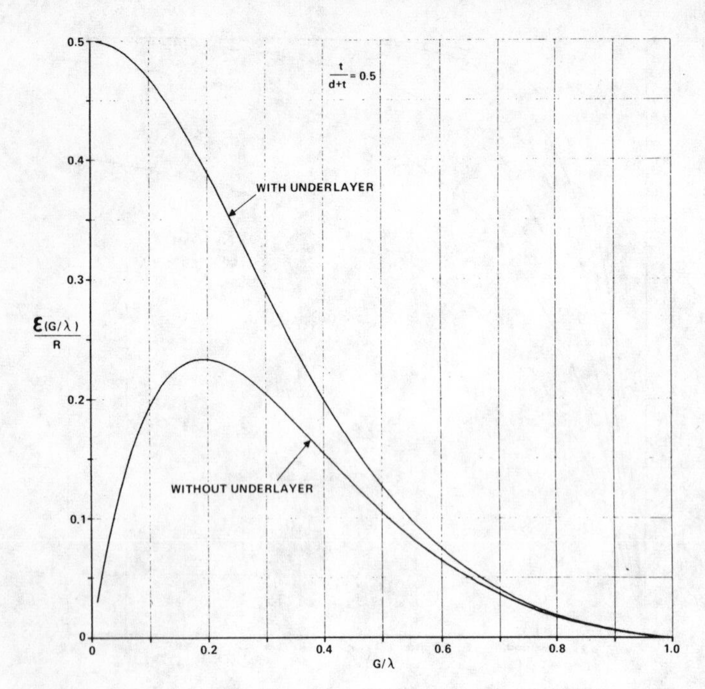

Fig. 7. Sine-wave response of gapped head with and without underlayer, up to the first gap-null at $G = \lambda$. The spacing $d = G/2$, thickness $t = G/2$.

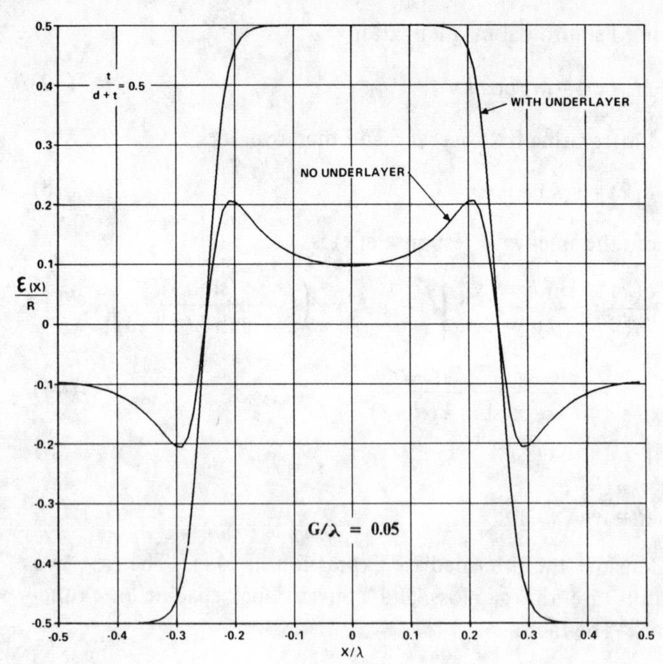

Fig. 8. Output waveforms for square-wave magnetization, with and without underlayer $d = t = G/2$, $G/\lambda = 0.05$ (λ is repeat distance of magnetization).

the dc response. At very short wavelength, the loss approaches the exponential form of a head without underlayer.

B. Square-Wave Response

A square-wave magnetization with repeat distance λ is composed of a Fourier series of cosine terms

$$M_y(x) = M_r \sum_{n \text{ odd}} \frac{4}{\pi n} (-1)^{(n-1)/2} \cos nk_0 x \tag{50}$$

where $k_0 = 2\pi/\lambda$ is the wave vector of the lowest frequency component. The square-wave response is found by using (50) for M_y in (41) and can be written down by inspection from (44):

$$\epsilon^{SQ}(x, k_0) = R \sum_{n \text{ odd}} \frac{4}{\pi n} (-1)^{(n-1)/2} \frac{\sin nk_0 g}{nk_0 g} \cdot \frac{\sinh nk_0 t}{\sinh nk_0(d+t)} \cos nk_0 x. \tag{51}$$

This has a maximum amplitude at $x = 0$, for all k_0, d, and t, given by

$$\epsilon^{SQ}(k_0) = R \sum_{n \text{ odd}} \frac{4}{n\pi} (-1)^{(n-1)/2} \frac{\sin nk_0 g}{nk_0 g} \frac{\sinh nk_0 t}{\sinh nk_0(d+t)}. \tag{52}$$

Using the identity

$$\frac{\pi}{4} = 1 - \frac{1}{3} + \frac{1}{5} - \frac{1}{7} + \cdots, \tag{53}$$

it is seen that the dc response of the head with underlayer is the same for both square-wave and sine-wave magnetization:

$$\frac{\epsilon^{SQ}(k_0 \to 0)}{R} = \frac{t}{d+t}. \tag{54}$$

The output from a head without underlayer, ϵ_{NU}^{SQ}, can be found from (50) and the first term in (44):

$$\epsilon_{NU}^{SQ}(x, k_0) = R \sum_{n \text{ odd}} \frac{4}{\pi n} (-1)^{(n-1)/2} \frac{\sin nk_0 g}{nk_0 g} e^{-nk_0 d} \cdot (1 - e^{-nk_0 t}) \cos nk_0 x. \tag{55}$$

In contrast with (51), it has a maximum at $x = 0$ only for short wavelengths. Output waveforms from square-wave magnetization, for heads with and without underlayer, at $d = G/2$ and $t = G/2$, are given in Figs. 8-10 for $G/\lambda = 0.05$, 0.1, and 0.2, respectively. Because the output from the head with underlayer does not dip between transitions, as does the conventional head at long wavelengths, it may be more immune to false zero-crossings and hence support an encoding method requiring a larger bandwidth. The maximum square-wave amplitude versus $(\text{wavelength})^{-1}$ is given in Fig. 11. For the head with underlayer, (52) is used. For the head without underlayer, the dc response is finite, as expected from an isolated transition. The dashed curve gives $\epsilon_{NU}^{SQ}(k_0, x = 0)$, showing the drop in output between transitions at long wavelength. The square-wave response is greater at all wavelengths than the corresponding sine-wave response (Fig. 7).

The output for an isolated transition is given in the Appendix. Simple expressions are derived for the special cases of zero gap-length and zero separation. For the general case, two analytical approximations to the waveform $\epsilon(x)$ are presented.

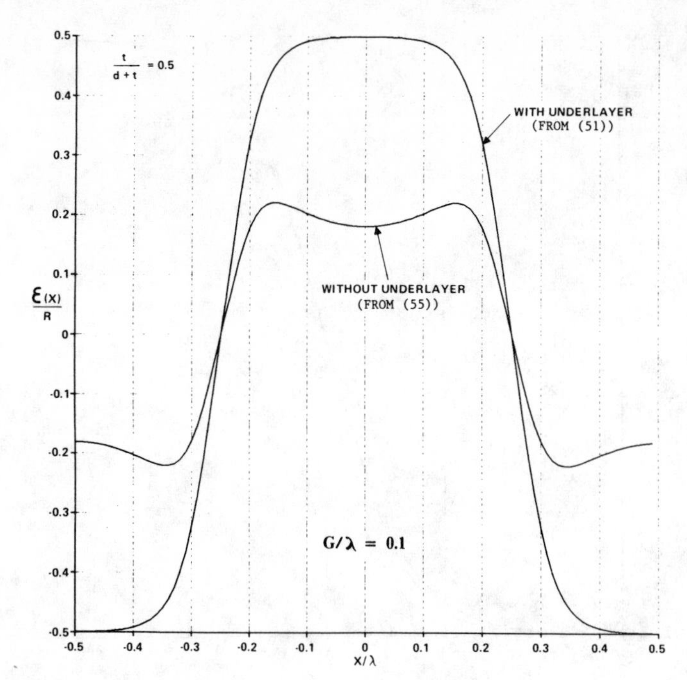

Fig. 9. Output waveforms for square-wave magnetization, with and without underlayer. $d = t = G/2$, $G/\lambda = 0.1$.

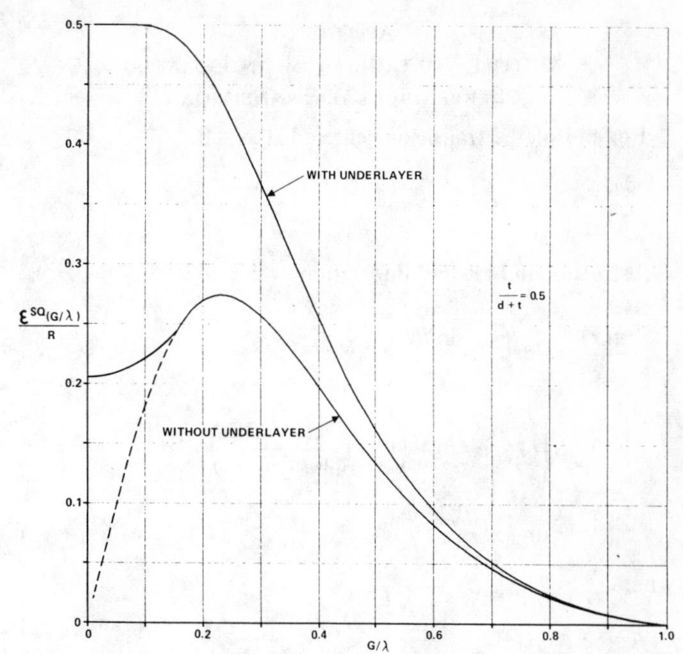

Fig. 11. Maximum square-wave amplitude of gapped head with and without underlayer, up to the first gap-null at $G = \lambda$. The spacing is $d = G/2$ and the medium thickness is $t = G/2$. Maximum response of the head with underlayer(52) always occurs at $x = 0$. For $G/\lambda > 0.17$ the maximum response of the head without underlayer occurs at $x = 0$. Dashed line for $G/\lambda < 0.17$ gives the output at $x = 0$ (between transitions).

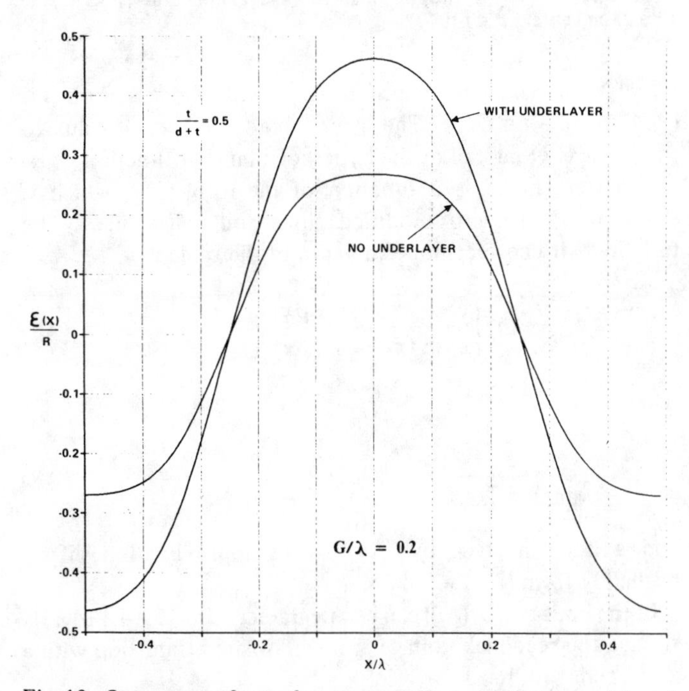

Fig. 10. Output waveforms for square-wave magnetization, with and without underlayer. $d = t = G/2$, $G/\lambda = 0.2$.

VII. Summary

A solution of Maxwell's equations in terms of spectral components has been given for a gapped recording head with a permeable underlayer. The effects due to eddy currents are unimportant at the wavelengths and frequencies of interest in digital recording. An infinite permeability underlayer is shown to produce fields that give superior amplitude and resolution in readback from a perpendicularly oriented medium, due to a sharper field gradient $\partial H_y/\partial x$ at the gap center. Using reciprocity, the sine-wave response from a medium of finite thickness is given in a simple form, allowing interpretation in terms of readback from a reduced pole density (of the medium) at the underlayer. This pole density is the fraction

$$\frac{\sinh kd}{\sinh k(d+t)} \tag{56}$$

of the pole density at the top surface. The spacing loss is not given by Wallace's exponential factor alone. Instead, the spacing and thickness losses are combined into the factor

$$\frac{\sinh kt}{\sinh k(d+t)}. \tag{57}$$

The square-wave response, appropriate to perpendicular recording, is given as a superposition of sine-waves. The square-wave response is shown to be larger than the sine-wave for finite wavelength. At low frequency, the waveform from a square-wave magnetization *without* underlayer shows a dip which, because of possible false zero crossings, could limit the bandwidth of the encoding method. There is no such dip in the response *with* an underlayer. Either of two approximate analytical functions, which are given for the readback from an isolated transition, can be superimposed to determine readback from an arbitrary set of transitions.

If the underlayer is not saturated during writing, as may occur with a thick underlayer and a thin-film gapped head, the written transitions will probably be formed behind the trailing pole. Operation of a recording system under these conditions merits inquiry.

APPENDIX
OUTPUT VOLTAGE FROM AN ISOLATED PERPENDICULAR TRANSITION

For an isolated transition centered at $x' = 0$,

$$\frac{\partial M_y}{\partial x'} = 2M_r \delta(x'). \tag{A1}$$

Substitution in (41) and integration over x' gives

$$\frac{\epsilon(x)}{R} = \frac{2}{\pi} \int_0^\infty \frac{\sin kg}{kg} e^{-kd} \cdot \left[(1 - e^{-kt}) + \left(1 - \frac{\sinh kd}{\sinh k(d+t)}\right) e^{-kt}\right] \cdot \sin kx \frac{dk}{k} \tag{A2}$$

where

$$R = wvN\eta\mu_0 M_r.$$

The first term in brackets gives the output from a gapped head without underlayer and can be integrated directly, using [9, (3.947.2)]:

$$\frac{\epsilon_{NU}(x)}{R} = f(d) - f(d+t)$$

where

$$f(u) = \frac{1}{\pi g}\left[g \tan^{-1} \frac{2ux}{u^2 + g^2 - x^2} + x \tan^{-1} \frac{2ug}{u^2 + x^2 - g^2} + \frac{u}{2} \ln \frac{u^2 + (g-x)^2}{u^2 + (g+x)^2}\right]. \tag{A3}$$

The voltage $\epsilon(x)/R$ from a gapped head with underlayer (A2) can be approximated by noting that 1) the asymptotic value ($x \to \pm\infty$) is $t/(d+t)$ and 2) the slope at $x = 0$ is proportional to the slope of H_y/H_g. Differentiating (36b) and using [9, (3.893.1) and (4.132.2)],

$$s \equiv 2g \frac{\partial}{\partial x}\left(\frac{H_y}{H_g}\right)\Bigg|_{x=0} = \frac{G}{h} \frac{\sinh \pi G/2h}{\cosh \pi G/2h - \cos \pi y/h}. \tag{A4}$$

Because the asymptotic value of H_y/H_g is $G/2h$, the slope s' of $\epsilon(x)/R$ at $x = 0$ is related to s by

$$s' \cong \left(\frac{t}{d+t}\right)\left(\frac{2h}{G}\right) s(y = d + t/2) = \frac{2t}{G} s(y = d + t/2). \tag{A5}$$

The slope is evaluated at the medium center ($y = d + t/2$) and $h = d + t$.

Choosing a hyperbolic tangent to approximate $\epsilon(x)/R$, and matching both asymptotic value $t/(d+t)$ at $x = \infty$ and slope s' at $x = 0$,

$$\frac{\epsilon(x)}{R} \cong \frac{t}{d+t} \tanh\left[\frac{2(d+t)sx}{G^2}\right]. \tag{A6}$$

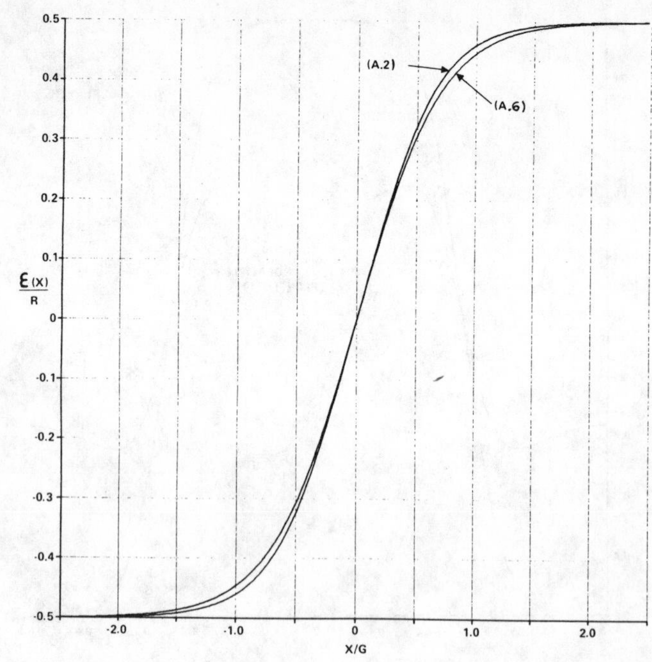

Fig. 12. Output from isolated transition (A2) and an approximation based on tanh function (A6) with the same asymptotic magnitude and slope at $x = 0$ proportional to field gradient $\partial H_y/\partial x$ at the medium center. $d = t = G/2$.

Fig. 12 compares this approximation with the actual output (A2), for $d = t = G/2$. The error (about 3 percent) is due to the relatively long tail of the hyperbolic tangent function.

A better choice is a function of the form (38), which is exact at $y = h$, with modified slope and magnitude. The function with correct slope and asymptotic value is

$$\frac{\epsilon(x)}{R} \cong \frac{2t}{\pi G} \ln\left[\frac{\cosh(\pi G/4h + \beta x)}{\cosh(\pi G/4h - \beta x)}\right], \tag{A7}$$

where

$$\beta = \frac{\pi}{2} \frac{s}{\tanh \pi G/4h} \tag{A8}$$

and s is again given by (A4). This approximation differs negligibly from (A2) when $t < d$.

In the limit $g \to 0$, (A2) can be integrated, using [9, (4.114.1)], to give the readback voltage from an isolated transition with a zero gap-length head:

$$\frac{\epsilon(x)}{R}\bigg|_{g=0} = \frac{2}{\pi} \tan^{-1}\left[\tan \frac{\pi t}{2(d+t)} \tanh \frac{\pi x}{2(d+t)}\right]. \tag{A9}$$

The asymptotic amplitude (for large x) is again given by $t/(d+t)$.

In the limit $d \to 0$, (A2) is integrable, using [9, (3.741.3)], to give the readback voltage at zero separation:

$$\frac{\epsilon(x)}{R}\bigg|_{d=0} = \begin{cases} x/g, & |x| < g \\ 1, & |x| > g. \end{cases} \tag{A10}$$

This is proportional to the assumed Karlquist potential across the gap, and is *independent* of thickness. By contrast, readback without an underlayer, given by (A3), is thickness dependent and is proportional to t for $t \ll g$.

Acknowledgment

If is a pleasure to acknowledge the contributions of Harold Anderson and Jess Rifkind, who encouraged this work by creating an extraordinarily stimulating environment at the Xerox Advanced Development Laboratory.

References

[1] B. J. Langland and P. A. Albert, "Recording on perpendicular anisotropy media with ring heads," *IEEE Trans. Magn.*, vol. MAG-17, pp. 2547–2549, 1981.

[2] R. I. Potter and I. A. Beardsley, "Self-consistent computer calculations for perpendicular magnetic recording," *IEEE Trans. Magn.*, vol. MAG-16, pp. 967–972, 1980.

[3] O. Karlquist, "Calculation of the magnetic field in the ferromagnetic layer of a magnetic drum," *Trans. Royal Inst. of Tech. Sweden*, vol. 86, pp. 2–27, 1954.

[4] O. Lopez, W. P. Wood, N. H. Yeh, and M. Jurisch, "Interaction of a ring head and double layer media-field calculations," presented at Third Joint INTERMAG-MMM Conf., 1982.

[5] K. Yamamori, R. Nishikawa, T. Asano, and T. Fujiwara, "Perpendicular magnetic recording performance of double-layer media," *IEEE Trans. Magn.*, vol. MAG-17, pp. 2538–2540, Nov. 1981.

[6] R. L. Wallace, "The reproduction of magnetically recorded signals," *Bell Sys. Tech. J.*, vol. 30, pp. 1145–1173, 1951.

[7] J. Toda, K. Kobayashi, and M. Hiyane, "A thin film head for perpendicular magnetic recording," presented at Third Joint INTERMAG-MMM Conf., 1982.

[8] R. L. Stoll, *The Analysis of Eddy Currents.* Oxford: Clarendon, 1974, ch. 4.

[9] I. S. Gradshteyn and I. M. Ryzhik, *Table of Integrals, Series, and Products.* New York: Academic, 1965.

[10] A. S. Hoagland, *Digital Magnetic Recording.* New York: John Wiley, 1963.

Part VII
Noise

Maximum Signal-to-Noise Ratio of a Tape Recorder

J. C. MALLINSON

***Abstract*—Using the Wiener autocorrelation theorem, the noise power spectrum of the pole strength in a thin lamina of an erased tape is shown to be approximately white. The noise power spectrum of the reproduce head voltage is calculated for a thick tape and compared with the signal power. The wide-band signal-to-noise ratio of a tape recorder equalized flat is deduced and expressed in very simple forms, which are inversely dependent upon the square of a bandwidth. Notably, in this special case the wide-band result is independent of reproduce head-to-tape spacing. Numerical examples demonstrate that this simple theory yields results in excellent agreement with practice.**

Introduction

THE signal-to-noise ratio (SNR) of a tape recorder is, with the possible exception of the dropout behavior, the most important factor governing its utility as an information storage system. The maximum possible SNR, which occurs when the principal noise source in the system is the tape itself, depends naturally not only upon the fundamental parameters of the tape but also upon the manner of its use. The discussions of SNR given previously [1], [2], though correct, seem to be needlessly complex. Further, the results are not in forms readily usable by the system designer. In the present paper the entire problem is reworked in a simple direct manner using the Wiener autocorrelation theorem.

It is shown that the wide-band SNR may be expressed in very simple forms which yield values in exceptionally close agreement with experiment. Several new relationships of practical significance are derived and discussed. Further, since all the important expressions are derived from first principles, it is believed that the work is not without pedagogic merit.

Initial Considerations

Whereas the signal in a tape recorder relates to the mean magnetization of the tape particles, the noise arises from the deviations from the mean of the magnetization. In an erased tape the major source of these deviations is the randomness of the particle magnetization directions. We shall assume that only two directions exist, positive and negative, which are occupied at random.

As the tape becomes magnetized and above directional randomness decreases, one might expect the noise to

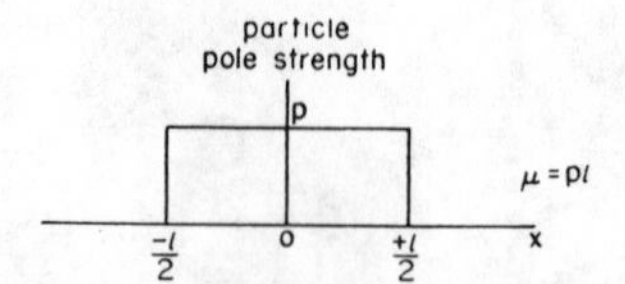

Fig. 1. Particle pole strength.

decrease. In fact, it increases somewhat, probably due to nonuniform particle packing effects. A noise which depends upon the signal (modulation noise) is neither stationary nor additive. However, since in the best tapes the noise increase is slight (≈3–4 dB), we shall assume that the noise is stationary and additive at all signal levels.

Tape Magnetization Statistics

We seek first the autocorrelation function (ACF), taken in the direction of head-to-tape motion x, of the pole strength[1] in a lamina, of width w and thickness δy, of an erased oriented particulate tape. Suppose the single-domain particles are identical, have dipole moment $\mu = pl$ (see Fig. 1), and are at a density n. The pole strength of the lamina at longitudinal position x is

$$P(x) = \sum_i b_i p_i(x) \tag{1}$$

where $b_i = \pm 1$ at random. The ACF is, by definition for stationary random processes [3, p. 258],

$$\mathrm{ACF}(x') = \lim_{x\to\infty} \frac{1}{x} \int_{-x/2}^{+x/2} P(x)P(x-x')\,dx \tag{2}$$

$$= \lim_{x\to\infty} \frac{1}{x} \int_{-x/2}^{x/2} \sum_i b_i p_i(x) \sum_j b_j p_j(x-x')\,dx \tag{3}$$

$$= nw\delta y \int_{-\infty}^{+\infty} p(x)p(x-x')\,dx \tag{4}$$

since

$$b_i b_j = \begin{cases} 1, & \text{if } i = j \\ 0, & \text{if } i \neq j. \end{cases}$$

Since, on the average, the particles only correlate with themselves, the lamina pole strength ACF turns out to be simply the sum of the individual particle pole strength ACFs, each of which is equal to $p^2(l - |x'|)$ as shown in Fig. 2. According to the Wiener theorem the noise power

Manuscript received March 19, 1969. Paper 2.7, presented at the 1969 INTERMAG Conference, Amsterdam, The Netherlands, April 15–18.

The author is with the Research and Advanced Technology Division, Ampex Corporation, Redwood City, Calif. 94303.

[1] The pole strength is defined by $P(x) = \int_A M(x)\,dA$, where $M(x)$ is the magnetization and A is the cross-sectional area.

Reprinted from *IEEE Trans. Magn.*, vol. MAG-5, pp. 182–186, Sept. 1969.

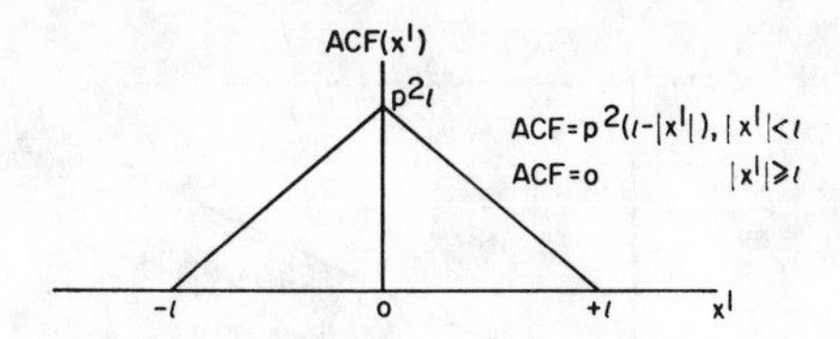

Fig. 2. Autocorrelation function of particle pole strength.

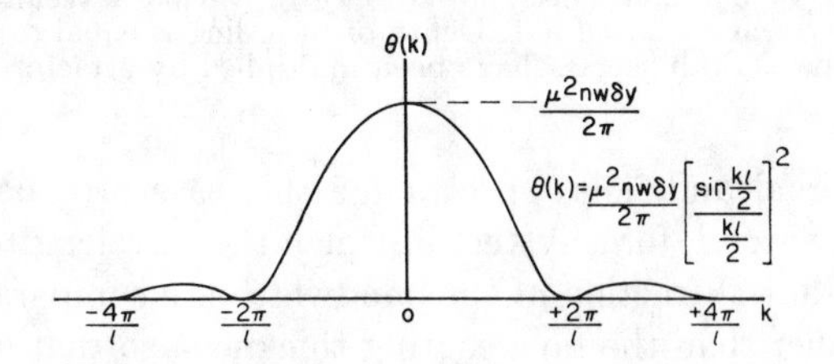

Fig. 3. Noise power spectrum of lamina pole strength.

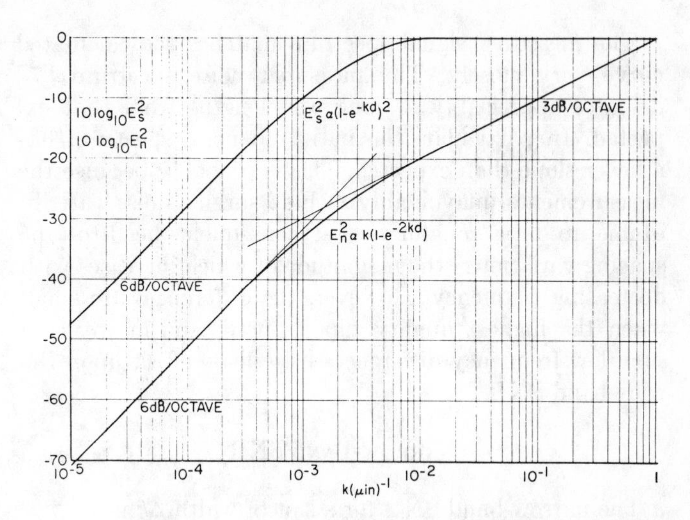

Fig. 4. Relative signal and noise power spectra versus wavenumber k for a 400-μin coating. No head-to-tape spacing effect is shown since it would change both curves equally.

spectrum is given by the Fourier cosine transform of the ACF [3, p. 56–59]. Thus the noise power spectrum of the lamina pole strength is

$$\theta(k) = \frac{1}{2\pi}\int_{-\infty}^{+\infty} nw\delta y p^2(l - |x'|) \cos kx' \, dx' \quad (5)$$

$$= \frac{\mu^2 nw\delta y}{2\pi}\left[\frac{\sin (kl/2)}{kl/2}\right]^2. \quad (6)$$

This function is plotted in Fig. 3. Note the important result that, for wavelengths λ substantially larger than the individual particle length, the lamina pole strength noise power spectrum is flat. This white spectrum approximation is assumed hereafter.

Output Noise Power Spectra (NPS)

Having defined the statistics, we proceed to compute the reproduce head voltage NPS. Providing gap losses may be neglected, the reproduce head exhibits a linear voltage transfer function $4\pi V|k|e^{-|k|y}$. The fact that this is true may be seen immediately since

$$\int_a^{a+d} 4\pi V|k|e^{-|k|y}\,dy = 4\pi V(1 - e^{-|k|d})e^{-|k|a} \quad (7)$$

which is the familiar Wallace output voltage spectrum [4]. To compute the output voltage NPS, we multiply the lamina pole strength noise power spectrum by the reproduce head power transfer function and integrate through the tape thickness. This operation is, in physical terms, allowing for the fact that the reproduce head only senses a wavelength dependent, limited volume of tape adjacent to the gap.

Thus

$$E_n^2(k) = \int_a^{a+d} \frac{\mu^2 nw\delta y}{2\pi}[4\pi V|k|e^{-|k|y}]^2 \quad (8)$$

$$= 4\pi\mu^2 nwV^2|k|(1 - e^{-2|k|d})e^{-2|k|a} \quad (9)$$

a result obtained by both Daniel [1] and Stein [2]. A similar development using the transfer function for a nondifferentiating head ($4\pi e^{-|k|y}$) leads to the output flux NPS.

$$\Phi_n^2(k) = \frac{E_n^2(k)}{V^2k^2} = \frac{4\pi\mu^2 nw(1 - e^{-2|k|d})e^{-2|k|a}}{|k|}. \quad (10)$$

If expressions (9) and (10) are integrated over an infinite bandwidth despite the comments following (6) and the onset of reproduce gap losses, the results are as follows:

$$\int_{-\infty}^{\infty} E_n^2(k)\,dk = 4\pi\mu^2 nwV^2\left\{\frac{d[a + (d/2)]}{a^2(a+d)^2}\right\} \quad (11)$$

$$\int_{-\infty}^{\infty} \Phi_n^2(k)\,dk = 8\pi\mu^2 nw \log_e\left(\frac{a+d}{a}\right) \quad (12)$$

as given by Mee [5]. They represent merely upper bounds to the total noise power. It will be evident that should exact results be needed they could be computed with little difficulty.

Output Signal Power Spectrum (SPS)

Suppose that, perhaps because of the need to minimize distortion, the sinusoidal signal magnetization recorded on the tape is only at a fraction f of the maximum amplitude possible. Further suppose, perhaps because of the need to minimize short wavelength record process losses, the tape is only recorded upon to a limited depth $d' \leq d$. Apparently by inspection of (7) the output signal power spectrum is

$$E_s^2(k) = \tfrac{1}{2}[4\pi\mu nwfV(1 - e^{-|k|d'})e^{-|k|a}]^2. \quad (13)$$

It will be noted the head-to-tape spacing dependence of both the SPS and NPS is identical. This occurs because the same physical laws govern both signal and noise of the same frequency. The two spectra are shown in Fig. 4.

The measured signal spectrum matches the calculated curve very closely. The measured noise spectrum [7] deviates appreciably at long wavelengths from that expected. In particular, the measured noise spectrum has a lower slope than expected. This is probably because the measurements unavoidably include surface noise (attributable to tape roughness and consequent head-to-tape spacing variations) the magnitude of which increases with decreasing frequency. However, the differences are small when the highest quality tape is used, and in any case such low frequency differences have little effect upon the wide-band SNR.

Narrow-Band SNR

The narrow-band SNR for a slot of width Δk is

$$\mathrm{SNR}_{\mathrm{narrow}} = \frac{2\pi n w f^2 (1 - e^{-|k|d'})^2}{|k|(1 - e^{-2|k|d})\Delta k}. \tag{14}$$

It is, of course, independent of head-to-tape spacing. The adverse effects of nonsaturation and partial penetration recording are evident; both reduce the SNR because, whilst only a limited number of particles contribute to the signal, all still contribute to the noise.

Wide-Band SNR

Since the signal and noise power spectra are not identical, the wide-band SNR depends upon the reproduce system equalization. Generally, wide-band SNRs will also depend upon the head-to-tape spacing. A simple case, of particular interest because of its widespread use, occurs when the output signal is equalized flat. To achieve this, the power transfer function of all parts of the reproduce system after the head must be the reciprocal of the signal power spectrum given by (13).

Note that this particular equalization makes the equalized noise power spectrum independent of head-to-tape spacing. An important consequence is that, in this special case, the wide-band SNR is independent of reproduce head-to-tape spacing. Out-of-contact playback need not entail a loss in SNR provided other noises in the system are kept below the (attenuated) tape noise. The onus is on the system designer.

The wide-band SNR for such systems is customarily defined to be the equalized signal power divided by the integrated noise power in the system bandwidth. That is,

$$\mathrm{SNR}_{\mathrm{wide}} = \left[\int_{|k|_{\min}}^{|k|_{\max}} \frac{(1 - e^{-2|k|d})\,|k|\,d\,|k|}{2\pi n w f^2 (1 - e^{-|k|d'})^2}\right]^{-1}. \tag{15}$$

Before evaluating this expression, two simplifications may be mentioned. First, if we consider only full coating depth recording ($d' = d$), then dropping the modulus signs,

$$\mathrm{SNR}_{\mathrm{wide}} = 2\pi n w f^2 \left[\int_{k_{\min}}^{k_{\max}} k \coth\frac{kd}{2}\, dk\right]^{-1}. \tag{16}$$

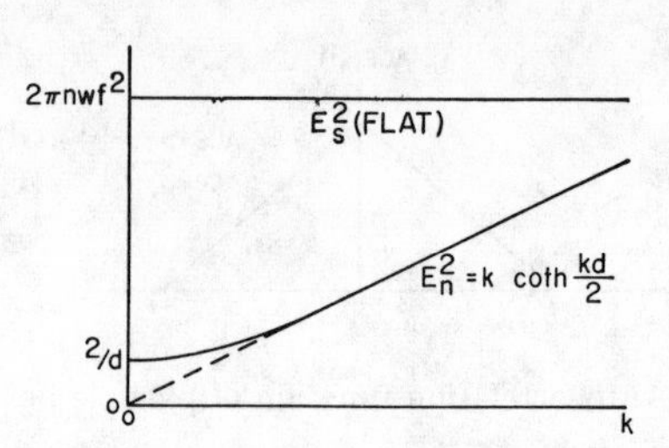

Fig. 5. Signal and noise power spectra versus wavenumber k in a system equalized flat. Depth of recording is equal to coating thickness. Both spectra have been multiplied by a factor $2\pi nwf^2$.

The signal and noise spectra for this case are shown in Fig. 5. Second, for a system in which the wavelengths over a substantial fraction of the bandwidth are comparable to or smaller than the tape coating thickness, so that $kd \gg 1$ and $\coth kd/2 \approx 1$, then

$$\mathrm{SNR}_{\mathrm{wide}} \approx 4\pi n w f^2 [k_{\max}^2 - k_{\min}^2]^{-1}. \tag{17}$$

This form may be compared with that resulting from the common but erroneous assumption that the tape noise is white in which case $\mathrm{SNR}\ \alpha (k_{\max} - k_{\min})^{-1}$. In a system equalized flat, the NPS rises at approximately 3 dB per octave, and consequently doubling the bandwidth actually entails a loss in SNR of about 6 dB rather than 3 dB.

It will be shown below that the extremely simple form of (17) does indeed closely approximate measured SNRs. It should be noted that the tape speed V, the head-to-tape spacing a, the coating thickness d, and the dipole moment μ do not appear; they do not have an important effect upon the maximum SNR.

The best tapes obviously yield the highest product nf^2. Magnetostatic interparticle interactions, which are rather poorly understood, control the distortion limit f and consequently n and f are not independent variables. No simple theory giving the functional dependence of f on n can presently be given.

Numerical Examples

We consider first the case of a 400-Hz–1.5-MHz, 120-in/s, 50-mil-trackwidth wide-band analog recorder equalized flat. The record gap length (150 μin) used is known to be noncritical, since the adjustments of the input currents largely compensate for differing gap lengths. Both the reproduce gap length (25 μin) and the average γ-Fe_2O_3 particle length (about 20 μin) are much smaller than the minimum wavelengths occurring (80 μin). A–c bias is used at a level which yields the maximum short wavelength output. If a head-to-tape spacing of 20 μin is assumed, the unequalized signal spectrum matches that expected for a partial penetration depth of about 75–100 μin. The signal input level is adjusted so that no more than 1-percent third harmonic distortion exists at long wavelengths. Under similar conditions, the rms remanent flux in audio tapes has been found to be about 200 nWb/m of trackwidth which is equivalent to a peak magnetization of about 250 gauss [7]. Since the maximum remanence of

TABLE I

$10\log_{10}\int_0^u \frac{S(1-e^{-2S})dS}{(1-e^{-\alpha S})^2}$ FOR SOME VALUES OF u AND α

u	$\alpha = 0.2$	$\alpha = 0.4$	$\alpha = 0.6$	$\alpha = 0.8$	$\alpha = 1.0$
5	20.0	15.5	13.6	12.5	12.0
10	22.1	18.9	17.9	17.5	17.3
20	25.2	23.6	23.3	23.1	23.1
40	29.7	29.2	29.1	29.1	29.0
80	35.2	35.1	35.1	35.1	35.1
160	41.1	41.1	41.1	41.1	41.1

γ-Fe_2O_3 analog tapes is about 1250 gauss, the distortion limit f is taken to be 0.2. The tape (Ampex 771) of coating thickness 400 μin, contains acicular γ-Fe_2O_3 particles of dimensions 20 by 4 by 4 μin (i.e., 5000 by 1000 by 1000 Å) which are packed at one third by volume. The number of particles per cubic microinch n is therefore about 10^{-3}.

The exact SNR given in (15) may be written

$$\text{SNR}_{\text{wide}} \approx 2\pi n w f^2 \left[\frac{1}{d^2}\int_0^u \frac{S(1-e^{-2S})\,dS}{(1-e^{-\alpha S})^2}\right]^{-1} \tag{18}$$

where

$$u = k_{\max} d, \quad \alpha = d'/d.$$

In the present case, substituting numbers,

$$\text{SNR}_{\text{wide}} \approx 2\cdot 10^6 \left[\int_0^u \frac{S(1-e^{-2S})\,dS}{(1-e^{-\alpha S})^2}\right]^{-1} \tag{19}$$

with

$$u = 2\pi d/\lambda_{\min} \approx 30, \quad \alpha = 0.4.$$

The integral has been evaluated numerically and the results are tabulated in Table I. Consequently, (19) may be written

$$10\log_{10}(\text{SNR})_{\text{wide}} = 10\log_{10}(2\cdot 10^6) - 26.6$$
$$= 63 - 26.6 = 36.4 \text{ dB}$$

The simple approximate form given in (17) yields

$$10\log_{10}(\text{SNR})_{\text{wide}} = 10\log_{10}(4\pi n w f^2 k_{\max}^{-2})$$
$$= 10\log_{10}(4000) = 36 \text{ dB}$$

which is less than 1 dB different from the exact result. Experimentally, if due care is taken to minimize other noises (mainly those due to reproduce head eddy currents) and to maintain the head efficiency at the upper frequencies, wide-band rms-signal-to-rms-noise ratios of 34–35 dB have been measured in excellent agreement with the above theory.

As a second example we consider briefly a 40-Hz–15-kHz, 7.5-in/s, 80-mil-trackwidth professional audio recorder. Such machines use both variable preequalization (of the record current) and fixed postequalization whereas the above theory considered only variable postequalization. It might seem, therefore, that the theory is not directly applicable. However, it turns out that direct application of (15) and (17) in fact does yield good numbers. This coincidence is related to the following considerations. The better quality tapes need little preequalization and thus produce an output spectrum close to that given by (7); the poorer tapes have considerable preequalization applied but again they yield the same output spectrum; and the fact that there is not much difference between the noise spectra of the different tapes.

To proceed with the calculation then we note that whereas the distortion limit is the same as in the previous example, now the depth of recording is equal to the standard γ-Fe_2O_3 coating thickness 400 μin. In this case ($u = 5$, $\alpha = 1.0$), (15) and (17) yield 10 log SNR_{wide} values of 54 and 55 dB, respectively, which values compare favorably with the 56–57 dB usually measured on such half-track audio machines.

The uncertain factors in these calculations are, of course, the partial penetration depth d' and the distortion factor f. Whereas the calculated SNR_{wide} is not sensitively dependent upon the exact value of the penetration depth, it does depend critically upon the distortion factor used. The value adopted here (0.2) is believed to be quite accurate and typical of modern analog tapes. However, even if the distortion factor is regarded simply as an adjustable parameter, the valuable fact remains that the theory, with $f = 0.2$, yields results in such excellent agreement with practice.

The above theory does not consider the effects of magnetostatic interactions which, particularly is nonuniformly packed tapes, will give rise to modulation noise. The excellent agreements found using the above simple theory indicate, however, that at least in the case of distortion limited recorders where, perforce, the signal level and tape magnetization is low, the effect of modulation noise upon SNR_{wide} is small.

NOMENCLATURE

A	cross-sectional area of tape (normal to head-tape motion)
a	head-to-tape spacing
b	dimensionless factor equal to ± 1
d	tape coating thickness
d'	depth of recording ($d' \leq d$)
E_n	reproduce head-noise voltage (k domain)
E_s	reproduce head-signal voltage (k domain)
f	ratio of signal to maximum possible signal
k	wavenumber ($2\pi/\lambda$)
$k_{\max}$	maximum wavenumber
$k_{\min}$	minimum wavenumber
l	magnetic particle length
M	tape lamina longitudinal magnetization (x domain)
n	number of particles per unit volume
p	magnetic particle pole strength (x domain)
P	tape lamina pole strength (x domain)

S	dimensionless factor (kd)
u	dimensionless factor ($k_{max}d$)
V	head-to-tape relative velocity
w	trackwidth
x	tape longitudinal coordinate
x'	offset tape coordinate
y	tape normal coordinate
α	dimensionless factor (d'/d)
λ	wavelength
μ	magnetic particle dipole moment (pl)
θ	lamina pole strength noise power (k domain).

References

[1] E. D. Daniel, "A basic study of tape noise," December 1960 (unpublished).

[2] I. Stein, "Analysis of tape noise," *IRE 1962 Internatl. Conv. Rec.* vol. 10, pt. 7, pp. 42–65.

[3] Y. W. Lee, *Statistical Theory of Communications.* New York: Wiley, 1960.

[4] R. L. Wallace, "The reproduction of magnetically recorded signals," *Bell Sys. Tech. J.*, pp. 1145–1173, October 1951.

[5] C. D. Mee, *The Physics of Magnetic Recording.* Amsterdam: North-Holland Publishing Company, 1964, p. 131.

[6] P. Smaller, "Factors affecting the wide-band SNR of direct recording systems," October 1963 (unpublished).

[7] J. G. McKnight, "The measurement of medium wavelength flux on the magnetic tape record," *J. Audio Engrg. Soc.* (to be published).

Statistical Analysis of Signal and Noise in Magnetic Recording

L. THURLINGS, MEMBER, IEEE

Abstract–**A general theory is developed for the power spectrum of the induced voltage at the read head in magnetic recording, originating from the particulate nature of the tape. The theory is general in that it yields the signal power as well as the noise spectrum, including the effects of clustered particulate media. The statistics of all relevant parameters such as particle length, particle magnetic moment, and partial penetration of the signal into the layer have been taken into account. Experiments on nonoriented Fe_2O_3 showed a discrepancy at short wavelengths of the order of 10 dB. This is qualitatively explained by particle interaction mechanisms.**

I. Introduction

NOISE in magnetic recording is caused by many sources. The noise due to the tape can be roughly divided into two categories, i.e., noise related to its magnetic properties and noise related to its mechanical properties. The latter arises from tape vibrations, which introduce frequency modulation of the signal, and from tape surface roughness which cause head-to-tape distance variations [1]-[6]. The magnetic characteristics of the particulate tape material introduce background noise and modulation noise. The background noise is directly related to the fact that the magnetic material consists of discrete particles which are permanently magnetized [7]-[12]. Thus, even when the tape is perfectly demagnetized so that the bulk remanence is zero, each particle still induces a voltage in the read head. When a signal has been recorded in particulate media modulation noise arises [1]-[6], [9], [11], and [13]. This is explained by the fact that the particles are not positioned at random but exhibit some degree of particle agglomeration which enlarges the noise flux.

The mechanism of the background noise arising from random particulate media is quite well understood [7]-[10]. However, little information is available about the power spectrum resulting from clustered media [11], [14], and [15]. Furthermore, it is well-known that the signal behavior as determined by Wallace [16] and Westmijze [17] is based on the assumption that in any point within the magnetic medium the magnetization can be defined. This is not the case in a particulate medium, so we will determine the signal power for such a particulate medium taking into account the statistics of all the relevant parameters. Hence, we have the general problem of establishing the power spectral density for noise and signal for a particulate magnetic medium. The aim of the study, then, is to investigate the effects of the statistics of the parameters involved on the signal and noise power. Important mechanisms, such as particle orientation and particle clustering, can be considered as well. We make the following assumptions.

1) The magnetic state of each particle is modeled by a bar magnet (dipole) with infinitesimal cross section but with finite length l (see Nomenclature). This, however, implies that the particle has no volume. Through the magnitude of the magnetic moment, which is related to the volume of the particle, the volume of the particle is introduced in the model. Hence, in our model it is possible that the "particles" overlap each other.

2) It is supposed that the magnetic moments of the particles do not change during the passage of the read head, although it is well-known that the reversible permeability alters the signal output [17].

3) During the reading process the distance from a particle to the head remains constant. This implies that the tape surface is ideally smooth.

4) The magnetic field of the head is represented by the Karlqvist expression [20]:

$$H_x = \frac{1}{\pi g}\left\{\arctan\frac{x_0 + (g/2)}{y_0} - \arctan\frac{x_0 - (g/2)}{y_0}\right\} \tag{1.1.a}$$

$$H_y = -\frac{1}{2\pi g}\ln\frac{(x_0 + (g/2))^2 + y_0^2}{(x_0 - (g/2))^2 + y_0^2}. \tag{1.1.b}$$

Starting from these assumptions we shall investigate the following cases.

A. AC-Erased Noise of a Uniform Medium

The medium is erased by means of a slowly decreasing ac field so that the bulk remanence is zero. All statistical variables are assumed to be independent, and the position of the particles is independently random (this will be referred to as uniform medium). Then we obtain a general formula, which includes the result obtained by Mann [7], Daniel [8], and Mallinson [10].

B. Harmonic Signal in a Uniform Medium

Here it is assumed, as in A, that the particles have random positions (no clusters), but that the polarity of the magnetic moments is related to the position of the particles so that a sinusoidal signal is recorded.

C. DC Noise of a Clustered Medium

We assume a medium with a typical cluster structure and apply a dc field to the medium. The spectrum thus obtained is known as the "dc-modulation noise spectrum."

Manuscript received June 1, 1979; revised December 21, 1979. This paper was presented at the Second Joint INTERMAG-MMM Conference, New York, NY, July 17-20, 1979.

The author is with Philips Research Laboratories, Eindhoven, The Netherlands.

Reprinted from *IEEE Trans. Magn.*, vol. MAG-16, pp. 507-513, May 1980.

D. Harmonic Signal in a Clustered Medium

A clustered medium is supposed, and a harmonic signal has been recorded so that the polarity of the magnetic moments depends on their position as in B.

II. GENERAL THEORY[1]

Suppose a particulate tape passes the read head, and assume that no mechanical noise sources are involved; thus the tape velocity is constant and the tape surface is ideally smooth. Further, each particle is assumed to be single domain so that the read process may be seen as the induction of a large amount of pulses arising from the individual pulse which each particle induces in the read head. As mentioned above, it is also assumed that the magnetic state of each particle does not change during the read process. This read process is in fact a stochastic process with the induction voltage as a stochastic variable. Assuming that the process is stationary the power spectrum of the read voltage may be written as [18], [19]

$$G(k) = \lim_{vT \to \infty} \frac{1}{2T} E\{|S_t(k)|^2\} \tag{2.1}$$

where $2T$ is the time interval of observation, k is the wavenumber $2\pi/\lambda$, v is the tape velocity, $S_t(k)$ is the Fourier transform of the read voltage of the particles passing the read head, and $E\{\cdot\}$ stands for the statistical expectation. The read voltage is the sum of the individual induction voltages of the N_T particles passing the read head in the time interval $2T$, so

$$S_t(k) = \sum_{i=1}^{N_T} S_i(k) \tag{2.2}$$

where $S_i(k)$ is the Fourier transform of the induction voltage of one particle. Thus, to compute the power spectrum we have to determine the expectation of the square of the absolute value of the sum of the individual Fourier transforms. We further note that (2.1) holds for all frequencies, whereas in practice the power spectral density has positive frequencies only. Therefore, we have to reckon with [18],

$$P(k) = 2 \cdot G(k), \quad k \geqslant 0. \tag{2.3}$$

Before evaluating (2.1) for the five cases mentioned in the introduction we first have to determine the expression for the individual Fourier transform $S_i(k)$. For that purpose the particle is represented by a bar magnet of finite length l, infinitesimal cross-section, position x, y on the tape coordinates, and with orientation (θ, φ), see Fig. 1. The Fourier transform of the induction voltage of one particle is defined as

$$S(k) = \frac{1}{v} \int_{-\infty}^{\infty} e(\xi)\, e^{-jk\xi}\, d\xi, \tag{2.4}$$

and the induction voltage itself can be found by means of the reciprocity theorem [17]

$$e(\xi) = -nv \frac{d}{d\xi} \left\{ \mu_0 \iiint M \cdot H\, dx_0\, dy_0\, dz_0 \right\}, \tag{2.5}$$

[1]SI units are used throughout this paper, with the magnetization M defined as $M = (B/u_0) - H$.

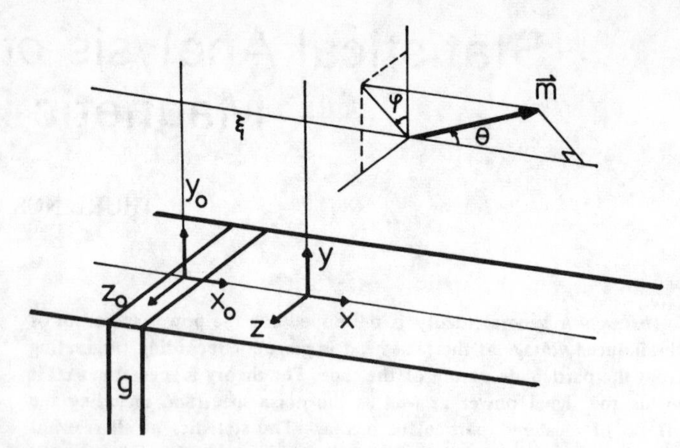

Fig. 1. Coordinate system for magnetic dipole passing read head.

where H would be the field of the read head if it were excited by a unit current, the displacement of the tape $\xi = v \cdot t$, and M represents the magnetization of the dipole described on the coordinate system (x_0, y_0, z_0). This results in

$$S(k) = n \cdot \eta \cdot \mu_0 m_0 e^{-|k|y} \cdot e^{jkx} \cdot \frac{\sin k(g/2)}{k(g/2)} \cdot \frac{e^{-|k|l \sin\theta \cos\varphi} \cdot e^{jkl\cos\theta} - 1}{l} \tag{2.6}$$

where n is the number of turns, η is the read head efficiency, μ_0 is the permeability, m_0 is the dipole moment, l is the length of the particle, (θ, φ) is the orientation of the particle, $0 \leqslant \varphi < 2\pi$, $0 \leqslant \theta \leqslant \pi$, and x, y is the position of the particle.

Now we have to recognize that when computing (2.1) all the variables describing the condition of the particles in (2.6) are in fact stochastic ones. This means that it is not the actual value which is relevant but merely the statistical magnitudes. Expression (2.6) is a complex function, thus

$$S(k) = {}^rS + j \cdot {}^jS \tag{2.7}$$

where ${}^rS$ is by definition the real part of $S(k)$, and ${}^jS$ is the imaginary part. Then (2.1) can be rewritten as

$$G(k) = \lim_{vT \to \infty} \left\{ \frac{1}{2T} E\left\{ \left(\sum_{i=1}^{N_T} {}^rS_i \right)^2 \right\} + \frac{1}{2T} E\left\{ \left(\sum_{i=1}^{N_T} {}^jS_i \right)^2 \right\} \right\}. \tag{2.8}$$

Now the evaluation of this equation depends on the statistical properties of the medium under consideration. We shall first treat the case where the particle position x is arbitrary. This will then reveal the well-known background noise spectrum.

III. AC-ERASED NOISE OF A UNIFORM MEDIUM

Imagine that a tape with thickness d, trackwidth w, and head-to-tape distance a is passing the read head. Assume that all the statistical variables are mutually independent. Now each term of (2.8) can be written as

$$E\left\{\left(\sum_{i=1}^{N_T} S_i\right)^2\right\} = \sum_{i=1}^{N_T} E\{S_i^2\} + 2\sum_{j=1}^{N_T}\sum_{i=j+1}^{N_T} E\{S_i \cdot S_j\}$$
$$= N_T \cdot E\{S^2\} + N_T(N_T - 1)E\{S\}^2. \tag{3.1}$$

With this expression we can now compute the power spectrum of (2.8) by treating the real and imaginary part separately. According to (2.6) S is a product of functions of different stochastic variables which are assumed to be independent. The same statistical rules as with (3.1) can now be applied. Then, with the assumption that the particles are randomly distributed in the longitudinal tape direction, i.e.,

$$f(x) = \frac{1}{2vT} \tag{3.2}$$

which satisfies

$$\int_{-vT}^{vT} f(x)\,dx = 1 \tag{3.3}$$

we obtain, after applying the limit $vT \to \infty$:

$$\lim_{vT\to\infty} \frac{N_T}{2T} E\{\cos^2 kx\} = \lim_{vT\to\infty} \frac{N_T}{2T}\int_{-vT}^{vT} \cos^2 kx \cdot \frac{1}{2vT}\,dx = \frac{\tilde{N}}{2} \tag{3.4.a}$$

$$\lim_{vT\to\infty} \frac{N_T}{2T} E\{\sin^2 kx\} = \lim_{vT\to\infty} \frac{N_T}{2T}\int_{-vT}^{vT} \sin^2 kx \cdot \frac{1}{2vT}\,dx = \frac{\tilde{N}}{2} \tag{3.4.b}$$

where

$$\tilde{N} = N \cdot w \cdot v \cdot d. \tag{3.5}$$

N is the number of particles per unit volume, w is the trackwidth, v is the tape speed, and d is the thickness. Now it follows that the last term in expression (3.1) disappears, which results in

$$G(k) = \lim_{vT\to\infty} \frac{N_T}{2T}\{E\{{}^rS^2\} + E\{{}^jS^2\}\}. \tag{3.6}$$

For simplicity we now assume that the magnetic moment of a particle can be written as

$$m_0 = \sigma_s \cdot V \cdot s_m \tag{3.7}$$

where σ_s is the bulk saturation magnetization of the magnetic material, V is the volume of the particle, and s_m is the sign of m_0 with respect to the x axis. This implies that now the angle θ is restricted to $0 \leqslant \theta \leqslant \pi/2$. This leads to the general expression for the background noise:

$$P(k) = 2 \cdot \tilde{N}\,[E\{A^2 \cdot s_m^2\} + E\{B^2 \cdot s_m^2\}], \tag{3.8}$$

where

$$A = -CVe^{-ky}\,\frac{e^{-kl\sin\theta\cos\varphi}\cdot\cos(kl\cos\theta) - 1}{l}, \tag{3.9.a}$$

$$B = -CVe^{-ky}\,\frac{e^{-kl\sin\theta\cos\varphi}\cdot\sin(kl\cos\theta)}{l}, \tag{3.9.b}$$

and

$$C = n\mu_0\sigma_s\eta\,\frac{\sin k(g/2)}{k(g/2)}. \tag{3.9.c}$$

This further reduces to

$$P(k) = 2Nw\,dv\,C^2 \cdot E\{s_m^2\}\cdot E\{V^2\}\cdot E\{e^{-2ky}\}$$
$$\cdot E\left\{\frac{e^{-2kl\sin\theta\cos\varphi} - 2e^{-kl\sin\theta\cos\varphi}\cdot\cos(kl\cos\theta) + 1}{l^2}\right\}. \tag{3.10}$$

In deducing this result we have assumed that all the stochastic variables are independent and that the density function of the variable x is uniform. So far no assumptions have been made concerning the density functions of the other variables. They can be treated separately. By definition, the sign s_m of the magnetic moment can only have the value ± 1. Let us assume that the probability that a magnetic moment is positive is $p \cdot ds_m$, independent of the position, then the density function of s_m is

$$f(s_m) = p \cdot \delta(s_m - 1) + (1 - p)\cdot\delta(s_m + 1), \tag{3.11}$$

where δ is the Dirac function.

If $p = \frac{1}{2}$ we have equal amounts of positively and negatively magnetized particles and the remancence is zero. For $p \neq \frac{1}{2}$ the medium is dc magnetized to a magnetization level $2p - 1$. Now we obtain

$$E\{s_m^2\} = \int_{-\infty}^{\infty} s_m^2 \cdot f(s_m)\,ds_m = 1. \tag{3.12}$$

This result is independent of p, which implies that in a medium where the particles have arbitrary x positions and the sign s_m is uncorrelated with all the other variables, the ac-erased and the dc-erased noise levels are the same. This is obvious because there is no mechanism, such as clustering, that could alter the noise level.

The expectation of the particle volume with density function $f(V)$ is

$$E\{V^2\} = \int_0^{\infty} V^2 f(V)\,dV = \overline{V}^2 + \sigma_v^2 \tag{3.13}$$

where $\overline{V}$ is the average particle volume, and σ_v^2 is the variance.

When a uniform packing density is assumed the density function for y is constant:

$$f(y) = \frac{1}{d}, \tag{3.14}$$

and the expectation of e^{-2ky} becomes

$$E\{e^{-2ky}\} = \int_a^{a+d} e^{-2ky} \cdot \frac{1}{d}\, dy = \frac{e^{-2ka}(1 - e^{-2kd})}{2kd}. \tag{3.15}$$

This is essentially the noise read function, expressing the distance and thickness loss [7]-[11].

The last term of (3.10) can be treated along the same lines, though the expressions become rather involved and have to be obtained numerically. We shall proceed here for the case that $kl < 1$. Then it turns out that

$$E\left\{\frac{e^{-2kl \sin\theta \cos\varphi} - 2e^{-kl \sin\theta \cos\varphi} \cdot \cos(kl \cos\theta) + 1}{l^2}\right\}$$

$$= k^2 \cdot E\{\sin^2\theta \cdot \cos^2\varphi + \cos^2\theta\}. \tag{3.16}$$

Here the influence of the finite particle length is assumed to be negligible. Considering the case of random particle orientation the density function becomes, with $0 \leqslant \theta \leqslant \pi/2$,

$$f(\theta, \varphi) = \frac{1}{2\pi} \tag{3.17}$$

so that

$$E\{\sin^2\theta \cdot \cos^2\varphi + \cos^2\theta\} = \tfrac{2}{3}. \tag{3.18}$$

The noise power spectrum for this case is

$$P(k) = N \cdot wv \left\{n\mu_0 \eta \overline{m} \frac{\sin k(g/2)}{k(g/2)}\right\}^2 \left(1 + \frac{\sigma_v^2}{\overline{V}^2}\right)$$

$$\cdot e^{-2ka}(1 - e^{-2kd}) \cdot \frac{2}{3} \cdot k \tag{3.19}$$

where

$$\overline{m} = \sigma_s \cdot \overline{V}. \tag{3.20}$$

However, when the particles are perfectly aligned with the longitudinal tape direction the expectation in the right side of (3.16) equals one, indicating that the noise increases due to alignment of the particles although the increase is only 1.8 dB [7], [8].

IV. Harmonic Signal in a Uniform Medium

In the preceding section it was assumed that—apart from dc magnetization—no signal was recorded in the medium. Now we shall derive the power spectrum for the case in which a sinusoidal signal is recorded. To this end we assume that the sign s_m of the magnetic moments depends on the position x. All the other variables are again statistically independent. Now the expectations concerning the variables x and s_m cannot be treated separately but have to be computed from the joint density function

$$f(s_m, x) = \frac{1}{4vT}[1 + \sin k_0 x]\,\delta(s_m - 1)$$

$$+ \frac{1}{4vT}[1 - \sin k_0 x] \cdot \delta(s_m + 1), \tag{4.1}$$

where $k_0 = 2\pi/\lambda_0$, and λ_0 is the wavelength of the recorded signal. For simplicity it is assumed here that in the peaks of the sine wave the medium is in saturation. The computation of the power spectrum starts again from (2.8), and with (3.1) we obtain the following relations:

$$\lim_{vT\to\infty} \frac{N_T(N_T - 1)}{2T} E\{s_m \cdot \sin kx\}^2$$

$$= \frac{\pi}{v} \cdot \frac{\tilde{N}^2}{2}[\delta(k - k_0) + \delta(k + k_0)] \tag{4.2.a}$$

$$\lim_{vT\to\infty} \frac{N_T}{2T} E\left\{s_m^2 \begin{matrix}\cos^2 kx \\ \sin^2 kx\end{matrix}\right\} = \frac{\tilde{N}}{2} \tag{4.2.b}$$

where $\tilde{N}$ is defined by (3.5). Hence, the general expression for the total power spectrum is

$$P(k) = 2\tilde{N}[E\{A^2\} + E\{B^2\}] + \frac{\pi}{v} \cdot \tilde{N}^2 [E\{A\}^2 + E\{B\}^2]$$

$$\cdot \delta(k - k_0), \tag{4.3}$$

where A and B are defined by (3.9).

Apart from s_m, which has now been used to constitute the signal, the first term is equal to the background noise expressed by (3.8), and the second term is the signal power which obeys the well-known expression for the wavelength losses

$$E\{e^{-ky}\} = \frac{e^{-ka}(1 - e^{-kd})}{kd} \tag{4.4}$$

as originally found by Wallace [16]. This illustrates the generality of our analysis in that it is capable of determining both the noise and the signal behavior of particulate magnetic recording media.

V. DC Noise of a Clustered Medium

When a tape is subjected to a dc field it is well-known that the noise power can be much higher than the ac-erased noise. For a smooth tape this is explained by particle clumping. Although this effect is understood intuitively and experimental evidence exists, there is no known straightforward analytical formulation as exists in the case of ac-erased tape. Therefore, a model will be proposed here which yields the power spectrum for the case where a clustered particulate medium is subjected to a dc field.

Suppose a particulate medium is passing the read head. The average number of particles passing the head in the time interval $2T$ is again N_T. Now we assume

$$N_T = K \cdot M_T, \tag{5.1}$$

where M_T is the average number of clusters passing the head per time interval $2T$, and K is the average number of particles per cluster. Further assume that the clusters are randomly distributed, described by coordinate u_1, but the position u_2 of the particles within a cluster is subject to certain statistics described by its density function $f(u_2)$. Now the stochastic variable x can be separated

$$x = u_1 + u_2 \tag{5.2}$$

with

$$f(u_1) = \frac{1}{2vT} . \tag{5.3}$$

Equation (2.8) can now be rewritten as

$$G(k) = \lim_{vT\to\infty} \frac{1}{2T} [N_T \cdot E\{{}^rS_{ij}^2\} + N_T \cdot E\{{}^jS_{ij}^2\} + N_T(K-1)\cdot E\{{}^rS_{ij}{}^rS_{il}\}_{j\neq l} + N_T(K-1) \cdot E\{{}^jS_{ij}{}^jS_{il}\}_{j\neq l} + N_T(N_T-K)E\{{}^rS_{ij}{}^rS_{lk}\}_{i\neq l} + N_T(N_T-K)E\{{}^jS_{ij}{}^jS_{lk}\}_{i\neq l}] \tag{5.4}$$

where the first index of S denotes the cluster, and the second one denotes the particle in the cluster.

The first two terms of (5.4) concern the autocorrelation of the magnetic moments, the third and fourth terms concern the correlation of the particles within the same cluster, and the last two terms concern the correlation among different clusters. Again it is assumed that all the variables involved are independent. We shall not go through the elaborate deduction of (5.4), for it is essentially the same as in the foregoing sections. We only remark here that the statistical interpretation of clusters is that the particles in a given cluster have the same coordinate u_1 which reveals the third and fourth term of (5.4). Hence, the general formulation of the power spectrum is

$$P(k) = 2\tilde{N} [E\{A^2 \cdot s_m^2\} + E\{B^2 \cdot s_m^2\}] + 2\tilde{N}(K-1)[E\{A \cdot s_m\}^2 + E\{B \cdot s_m\}^2] \cdot (E\{\cos ku_2\}^2 + E\{\sin ku_2\}^2). \tag{5.5}$$

The first part of (5.5) is exactly the background noise of (3.8). However, a second noise contribution exists which is proportional to the average number of particles K per cluster, and the wavelength-dependent part is described by the expectation of the sine and cosine functions of u_2. When $f(u_2)$ takes a Gaussian form (Fig. 2(a)) we obtain

$$E\{\cos ku_2\} = \sqrt{\frac{\alpha}{\pi}} \int_{-\infty}^{\infty} \cos ku_2 \cdot e^{-\alpha u_2^2}\, du_2 = e^{-k^2/4\alpha}. \tag{5.6}$$

Here the clusters have infinite dimensions. Clusters which are restricted to some average dimension L (Fig. 2(b)), yield

$$E\{\cos ku_2\} = \frac{1}{L}\int_{-L/2}^{L/2} \cos ku_2\, du_2 = \frac{\sin kL/2}{kL/2}. \tag{5.7}$$

Now with (5.6) as an example the power spectrum becomes

$$P(k) = \text{background noise} + 2\tilde{N}(K-1)C^2E\{V\}^2 \cdot E\{s_m\}^2 E\{e^{-ky}\}^2 \cdot e^{-k^2/2\alpha} \cdot \left(E\left\{\frac{e^{-kl\sin\theta\cos\varphi}\cdot\cos kl\cos\theta - 1}{l}\right\}^2 + E\left\{\frac{e^{-kl\sin\theta\cos\varphi}\cdot\sin kl\cos\theta}{l}\right\}^2\right). \tag{5.8}$$

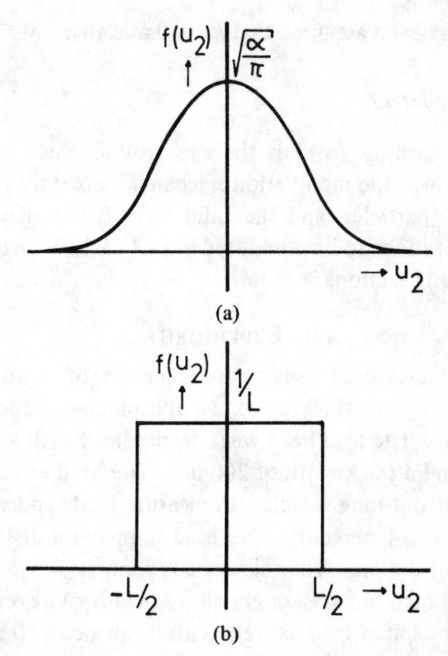

Fig. 2. Density function of cluster. (a) Gaussian distribution. (b) Uniform distribution.

From this expression it is evident that when the medium is ac erased, the modulation noise disappears because $E\{s_m\}$ is zero in that case. On the other hand when the tape is poorly ac erased so that a small dc component in the magnetization is left behind, the erased noise is increased due to the modulation noise.

When the tape is dc magnetized and the particles have random orientation we obtain for the case that $kl < 1$:

$$P(k) = \text{background noise} + \frac{K-1}{2d} \cdot Nwv\overline{V}^2 \cdot C^2 \cdot e^{-2ka}(1-e^{-kd})^2 \cdot e^{-k^2/2\alpha}. \tag{5.9}$$

Observe that the modulation noise term behaves as a signal term in that the wavelength losses are expressed by the same relations as for signal magnetization.

VI. Harmonic Signal in a Clustered Medium

Due to the clumping of the particles, noise sidebands occur in the power spectrum in addition to the signal when a harmonic signal is recorded in the medium. We shall treat this case by considering the cluster model of the preceding section and assume that a harmonic signal is recorded with the statistics described by (4.1). To compute the power spectrum we can again start from (5.4) using the assumptions (5.2) and (5.3). Because the derivation is essentially the same as in the foregoing we shall not go through the extensive deduction but merely state the result

$$P(k) = 2\tilde{N}[E\{A^2\} + E\{B^2\}] + \tilde{N}(K-1)[E\{A\}^2 + E\{B\}^2](E\{\cos(k+k_0)u_2\}^2 + E\{\cos(k-k_0)u_2\}^2) + \tilde{N}^2[E\{A\}^2 + E\{B\}^2] \cdot \frac{\pi}{v}\cdot\delta(k-k_0), \tag{6.1}$$

where it has been assumed that $k_0 \neq 0$ and that

$$f(-u_2) = f(u_2). \tag{6.2}$$

The first term in (6.1) is the background noise power, the second term is the modulation sidebands caused by the clumping of the particles, and the third term is the signal power. The expectations to be computed in (6.1) follow directly from the foregoing sections.

VII. Experimental

We have measured noise power spectra of a nonoriented Fe_2O_3 layer with thickness 5 μm and pigment concentration 1.93 g/cm^3. The read head was a ferrite head with a gaplength of 1 μm and a trackwidth of 200 μm. The head efficiency was estimated from long wavelength measurements and was found to be about 54 percent. The head-to-medium distance is of the order of 0.3 μm. The velocity v is 5 cm/s.

First the medium was ac erased by means of a conventional erase head excited by a current with frequency 250 kHz. The power spectrum thus obtained is shown in Fig. 3, curve (a). The spectrum increases slightly, but above 1.5 kHz it decreases considerably. Curve (b) shows the theoretical curve calculated with (3.17), assuming that $\sigma_v = 0$. This curve is fitted at long wavelengths to the experimental curve. Thus, we obtain the average dipole moment m_0 which appears to be 0.2×10^{-14} (A $\cdot$ m^2) and the average number of particles $N = 7 \times 10^{19}$ (m^{-3}). Fitting at long wavelengths is reasonable, because according to the theory none of the relevant mechanism such as the possible correlations among the magnetic moments or even the simple mechanisms of distance and spacing losses contribute significantly to the power spectrum at long wavelengths. The difference of the order of 20 dB at short wavelength is too large to be accounted for by the present theory. Even a possible misinterpretation in the head-to-medium distance, which is the most sensitive parameter, cannot give such a big result. However, it may be argued that particle interaction can cause such enormous effects. For instance, the hypothetical case of a tape where each particle has a neighboring particle which has the same orientation and the same position (x, y) but which is magnetized antiparallel after ac erasing due to the magnetic interaction, will have no noise at all because the flux from one particle is canceled by the flux of its neighboring particle. In practice, the same sort of interaction mechanism may occur, resulting in a decrease of the ac-noise power.

The second experiment concerns the dc-noise power spectrum. A large direct current was supplied to the erase head in order to saturate the magnetic layer. This noise power spectrum, Fig. 3, curve (c), which is much higher than the erased noise, increases with frequency to about 5 kHz and then strongly decreases. The theoretical dc-noise power spectrum is calculated with (5.8) and fitted as well as possible to the experimental one at long wavelengths. This yields an average number of particles per cluster K of the order of 12. This number seems reasonable and is comparable to the value determined by Su and Williams [13]. At very short wavelengths the theoretical dc-noise power approaches the theoretical ac-noise power. However, the experimental dc-noise spectrum shows a much steeper decrease than that of the theoretical spectrum. This can possibly be explained by the fact that the medium is in saturation remanence, which can imply that a certain amount of particles are magnetized antiparallel to other particles. If this magnetization state is correlated with the size of the cluster, i.e., only the smaller ones have particles which are magnetized antiparallel, it is evident that there is a strong decrease at short wavelengths of the dc noise.

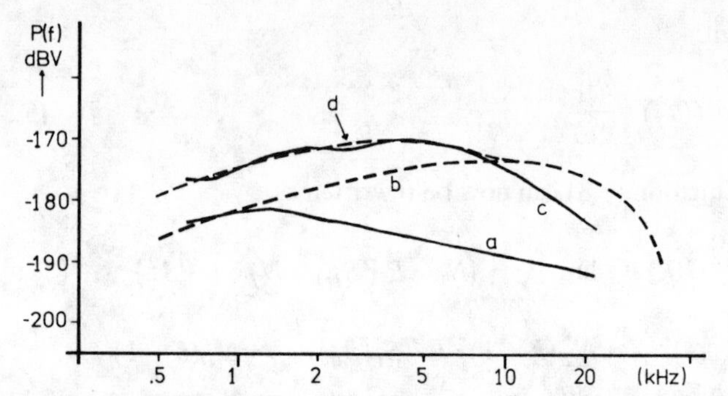

Fig. 3. Head-induced power spectral density for ac-erased and dc-erased Fe_2O_3 magnetic layer. Full line: experiment. Broken line: theory.

Discussion and Conclusion

The noise power spectrum for the ac-erased tape is well-known in the literature. However, a general theory based on classical statistics, which incorporates noise as well as signal behavior and which can deal with clustered media, is lacking. In this paper we have derived such a theory. First we derived a new formulation for the ac-erased case, which has a general character and includes the well-known random noise spectrum of Mann [7]. We then considered the case where a sinusoidal signal was recorded. It appeared that no modulation noise existed, as it should be, for there are no possible sources that can cause the modulation noise. Next a clustered medium was considered, and a general expression for the dc-modulation noise was derived. It was observed that the modulation noise term is proportional to the number of particles per cluster, and the wavelength dependence is similar to the wavelength losses, as in the case of signal magnetization. The latter is explained by the fact that this noise source originates from the cross correlation of the magnetic moments, whereas the background noise originates from the auto correlation of the magnetic moments. Further, the harmonic magnetization in a clustered medium was considered, and the power spectrum was found to consist of three fundamental contributions: the signal power, background noise, and modulation sidebands.

Verification of the power spectrum for the case of ac-erased and dc-erased tape, taking into account the tolerances of the parameters involved, revealed that there is a large discrepancy between the experimental and theoretical noise power spectrum. This is possibly caused by the particle interaction, which can theoretically reduce the net flux emanating from the tape.

NOMENCLATURE

a	Head to medium distance (m).
A	Auxiliary variable defined by (3.9).
B	Auxiliary variable defined by (3.9).
C	Auxiliary variable defined by (3.9).
d	Tape thickness (m).
E	Statistical expectation.
f	Density function.
g	Gaplength (m).
G	Power spectrum $-\infty < k < \infty$.
H_x	Longitudinal component of the Karlqvist head field.
H_y	Transversal component of the Karlqvist head field.
j	$\sqrt{-1}$.
k	Wavenumber $k = 2\pi/\lambda$ (m^{-1}).
k_0	Wavenumber of recorded signal (m^{-1}).
K	Average number of particles per cluster.
l	Length of a particle (m).
L	Parameter for uniform clusters.
m_0	Magnetic moment of a particle ($A \cdot m^2$).
$\overline{m}$	Average magnetic moment of the particles ($A \cdot m^2$).
M_T	Average number of clusters passing the read head in the time interval $2T$.
n	Number of turns.
N	Number of particles per unit volume (m^{-3}).
$\tilde{N}$	Number of particles passing the read head per unit time (s^{-1}).
N_T	Number of particles passing the read head during the time interval $2T$.
p	Percentage of particles with positive polarity.
P	Power spectrum $k \geqslant 0$ (V^2/Hz).
s_m	Sign of the magnetic moment of a particle with respect to the x axis.
S	Fourier transform of the induction voltage of a particle passing the read head.
${}^rS$	Real part of the Fourier transform S.
${}^iS$	Imaginary part of the Fourier transform S.
t	Time (s).
$2T$	Time interval of observation (s).
u_1	Position of the clusters in the tape in the longitudinal direction (m).
u_2	Position of the particles in the cluster in the longitudinal direction (m).
v	Tape velocity (m/s).
V	Volume of a particle (m^3).
$\overline{V}$	Average of the particle volume (m^3).
w	Track width (m).
(x, y)	Position of a particle in a coordinate system associated with the tape.
(x_0, y_0, z_0)	Coordinate system associated with the head.
α	Parameter for Gaussian clusters.
δ	Dirac function.
η	Efficiency of the read head.
(θ, φ)	Orientation of a particle.
λ	Wavelength (m).
λ_0	Wavelength of recorded signal (m).
μ_0	Permeability $4\pi \times 10^{-7}$ ($V \cdot s/A \cdot m$).
ξ	Displacement of the tape $\xi = v \cdot t$ (m).
σ_s	Bulk saturation magnetization of the magnetic material (A/m).
σ_v	Variance of the particle volume (m^3).

REFERENCES

[1] R. L. Price, "Modulation noise in magnetic tape recordings," *IRE Trans. Audio*, vol. 6, pp. 29–40, Mar. 1958.

[2] P. Smaller, "The noise in magnetic recording which is a function of tape characteristics," *J. Audio Eng. Soc.*, vol. 7, pp. 196–202, Oct. 1959.

[3] D. F. Eldridge, "D-C and modulation noise in magnetic tape," *Trans. Comm. Electr.*, vol. 83, pp. 585–588, Sept. 1964.

[4] E. G. Trendell, "The measurement and subjective assessment of modulation noise in magnetic recording," *J. Audio Eng. Soc.*, vol. 17, pp. 644–653, Dec. 1969.

[5] E. D. Daniel, "Tape noise in audio recording," *J. Audio Eng. Soc.*, vol. 20, pp. 92–99, Mar. 1972.

[6] E. D. Daniel, "A preliminary analysis of surface induced tape noise," *Trans. Comm. Electr.*, vol. 83, pp. 250–253, May 1964.

[7] P. A. Mann, "Das rauschen eines magnettonbandes," *Arch. Elektris. Uebertragung*, vol. 11, pp. 97–100, Mar. 1957 (in German).

[8] E. D. Daniel, "A basic study of tape noise," Ampex Res. Rep. AEL-1, Dec. 1960.

[9] I. Stein, "Analysis of noise from magnetic storage media," *J. Appl. Phys.*, vol. 34, pp. 1976–1990, July 1963.

[10] J. C. Mallinson, "Maximum signal-to-noise ratio of a tape recorder," *IEEE Trans. Magn.*, vol. MAG-5, pp. 182–186, Sept. 1969.

[11] S. Satake and J. Hokkyo, "A theoretical analysis of the erased noise of magnetic tape," in *IECE Techn. Group Meeting of Magn. Rec.*, Japan, vol. MR 74-23, pp. 39–49, 1974 (in Japanese).

[12] H. U. Ragle and P. Smaller, "An investigation of high-frequency bias-induced tape noise," *IEEE Trans. Magn.*, vol. MAG-1, pp. 105–110, June 1965.

[13] J. L. Su and M. L. Williams, "Noise in disk data-recording media," *IBM J. Res. Devel.*, vol. 18, pp. 570–575, Nov. 1974.

[14] P. A. Mann, "Ueber das modulationsrauschen eines magnettonbandes," *Arch. Elektris. Uebertragung*, vol. 15, pp. 18–24, Jan. 1961, (in German).

[15] P. G. Rothe, "Einige Bemerkungen zum Modulationsrauschen eines Magnettonbandes," *Arch. Elektris. Uebertragung*, vol. 16, pp. 535–542, Nov. 1962, (in German).

[16] R. L. Wallace, "The reproduction of magnetically recorded signals," *Bell Syst. Techn. J.*, vol. 30, pp. 1145–1173, Oct. 1951.

[17] W. K. Westmijze, "Studies on Magnetic Recording," *Philips Res. Depts.*, vol. 8, pp. 148–366, Apr. 1953.

[18] J. H. Laning and R. H. Battin, *Random Processes in Automatic Control.* New York: McGraw-Hill, 1956, Chapt. 3.

[19] A Papoulis, *Probability, Random Variables and Stochastic Processes.* Tokyo: McGraw-Hill, 1965, Chapt. 10.

[20] D. Karlqvist, "Calculation of the magnetic field in the ferromagnetic layer of a magnetic drum," *Trans. Royal Inst. of Techn.*, Stockholm, vol. 86, pp. 3–27, 1954.

A Theoretical Analysis of Modulation Noise and dc Erased Noise in Magnetic Recording

K. Tarumi

Fachbereich Physik, Universität Bremen, Fed. Rep. Germany

Y. Noro

Consumer Products Research Center, Hitachi Ltd. 292 Yoshida-machi Totsuka-ku, Yokohama, 244, Japan

Received 26 January 1982/Accepted 17 April 1982

Abstract. The modulation noise and the special case, dc erased noise in magnetic recording are discussed theoretically, taking account of two main causes at the same stage, i.e. the magnetic properties such as the inhomogeneous distribution of the particles on the tape, and the mechanical properties such as the head-to-tape space variations arising from the tape surface roughness. We derive the signal as well as the noise power spectrum of the induced voltage at the read head. This is done by generalizing the theory of the signal behaviour. The theoretical result turned out to agree quite well with the experimental one. The modulation noise power spectrum can be interpreted to consist of two parts, i.e. the steep peak due to the mechanical causes near the recorded signal wavenumber and the broad peak due to magnetic causes.

PACS: 02.50, 07.55, 41.10D, 75.90

The signal-to-noise ratio (S/N) of a tape is the most important factor in its application as an information storage system in magnetic recording. The mechanism of the signal is quite well understood [1–3], however some problems are still unsolved about the noise. The causes of the noise in the magnetic tape can be classified into two categories, i.e. causes due to its magnetic properties and causes due to its mechanical properties.

Some authors [4–7] have investigated the noise originating from the former causes. Stein [4] and Mallinson [5] analyzed the case where the magnetic particles are distributed discretely but homogeneously. Thurlings [7] has demonstrated very recently that indeed the background noise can be quite well understood by the theories of Stein and Mallinson. But the modulation noise arising in the case that a signal has been recorded, should be explained by the fact that the particles exihibit some inhomogeneous distribution such as agglomeration. Thurlings discussed the noise power spectrum at the read head. originating from the particulate nature including the effects of clustered particulate media. The modulation noise and the special case dc erased noise will be discussed in our paper, too, but we will take into account the inhomogeneous distribution of the particles in a more conventional manner. It will be shown that our result agrees with that of Thurlings's in the essential term.

The noise related to the mechanical properties of the tape originates from the tape surface roughness which cause head-to-tape space variations. Some authors [8, 9] have investigated the dc erased noise due to the surface asperities of the tape. Iijima [9] derived the dc erased noise power spectrum related to the shape and the distribution of the asperities. However there seems to be superfluous complexities to evaluate the parameters concerning the shape and the distribution of the asperities. Furthermore he didn't consider the modulation noise for non-zero wavenumber. We shall discuss also that general case. We take account of mechanical cause of the noise by the way that the surface roughness is assumed to be mainly related to a

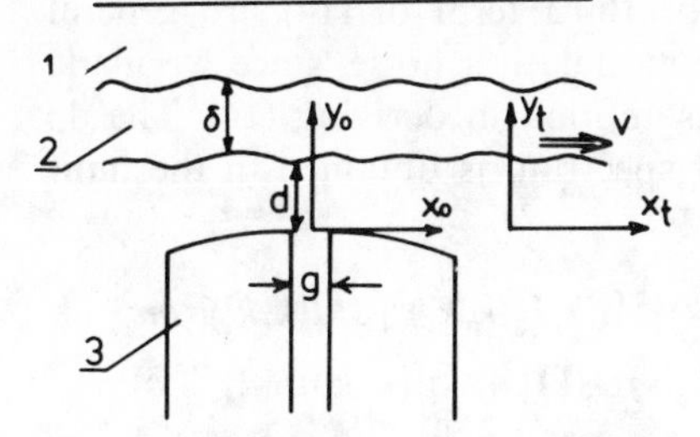

Fig. 1. The read process (1 is the tape base, 2 is the magnetic layer of the tape, 3 is the read head)

lack of magnetization on the tape and not, for example, to the plastic binder.

No one seems to have discussed the tape noise taking into consideration the two main causes at the same stage. In this paper we develop a theory for the power spectrum of the induced voltage at the read head taking account of both main causes. It will be demonstrated that the modulation noise and the dc erased noise as well as the signal can be treated by our theory.

In Sect. 1 the model is explained, and conventional stochastic variables are introduced. In Sect. 2 the expression for the signal and the noise power spectrum are obtained. Furthermore handy forms of the power spectrum are derived by the help of an Ansatz. Section 4 summarizes our results.

1. Model

The read process is shown in Fig. 1. We use (x_t, y_t) for the coordinates on the tape, and (x_0, y_0) for the coordinates fixed on the read-head gap. (x_t, y_t) and (x_0, y_0) obey the equation

$$x_t = x_0 - vt$$
$$y_t = y_0, \tag{1}$$

where v is the relative velocity between the head and the tape.

The read head-detected flux Φ is described by the reciprocity theorem [2],

$$\Phi = \frac{K}{i} \int_{-\infty}^{\infty} dx_0 \int_{d}^{d+\delta} dy_0 \mathbf{H} \cdot \mathbf{M} \tag{2}$$

where K is the constant which includes the number of turn, the read head efficiency and the permeability μ_0, d is the head-to-tape spacing, δ the tape thickness, $\mathbf{H}$ the magnetic field of the Karlqvist read head field excited by a unit current i, and $\mathbf{M}$ the magnetic moment which has been recorded on the tape.

The track width is assumed to be unity for (2).

H is to be represented by the Karlqvist expression [10], namely

$$H_x = \frac{i}{\pi g} \{\arctan[(g/2 + x_0)/y_0] + \arctan[(g/2 - x_0)/y_0]\}$$
$$H_y = \frac{i}{2\pi g} \ln\{[(x_0 + g/2)^2 + y_0^2]/ \cdot [(x_0 - g/2)^2 + y_0^2]\}, \tag{3}$$

where g is the gap length of the read head.

The remanent magnetic moment $\mathbf{M}$ is assumed to be of the form

$$\mathbf{M}(x_t, y_t) = m(x_t, y_t)\mathbf{S}(x_t, y_t), \tag{4}$$

where $\mathbf{S}$ represents the signal pattern recorded on the tape. Now m is not a constant, but it varies around an average value m_{av} because of the inhomogeneous density and dispersion of the magnetic properties, i.e.

$$m(x_t, y_t) = m_{av} + \Delta m(x_t, y_t). \tag{5}$$

The head-to-tape spacing is not constant either, but it varies around an average d_{av} due to the tape surface roughness which is assumed to be related to a lack of magnetization on the tape, i.e.

$$d(x_t) = d_{av} + \Delta d(x_t), \tag{6}$$

Δm and Δd are stochastic variables. We assume that they satisfy the following conditions

$$\langle \Delta m(x_t, y_t) \rangle = 0 \quad \langle \Delta d(x_t) \rangle = 0 \tag{7}$$

Δm and Δd are stationary,

where $\langle \ldots \rangle$ stands for the average; stationary means that $\langle \Delta d(x_t) \Delta d(x_t - X) \rangle$ is a function only of X.

We introduce the Fourier transformation to the wave space (k-space) by the equations,

$$\Delta m(\mathbf{x}_t) = \int_{-\infty}^{\infty} d\mathbf{k} \Delta M(\mathbf{k}) \exp\{\mathrm{j}[\mathbf{k} \cdot \mathbf{x}_t + \theta_m(\mathbf{k})]\}$$
$$\Delta d(x_t) = \int_{-\infty}^{\infty} dk \Delta D(k) \exp\{\mathrm{j}[k x_t + \theta_d(k)]\}, \tag{8}$$

where $\mathbf{x}_t = (x_t, y_t)$. θ_m and θ_d are the stochastic variables in the k-space.

θ_m and θ_d have to be determined so that the conditions (7) are satisfied. It is found that the stochastic variables θ_m, θ_d obey the following equations (Appendix)

$$\langle \exp[\mathrm{j}\theta_m(\mathbf{k})] \rangle = 0 \langle \exp[\mathrm{j}\theta_d(k)] \rangle = 0$$
$$\langle \exp[\mathrm{j}\theta_m(\mathbf{k}_1)] \exp[-\mathrm{j}\theta_m(\mathbf{k}_2)] \rangle = \delta(\mathbf{k}_1 - \mathbf{k}_2) \tag{9}$$
$$\langle \exp[\mathrm{j}\theta_d(k_1)] \exp[-\mathrm{j}\theta_d(k_2)] \rangle = \delta(k_1 - k_2).$$

2. Theory and Discussion

After introducing the following transformation of the coordinates

$$\begin{aligned} x_0 &= x_1 \\ y_0 - \Delta d(x_t) &= y_1 \end{aligned} \quad (10)$$

(1), (2), (4)–(6), and (10) give the induced voltage at the read head, $e(t) = -d\Phi/dt$.
Expanding it up to the first order of $\Delta m, \Delta d$, we find (Appendix)

$$\begin{aligned} e(t) = -K \int_{-\infty}^{\infty} dx_1 \int_{d_{av}}^{d_{av}+\delta} dy_1 \\ \cdot [\mathbf{H}(x_1, y_1) \cdot \frac{\partial}{\partial t} \mathbf{S}(x_1 - vt, y_1) m_{av} \\ + \mathbf{H}(x_1, y_1) \cdot \frac{\partial}{\partial t} [\mathbf{S}(x_1 - vt, y_1) \Delta m(x_1 - vt, y_1)] \\ + \frac{\partial}{\partial y_1} \{\mathbf{H}(x_1, y_1) \cdot \frac{\partial}{\partial t} [\mathbf{S}(x_1 - vt, y_1) \\ \cdot \Delta d(x_1 - vt)] m_{av}\}]. \end{aligned} \quad (11)$$

The first term is the induced voltage of the signal. The second term represents the induced voltage of the noise due to the variation Δm, and the third term reveals the induced voltage due to the variation Δd.
Suppose the sinusoidal signal of wave number k_c is recorded on the tape, $\mathbf{S}(x, y)$ can be described generally by the form,

$$\mathbf{S}(x, y) = \exp(jk_c x)\mathbf{S}(y). \quad (12)$$

Noting that the power spectrum $P(k)$ is given by the Fourier transformation of $\langle e(t)e^*(t-\tau)\rangle$ i.e.

$$P(k) = \int_{-\infty}^{\infty} d\tau \exp(jkv\tau)\langle e(t)e^*(t-\tau)\rangle, \quad (13)$$

from (8), (9), and (11)–(13) we get the power spectrum (Appendix)

$$\begin{aligned} P(k) = v[Km_{av}k_c L_g(k_c)\mathbf{I}_1]^2 \delta(k-k_c) \\ + v[Km_{av}kL_g(k)]^2 \int dk_y \mathbf{I}_2(k \cdot k_y)^2 \\ \cdot |\Delta M(k-k_c, k_y)|^2/m_{av}^2 + v[Km_{av}kL_d(k)L_g(k)]^2 \\ \cdot [\exp(-k\delta)\mathbf{S}(d_{av}+\delta) - \mathbf{S}(d_{av})]^2 |\Delta D(k-k_c)|^2 \end{aligned} \quad (14)$$

with

$$L_g(k) = \sin(gk/2)/(gk/2)$$

$$L_d(k) = \exp(-kd_{av})$$

$$\mathbf{I}_1(k) = \int_{d_{av}}^{d_{av}+\delta} dy \mathbf{S}(y) \exp(-ky)$$

$$\mathbf{I}_2(k_1, k_2) = \int_{d_{av}}^{d_{av}+\delta} dy \mathbf{S}(y) \exp(-k_1 y + jk_2 y).$$

The second and the third term of (14) are general expressions for the modulation noise, since we made no questionable assumption in deriving (14). The dc erased noise power spectrum is obtained in the limit $k_c \to 0$, i.e.

$$\begin{aligned} P_{dc}(k) = v[\kappa m_{av}kL_g(k)]^2 \int dk_y \mathbf{I}_2(k, k_y)^2 |\Delta M(k, k_y)|^2/m_{av}^2 \\ + v[\kappa m_{av}kL_d(k)L_g(k)]^2 [\exp(-k\delta)\mathbf{S}(d_{av}+\delta) \\ - \mathbf{S}(d_{av})]^2 |\Delta D(k)|^2. \end{aligned} \quad (15)$$

In order to get the noise spectrum in a handy form, we make the following assumptions.

Ansatz 1

We put

$$\mathbf{S}(y) = \hat{\mathbf{x}} \quad (16)$$

where $\hat{\mathbf{x}}$ represents the unit vector of $\mathbf{x}$. Equation (16) was introduced by Westmijze [2], when he derived the signal power spectrum.

Ansatz 2

We assume the correlation function in the real space of Δm and Δd to be of the form,

$$\begin{aligned} \langle \Delta m(\mathbf{x}_t)\Delta m(\mathbf{x}_t - \mathbf{x})\rangle = \langle \Delta m^2 \rangle \exp(-\mathbf{x}^2/l_m^2) \\ \langle \Delta d(x_t)\Delta d(x_t - X)\rangle = \langle \Delta d^2 \rangle \exp(-X^2/l_d^2), \end{aligned} \quad (17)$$

where $\langle \Delta m^2 \rangle$ represents the average value of the variation of Δm, and l_m is the correlation length of the variation Δm. l_m will be found to be interpreted as the size of the cluster. $\langle \Delta d^2 \rangle$ represents the average value of the variation of Δd and l_d is the correlation length of the variation Δd. $\langle \Delta d^2 \rangle$ and l_d can be evaluated by the surface roughness' data of the tape, i.e. they are related to the Fourier transformation of the surface roughness' curve by the Wiener-Khintchenés theorem.
By the help of Ansatz 1 and Ansatz 2 $P(k)$ is found to have the form, (Appendix)

$$\begin{aligned} P(k) = A(k_c)\{[1 - \exp(-k\delta)]^2 \delta(k-k_c) \\ + A(k)k \exp(-k\delta)(\langle \Delta m^2 \rangle/m_{av}^2)l_m^2 \\ \cdot \exp[-l_m^2(k-k_c)^2/4](1/\sqrt{\pi}) \int_0^{\delta/l_m} \\ \cdot du \sinh\{l_m k[(\delta/l_m) - u]\} \exp(-u^2) \\ + A(k)[1 - \exp(-k\delta)]^2 (l_d/2\sqrt{\pi})\langle \Delta d^2 \rangle k^2 \\ \cdot \exp[-l_d^2(k-k_c)^2/4], \end{aligned} \quad (18)$$

where

$A(k)$ is $v[Km_{av}L_g(k)L_d(k)]^2$.

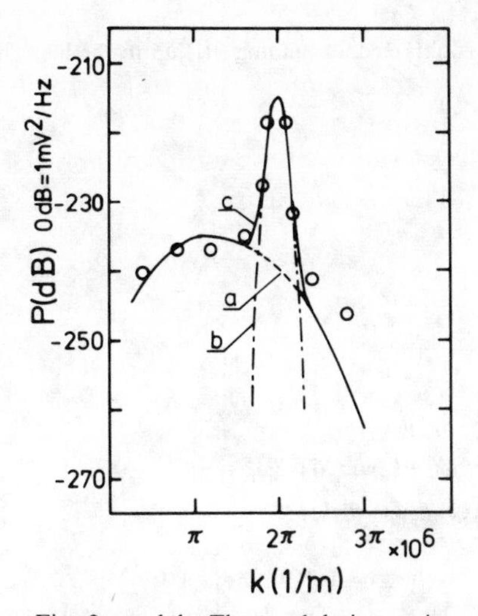

Fig. 2a and b. The modulation noise power spectrum P vs the wavenumber k. (a) denotes the theoretical curve due to the magnetic causes, and (b) the theoretical curve due to the mechanical causes. The curve c is the summation of a and b. The plots are the experimental results, obtained under the conditions: $g=0.4\,\mu\text{m}$, $\delta=4\,\mu\text{m}$, $k_c=2\pi\times 10^6\,(1/\text{m})$, $d_{av}=0.15\,\mu\text{m}$. (d_{av} is evaluated by the wavenumber dependency of the signal)

The first term is the signal term of the recorded wave length k_c, which is the same result as that of Westmijze [2]. The second term represents the modulation noise due to the inhomogeneous density and dispersion of the magnetic properties. The third term reveals the modulation noise due to the head-to-tape spacing variation arising from the tape surface roughness. The dc erased noise power spectrum is easily obtained in the limit $k_c \to 0$ as follows,

$$
\begin{aligned}
P_{dc}(k) &= A(k)k\exp(-k\delta)(\langle \Delta m^2\rangle/m_{av}^2) \\
&\cdot l_m^2 \exp(-l_m^2k^2/4)(1/\sqrt{\pi}) \int_0^{\delta/l_m} \\
&\cdot du \sinh\{l_m k[(\delta/l_m)-u]\}\exp(-u^2) \\
&+ A(k)[1-\exp(-k\delta)]^2(l_d/2\sqrt{\pi})\langle \Delta d^2\rangle \\
&\cdot k^2\exp(-l_d^2k^2/4).
\end{aligned}
\tag{19}
$$

The remarkable progress in magnetic recording in recent years enables the use of a very short wavelength region. For example, the wavelength reaches the order of 1 µm in case of home-used video tape recorder. In this wavelength region the integral of the second term of (18),

$$
\int_0^{\delta/l_m} du \sinh\{l_m k[(\delta/l_m)-u]\}\exp(-u^2) \tag{20}
$$

can be calculated further. The part $\sinh\{l_m k[(\delta/l_m)-u]\}$ of the integrand changes from $\sinh(k\delta)$ to zero in the integration region, which is, in a rough approximation, compared with the change of $\exp(-u^2)$. (The usual tape thickness is $\delta \sim 4\,\mu\text{m}$.)

Equation (20) can be estimated roughly as follows.

$$
\begin{aligned}
&\int_0^{\delta/l_m} du \sinh\{l_m k[(\delta/l_m)-u]\}\exp(-u^2) \\
&\sim C\int_0^{\delta/l_m} du \sinh\{l_m k[(\delta/l_m)-u]\} = \frac{C}{l_m k}[1-\cosh(k\delta)].
\end{aligned}
\tag{21}
$$

where C is a constant and independent of k. Therefore the whole second term of (18) is of the form,

$$
\begin{aligned}
&v(Km_{av}L_gL_d)^2C[1-\exp(-k\delta)]^2(\langle \Delta m^2\rangle/m_{av}^2)(l_m/2\sqrt{\pi} \\
&-\exp[-l_m^2(k-k_c)^2/4].
\end{aligned}
\tag{22}
$$

Thurlings [7] investigated the modulation noise taking into consideration the effect of clustered particulate media. He derived the modulation noise power spectrum to be of the form,

$$
v(K'L_gL_d)^2[1-\exp(-k\delta)]^2(1/2\alpha)\exp[-(k-k_c)^2/2\alpha],
$$

where K' is the constant, and α is the parameter concerning the particle distribution in the cluster. $1/\sqrt{2\alpha}$ corresponds to the size of the cluster. Our result agrees with Thurlings's apart from constants originating from the different models. And l_m introduced in Ansatz 2 can be interpretated as the size of the cluster.

It should be noted by the comparison of (22) and the third term of (18) that the wave number dependencies of the magnetic and mechanical contributions differs by a factor k^2. It is interesting that the variation Δm contributes to the noise power in the form of the ratio $\langle \Delta m^2\rangle/m_{av}^2$; while the variation Δd contributes to the noise power in the direct form $\langle \Delta d^2\rangle$.

In Fig. 2 a theoretical result of the modulation noise power spectrum is shown using (18) and (21). $l_d=5\,\mu\text{m}$ is evaluated from the tape surface roughness by assuming that the tape surface roughness is related to the variation of Δd. $l_m=0.6\,\mu\text{m}$ and $\langle \Delta d^2\rangle/C(\langle \Delta m^2\rangle/m_{av}^2) = 6\times 10^{-14}$ are obtained by a fitting procedure. An experimental result is plotted, too. Although we made a quite rough estimation concerning the parameters, our theoretical result turn out to agree quite well with the experimental one. The modulation noise power spectrum is interpreted to consist of two parts, i.e. the steep peak due to the mechanical causes near the recorded signal wave number and the broad peak due to the magnetic causes. In this paper we reported briefly on the application of our theory to experiments. A further detailed report on that is now under study.

Conclusion

The modulation noise and the special case, dc erased noise were discussed taking account of two main

causes, i.e. the magnetic properties such as the inhomogeneous density and dispersion of the particles of the tape, and the mechanical properties such as the head-to-tape spacing variation arising from the tape surface roughness. We derived not only the signal but also the noise power spectrum. This is done by generalizing the theory of the signal behaviour [1–3]. The noise power spectrum due to the magnetic causes turned out to agree with the result of Thurlings's [7] who has reported a theory for the power spectrum originating from the particulate nature of the tape including the effects of clustered particulate media. We considered the effects of clustered media in a different manner introducing convenient stochastic variables. By our theory also the noise due to mechanical causes could be treated. The derived noise power spectrum due to the mechanical causes turned out to differ from that due to the magnetic causes by a factor of the square of the wavenumber. We reported briefly on the application of our theory to experiments and showed that our theoretical result turn out to agree quite well with the experimental one.
This paper is devoted almost to the theoretical result. A further application of our theory to experimental results is now under way.

Acknowledgements. The authors are very grateful to Prof. Akizuki of Waseda University, Tokyo, for the helpful discussion concerning the introduction of the stochastic variables. They are also thankful to Prof. Schwegler of Bremen University and Dr. Nagai of Waseda University Tokyo.

Appendix A

Derivation of (9)

We shall show that (7) is satisfied with the condition (9).

$$\langle \Delta m(\mathbf{x}_t)\rangle = \langle \int d\mathbf{k}\,\Delta M(\mathbf{k})\exp[\mathrm{j}(\mathbf{k}\cdot\mathbf{x}_t+\theta_m(\mathbf{k})]\rangle$$
$$= \int d\mathbf{k}\,\Delta M(\mathbf{k})\exp(\mathrm{j}\mathbf{k}\cdot\mathbf{x}_t)\langle\exp(\mathrm{j}\theta_m(\mathbf{k}))\rangle = 0 \tag{A1}$$

$$\langle \Delta m(\mathbf{x}_t)\Delta m^*(\mathbf{x}_t-\mathbf{X})\rangle$$
$$= \langle \int d\mathbf{k}d\mathbf{k}'\,\Delta M(\mathbf{k})\Delta M^*(\mathbf{k}')\exp[\mathrm{j}(\mathbf{k}\cdot\mathbf{x}_t+\theta_m(\mathbf{k})]$$
$$\cdot\exp[-\mathrm{j}(\mathbf{k}'\cdot(\mathbf{x}_t-\mathbf{X})+\theta_m(\mathbf{k}')]\rangle$$
$$= \int d\mathbf{k}d\mathbf{k}'\,\Delta M(\mathbf{k})\Delta M^*(\mathbf{k}')\exp[\mathrm{j}(\mathbf{k}-\mathbf{k}')\cdot\mathbf{x}_t+\mathrm{j}\mathbf{k}'\cdot\mathbf{X}]$$
$$\cdot\langle\exp[\mathrm{j}\theta_m(\mathbf{k})]\exp[-\mathrm{j}\theta_m(\mathbf{k}')]\rangle$$
$$= \int d\mathbf{k}d\mathbf{k}'\,\Delta M(\mathbf{k})\Delta M^*(\mathbf{k}')\exp[\mathrm{j}(\mathbf{k}-\mathbf{k}')\cdot\mathbf{x}_t+\mathrm{j}\mathbf{k}'\cdot\mathbf{X}]\delta(\mathbf{k}-\mathbf{k}')$$
$$= \int dk|\Delta M(\mathbf{k})|^2\exp(\mathrm{j}\mathbf{k}\cdot\mathbf{X}). \tag{A2}$$

Equation (A2) is the function only of X, so Δm is stationary. The calculation of Δd is almost same.

Appendix B

Derivation of (11)

Equations (1), (2), and (4)–(6) give

$$\Phi = K\int_{-\infty}^{\infty} dx_1 \int_{d_{av}}^{d_{av}-\delta} dy_1\{H(x_1, y_1+\Delta d(x_1-vt))$$
$$\cdot S(x_1-vt, y_1+\Delta d(x_1-vt))$$
$$\cdot[m_{av}+\Delta m(x_1-vt, y_1+\Delta d(x_1-vt))]\}. \tag{B1}$$

The induced voltage $e(t)$ is obtained, expanding (B1) up to the first order of Δm and Δd.

Appendix C

Derivation of (14)

Equation (8), (11), and (12) give

$$e(t) = K\int_{-\infty}^{\infty} dx_1 \int_{d_{av}}^{d_{av}+\delta} dy_1$$
$$\cdot[\mathbf{H}(x_1,y_1)\cdot\mathbf{S}(y_1)\mathrm{j}x_1 v\exp[\mathrm{j}k_c(x_1-vt)]m_{av}$$
$$+\int dk_x\mathbf{H}(x_1,y_1)\cdot\mathbf{S}(y_1)\mathrm{j}(k_c+k_x)v\exp[\mathrm{j}k_c(x_1-vt)]$$
$$\Delta M(k)\exp[\mathrm{j}k_x(x_1-vt)+k_y y_1+\theta_m(k)]$$
$$+\int dk_y\frac{\partial}{\partial y_1}\mathbf{H}(x_1,y_1)\cdot\mathbf{S}(y_1)\mathrm{j}(k_c+k_x)v\exp[\mathrm{j}k_c(x_1-vt)]$$
$$m_{av}\Delta D(k_x)\exp[\mathrm{j}k_x(x_1-vt)+\theta_d(k_x)]]. \tag{C1}$$

The integration of x_1 in (C1) is easily done.

$$\int dx H(x,y)\exp(\mathrm{j}kx) = (\hat{\mathbf{x}}+\mathrm{j}\hat{\mathbf{y}})L_g(k)\exp(-ky). \tag{C2}$$

After substitution of (C2) into (C1), $\langle e(t)e^*(t-\tau)\rangle$ is obtained to have the form

$$\langle e(t)e^*(t-\tau)\rangle = [Kvm_{av}k_cL_g(k_c)I_1(k_c)]^2\exp(-\mathrm{j}k_cv\tau)$$
$$+\int dk(Kv(k_c+k_x)L_g(k_c+k_x)I_2(k_c+k_x,k_y)$$
$$\cdot|\Delta M(k)|)^2\exp[-\mathrm{j}(k_c+k_x)v\tau]$$
$$+\int dk_x\{Km_{av}(k_c+k_x)L_g(k_c+k_x)L_d(k_c+k_x)$$
$$\cdot\exp[-(k_c+k_x)\delta]\}^2|S(d_{av}+\delta)-S(d_{av})|^2$$
$$\cdot|\Delta D(k_x)|^2\exp[-\mathrm{j}(k_c+k_x)v\tau]. \tag{C3}$$

The Fourier transformation of (C3) gives (14).

Appendix D

Derivation of (18)

I_1^2 and I_2^2 are obtained according to the definition (14) to be of the form,

$$\mathbf{I}_1^2 = \exp(-2kd_{av})\frac{1}{k^2}[1-\exp(-k\delta)]^2$$

$$\mathbf{I}_2^2 = \exp(-2k_1d_{av})\frac{1}{k_1^2+k_2^2} \tag{D1}$$
$$\cdot[\exp(-2k_1\delta)-2\exp(-k_1\delta)\cos(k_2\delta)+1].$$

Noting that

$$\int_0^{\infty} dx\exp(-a^2x^2)/(x^2+b^2) = (\sqrt{\pi}/b)\exp(a^2b^2)\int_{ab}^{\infty}du\exp(-u^2)$$

$$\int_0^{\infty} dx\exp(-a^2x^2)\cos(cx)/(x^2+b^2)$$
$$= (\sqrt{\pi}\;b)\cosh(bc)\exp(a^2b^2)\int_{ab}^{\infty}du\exp(-u^2) \tag{D2}$$
$$+(\sqrt{\pi}/b)\int_0^{(c/2a)}du\sinh[(2au-c)b]\exp(-u^2),$$

we get (18).

Nomenclature

$A(k)$	variable defined by $v(Km_{av}L_gL_d)^2$
d	head-to-tape spacing [m]
d_{av}	average value of d [m]
Δd	variance of d [m]
ΔD	variable defined by (8)
$e(t)$	induced voltage at the read head [V]
g	gap length [m]
$\mathbf{H}$	the Karlqvist head field excited by a current unit i [A/m]
i	a curren unit [A]
j	$\sqrt{-1}$
$k=\|\mathbf{k}\|$	wavenumber [m^{-1}]
k_c	wavenumber of recorded signal [m^{-1}]
K	constant which includes the number of turn of head, the read head efficiency and the permeability μ_0 [Vs/Am]
l_d	correlation length of the variation Δd [m]
l_m	correlation length of the variation Δm [m]
$L_d(k)$	spacing loss defined by $\exp(-kd_{av})$
$L_g(k)$	gap loss defined by $\sin(gk/2)/(gk/2)$
$\mathbf{M}$	magnetization of the tape [A/m]
ΔM	variable defined by (8)
m	amplitude of the magnetization of the signal pattern [A/m]
m_{av}	average value of m [A/m]
Δm	variance of d [A/m]
$P(k)$	power spectrum [V^2/Hz]
$P_{dc}(k)$	noise power spectrum [V^2 Hz]
$\mathbf{S}$	signal pattern of the tape
v	relative velocity between the head and the tape [m s]
$\mathbf{x}_t=(x_t, y_t)$	coordinate on the tape
(x_0, y_0)	coordinate fixed on the read head
(x_1, y_1)	coordinate defined by (10)
δ	tape thickness [m]
θ_d	random phase variable defined by (8)
θ_m	random phase variable defined by (8)
μ_0	permeability [Vs Am]
Φ	read-head detected flux [Wb]

References

1. R.L. Wallance: Bell. Syst. Techn. J. **30**, 1145–1173 (1951)
2. W.K. Westmijze: Philips Res. Dep. **8**, 148–366 (1953)
3. B.K. Middleton: IEEE. Trans. MAG-**11**, 1170 (1975)
4. I. Stein: J. Appl. Phys. **34**, 1976–1990 (1963)
5. J.C. Mallinson: IEEE Trans. MAG-**5**, 182–186 (1969)
6. S. Satake, J. Hokkyo: IECE Techn. Group Meeting of Magn. Rec., Japan, MR **74–23**, 39–49 (1974)
7. L. Thurlings: IEEE Trans MAG-**16**, 507–511 (1980)
8. E.D. Daniel: IEEE Trans. CE-**83**, 250–253 (1964)
9. T. Iijima, S. Hosokawa: IECE Tech. Group Meeting of Magn. Rec., Japan, MR **78–33**, 21–30 (1978)
10. D. Karlqvist: Trans. R. Inst. Techn. (Stockholm) **86**, 3–27 (1954)

Part VIII
Codes

P. A. Franaszek

Sequence-state Methods for Run-length-limited Coding

Abstract: Methods are presented for the encoding of information into binary sequences in which the number of ZEROS occurring between each pair of successive ONES has both an upper and a lower bound. The techniques, based on the state structure of the constraints, permit the construction of short, efficient codes with favorable error-propagation-limiting properties.

1. Introduction

This paper presents a study of a class of codes that are of interest in connection with a number of digital recording and communication techniques. The codes are such that coder output sequences are binary, and have the property that two consecutive ONES are separated by at least d but no more than k ZEROS. The parameter d may be used to control interference between recorded transitions in saturation recording, or to limit spectrum spread in frequency-shift keying. The parameter k imposes a bound on the maximum transition spacing, a dimension that must be specified in most systems that employ self-clocking.

Problems connected with such run-length-limited codes have received considerable attention.[1-6] Application of the codes, particularly to magnetic recording, was discussed by Melas,[1] Kautz,[2] Tang[3] and Gabor.[4] Properties of the output sequences were studied by Melas.[1] Asymptotically optimal coding techniques whose complexity grows linearly with the code word length were described by Kautz[2] and Tang.[3] In addition, a number of short codes and state-oriented coding techniques were presented by Freiman and Wyner,[6] Gabor[4] and Tang.[5]

The approach taken in this paper is based on the use of finite-state machines as models of the run-length-limited sequence constraints. The analysis is thus applicable to any constraints that can be described in this form. Algorithms presented in a recent report[7] are used to construct synchronous (fixed-rate) codes that are optimal in the sense that the maximum word length is minimized for a given bit-per-symbol value. Word lengths of fixed- and of variable-length codes in this class are compiled for a number of (d, k) constraints. The results indicate that varying the word length frequently yields codes that are shorter and easier to implement. The problem of error propagation, which arises in state-dependent and variable-length coding,[8] is studied. It is shown that one can always limit error propagation in fixed-length (d, k) codes by a proper assignment of message digits to code words. A method for constructing error-propagation-limiting, variable-length (d, k) codes is described, the method being valid for the more general case of constrained sequences with finite memory. Finally, some examples of code construction are included to illustrate the methods.

The author is located at the IBM Thomas J. Watson Research Center, Yorktown Heights, New York.

2. Run-length-limited sequence constraints

In some applications it is desirable to impose minimum and maximum distances between transitions or pulses in a signal. Digital magnetic recording is an example. The recording medium is typically partitioned into intervals of length T. Information is stored in these intervals by means of the presence or absence of transitions between saturation levels. Intervals in which transitions occur are assigned the value ONE. If no transition occurs, the interval is assigned the value ZERO. The separation between such transitions must be sufficiently great to limit interference to acceptable levels. If clocking is to be derived from the recorded data, an additional requirement is that transitions occur frequently enough to provide adequate energy for the timing circuits.

Suppose that Q and V are respectively the minimum and maximum tolerable distances between transitions. A possible approach to the signal design problem is to require that the recorded sequence be (d, k) *run-length-limited* with $(d + 1)T \geq Q$ and $(k + 1)T \leq V$. That is, at least $(d + 1)$ but no more than $(k + 1)$ intervals occur between each pair of successive transitions. If Q and V are fixed, an increase in $(d + 1)$ (i.e., a decrease in T)

Reprinted with permission from *IBM J. Res. Develop.*, vol. 14, pp. 376–383, July 1970.

will produce an increase in the amount of information that can be stored, and a decrease in the level of tolerable degradation. Thus, the choice of d and k is a function of the interference and of the circuit quality in a given system.

The (d, k) sequence constraints may be represented by a finite-state sequential machine. Figure 1 illustrates a possible state-transition diagram. There are $(k + 1)$ states, $(\sigma_1, \sigma_2, \cdots, \sigma_{k+1})$. Transmission of a ZERO takes the sequence from state σ_i to state σ_{i+1}. Transmission of a ONE takes the sequence to σ_1. That is, state σ_r, $r = 1, 2, \cdots, k + 1$, indicates that $(r - 1)$ ZEROS have occurred since the last ONE. A ONE may be transmitted only when the sequence occupies states $\sigma_{d+1}, \cdots, \sigma_{k+1}$. If the sequence occupies state σ_{k+1}, then only a ONE may be transmitted.

The skeleton transition matrix, which gives the number of ways of going from state σ_i to state σ_j, is given by the $(k + 1) \times (k + 1)$ array:

$$\mathbf{D} = \{d_{ij}\}, \tag{1}$$

where

$$d_{i1} = 1, \quad i \geq d + 1; \tag{2}$$

$$d_{ij} = 1, \quad j = i + 1; \tag{3}$$

$$d_{ij} = 0, \quad \text{otherwise.} \tag{4}$$

For example, the $\mathbf{D}$ matrix for $(d, k) = (2, 4)$ is

$$\mathbf{D} = \begin{bmatrix} 0 & 1 & 0 & 0 & 0 \\ 0 & 0 & 1 & 0 & 0 \\ 1 & 0 & 0 & 1 & 0 \\ 1 & 0 & 0 & 0 & 1 \\ 1 & 0 & 0 & 0 & 0 \end{bmatrix}. \tag{5}$$

The above representation is related to the input-restricted channels first studied by Shannon.[9] The finite-state machine model permits the computation of the channel capacity, defined as the number of bits per symbol that may be carried by the sequence. This quantity is given[9] by the base-two logarithm of the largest real root of

$$\det [d_{ij} Z^{-1} - \delta_{ij}] = 0. \tag{6}$$

For large values of k it may be more convenient to obtain the channel capacity from[8]

$$C \approx \frac{1}{n} \log_2 \sum_{ij} d^n_{ij}, \tag{7}$$

where

$$d^n_{ij} \equiv [\mathbf{D}^n]_{ij}. \tag{8}$$

If the value of n is a power of two, then $\mathbf{D}^n$ may be obtained from $\mathbf{D}$ by a total of $\log_2 n$ matrix multiplications.

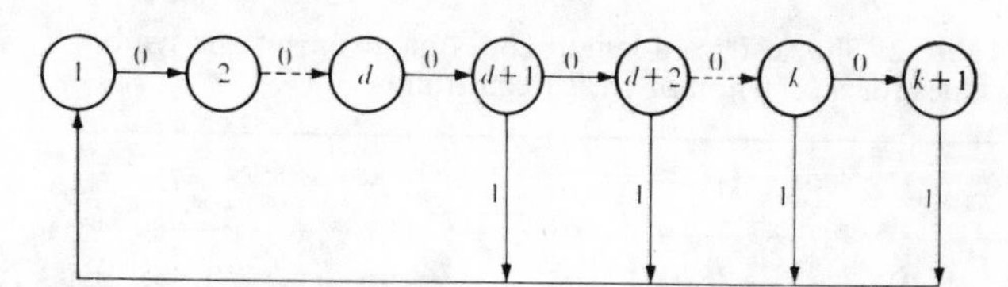

Figure 1 State-transition diagram for a (d, k) sequence.

Table 1: Channel capacities for a selection of (d, k) constraints.

d	k	*Channel capacity in bits/symbol*
0	1	0.694
0	2	0.879
0	3	0.947
1	3	0.552
1	4	0.617
1	5	0.651
1	7	0.679
2	5	0.465
2	8	0.529
2	11	0.545
3	7	0.406
3	11	0.452
3	15	0.462
4	9	0.362
4	14	0.397
5	12	0.337
5	17	0.356
6	13	0.301
6	17	0.318
6	20	0.324
7	15	0.279
7	23	0.298
8	17	0.260
8	26	0.276

The required value of n may be estimated from the following bound for the capacity[7]:

$$\frac{1}{n + k} [\log_2 \sum_{ij} d^n_{ij} - 2 \log_2 (k + 1)] \tag{9}$$

$$\leq C \leq \frac{1}{n} \log_2 \sum_{ij} d^n_{ij}.$$

Values of C for a number of (d, k) constraints were obtained by computer calculation. These values are given in Table 1.

A property of the above sequence constraints is that the channel is of finite memory. That is, the channel state may be uniquely determined from a finite number of previously transmitted symbols. More precisely, if the channel occupies state σ_r, $r = 1, 2, \cdots, k + 1$, (we use the numbering of Fig. 1), then the channel state may

Table 2: Shortest fixed-length codes of given bit-per-symbol values for a selection of (d, k) constraints.

d	k	α	N	E
0	1	3	5	0.865
0	2	4	5	0.910
0	3	9	10	0.950
1	3	1	2	0.906
1	5	6	10	0.922
1	7	6	10	0.884
2	5	4	10	0.860
2	8	11	22	0.945
2	11	8	16	0.917
3	7	46	115	0.985
3	11	8	20	0.885
3	15	8	20	0.866
4	9	9	27	0.921
4	14	12	33	0.917
5	12	9	30	0.890
5	17	15	45	0.843
6	13	12	44	0.906
6	17	9	33	0.858
6	20	15	50	0.926
7	15	9	36	0.896
7	23	7	28	0.839
8	17	27	108	0.962
8	26	12	48	0.906

α = Number of bits per word
N = Word length
$E = [\alpha/N]/C$, the code efficiency

be identified through knowledge of the r previously transmitted symbols. This property is used to limit error propagation in variable-length codes.

3. Fixed-length binary *(d, k)* codes

• *Discussion*

Suppose one wishes to map binary information onto a (d, k) sequence with a code of fixed length. The message sequence is partitioned into blocks of length α, and such blocks mapped by the code onto words composed of N channel symbols. A code may be state dependent, in which case the choice of the word used to represent a given binary block is a function of the channel state, or the code may be state independent. State independence implies that code words can be freely concatenated without violating the sequence constraints.[6] It is an additional restriction that, in general, leads to codes that are longer than state-dependent codes for a given bit-per-symbol value. In some cases state independence may yield advantages in error propagation limitation, since channel words may be decoded without knowledge of the state. However, state-independent decoding may be achieved for any fixed-length (d, k) code, as is indicated by the theorem below. Since coder and decoder complexity tends to increase exponentially with the code word length, it is usually advantageous to search for the shortest existing code without regard to state independence, and to achieve state-independent decodability by a proper assignment of message digits to code words.

Let S denote the states of a finite-state machine that represents a (d, k) sequence. Let $W(\sigma_i)$ denote the set of words that may be transmitted when the sequence occupies the initial state σ_i.

Theorem 1: Given a set of states $S' = \{\sigma_i\} \subset S$ and a class of associated word sets $\{W(\sigma_i)\}$ such that each $W(\sigma_i)$ contains at least 2^α words, it is possible to assign binary blocks b_r, $r = 1, 2, \cdots, 2^\alpha$, to the word sets $\{W(\sigma_i)\}$ so that there is a unique inverse mapping.

Proof: A word w that is allowable from at least one state (i.e., it may be obtained from that state by an appropriate path through the state transition diagram) is not allowable from another state either because there is a ONE too close to the beginning of the word, or because there are too many ZEROS before the first ONE. Suppose w is allowable from state σ_n and σ_{n+m}, where $m > 0$. Then the allowability from σ_n implies that the first ONE is not too close to the beginning for w to be allowable from $\sigma_{n+1}, \sigma_{n+2}, \cdots, \sigma_{n+m-1}$. The allowability of w from σ_{n+m} implies that it does not have too many ZEROS in the prefix to be allowable from $\sigma_{n+1}, \sigma_{n+2}, \cdots, \sigma_{n+m-1}$. Thus,

$$w \in W(\sigma_n) \cap W(\sigma_{n+m}) \Rightarrow w \in W(\sigma_{n+j}),$$

$$j = 1, 2, \cdots, m-1, \quad (10)$$

which is a sufficient condition[8] for the theorem to hold.

• *Fixed-length codes of minimum length*

A quick method for determining whether there exists a code with word length N and α bits per word is given in Ref. 7. It is a recursive search technique for determining the existence of a set of *principal states* through operations on the **D** matrix. These are the states from each of which there exists a sufficient number, 2^α, of paths terminating at other principal states. The existence of a set of principal states is a necessary and sufficient condition for the existence of a code with the given values of α and N, and in the case of (d, k) sequences, is a necessary and sufficient condition for the existence of a state-independently decodable code. A number of (d, k) combinations and the length of the shortest existing fixed-length codes for a given ratio of α/N were computed. These are listed in Table 2, which may be used as an aid in construction of both state-dependent and state-independent codes. In the latter case the Table indicates a lower bound on the code-word length for a given bit-per-symbol value. The systematic construction of a state-dependent code is illustrated next.

• *Code construction*

The words available for encoding are the paths of length N connecting the principal states. These may be obtained from the Nth power of the $(k+1) \times (k+1)$ channel-transition matrix

$$\mathbf{A} = \{a_{ij}\}, \tag{11}$$

where a_{ij} represents the sequence digit corresponding to a transition from state σ_i to σ_j. If it is not possible to go from σ_i to σ_j with one digit (i.e., if $d_{ij} = 0$), then $a_{ij} = \emptyset$, where $\emptyset$ denotes the null symbol. Powers of $\mathbf{A}$ are formed by the operations of disjunction, $+$, and concatenation. The concatenation of the null symbol $\emptyset$ with any symbol results in $\emptyset$.

As an example, consider the transition matrix

$$\mathbf{A} = \begin{bmatrix} 1 & 0 \\ 1 & \emptyset \end{bmatrix}, \tag{12}$$

which corresponds to $(d, k) = (0, 1)$.

The outputs of length two are given by

$$\mathbf{A}^2 = \begin{bmatrix} 11 + 01 & 10 \\ 11 & 10 \end{bmatrix}. \tag{13}$$

Code words in $W(\sigma_i)$ may be obtained from $\sum_j [\mathbf{A}^N]_{ij}$, where the operation, $+$, is taken over the principal states.

A state-independently decodable code may be obtained as follows[8]:

1) List the principal states $\sigma_{m_1}, \sigma_{m_2}, \cdots, \sigma_{m_q}$, where $m_1 < m_2 < \cdots < m_q$.
2) Assign a binary word of length α to each of the 2^α words from $W(\sigma_{m_1})$.
3) If $w_x \in W(\sigma_{m_i}) \cap W(\sigma_{m_j})$, where $i < j$, assign the same binary word to w_x for coding from σ_{m_i} as from σ_{m_j}. If $w_x \in W(\sigma_{m_j})$ but $w_x \notin W(\sigma_{m_i})$, assign the binary word to a $w_y \in W(\sigma_{m_i})$ such that $w_y \notin W(\sigma_{m_j})$.

Theorem 1 assures the success of the procedure.

• *Example*

Consider $(d, k) = (1, 3)$. $\mathbf{D}$ and $\mathbf{D}^2$ are given by

$$\mathbf{D} = \begin{bmatrix} 0 & 1 & 0 & 0 \\ 1 & 0 & 1 & 0 \\ 1 & 0 & 0 & 1 \\ 1 & 0 & 0 & 0 \end{bmatrix}, \tag{14}$$

$$\mathbf{D}^2 = \begin{bmatrix} 1 & 0 & 1 & 0 \\ 1 & 1 & 0 & 1 \\ 1 & 1 & 0 & 0 \\ 0 & 1 & 0 & 0 \end{bmatrix}. \tag{15}$$

Table 2 shows that a bit-per-symbol value of $\frac{1}{2}$ represents over 90 percent of the channel capacity.

Use of the search algorithm described in the subsection on fixed-length codes of minimum length indicates that the shortest existing code with this bit-per-symbol value has the parameters $\alpha = 1$, $N = 2$. The principal states are σ_1, σ_2, σ_3. Code words may be obtained from

$$\mathbf{A}^2 = \begin{bmatrix} 01 & \emptyset & 00 & \emptyset \\ 01 & 10 & \emptyset & 00 \\ 01 & 10 & \emptyset & \emptyset \\ \emptyset & 10 & \emptyset & \emptyset \end{bmatrix}. \tag{16}$$

The alphabets available for encoding are the word sets associated with the principal states:

$$W(\sigma_1) = \begin{cases} 01 \\ 00 \end{cases}, \tag{17}$$

$$W(\sigma_2) = \begin{cases} 01 \\ 10 \end{cases}, \tag{18}$$

$$W(\sigma_3) = \begin{cases} 01 \\ 10 \end{cases}. \tag{19}$$

A state-independently decodable code may be constructed with the following binary assignments:

Binary symbols	*State*	*Channel code words*
1	σ_1	01
	σ_2	01
	σ_3	01
0	σ_1	00
	σ_2	10
	σ_3	10

This code is well known and is usually referred to as MFM (modified frequency modulation). It is of interest to note that it results almost automatically from the application of the above techniques to the $(d, k) = (1, 3)$ sequence.

4. Variable-length synchronous binary codes

• *General*

The codes discussed in the previous section have the property that the code may be a function of the state occupied by the sequence. This permits a minimization of word length over the class of fixed-length codes with a given bit-per-symbol value. In this section, an additional degree of freedom is introduced by allowing the word length to vary. The result is often a significant decrease in the required code word length. The rate of information

Table 3: Shortest variable-length codes of given bit-per-symbol values for a selection of (d, k) constraints.

d	k	α	N	M	E
2	8	1	2	4	0.945
2	11	1	2	4	0.917
3	7	2	5	8	0.985
3	11	2	5	3	0.885
3	15	2	5	3	0.866
4	9	1	3	3	0.921
4	14	4	11	3	0.917
5	12	3	10	3	0.890
5	17	1	3	6	0.843
6	13	3	11	3	0.906
6	17	3	11	2	0.858
6	20	3	10	4	0.926
7	15	1	4	3	0.896
7	23	1	4	3	0.839
8	17	1	4	8	0.962
8	26	1	4	5	0.906

α = number of bits per word
N = word length
NM = maximum word length

transmission is kept constant by fixing the bit-per-symbol value for each word. That is, words of length N carry half as many bits as those of length $2N$. Algorithms described in a recent report[7] are used to construct variable-length fixed-rate (synchronous) codes that are optimal in the sense that the maximum word length is minimized. Procedures are developed to find codes with error-propagation-limiting properties.

- *Variable-length codes of minimum length*

A recursive search and optimization method to determine whether there exists a variable-length code with a given bit-per-symbol value and maximum word length is described in Ref. 7. To apply the procedure, a *basic channel word length* N and number α of bits per N channel symbols are chosen, along with a maximal word length MN. Words may be of length jN, $j = 1, 2, \cdots, M$. The procedure involves operations on powers of the $\mathbf{D}$ matrix. If the search is successful, the results are a set of states from which coding may be performed (termed the *principal states*) and, for each such state σ_i, a set of paths that maximizes the quantity

$$\Psi(\sigma_i) = L_1(\sigma_i) + 2^{-\alpha} L_2(\sigma_i) + \cdots + 2^{-\alpha(M-1)} L_M(\sigma_i), \tag{20}$$

where $L_j(\sigma_i)$ is the number of available code words of length jN. It is required that $\Psi(\sigma_i) \geq 2^{\alpha}$ for a code to exist. Terminal states are specified for the paths. These terminal states are in turn principal states, so that channel words always lead to states from which coding is possible.

The shortest maximum word lengths for a number of (d, k) constraints and bit-per-symbol values were computed with the above technique and are shown in Table 3. A comparison of the variable-length codes with the fixed-length codes of Table 2 shows that the extra degree of freedom in variable-length coding often yields a very significant decrease in word length. For example, the $(d, k) = (4, 9)$ constraints, with a bit-per-symbol value of $\frac{1}{3}$, result in fixed-length codes with a minimum of 9 bits per word. Thus, at least $2^9 = 512$ words are required for fixed-length coding, while a variable-length code (illustrated in an example given later in this section) exists with a maximum word length corresponding to 3 bits, and which has a total of 6 words. Since coding is done by table look-up, the advantage of the variable-length code is clear. Further benefits resulting from the shorter word length are discussed below.

- *Error propagation limitation*

Consider the fixed-length codes of Section 3. It was shown that if a fixed-length code exists, then it is possible to code so that decoding can be performed without a knowledge of the state occupied by the channel at the beginning of the transmitted word. This property implies that an error in the detection of a symbol in a given word does not affect the decoding of the next word. However, the property is not sufficient to limit error propagation in codes of variable word length since an error in detection may cause the decoder to treat the received symbol as part of a word of length different from that which was transmitted, possibly resulting in serious error propagation. Thus, error propagation in variable-length codes may result from improper blocking of the received sequence into words (misframing), as well as from code state dependency.

Suppose it were possible to mark word endings with, for example, a special sequence of symbols. Misframing as a result of an error in detection could then be eliminated after the correct reception of at most one word. For channels with finite memory [of which the (d, k) constraints are an example] it is possible to achieve essentially this result by an appropriate choice of code paths, as shown below. Since the method depends in part on identification of the sequence state, it overcomes error propagation due to state dependence as well as to misframing.

Let S_a be the subset of the principal states in which code words actually terminate. Consider a code such that code words are *required* to terminate when a state that is a member of S_a is entered after rN symbols, where r is an integer. Since words may end only after an integer multiple of N symbols, this restriction is sufficient to permit the decoder to determine word endings by tracking the state of the sequence. Knowledge of the state also resolves ambiguities resulting from code-state dependence.

Let L be the maximum number of channel symbols that are required to identify such a state. In general, L is less than the channel memory, $\mathfrak{M}$. The decoder may identify the first word termination occurring after at most L correctly received symbols. Note that this is equivalent to marking word endings with any one of a number of special sequences, some of which may be longer than code words. An advantage of this approach over that of using a single sequence to mark word terminations is that it may result in a larger number of available code words of a given length, and thus in a shorter code for a given bit-per-symbol value.

- *A search routine for a class of error-propagation-limiting codes*

In order to implement the error-propagation-limiting (EPL) procedure discussed in the previous section, it is necessary to find a set of states S_e (termed an EPL set) and paths connecting them that have the following properties:

1) The set S_e is a subset of S_p, the set of principal states. This is due to the necessity of starting words from principal states.
2) A word that enters a state $\sigma_i \in S_e$ after rN symbols, $r = 1, 2, \cdots, M$, terminates there. All words terminate in states that are members of S_e.
3) Associated with each $\sigma_i \in S_e$ there is a set of words of length rN, $r = 1, 2, \cdots, M$, such that

$$\Psi(\sigma_i) = L_1(\sigma_i) + 2^{-\alpha} L_2(\sigma_i) + \cdots + 2^{-\alpha(M-1)} L_M(\sigma_i) \geq 2^{\alpha}. \tag{21}$$

It follows from these properties that there may be instances when one or more states must be eliminated from S_p in order to form a set S_e, if it exists.

Let V_p^r be an r-tuple of states that may be elided from S_p while the possibility of encoding is maintained. That is, from each state $\sigma_i \in (S_p - V_p^r)$ there is available a sufficient number of words terminating in states belonging to $(S_p - V_p^r)$ to satisfy Eq. (21). Such subsets of the principal state set are termed *principal subsets*. Let Ω^r be the set of V_p^r r-tuples.

The set Ω' can be found by using the optimization procedure of Ref. 7 to determine whether $(S_p - \sigma_i)$ is a principal subset of states for each $\sigma_i \in S_p$. The dimensionality of the search for all principal subsets of S_p may be reduced by noting that if an r-tuple of states is a member of Ω^r, then all combinations of $(r - 1)$ members of the r-tuple are members of Ω^{r-1}.

Each principal subset S_p^a may be tested to determine whether it forms an EPL set. The subset S_p^a is an EPL set if, for each $\sigma_i \in S_p^a$, $\Psi_e(\sigma_i) \geq 2^{\alpha}$.

$$\Psi_e(\sigma_i) = \sum_{\sigma_k \in \mathcal{S}_1} d_{ik}^N + 2^{-\alpha} \sum_{\sigma_e \in \mathcal{S}_1} \sum_{\sigma_m \in \mathcal{S}_2} d_{ie}^N d_{em}^N + \cdots + 2^{-\alpha(M-1)} \sum_{\sigma_e \in \mathcal{S}_1} \cdots \sum_{\sigma_q \in \mathcal{S}_{M-1}} \sum_{\sigma_r \in \mathcal{S}_M} d_{ie}^N \cdots d_{qr}^N, \tag{22}$$

where the $\mathcal{S}_r \equiv S_p^a[rN, \sigma_i]$ are those members of S_p^a that are reachable from σ_i with rN symbols.

The available code words can be obtained from the channel-transition matrix **A**. Let $W_r(\sigma_i)$ be the words of length rN available from state σ_i. Using the operations defined in Section 3 we construct the words

$$W_1(\sigma_i) = \sum_{\sigma_q \in S_p^a} a_{iq}^N$$

$$W_2(\sigma_i) = \sum_{\sigma_q \in S_p^a} \sum_{\sigma_l \in S_p^a} a_{iq}^N a_{ql}^N$$

$$\vdots$$

$$W_{MN}(\sigma_i) = \sum_{\sigma_q \in S_p^a} \sum_{\sigma_l \in S_p^a} \cdots \sum_{\sigma_n \in S_p^a} a_{iq}^N a_{ql}^N \cdots a_{mn}^N. \tag{23}$$

Examples of codes for $(d, k) = (4, 9)$ and $(d, k) = (7, 15)$ are given below.

- *Example: a variable-length code for* $(d, k) = (4, 9)$

Consider the sequence with $(d, k) = (4, 9)$. There are ten states: $\sigma_1, \sigma_2, \cdots, \sigma_{10}$. Using the procedure discussed in the second subsection of Section 4 a set of principal states can be found for coding with $\alpha = 1$, $N = 3$ and $M = 3$. These are σ_4, σ_5, σ_6 and σ_7. In this case, S_p is an EPL set. Moreover, $\{\sigma_4, \sigma_5, \sigma_6, \sigma_7\}$ is the only principal subset of states; elimination of one state from this subset eliminates the possibility of coding with the above α, N and M parameters. Channel words corresponding to the EPL paths for the principal states, obtained from Eq. (23), are shown in Fig. 2.

Table 4 shows an assignment of binary digits to the channel words such that decoding may be performed without recourse to state information as long as there are no errors.

Words that enter principal states after 3, 6 or 9 symbols terminate there. Word endings may be found through state identification. Since n symbols are required to identify σ_n, the correct reception of any word except 000 is sufficient to determine word termination. In the latter case, this can be done after reception of the next word, so that at worst, code words corresponding to four bits must be correctly received in order to identify word termination after an error in detection.

Table 2 shows that fixed-length coding with the same bit-per-symbol value as in the current example would require channel words of minimum length 27, each representing 9 bits. Here table look-up coding and decoding

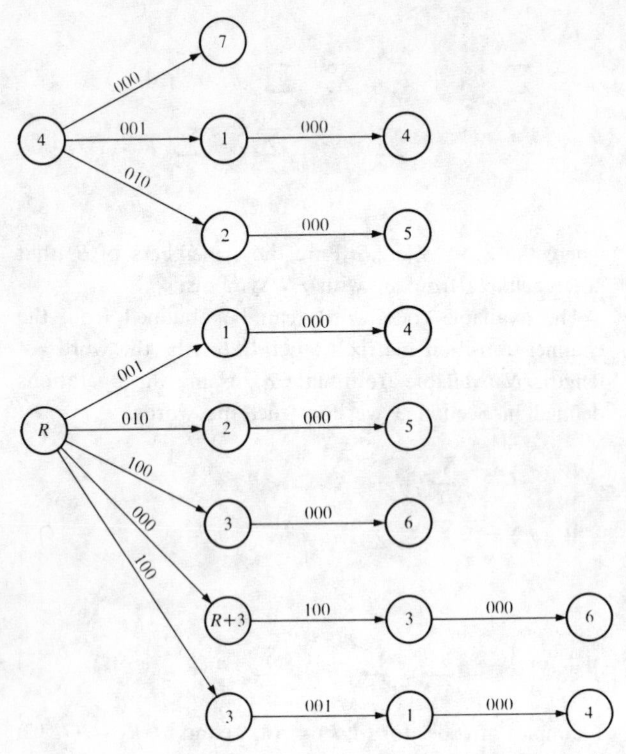

Figure 2 EPL coding paths for $(d, k) = (4, 9)$, $\alpha = 1$, $N = 3$ and $M = 3$. The principal states are σ_4, σ_5, σ_6, σ_7; $R = 5$, 6 and 7 for the paths corresponding to σ_5, σ_6 and σ_7, respectively.

might not be practical. Moreover, a single error in detection could result in the incorrect decoding of nine bits.

• *Example: Variable-length code for* $(d, k) = (7, 15)$

The $(d, k) = (7, 15)$ channel has sixteen states. A variable-length code with $\alpha = 1$, $N = 4$ and $M = 3$ exists, as can be seen in Table 3. The principal states for such encoding are $\sigma_5, \sigma_6, \cdots, \sigma_{13}$. These do not form an EPL set. However, there exist principal subsets that have the EPL property. The smallest such set is $\{\sigma_5, \sigma_6, \sigma_7, \sigma_8, \sigma_{10}, \sigma_{11}, \sigma_{13}\}$. A code with these as terminal states can be formed as shown in Table 5.

Word termination may be detected by channel state identification. That is, words terminate when a state in the principal subset is entered after rN symbols, where r is an integer. Once the detected sequence has been partitioned into words, decoding may proceed without further recourse to state information. Misframing as a result of an error in detection is corrected after the detection of at most 16 channel symbols, which correspond to 4 bits.

5. Conclusions

Methods have been described for the systematic construction of run-length-limited codes. The techniques are based on modeling the sequence constraints by finite-state machines, and are thus applicable to any constraints that can be described in this form. It was shown that varying the code word length often yields codes with superior error-propagation-limiting properties that are simpler and easier to implement than the corresponding codes of fixed word length.

Table 4 A $(d, k) = (4, 9)$ code for $\alpha = 1$, $N = 3$ and $M = 3$.

State	*Binary symbols*	*Channel code words*
σ_4	0	000
	10	001 000
	11	010 000
$\sigma_5, \sigma_6, _7$	10	001 000
	11	010 000
	01	100 000
	001	000 100 000
	000	100 001 000

Table 5 A $(d, k) = (7, 15)$ code for $\alpha = 1$, $N = 4$ and $M = 3$.

State	*Binary symbols*	*Channel code words*
σ_5	00	0001 0000
	010	0000 0100 0000
	011	0000 1000 0000
	11	0000 0000
	100	0000 0001 0000
	101	0000 0010 0000
σ_6, σ_7	1	0000
	00	0001 0000
	01	0010 0000
$\sigma_8, \sigma_{10}, \sigma_{11}, \sigma_{13}$	00	0001 0000
	01	0010 0000
	10	0100 0000
	11	1000 0000

Acknowledgment

The programming support of G. Krajcsik is gratefully acknowledged.

References

1. C. M. Melas, "Quelques Proprietes des Codes Binaires avec Symbols Consecutifs," IEEE International Symposium on Information Theory, 1967.
2. W. H. Kautz, "Fibonacci Codes for Synchronization Control," *IEEE Trans. Info. Theory* **IT-11,** 284 (1965).

3. D. T. Tang, "Run Length Limited Codes," IEEE International Symposium on Information Theory, 1969.
4. A. Gabor, "Adaptive Coding for Self-clocking Recording." *IEEE Trans. Elect. Computers* **EC-16,** 866 (1967).
5. D. T. Tang, "Practical Coding Schemes with Run-length Constraints," IBM Research Report RC-2022, Thomas J. Watson Research Center, Yorktown Heights, New York, 1968.
6. C. V. Freiman and A. D. Wyner, "Optimum Block Codes for Noiseless Input Restricted Channels," *Info. and Control* **7,** 398 (1964).
7. P. A. Franaszek, "On Synchronous Variable Length Coding for Discrete Noiseless Channels," *Info. and Control* **15,** 155 (1969).
8. P. A. Franaszek, "Sequence-state Coding for Digital Transmission," *Bell System Tech. J.* **47,** 143 (1968).
9. C. E. Shannon, "A Mathematical Theory of Communication," *Bell System Tech J.* **27,** 379 (1948).

Received October 9, 1969

AN OPTIMIZATION OF MODULATION CODES IN DIGITAL RECORDING

T. Horiguchi and K. Morita,
Central Research Lab.,
Nippon Electric Co., Ltd., Kawasaki Japan

ABSTRACT

A few new modulation codes(or run-length-limited codes) in digital magnetic recording is presented. Linear density limits of the new codes and the existing codes are evaluated for a typical recording channel. One of the new codes shows the highest linear density limit among the existing codes and is improved over, for example, MFM and 4/5-rate-NRZI codes by about 20% to 30% in the density limit for the channel. A method for constructing the new codes is presented.

1. INTRODUCTION

Various modulation codes are used in digital recording: FM, MFM and 4/5-rate-NRZI codes. They are in general categorized into run-length-limited codes (RLL codes). RLL codes are such that any two consecutive ones in coded binary sequence are separated by at least d zeros but no more than k zeros. The constraint d is used to control pulse crowding effects. The constraint k is used to provide a self-clocking ability. To meet the (d,k) constraint, m bits of data are mapped into n bits of code, on the average, where $n > m$. According to Franaszek [1], RLL codes are denoted by (d,k;m,n;r). Codes with r=1 are fixed length codes. Codes with $r > 1$ are called variable-length fixed rate codes. In these codes, a data sequence is partitioned into blocks of length m, and j consecutive blocks are mapped into a binary code word of length jn, varying j between 1 and r to achieve the (d,k) constraint. The resulting coded binary sequence is then converted into a waveform using NRZI rules, i.e., a transition for one and no transition for zero. The advantages of variable length codes over fixed length codes are seen in [1]. Existing codes, extracted from [1-4] are presented in Table I. Provided that a data sequence of rate 1/T bits/sec is encoded into a (d,k;m,n;r) code sequence, (m/n)T sec may be allocated to one code bit. The time is called detection window $T_w(=(m/n)T)$: In demodulation, detection of the transition is done in the time. Minimum and maximum interval between transitions are $T_{min}=(m/n)(d+1)T$ sec and $T_{max}=(m/n)(k+1)T$ sec. The degree to which the effects of signal distortion can be reduced depends on the following parameters:

- Code rate R=m/n. Codes with a high R are less sensitive to timing jitter caused by noise and peak-shift.
- Minimum transition interval M=(m/n)(d+1). Codes with high M may provide read-back signals with a high signal-to-noise ratio(S/N).
- P=(k+1)/(d+1) ratio. Codes with high P tend to have a large peak-shift, poor self-clocking ability and a large spectral DC component.

Hence, codes with high R, high M and low P are desirable to maximize the linear density for a given recording channel, while maintaining satisfactory data reliability. However, realizable codes with high R have low M and codes with high M result in low R, as shown in Table I. That is, M is inversely proportional to R. Hence, the codes maxmizing density may be selected by compromise between M, R and P. Here, the linear density limit, defined by reference [5], can be used to indicate the degree to which a code is suitable for maximizing density in the recording channel. We calculated the linear density limits of all the codes in Table I. The results are shown in Table I. The calculating method are the same as are written in reference [5]. The conditions for the calculations, i.e., the channel characteristics and the recording system are assumed to be the same as are written in reference [5]. In the calculations, all the codes with P=2 in Table I are assumed to have the same peak-shift vs. recording frequency curve of MFM code as is written in [5]. And, all the codes with P=3 are assumed to have the same peak-shift vs. recording frequency curve of 4/5-rate NRZI code as is written in reference [5]. MFM code and 4/5-rate NRZI code have P=2 and P=3, respectively. The codes with higher density limits may be more suitable for maximizing linear density in the recording channel. As is seen from Table I, the density limits of the existing codes are comparatively low. This is because the existing codes have either high M and low R or low M and high R. So, it was assumed that codes with medium R ($1/2 < R \leq 2/3$), medium M ($1 < M \leq 4/3$ or d=1) and as low a P as possible, will match the channel characteristics and have higher density limits. The new codes with the above parameters are constructed by a method presented in Section 2. They have comparatively higher density limits for the channel, as shown in Table I.

TABLE I

Parameters and linear density limis of new codes and existing codes with $2 \leq P \leq 4$.
D: Linear density limit (Bits/mm).

d	k	m	n	r	M	R	D	Note
0	2	4	5	1	0.8	0.8	200	4/5-rate NRZI
0	3	9	10	1	0.9	0.9	below 231	
1	3	1	2	1	1.0	0.5	185	MFM
2	8	1	2	4	1.5	0.5	217	
3	7	2	5	8	1.6	0.4	211	
4	14	4	11	3	1.8	0.36	below 185	[1]
5	17	1	3	6	2.0	0.33		
6	17	3	11	2	1.9	0.27		
7	15	1	4	3	2.0	0.25	152	
8	17	1	4	8	2.5	0.25	173	
2	5	2	5	?	1.2	0.4	173	[2] Tang's
0	1	2	3	?	0.66	0.66	169	[3]
1	7	8	13	1	1.23	0.62	below 221	[4]
1	6	2	3	5	1.33	0.66	about 238	New Codes
1	7	2	3	2	1.33	0.66	below 238	
1	5	7	11	2	1.27	0.64	235	
1	5	5	8	2	1.25	0.63	228	
1	5	8	13	1	1.23	0.62	221	
1	4	3	5	2	1.20	0.60	about 224	

2. CODE CONSTRUCTION

The new (d,k;m,n;r) codes are listed in Table I. In the following, a code construction method and a few examples of the new codes will be presented.

(1,7;2,3;2) code. Table II shows the coding rule for this code. A data sequence is partitioned into blocks of length 2(=m). Blocks of data 01, 10 and 11 are encoded into corresponding words of length 3(=n), shown in coding rule S_1 of Table II. Blocks of data 00 cannot be encoded by S_1. So, they are encoded together with the succeeding block into corresponding words of length 6(=nr), shown in coding rule S_2 of Table II. Hence, all data sequences can be encoded, as shown in the following example.

```
                        (Time ——▶ )
Data sequence   :1 0 :1 1 :0 0 1 0 :1 0 :0 1
Coded sequence  :0 1 0:1 0 1:0 0 0 0 0 0:0 1 0:1 0 0
```

As in the example, a value of X in the words is set as either 1 or 0, according to the following rule, when

Reprinted from *IEEE Trans. Magn.*, vol. MAG-12, pp. 740-742, Nov. 1976.

the words are concatenated: X=Complement of a value of the code bit preceeding the bit X. The resultant coded sequence achieves constraint d=1 and k=7.

TABLE II. Coding rule for (1,7;2,3;2) code

	Data	Word		Data	Word
	01	X00		0001	X00001
S_1	10	010	S_2	0010	X00000
	11	X01		0011	010001
				0000	010000

In decoding, a coded sequence may be partitioned into blocks of length 3. The decoder checks to determine whether two consecutive blocks coincide with a word in S_2. If coincidence occurs, the two blocks will be decoded into the corresponding data in S_2. Otherwise, the first block of the two blocks may be decoded into the corresponding data in S_1. Succeeding blocks may be decoded similarly. Checking for coincidence and decoding may be done ignoring X in the words, since the words in S_i are distinct from each other, excluding the X bit. (The encoder and decoder for this code may be operated by the flow-chart presented later in Fig.1.) In the above decoding scheme, if the blocks are treated as a word whose length differs from that which was transmitted, incorrect decoding results. It should be proved that all coded sequences can be decoded into original data sequences. Let Ci be the set of words in coding rule S_i as shown in Eqs. (1) and (2). Define C_iC_j---C_k as a set given by cocatenating all the words in C_i, C_j, ---, C_k in this order. For example, set C_1C_2 can be calculated as in Eq. (3). When any sequence in C_p does not coincide with any sequence which consists of the first 3•p bits of a sequence in set C_iC_j---C_k, independently of a value of X, notation $(C_iC_j---C_k)\cap C_p=\phi$ is used, where ϕ is the empty set. For example, $(C_1C_2)\cap C_2=\phi$ from Eqs. (2) and (3). Similarly, $(C_1C_1)\cap C_2=\phi$. Moreover, it is clear that $(C_1C_iC_j---)\cap C_2=\phi$ for all i, j,---.

$$C_1=\left\{\begin{matrix}X00\\010\\X01\end{matrix}\right\} \quad (1) \qquad C_1C_2=\left\{\begin{matrix}X00\\010\\X01\end{matrix}\right\}\left\{\begin{matrix}X00001\\X00000\\010001\\010000\end{matrix}\right\}=\left\{\begin{matrix}X00100001\\100000\\\text{"}\ 010001\\010000\\010100001\\100000\\\text{"}\ 010001\\010000\\X01000001\\000000\\\text{"}\ 010001\\010000\end{matrix}\right\} \quad (3)$$

$$C_2=\left\{\begin{matrix}X00001\\X00000\\010001\\010000\end{matrix}\right\} \quad (2)$$

It will be shown later that equation $(C_1C_iC_j---)\cap C_2=\phi$ gives the sufficient condition for the correct decodability of this code. That is, all the coded sequences can be correctly decoded into original data sequences.

● Code construction. First, it is necessary to find as many distinct length 3 words in S_1 as possible, which achieve d=1 and a small k-constraint when they are freely concatenated. Three words shown in S_1 of Table II satisfy the above conditions. Next, there must be four distinct length 6 words in S_2 which achieve d=1 and a small k-constraint, and satisfy the condition $(C_1C_i)\cap C_2=\phi$, for all data to be encoded by the code with r=2. They are obtained as follows. If words X00 and X00 in S_1 are concatenated in this order, sequence X00100 results. Here, if the coding rule of X is violated, sequence X00000 results. X00000 may be a word in S_2. The other three words in S_2 are obtained, similarly. Next, data will be allocated to each word. This allocation can be arbitrarily accomplished, but there may be an optimum allocation according to some criterion. Lastly, error propagation property of this code should be checked by the method described later. It is shown that a single error in code bit detection results in incorrect decoding of 6 data bits. That is, error propagation is limited in this code.

<u>(d,k;m,n;r) code</u>. It is assumed that the (d,k;m,n;r) codes in the present discussion have structures similar to that of the (1,7;2,3;2) code described previously. That is, the (d,k;m,n;r) code will have r coding rules, i.e., S_1, S_2, ---, S_r. Coding rule S_i is a one-to-one correspondence table between data of length i·m and words of length i.n, such as the coding rules S_1 and S_2 shown in Table II. Encoding and decoding for the (d,k;m,n;r) code may be performed using the coding rules S_r, S_{r-1}, ---, S_1 in this priority order by the flow-chart shown in Fig.1. Here, it should be noted that, in the above encoding scheme, codes in which concatenations of some words are inhibited can be constructed. But, in the present discussion, codes in which all the words can be freely cocatenated each other are assumed, for simplicity. For example, all the words in the (1,7;2,3;2) code shown in Table II can be freely concatenated. The words in the (d,k;m,n;r) should satisfy the following conditions.

● Condition for correct decoding. Let a data sequence be encoded using coding rules S_i, S_j, S_k, --- in this order. A sequence in set $C_iC_jC_k$--- is then transmitted, where C_i is a set of words of length i.n in coding rule S_i and set $C_iC_jC_k$--- can be calculated in the same way as in Eq. (3). That is, a word in C_i(or S_i) is transmitted first, a word in C_j is transmitted second, etc.. (Note here that number series i, j, k,--- can be any combination out of 1, 2, --, r, but is unique for a given data sequence.)
The decoder will receive the transmitted sequence in set $C_iC_jC_k$---, and check to determine if the first p.n bits of the sequence coincide with a word in C_p(or S_p). This checking for coincidence is accomplished in the priority order of p=r, r-1, ---, 1, as shown in Fig.1. If the first coincidence occurs at p=y, the first y·n bits of the sequence will be decoded as a word in C_y. Hence, the coincidence must not occur at p=r, r-1, ---, i+1, for the first i·n bits of the sequence in set $C_iC_jC_k$--to be correctly decoded as a word in C_i. Clearly, this must hold for all sequences in $C_iC_jC_k$--- and for all i, j, k, ---. Thus, Eq. (4) must be satisfied for correct decoding:

$$(C_iC_jC_k---)\cap C_p=\phi,\ i=1,2,---,r-1,\ p=i+1,i+2,---,r, \text{ and for all } j,k,---. \quad (4)$$

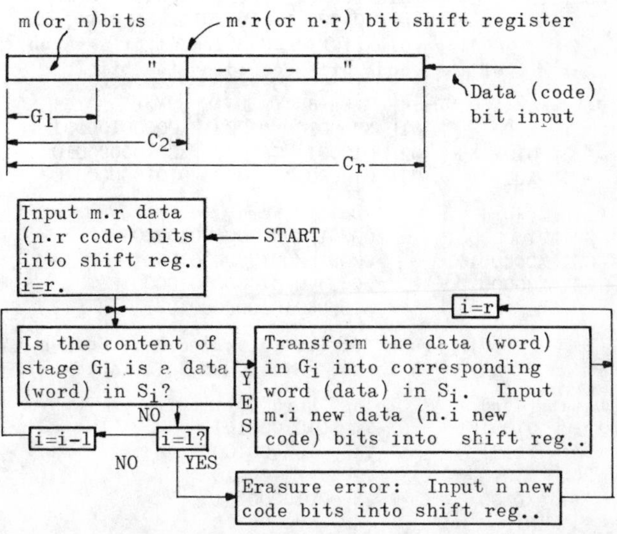

Fig. 1. Encoding/decoding flow-chart for (d,k;m,n;r) code. In decoding, the words in the parenthesises are used.

That is, the first p.n bits of any sequence in set $C_iC_jC_k$--- do not coincide with any word in set C_p. Reversely, it will be shown that if Eq. (4) is satisfied, all sequences in set $C_iC_jC_k$--- can be correctly decoded for all i, j, k, --, by the flow-chart in Fig. 1.

• Condition for error propagation limitation.

Assume that a sequence in set ---$C_aC_iC_jC_k$--- is transmitted and that the decoder receives the sequence from the middle, i.e., from and including the last q·n bits of the transmitted word in C_i, where $1 \leq q \leq i-1$. Denote $R_q(C_i)$ as a set of the sequences each of which consists of last q.n bits of each word in C_i. For example,

$$C_3=\begin{Bmatrix}010000001\\ X00000101\\ X00000100\end{Bmatrix}, \text{ then } R_2(C_3)=\begin{Bmatrix}000001\\ 000101\\ 000100\end{Bmatrix},$$

where C_3 is a set of the (1,6;2,3;5) code shown in Table III. Namely, the decoder receives a sequence in set $R_q(C_i)C_jC_k$---. If the first (q+w).n bits of the sequence coincide with a word in C_{q+w}, the first w.n bits of the sequence in C_jC_k--- will be decoded together with the sequence in $R_q(C_i)$ as a word, resulting in incorrect decoding of the sequence in C_jC_k---, where q+w is any of q+1, q+2, --, r. Hence, for correct decoding of the sequence in C_jC_k--- , Eq. (5) must be satisfied. That is, the first p.n bits of any sequence in set $R_q(C_i)C_jC_k$--- do not coincide with any word in C_p, independently of a value of X in the words of C_p, such as X in the words shown in Table II.

$$(R_q(C_i)C_jC_k---)\cap C_p=\phi,\ i=2,3,---,r,\ q=1,2,---,i-1,\quad (5)$$
$$p=q+1,q+2,---,r, \text{ and for all } j,k,---.$$

Reversely, it will be shown that if Eq. (5) is satisfied, the first q.n bits of any sequence in $R_q(C_i)C_jC_k$-- will be decoded as a string of words, whose total length is q.n by the flow-chart in Fig.1, and then, next decoding will be performed on the sequence in C_jC_k---, resulting in correct decoding of the sequence in C_jC_k----. When errors in the code bit detections occur, another word which terminates in the middle of a transmitted word may be decoded. In that case, the decoder will receive the sequence from the middle of the transmitted word. This is equivalent to the above assumption. (Here, it was assumed, without loss of generality, that the errors did not occur from the middle of the word.)

Hence, if Eq. (5) is satisfied, the errors do not affect the decoding of the succeeding words, that is, error propagation is limited within the word in the middle of which the word termination occured. All the new codes in Table I satisfy Eqs. (4) and (5). Hence, they can be properly decoded into the original data sequences, with error propagation limited. The new codes are proved to have error propagation property that a single error in code bit detection results in incorrect decoding of (2·r-1)·m data bits, at most. Table III shows the coding rules for the (1,6;2,3;5) code. In the present attempts, it was found that about a (1,6;2,3;r) code that the code with r=3 does not exist, and the code with r=4 code exists. However, error propagation is not limited in the code with r=4. In the code with r=5, shown in Table III, error propagation is limited. That is, Table III code satisfies Eqs. (4) and (5). A single error in detection results in incorrect decoding of 12 data bits, at the worst case, in Table III code. Table IV shows the (1,4;3,5;2) code, which satisfies Eqs. (4) and (5). The (d,k;m,n;r) codes in Table I were constructed by the manner similar to that used in the (1,7;2,3;2) code construction: Words in S_1, S_2, ---, S_r may be obtained in that order, using violation of the coding rule of X. After the words are obtained, it should be verified that they satisfy Eqs. (4) and (5), and (d,k) constraint.

Authors would like to thank Mr. Kiji and Dr. Naito for helpful discussions.

TABLE III. Coding rule for (1,6;2,3;5) code. Symbols in data are Ø=00, 1=01, 2=10 and 3=11. Coding rule of X: X=Complement of a value of the code bit preceeding the bit X.

	Data	Word
S_1	1	X00
	2	010
	3	X01
S_2	Ø1	X00001
	Ø2	010001
	Ø3	010000
S_3	ØØ1	010000001
	ØØ2	X00000101
	ØØ3	X00000100
S_4	ØØØ1	X00000100001
	ØØØ2	X00100000010
	ØØØ3	010100000010
S_5	ØØØØ1	X00100000010000
	ØØØØ2	X00100000010001
	ØØØØ3	010100000010000
	ØØØØØ	010100000010001

TABLE IV. Coding rule for (1,4;3,5;2) code. Coding rule of X is the same as that in Table III.

	Data	Word		Data	Word
	Ø	01001		7Ø	0101000101
	1	X0101		71	" 00100
	2	01010		72	00010
S_1	3	X0100	S_2	73	X010000101
	4	01000		74	" 00100
	5	X0001		75	X001000101
	6	X0010		76	" 00100
				77	00010

REFERENCES

[1] P. A. Franaszek, "Sequence-state methods for run-length-limited coding", IBM J Res. Develop., Vol. 14, pp. 376, July 1970.

[2] H. Kobayashi, "A survey of coding schemes for transmission or recording of digital data", IEEE Trans. Commun. Technol., Vol. 19, pp. 1087, Dec. 1971.

[3] A. Gabor, "Adaptive coding for self-clocking recording", IEEE Trans. Electron. Comput., Vol. 16, pp. 866, Dec. 1967.

[4] P. Hodge, "High rate digital modulation/demodulation method and means", U. S. Patent, No. 3,852,687, Dec. 1974.

[5] T. Tamura, et al., "A coding method in digital magnetic recording", IEEE Trans. Magn., Vol. MAG-8, No 3, 1972.

Channel Capacity of Charge-Constrained Run-Length Limited Codes

KERMIT NORRIS, ASSOCIATE MEMBER, IEEE, AND DAN S. BLOOMBERG, MEMBER, IEEE

Abstract—**The methods of information theory are applied to run-length limited codes with charge constraints. These self-clocking codes are useful in several areas of information storage, including magnetic recording, where it may be desirable to eliminate the dc component of the frequency spectrum. The channel capacity of run-length limited codes, with and without charge constraints, is derived and tabulated. The channel capacity specifies the maximum ratio of data/message bits achievable in implementing these codes and gives insight into the choice of codes for a particular task. The well-known frequency modulation (FM) code provides a simple example of these techniques.**

I. Constrained Codes and Information Theory

On MAGNETIC media, a change from one polarity of magnetic saturation to its opposite is termed a "transition." When digital data are recorded, transitions are written or not written at regular discreet intervals along the media in accordance with some code or rule. Such transitions and lack of transitions are termed "encoded data ones and zeros," or merely "ones" and "zeros," respectively.

Imagine an encoder (Fig. 1) that accepts random binary data at one rate and outputs less random message bits at a higher rate. Without discussing the encoder in detail, it is possible to evaluate the information-handling capacity of the encoded message. In this paper we derive, from an information theory viewpoint, the characteristics of two broad classes of codes that are useful in magnetic recording: run-length limited (RLL) and charge-constrained run-length limited (CCRLL) codes.

RLL codes are those for which there are both a minimum (d) and a maximum (k) number of zeroes between ones. They are useful in magnetic recording because a) they are self-clocking if k is not too large; b) with increasing d they are able to reduce bit crowding; and c) with decreasing k they reduce shouldering problems which result in false zero-crossings in the differentiated waveforms.

Charge-constrained RLL codes (CCRLL) are a subset of the RLL codes that have no dc component in the recorded waveform. Charge increases or decreases one unit per message bit. As 1's reverse polarity, a running charge sum will result, whose value depends on the specific bit sequence. The running charge sum of a CCRLL code will never exceed, in absolute value, an integer (c). Because these codes have no dc component, they are useful when the signal is coupled through a transformer. They permit signal processing by integration, which for some noise spectra may be preferable to differentiation. They also reduce print-through problems on magnetic tape by reducing the extension of the stray dipole fields.

Manuscript received February 18, 1981; revised April 2, 1981.

The authors are with the Palo Alto Research Center, Xerox Corporation, El Segundo, CA 90245.

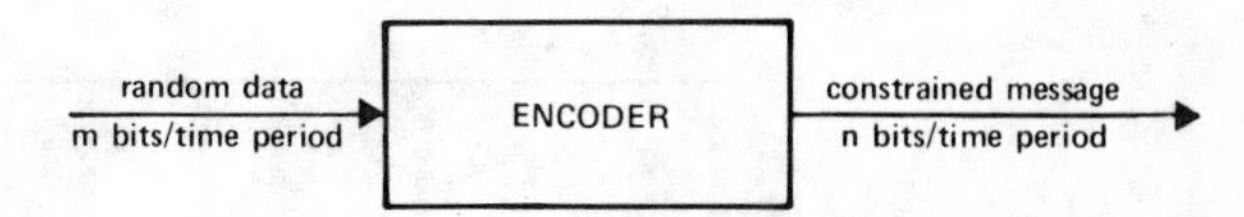

Fig. 1. Encoder for constrained code; $m < n$.

The RLL codes are characterized by five integers $(d, k; m, n; r)$, and the CCRLL codes are labeled by six integers $(d, k, c; m, n; r)$. The d, k (and c) specify constraints in the encoded message. The m, n, and r describe (not uniquely) the encoder implementation. For m original data bits, the encoded message contains n bits. The code rate is defined as the ratio m/n. The r value gives the number of "tables" that are used in the encoding/decoding process. For example, a (2, 11; 3, 6; 1) code uses a single table to encode 3 data bits into 6 message bits, taking the data 3 bits at a time. A (2, 7; 1, 2; 3) code uses 3 tables, taking data either 1, 2, or 3 bits at a time. Codes with $r = 1$ are called block or fixed length codes and are often simpler to implement than codes with larger r values. It is also generally believed that multiple table codes have more serious error propagation characteristics than $r = 1$ codes, but this is not always true.

As constraints are increasingly built into the encoded message, the average amount of information (or "surprise") contained in each bit decreases. The average information/bit in the constrained message relative to the data source is called (after Shannon [1]) the *channel capacity H* of a code. It is the upper limit of the ratio of the number of random data bits in a sequence to the number of message bits necessary to encode it averaged over all possible sequences. The channel capacity is less than one for all constrained codes. Because it is a theoretical upper limit, it is not usually obtained in the implementation of a particular code.

For CCRLL codes, the channel capacity depends only on the constraints d, k, and c. But the m and n values give the actual ratio of data/message bits in the encoded sequence. The *code rate m/n* can never be greater than the channel capacity; the ratio of code rate to channel capacity is the *efficiency* of the particular code. For example, the d, k, c constraints (1, 3, 4) describes a class of codes with a channel capacity of 0.5237. Proposed implementations all have $m/n = 1/2$ and an efficiency of 0.955. In practice an m/n ratio of two integers such as 1/2 or 2/3 is chosen. Those d, k, c codes with channel capacity falling at or above the chosen code rate are possible codes; others are not possible and need not be attempted. This paper

Reprinted from *IEEE Trans. Magn.*, vol. MAG-17, pp. 3452–3455, Nov. 1981.

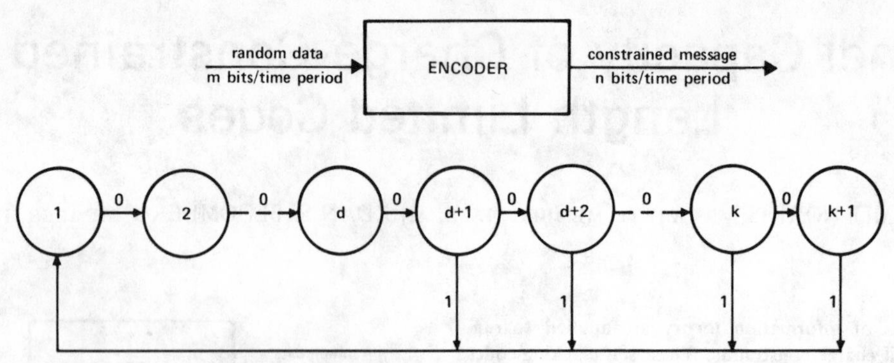

Fig. 2. State-transition diagram for RLL code.

will not discuss code implementation beyond the following remark. A weak trade-off is expected to exist between code performance and code complexity. By reducing k and c and increasing d, code performance will generally improve (smaller error rate, less peak shift). However, as efficiency is increased, the logic required for implementation tends to increase in complexity.

Specific CCRLL code algorithms have been studied by Patel [2], [3] and Hong and Ostapko [4]. A method for calculating the channel capacity of codes with only charge-constraint has been given by Chien [5]. In the next sections, derivations are given for the channel capacity of RLL and CCRLL codes. The RLL derivation is simpler and is given partly as an aid in understanding the CCRLL development. The channel capacities are found as the base 2 logarithm of the largest root of a polynomial and are tabulated for various CCRLL codes with $d \leqslant 4$ and $c \leqslant 11$.

II. The Channel Capacity of RLL Codes

The number of possible specific messages (PSM) of random binary data of length j bits is 2^j. For an RLL code, let the number of encoded PSM of length n be Q_n. For n significantly larger than the constraint length k, the ratio Q_{n+1}/Q_n should tend to a number z between 1 and 2. Thus

$$Q_n = z^n. \tag{1}$$

Since nH data bits can be represented by n bits of the encoded sequence,

$$2^{nH} = z^n. \tag{2}$$

Thus the channel capacity is related to z by

$$H = \log_2 z. \tag{3}$$

The state-transition diagram [6] (Fig. 2) shows how new PSM arise as n increases. The "one" state means a transition has just occurred; i.e., a 1 has appeared in the message sequence. The 1 must be followed by d 0's. At "$d + 1$" a choice is offered: return to state "one" with a 1 or continue to "$d + 2$" with another 0. Similar choices arise at "$d + 3$," "$d + 4$," $\cdots$ "k," but return to mandatory at "$k + 1$." Working backwards, if a 1 occurs at bit n, then the previous 1 occurred at bit $n - (d + 1)$ or $n - (d + 2)$ or $\cdots$ or $n - (k + 1)$. The number of PSM at bit n, Q_n, is then the sum of the PSM at these previous bits:

$$Q_n = Q_{n-(d+1)} + Q_{n-(d+2)} + \cdots + Q_{n-(k+1)}. \tag{4}$$

Using (1) and the geometric series relation

$$1 + r + r^2 + \cdots + r^{n-1} = (1 - r^n)/(1 - r), \tag{5}$$

z is found to be the largest root of the equation

$$z^{k+2} - z^{k+1} - z^{k-d+1} + 1 = 0. \tag{6}$$

This result was first derived by Shannon. For a particular d and k, the channel capacity $H = \log_2 z$ gives an upper bound to the ratio m/n which may be chosen.

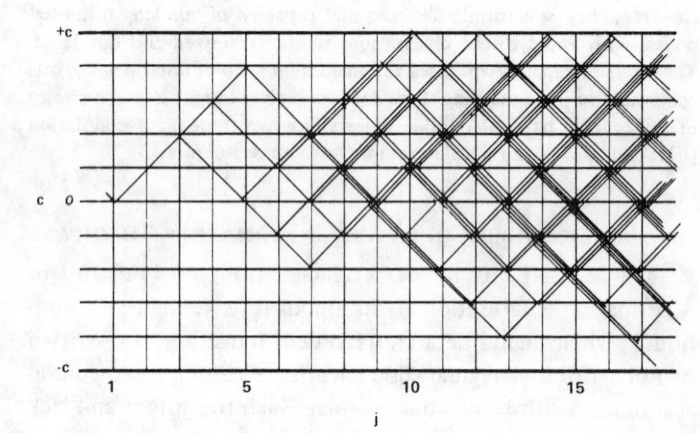

Fig. 3. Charge-trellis diagram showing growth from single upward transition in d, k, c sequence of 1, 3, 5.

III. The Channel Capacity of CCRLL Codes

For charge-constrained codes, the charge-trellis diagram (Fig. 3) is useful for counting PSM. Charge is plotted vertically against message bit and all possible branchings are shown. In order to permit run lengths of k zeroes, the total charge excursion $2c$ must satisfy

$$2c \geqslant \begin{cases} k + 1, & k \text{ odd} \\ k + 2, & k \text{ even.} \end{cases} \tag{7}$$

Let the number of PSM with upward and downward directed transitions at (c, j) be $U(c, j)$ and $D(c, j)$, respectively. For the $d, k, c = 1, 3, 5$ code in Fig. 3, inspection of the diagram gives

$$D(c,j) = U(c-2, j-2) + U(c-3, j-3) + U(c-4, j-4), \tag{8}$$

$$U(c,j) = D(c+2, j-2) + D(c+3, j-3) + D(c+4, j-4). \tag{9}$$

For sufficiently large j, symmetry about $c = 0$ yields

$$D(c,j) = U(-c, j). \tag{10}$$

As before, constancy of the information/bit is expressed by the geometric factor z:

$$D(c,j) = z^i D(c, j-i), \tag{11}$$

$$U(c,j) = z^i U(c, j-i), \tag{12}$$

where i is any integer. Substituting (11) and (12) into (8) and (9)

$$D(c,j) = z^{-2} U(c-2, j) + z^{-3} U(c-3, j) + z^{-4} U(c-4, j), \tag{13}$$

$$U(c,j) = z^{-2} D(c+2, j) + z^{-3} D(c+3, j) + z^{-4} D(c+4, j). \tag{14}$$

Applying (10) to (13) and dropping reference to j,

$$D(c) = z^{-2} D(-c+2) + z^{-3} D(-c+3) + z^{-4} D(-c+4). \tag{15}$$

Note that the use of (10) in (14) results in the same set of equations. Generalizing to any d, k,

$$D(c) = z^{-(d+1)} D(-c+d+1) + z^{-(d+2)} D(-c+d+2) \cdot + \cdots + z^{-(k+1)} D(-c+k+1), \quad c = -c, -c+1, \cdots c. \tag{16}$$

This is a set of $N = 2c + 1$ homogeneous linear equations. The requirement of a vanishing determinant gives an algebraic equation for z. The matrix with zero determinant has coefficients

$$A_{ij} = z^{-(i+j-N-1)} f(i+j-N-1) - \delta_{ij}, \quad 1 \leq i, j \leq N \tag{17}$$

where

$$f(p) = \begin{cases} 1, & \text{if } d+1 \leq p \leq k+1 \\ 0, & \text{otherwise} \end{cases}$$

$$\delta_{ij} = \begin{cases} 1, & \text{if } i = j \\ 0, & \text{otherwise.} \end{cases}$$

As before, the channel capacity is the logarithm of the largest root. The values of $D(c)$ for this root give relative probabilities of transitions occurring in these states for an ideal code and random data. The values of the channel capacities, found by varying z until the determinant vanishes, are given in Table I. The limiting value of the $c = \infty$ RLL code is also given.

The unconstrained charge result (6) can be obtained from (16) by letting the dimensionality become infinite. In this limit all charge states are equally likely and the set of linear equations degenerates to one equation,

$$1 = z^{-(d+1)} + z^{-(d+2)} + \cdots + z^{-(k+1)} \tag{18}$$

which is easily reduced to (6).

As a simple example, consider the biphase (or frequency modulation (FM)) code, with $d, k, c = 0, 1, 1$. The equation

$$\begin{vmatrix} -1 & 0 & 0 \\ 0 & -1 & z^{-1} \\ 0 & z^{-1} & z^{-2} - 1 \end{vmatrix} = 0 \tag{19}$$

which follows from (16) has $z = \sqrt{2}$ as the largest root. The channel capacity is exactly 1/2, which can also be seen from the actual (0, 1, 1; 1, 2; 1) code. There are two message bits for every data bit. Transitions always occur on the data cell boundaries; information is carried by the absence/presence of a transition in the middle of the cell corresponding to 0/1 data. Because half the bits carry random information (and the others carry none) the channel capacity is 1/2. One other set of codes (1, 3, 3) has a channel capacity of exactly 1/2. Known implementations are relatively simple, but they either suffer from infinite error propagation or require "infinite" look-ahead at the encoder.

It is expected that useful codes will have small integer ratios of m/n. By looking at state diagrams of particular $d = 2$ codes, Patel [2] has deduced that codes with d, k, c values of (2, 7, 8) and (2, 8, 7) have channel capacities slightly greater than 1/2. They are in fact the least constrained codes possible with $d = 2$ and $m/n = 1/2$. From Table I, (0, 1, 4), (1, 7, 10), (1, 8, 9), and (1, 11, 8) are the least constrained codes possible with $m/n = 2/3$; (1, 2, 6), (1, 3, 2), (2, 4, 8), (2, 5, 4), (3, 7, 12) and (3, 8, 7) can be constructed with $m/n = 2/5$; (1, 2, 2), (2, 4, 3), (3, 6, 5), (4, 8, 9), (4, 9, 7), and (4, 11, 6) with $m/n = 1/3$.

Code choice depends on characteristics of the entire recording channel. If bit crowding is the most severe limitation, a (3, 7, 12) code with 40 percent code rate might be preferable to a (1, 3, 4) code with 50 percent code rate. If shouldering problems occur, a code with a narrow bandwidth (small $(k+1)/(d+1)$), such as (2, 4, 8), should be considered. The timing window is proportional to code rate, which is disadvantageous for codes with small m/n. Other considerations include error propagation length, logic complexity, and whether or not an industry standard is established.

TABLE I
CHANNEL CAPACITIES OF CCRLL CODES

		c													
d	k	1	2	3	4	5	6	7	8	9	10	11	12	13	∞
0	1	.5000	.6358	.6662	.6778	.6834	.6866	.6885	.6898	.6907	.6914	.6919	.6922	.6925	.6942
0	2	--	.7664	.8244	.8468	.8578	.8640	.8678	.8704	.8722	.8734	.8744	.8751	.8757	.8792
0	3	--	.7925	.8704	.9012	.9165	.9252	.9306	.9342	.9367	.9386	.9399	.9410	.9418	.9468
0	4	--	--	.8832	.9120	.9380	.9486	.9552	.9596	.9627	.9650	.9667	.9680	.9690	.9752
0	5	--	--	.8858	.9256	.9460	.9578	.9652	.9702	.9738	.9763	.9783	.9798	.9810	.9881
0	6	--	--	--	.9273	.9488	.9614	.9694	.9747	.9786	.9811	.9834	.9851	.9864	.9942
0	7	--	--	--	.9276	.9497	.9627	.9710	.9766	.9806	.9836	.9858	.9875	.9888	.9971
0	8	--	--	--	--	.9499	.9632	.9717	.9774	.9815	.9845	.9868	.9886	.9900	.9986
0	9	--	--	--	--	.9500	.9633	.9719	.9777	.9819	.9849	.9873	.9891	.9905	.9993
1	2	--	.3471	.3822	.3931	.3978	.4003	.4018	.4027	.4034	.4038	.4041	.4044	.4046	.4057
1	3	--	.4248	.5000	.5237	.5341	.5396	.5428	.5449	.5463	.5473	.5480	.5486	.5490	.5515
1	4	--	--	.5391	.5746	.5905	.5989	.6039	.6702	.6093	.6109	.6121	.6129	.6136	.6175
1	5	--	--	.5497	.5947	.6153	.6263	.6328	.6371	.6400	.6421	.6436	.6448	.6457	.6509
1	6	--	--	--	.6020	.6260	.6391	.6470	.6522	.6557	.6582	.6601	.6615	.6626	.6690
1	7	--	--	--	.6039	.6305	.6451	.6540	.6599	.6639	.6668	.6689	.6706	.6718	.6793
1	8	--	--	--	--	.6321	.6477	.6574	.6638	.6682	.6713	.6737	.6755	.6769	.6853
1	9	--	--	--	--	.6325	.6488	.6590	.6657	.6704	.6738	.6763	.6783	.6798	.6888
1	10	--	--	--	--	--	.6492	.6597	.66666	.6715	.6751	.6777	.6798	.6814	.6909
1	11	--	--	--	--	--	.6493	.6600	.6671	.6721	.6758	.6785	.6806	.6823	.6922
1	12	--	--	--	--	--	--	.6601	.6673	.6724	.6761	.6789	.6811	.6828	.6930
2	3	--	.2028	.2625	.2757	.2807	.2832	.2845	.2853	.2859	.2863	.2866	.2868	.2869	.2878
2	4	--	--	.3471	.3777	.3893	.3950	.3981	.4001	.4013	.4022	.4029	.4034	.4038	.4057
2	5	--	--	.3723	.4199	.4384	.4475	.4526	.4557	.4578	.4593	.4603	.4611	.4617	.4650
2	6	--	--	--	.4366	.4614	.4737	.4807	.4851	.4879	.4899	.4914	.4925	.4933	.4979
2	7	--	--	--	.4418	.4718	.4870	.4956	.5011	.5047	.5072	.5091	.5105	.5115	.5174
2	8	--	--	--	--	.4761	.4935	.5036	.5099	.5142	.5172	.5194	.5210	.5223	.5293
2	9	--	--	--	--	.4774	.4965	.5077	.5148	.5196	.5230	.5255	.5274	.5288	.5369
2	10	--	--	--	--	--	.4977	.5097	.5174	.5227	.5264	.5291	.5312	.5328	.5418
2	11	--	--	--	--	--	.4980	.5107	.5188	.5244	.5283	.5313	.5335	.5352	.5450
2	12	--	--	--	--	--	--	.5110	.5195	.5253	.5295	.5325	.5352	.5369	.5471
2	13	--	--	--	--	--	--	.5111	.5198	.5258	.5301	.5333	.5357	.5376	.5485
3	4	--	--	.1903	.2101	.2162	.2188	.2202	.2210	.2215	.2219	.2221	.2223	.2225	.2232
3	5	--	--	.2434	.2902	.3049	.3112	.3146	.3166	.3178	.3187	.3193	.3197	.3200	.3218
3	6	--	--	--	.3224	.3464	.3570	.3625	.3658	.3679	.3694	.3704	.3711	.3716	.3746
3	7	--	--	--	.33329	.3660	.3807	.3885	.3932	.3962	.3982	.3996	.4007	.4015	.4057
3	8	--	--	--	--	.3746	.3929	.4029	.4088	.4127	.4153	.4172	.4185	.4196	.4251
3	9	--	--	--	--	.3774	.3990	.4107	.4179	.4224	.4257	.4279	.4296	.4309	.4376
3	10	--	--	--	--	--	.4017	.4149	.4230	.4283	.4320	.4346	.4366	.4381	.4460
3	11	--	--	--	--	--	.4025	.4170	.4259	.4318	.4359	.4388	.4410	.4427	.4516
3	12	--	--	--	--	--	--	.4179	.4274	.4338	.4382	.4414	.4438	.4456	.4556
3	13	--	--	--	--	--	--	.4182	.4282	.4349	.4396	.4430	.4455	.4475	.4583
4	5	--	--	.1278	.1662	.1747	.1779	.1794	.1803	.1808	.1812	.1814	.1816	.1817	.1823
4	6	--	--	--	.2271	.2480	.2559	.2597	.2618	.2631	.2639	.2644	.2647	.2649	.2669
4	7	--	--	--	.2478	.2822	.2955	.3019	.3055	.3078	.3092	.3103	.3110	.3115	.3142
4	8	--	--	--	--	.2975	.3162	.3254	.3306	.3338	.3360	.3374	.3385	.3393	.3432
4	9	--	--	--	--	.3030	.3267	.3386	.3453	.3496	.3523	.3543	.3557	.3568	.3620
4	10	--	--	--	--	--	.3316	.3458	.3540	.3592	.3626	.3650	.3667	.3681	.3746
4	11	--	--	--	--	--	.33336	.3496	.3591	.3650	.3690	.3719	.3739	.3755	.3833
4	12	--	--	--	--	--	--	.3514	.3619	.3686	.3731	.3763	.3786	.3804	.3894
4	13	--	--	--	--	--	--	.3520	.3633	.3706	.3756	.3791	.3817	.3837	.3937
5	6	--	--	--	.1313	.1451	.1493	.1511	.1521	.1527	.1530	.1533	.1534	.1536	.1542
5	7	--	--	--	.1713	.2054	.2160	.2206	.2230	.2244	.2252	.2257	.2260	.2262	.2281
5	8	--	--	--	--	.2318	.2499	.2578	.2620	.2644	.2660	.2670	.2676	.2680	.2709
5	9	--	--	--	--	.2415	.2672	.2786	.2847	.2883	.2906	.2922	.2933	.2941	.2979
5	10	--	--	--	--	--	.2755	.2903	.2983	.3030	.3061	.3081	.3096	.3107	.3158
5	11	--	--	--	--	--	.2786	.2965	.3063	.3121	.3159	.3185	.3204	.3217	.3282
5	12	--	--	--	--	--	--	.2996	.3109	.3178	.3222	.3253	.3275	.3292	.3369
5	13	--	--	--	--	--	--	.3007	.3134	.3212	.3263	.3298	.3323	.3342	.3432
6	7	--	--	--	.0934	.1217	.1279	.1303	.1314	.1320	.1324	.1327	.1328	.1329	.1335
6	8	--	--	--	--	.1691	.1850	.1910	.1939	.1954	.1962	.1967	.1970	.1972	.1993
6	9	--	--	--	--	.1863	.2133	.2237	.2287	.2315	.2332	.2342	.2349	.2353	.2382
6	10	--	--	--	--	--	.2268	.2418	.2492	.2533	.2559	.2576	.2587	.2594	.2633
6	11	--	--	--	--	--	.2320	.2516	.2614	.2669	.2703	.2726	.2741	.2750	.2804

* Ordinary RLL values (no charge constraint) are in right column.

ACKNOWLEDGMENT

The authors wish to thank Pat Kocsis for help in preparation of the manuscript.

REFERENCES

[1] C. E. Shannon, "A mathematical theory of communication," *Bell Syst. Tech. J.*, vol. 27, pp. 379-423, July 1948.

[2] A. M. Patel, "Zero-modulation encoding in magnetic recording," *IBM J. Res. Dev.*, vol. 19, pp. 366-378, July 1975.

[3] A. M. Patel, "Charge-constrained byte-oriented (0, 3) code," *IBM Tech. Disclosure Bull.*, vol. 19, pp. 2715-2724, Dec. 1976.

[4] J. Hong and D. L. Ostapko, "Codes for self-clocking, ac-coupled transmission: aspects of synthesis and analysis," *IBM J. Res. Dev.*, vol. 19, pp. 358-365, July 1975.

[5] T. M. Chien, "Upper bound on the efficiency of dc-constrained codes," *Bell Syst. Tech. J.*, vol. 49, pp. 2267-2287, Nov. 1970.

[6] P. A. Franaszek, "Sequence-state methods for run-length-limited coding," *IBM J. Res. Dev.*, vol. 14, pp. 376-383, July 1970.

Part IX
Equalization

Signal Equalization in Digital Magnetic Recording

GEORGE V. JACOBY, SENIOR MEMBER, IEEE

Abstract—The signal applied to the head in the writing process undergoes changes and distortions of various sources. Equalization must be applied to the read signal to help restore the input waveshape. The object is to improve timing accuracy which ultimately allows the use of increased bit density. An equalization method is presented in which amplitude and phase are controlled separately. The phase equalizer is adjustable, compensating for waveshape asymmetry. The networks are extremely simple and not sensitive to component tolerances. In continuous waveshape codes, such as phase recording or double frequency recording, the system is capable of compensating for considerable variations in the spacial parameters of recording. The equalization can also be used in nonreturn to zero (NRZ) code, where the ability to restore symmetry can be used to great advantage in eliminating errors.

Introduction

THE signal containing the information in digital magnetic recording suffers distortions of various origin, resulting in a degraded read signal that usually does not resemble the original information at all. Equalization or compensation must be applied to the read signal to remove these distortions. Then the timing of the equalized read signal can be detected accurately to recover or decode the original binary information. Equalization therefore, as an intermediate step in this signal processing chain, improves timing accuracy which, in turn, makes it possible to increase the bit density of the magnetic recording channel.

The most important types of signal distortion result from spacial parameter variations (also called "dropout" or separation losses) and pulse crowding, the inability of the recording channel, due to high bit density, to resolve individual pulses [1]. Imperfections in the write and erase process constitute another source of distortion resulting in timing error [2]. Yet another type of distortion appears in asymmetrical pulse shape [3], [4] caused by field pattern asymmetry, imperfections in the head, and spatial parameters that are different in forward and reverse motion. All these asymmetrical waveshape phenomena can be treated from a signal equalization standpoint as phase distortion.

The equalization method published in [5] is intended for slimming individual pulses (NRZ) by maintaining the original Gaussian pulse shape. This condition requires a gradual cutoff at the high-frequency end of the spectrum, resulting in a limited amount of pulse narrowing. Practically an improvement by a factor of two can be obtained in pulse width and consequently in bit density. The passive circuit providing the necessary amplitude response and maintaining linear phase simultaneously is rather complex and requires close tolerances. This method assumes a symmetrical read pulse from the head and has no means of compensating for asymmetry or phase error.

Manuscript received March 1, 1968; revised June 4, 1968. Paper 3.11, presented at the 1968 INTERMAG Conference, Washington, D. C., April 3–5.

The author is with the Electronic Data Processing Division, RCA, Camden, N. J.

Equalization of Continuous Waveshapes

The equalization method presented in this paper was developed originally for continuous waveshapes, i.e., phase recording or double frequency coding. The main object of the equalization of continuous rectangular waveshapes is to restore the original write current waveshape to a limited degree. The spectrum of a continuous signal in double frequency coding contains two fundamental frequencies f_0 and f_1 [Fig. 1(a)], where f_0 is the fundamental of the square wave for binary 0 and $f_1 = 2f_0$ is the fundamental for binary 1; it also equals the bit rate. In order to obtain a reasonably good approximation of this rectangular waveshape, it is enough to restore only one harmonic, besides the fundamentals, that is, the third harmonic of f_0. Because of the varying binary patterns, there are also low-frequency components present below f_0. It can be shown by Fourier analysis [6] that the amplitude of these decreases vary rapidly with decreasing frequency. It is enough to include only $0.25 f_1$, representing a worst case pattern of 11001100, etc. It is seen, therefore, that a limited bandwidth from $0.25f_1$ to $1.5f_1$ is adequate for the restoration of a double frequency coded rectangular wave.

A functional block diagram of the magnetic recording channel is shown in Fig. 2. The first block represents the ideal differentiating function of magnetic recording. The second block contains all the distortions, amplitude, and phase, resulting in actual imperfect reproduction of the input waveshape. The equalizer is composed of two parts: a passive network amplitude equalizer contributing also to the phase shift and an all-pass type adjustable phase equalizer having no effect on the amplitude. Due to the limited bandwidth requirements, the magnetic recording channel, including the equalizer, should have a flat frequency response only between $0.25f_1$ and $1.5f_1$. The behavior of the system outside this frequency band is of no practical interest.

The sinusoidal frequency response of the recording channel with and without equalization on a logarithmic scale is shown in Fig. 3. The curves *b* and *c* represent two conditions in which the spatial parameters are different; they will be discussed in detail later. Therefore, the amplitude equalizer is required to supply enough boost to make the frequency response flat between $0.25f_1$ and $1.5f_1$. This was done in case *b*, which will be referred to as flat equalization. The overall phase of the whole system should be either 0 or 180 degrees within the bandwidth from $0.25f_1$ to $1.5f_1$.

Reprinted from *IEEE Trans. Magn.*, vol. MAG-4, pp. 302–305, Sept. 1968.

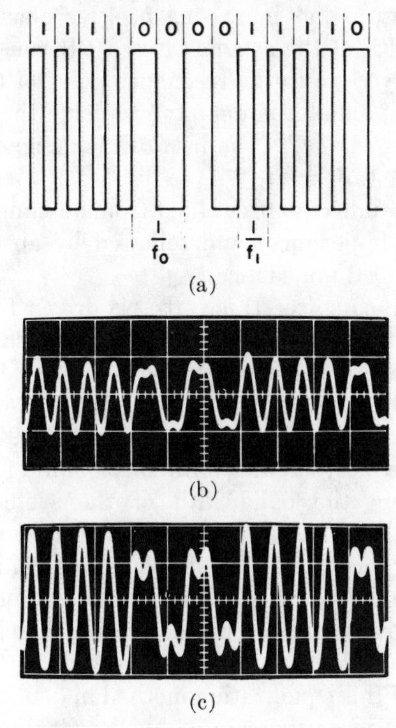

Fig. 1. Waveshapes in double frequency code. (a) Input—sequence of binary bits. (b) Equalized read signal in flat equalization. (c) Equalized read signal in overequalization.

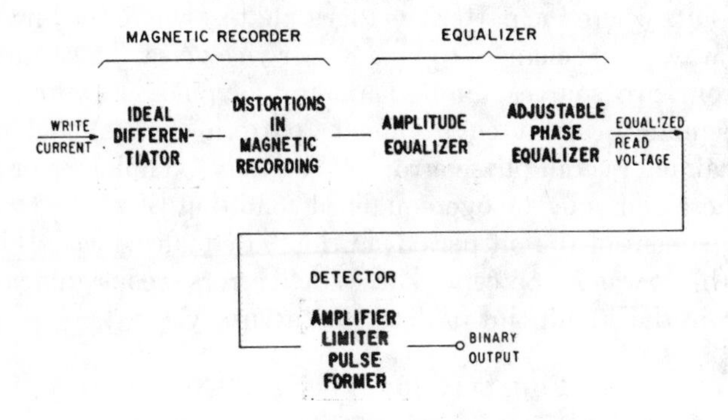

Fig. 2. Magnetic recording channel.

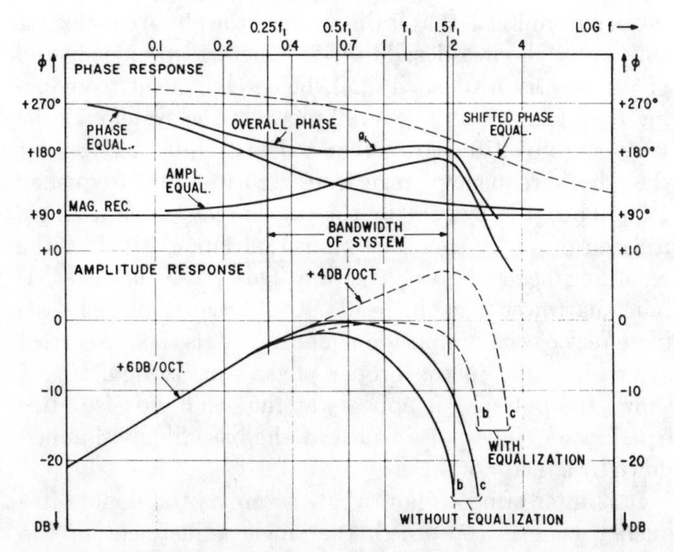

Fig. 3. Sinusoidal frequency response of magnetic recording channel.

The magnetic recording process contributes 90 degrees, due to the inherent differentiating function and an additional linear phase shift, which is equivalent to a constant delay. This linear phase shift has no effect at all on the waveshape; therefore it can be ignored, and it is not shown in Fig. 3. The amplitude and phase equalizer together must contribute the additional 90 degrees, thereby restoring the original phase relationship to each spectral component. Phase distortions in the channel, causing deviations from the ideal phase requirement, can be compensated for by adjusting the phase equalizer to the proper value. The adjustment can be visualized as shifting the curve of the phase equalizer along the frequency axis. This compensation is possible because the sum of the phase distortions and the equalizer's phase result in an overall 90 degrees plus linear phase relationship. Since linear phase is equivalent to a constant time delay, the phase distortions will not cause waveshape distortion. This is a very important and useful feature of the system, because it restores waveshape symmetry which is not possible in an equalizer built with fixed parameters.

The equalized waveshape, corresponding to the flat frequency response between $0.25f_1$ and $1.5f_1$ is shown in Fig. 1(b). It is easy to see that this is really an approximation of the original square wave input of Fig. 1(a), where the binary 1 waveshape is approximated by its fundamental and the binary 0 with the fundamental and the third harmonic term. The amplitude of the third harmonic is 33 percent of the fundamental; the same as in the Fourier spectrum of the binary 0 square wave. This, of course, is to be expected since the flat frequency response of the system did not change the relative amplitudes. It is possible to increase the third harmonic percentage and still maintain the desirable symmetrical waveshape. Th's is shown in Fig. 1(c), where the third harmonic is 66 percent of the fundamental. This waveshape corresponds to a frequency response, shown in Fig. 3 (curve *c*), where the equalized response has a +4 dB per octave slope in the bandwidth from $0.25f_1$ to $1.5f_1$. This type of response will be referred to as *overequaiization*.

As mentioned earlier, the spacial parameters for the two sets of curves (*b* and *c*) in Fig. 3 are different. Specifically, the head to medium separation in case *b* is twice as large as in case *c*. Curve *b*, therefore, represents considerable signal degradation or a partial dropout condition. It could also represent an increase in the bit density (decrease in the wavelength), compared to curve *c*, due to a different recording speed. Or it could represent a combination of these and other spatial parameters, the net effect of which results in an exponential drop in the frequency response. The advantage of overequalization is obvious. The amplitude equalizer should be designed such that it gives overequalization (curve *c* in Fig. 3) with the +4 dB per octave slope under the most favorable conditions of the recording parameters. When a change occurs in the spatial parameters and the frequency response drops to curve *b*, flat equalization will result. Only the relative amplitudes will be different, as seen in Fig. 1(b) and (c). The timing is equally

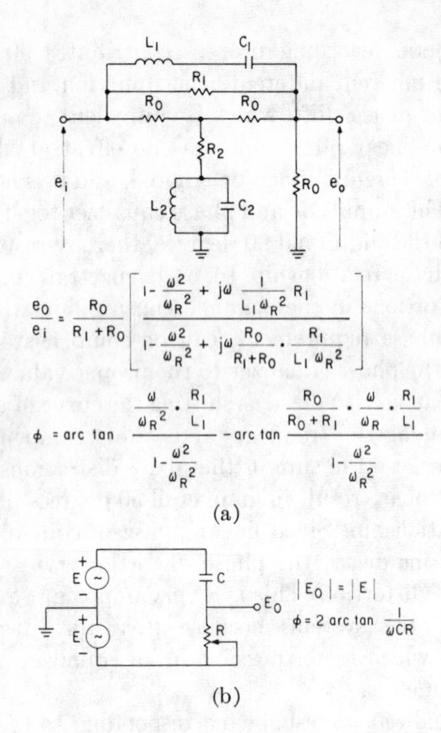

Fig. 4. Equalizer networks. (a) Amplitude equalizer, bridged-T network. (b) Phase equalizer network.

accurate in both cases. The result is not surprising because the timing accuracy is inherently connected with the bandwidth of the system, and this did not change, as seen in Fig. 3. Merely the slope of the amplitude response changed within the bandwidth. The phase did not change either, since spatial parameter variations result in an exponential type amplitude drop in the frequency response with an attendant linear phase relationship that can be ignored. Thus the system is inherently less sensitive to dropouts, to the extent indicated above.

The next step in the signal processing is the detection of the zero axis crossover points of the equalized read signal. The simplest way to do this is to perform amplification and limiting through several stages. Thus the shape of the original rectangular input is completely recovered. Then a pulse can be formed at each crossover point, resulting in one pulse for each binary 0 and two pulses for each binary 1. Detection and decoding will not be discussed in this paper. This particular method of crossover detection limits the maximum percentage value of the third harmonic in the binary 0 waveshape. If, e.g., 100 percent third harmonic were used this would result in a dip, touching the zero axis, causing a false crossover in the middle of the bit. Therefore, the 66 percent value represents a safe upper limit.

The essential elements of the equalizer are shown in Fig. 4. The transfer functions, in terms of amplitude and phase response, are given in the figure. The phase characteristic of the amplitude equalizer plays an important role in the selection of the circuit, since the phase equalizer must compensate for phase shifts both in the magnetic recorder and in the amplitude equalizer. For this reason, a bridged-T circuit was selected for the amplitude equalizer [Fig. 4(a)], because it has increasing positive phase shift over the bandwidth of the system (Fig. 3). Because the phase drops rapidly close to the resonance point of the network f_R, this point should be definitely outside the bandwidth of the system. Usually, the network is designed such that $f_R = 1.33 \times 1.5f_1 = 2f_1$.

The amplitude equalizer is fed from an emitter follower (low source impedance) and followed by an emitter follower (high load impedance).

The phase equalizer [Fig. 4(b)] is driven from a paraphase amplifier (differential amplifier) and has a positive decreasing phase characteristic with a range from 180 to 0 degrees (Fig. 3). Adding this to the phase characteristic of the bridged-T network and to the +90 degrees of the magnetic recorder, an overall phase shift of 180 degrees results within the bandwidth of the system (Fig. 3, curve a), thereby restoring the original phase relationship of the input spectrum. The simplicity and compactness of the whole equalizer network is its most outstanding advantage. Close tolerances are not required; reactive components with 5 percent accuracy are used.

The use of this equalization method in double frequency code resulted in an increase in bit density by a factor of three, as compared to the unequalized case. The waveshape is very clearly defined with sharp resolution, as can be seen in Fig. 1(b) and (c). A small timing error was measured at points where the pattern changes, that is, where the binary 1 waveshape changes to binary 0 or vice versa. This results from two sources: limited amount of pulse crowding in asymmetrical bit environment (pattern change) and remaining overall phase error in the system. Timing error at these points in the overequalized condition [Fig. 1(c)] was 5 percent of the bit period; in the flat equalized case [Fig. 1(b)] it was 7.5 percent. These small errors are insignificant from the standpoint of detection accuracy.

Equalization in NRZ Code

The equalization technique, specifically, the use of the bridged-T amplitude equalizer and the simple phase equalizer circuit, can be extended also to NRZ code. It can be seen from Fig. 3 that if the curve of the phase equalizer is shifted by about a decade toward the high frequency end of the spectrum (dashed line), being equivalent to reducing R of Fig. 4(b), the overall phase of the system will be very nearly +270 degrees (equivalent to +90 degrees) for the whole frequency range from zero to cutoff frequency $1.5f_1$. Thus the system in this mode plays back a single isolated pulse for each flux transition input, just like the recording head at low densities. This pulse, however, is much narrower than the head output because of the boosting effect of the amplitude equalizer. It also shows correct symmetry, due to the proper phase equalization. Fig. 5 shows the pulse, as it appears at the head and also after equalization. The pulse width at the base line is slimmed down by a factor of three.

It is interesting to note that, as far as the double frequency code is concerned, the single adjustment of the phase equalizer changes the overall phase of the system

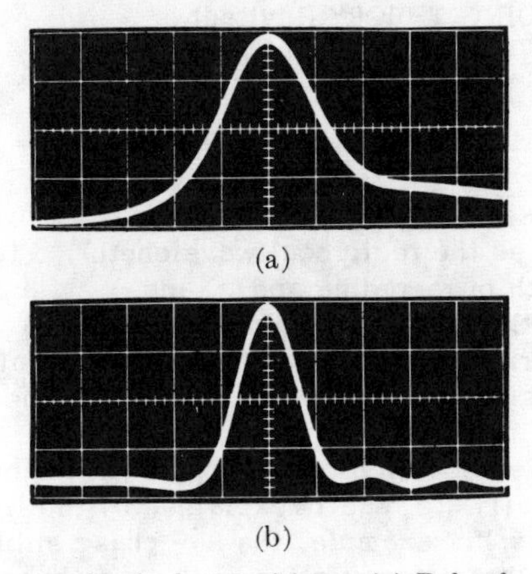

Fig. 5. Response to single transition. (a) Pulse from read head. (b) Pulse after equalization.

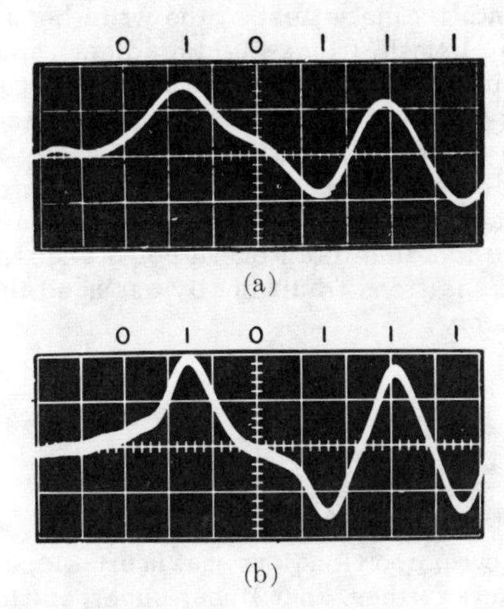

Fig. 6. Waveshapes in NRZ code. (a) Read signal from head. (b) Read signal after equalization.

within the bandwidth of interest (from $0.25f_1$ to $1.5f_1$) by 90 degrees without changing the amplitude response. Accordingly, the output waveshape corresponds to a sequence of either peaks or crossovers with the same resolution. This 90-degree phase shifting property of the phase equalizer is equivalent mathematically to the Hilbert transformation [7].

The fact that the system is able to reproduce individual narrow pulses, properly phase equalized down to zero frequency, is the key to its application in NRZ code. The pulse shape, as shown in Fig. 5(b), resembles a sin x/x function, resulting from the steep boost of the amplitude response of curve c in Fig. 3. This ringing is not objectionable in the phase recording or double frequency codes, because in the crossover waveshape [Fig. 1(b) and (c)] it will be clipped out during detection. However, since ringing at the base line is undesirable in NRZ, the amplitude equalizer network should be designed with much less boost than curve c (Fig. 3). The shape of the amplitude response will be similar to curve c without equalization, but shifted toward higher frequencies. This, of course, results in less pulse slimming, than shown in Fig. 5(b), but it has negligible ringing. Fig. 6 shows a sequence of NRZI binary bits before and after equalization. Besides narrowing the pulse, it is seen how well the symmetry of the 1 bit between the two 0 is restored, due to phase equalization. Indeed, the NRZ system is more sensitive to phase distortions, since waveshape asymmetry results in baseline shift which in severe form will cause erroneous bit detection. Phase equalization, therefore, is an absolute necessity in NRZ recording. Since ringing in the single pulse response is undesirable, the overequalization feature of the system cannot be utilized in NRZ recording. However, the network simplicity combined with the additional benefit of phase equalization can be used to great advantage.

Conclusion

The detrimental effects of pulse crowding in high-density magnetic recording can be improved by amplitude equalization. However, this alone is not enough because waveshape asymmetry is also present in the read head output. Pulse crowding and waveshape asymmetry have different origins but both have the same result: timing errors. Waveshape asymmetry can be corrected by a separate phase equalizer, giving great flexibility to the magnetic recording channel. The same system is equally applicable to NRZ and continuous recording codes, with the additional advantage of overequalization in the latter. The circuitry used in the equalizer is extremely simple. Elimination of timing errors enhances the reliability of the system or allows the use of increased bit densities.

Acknowledgment

The very valuable contributions of J. Gleitman during the whole course of development of this equalizer system are gratefully acknowledged. He also originated the use of the bridged-T circuit and the phase shifter network in the equalizer which resulted in simplification of design procedure. Contribution of J. R. Hall of the West Coast Division is also acknowledged for recognizing the importance of overequalization.

References

[1] A. S. Hoagland, *Digital Magnetic Recording*. New York: Wiley, 1963, pp. 130–140.
[2] A. V. Davies, "Effects of the writing process and crosstalk on the timing accuracy of pulses in NRZ digital recording," *IEEE Trans. Magnetics*, vol. MAG-3, pp. 217–222, September 1967.
[3] D. F. Eldridge, "Magnetic recording and reproduction of pulses," *IRE Trans. Audio*, vol. AU-8, pp. 42–57, March–April 1960.
[4] M. F. Barkouki and I. Stein, "Theoretical and experimental evaluation of RZ and NRZ recording characteristics," *IEEE Trans. Electronic Computers*, vol. EC-12, pp. 92–100, April 1963.
[5] H. M. Sierra, "Increased magnetic recording readback resolution by means of a linear passive network," *IBM J. Res. Develop.*, vol. 7, pp. 22–33, January 1963.
[6] C. L. Cuccia, *Harmonics, Sidebands and Transients in Communication Engineering*. New York: McGraw-Hill, 1952, pp. 310–315.
[7] R. M. Bracewell, *The Fourier Transform and Its Applications*. New York: McGraw-Hill, 1965, pp. 267–274.

A UNIFIED VIEW OF HIGH DENSITY DIGITAL RECORDING THEORY (Invited)

John C. Mallinson*

ABSTRACT

In an attempt to unify digital recording theory, the principal ideas concerning pulse response, linear superposition, medium properties, head geometry, channel codes, signal-to-noise ratio, post-equalization and achievable bit densities are discussed.

INTRODUCTION

Over the last decade, the lineal and areal densities achieved in digital magnetic recording have increased by several orders of magnitude and further increases seem likely.[1] Although these practical achievements have been accompanied by extensive and continuing theoretical studies, the overall status of high density digital recording theory remains rather unsatisfactory because it lacks cohesion. In particular, the relatively sparse attempts to apply the results of communications theory to the subject of magnetic recording have been inconclusive. In this paper, an attempt is made to collect the major theoretical precepts and to demonstrate that a logical prescription for system optimization can be established.

Topics discussed are the similarity between unbiased sine wave and digital recording, the pulse response and linear superposition, the effects of recording medium properties and head geometry, the choice of channel codes and their power spectral densities, the medium noise and intersymbol interference, optimum filters and system equalization philosophy, and finally, lineal and areal packing densities. Several important aspects of digital recording are not treated; the principal omissions are the incidence of dropouts, the closely allied topic of error detection and correction code selection and factors related to mechanical stability.

No claim is made to mathematical regour in this paper. Whilst some steps are demonstrably true, others are mere approximations which, in some cases, can only be justified by appealing to physical plausibility; some of the facts are, of course, simply not understood at the moment. The shorthand notation $e(x) \leftrightarrow E(k)$ indicates a Fourier transform pair.

THE RECORDING PROCESS

At short wavelengths, a magnetic recorder behaves as a low-pass filter and, consequently, almost indistinguishable output voltage waveforms result from sine wave or square wave input currents; this is due to both write process and reproduce process losses.[2] Since the spectrum observed at short wavelengths depends strongly upon the amplitude of the write current, it is usual to optimize the write current so that very nearly the maximum output is achieved at all short wavelengths. Under these conditions the complex spectrum of the output is well approximated by[3],

$$W(k) \;\alpha\; jS(k)\,(1-e^{-|k|d})\,e^{-|k|(a+f)} \qquad (1)$$

Manuscript received February 6, 1975.

*Ampex Corporation, Redwood City, CA 94063

where $j = \sqrt{-1}$, $S(k) = \pm 1$ for $k \geq 0$, $k < 0$, k is the wavenumber (2π times the reciprocal wavelength), d is the effective depth of recording and a and f are, respectively, the head-to-medium spacing and an empirical write process loss factor. Note that a + f may be considered as an effective head-to-medium spacing.

With the exception of the $\exp(-|k|f)$ factor, all the terms in Eq. (1) are readily explained by the linear reproduce process; for example, the 90° phase shift associated with the differentiation of the reproduce head flux leads to the factors $jS(k)$. The $\exp(-|k|f)$ factor, on the other hand, is due to the nonlinear write process and represents the net effect of the write head-medium spacing, gap-length, fringe-field gradient, and self-demagnetization; it is perhaps surprising that all these complicated phenomena may be represented accurately (see Fig. 1) by so simple a factor.

The simplest explanation of these facts seems to be that the short wavelength write process magnetizes the medium according to an overlapping sequence of the arctangent transitions traditionally assumed in digital recording theory,[4]

$$m(x) \;\alpha\; \tan^{-1}\frac{x}{f} \qquad (2)a$$

or

$$M(k) \;\alpha\; \frac{e^{-|k|f}}{jk} \qquad (2)b$$

where f is now proportional to maximum slope of the transition and, further, that linear superposition (LSP) is valid.

According to LSP, if an isolated input transition causes an output pulse e(x), then the composite output waveform is[5]

$$b(x) = \sum_{i=-\infty}^{\infty} a_i\, e(x - i\Delta)\ , \qquad (3)$$

where Δ is the bit (or transition) interval and a_i (= $\pm$ 1 or 0) is the transition polarity. Theoretically, LSP is expected to hold when the input transition rise time is less than or equal to the bit interval Δ, the density is not too high (say $<10^4$ reversals/cm), and self-demagnetizing fields are not large.[6]

The validity of LSP is a crucial issue in high density recording because it admits the application of linear post-equalization to the intersymbol interference problem discussed below. The precision with which the data in Fig. 1 follows the form $\exp(-|k|f)$ may be taken as an indirect confirmation of LSP. In FM video recorders, linear post-equalization is employed to achieve essentially distortion-free operation and this is also compelling evidence in support of LSP.

The write process loss factor f has been studied by many investigators. The simplest models merely set the peak self-demagnetizing field equal to the medium coercivity;[7] the most complex involve self-consistent iterative machine computations.[8] When applied to thin media very good agreement with experiment is obtainable

Reprinted from *IEEE Trans. Magn.*, vol. MAG-11, pp. 1166–1169, Sept. 1975.

rather easily in an analytic approximation which takes into account the spatial gradients of the write head fringing and the medium self-demagnetizing fields and the shape of the medium M-H loops.[9] For very small head-medium spacings ($a \approx 0.3\ \mu m$), this approach yields $f \approx 0.3 \mu m$ for optimized partial penetration recording on relatively thick γ-Fe_2O_3 and $f \approx 0.5\ \mu m$ for typical plated media; these values are consistent with the measured data (see Fig. 1).

Reproduction of the magnetization transition (2) is, of course, a purely linear process and the system transfer function is just (1) with $f=0$. The isolated output pulse expected is, therefore,

$$e(x) \propto \log \frac{(a+d+f)^2 + x^2}{(a+f)^2 + x^2} \qquad (4)a$$

$$E(k) \propto S(k) \frac{(1-e^{-|k|d})\, e^{-|k|(a+f)}}{k}, \qquad (4)b$$

in fair agreement with experiment. The most notable discrepancy in (4) is that $e(x)$ remains always positive and that, concomittantly, $E(k)$ has a non-zero dc ($k=0$) value. In actual fact, recorders cannot reproduce dc signals for several reasons, the most fundamental of which concerns the geometry of the reproduce head.[10] Just as the gap-length limits the short wavelength head response, so the pole-tip-length determines the long wavelength limit. With conventional head designs wavelengths as long as 1 - 2 cm are reproducible; the use of batch fabricated thin film heads will permit only narrow bandwidths to be reproduced.

CODES, NOISES AND EQUALIZERS

The bandpass nature of a recorder introduces many difficulties which may be discussed conveniently in terms of the missing long and short wavelengths respectively.

When random digital transitions are recorded, the absence of long wavelength response can cause the output waveform to be severely distorted; the terms "staircasing" or "wandering base-line" are often used to describe the effect. Channel coding can reduce the distortion by rearranging the transitions in such a way that very little dc or long wavelength power is present at the write head input. In Manchester (also called bi-phase or frequency doubling) coding, the average dc content is identically zero, the input power spectral density (PSD) being of the form,[11]

$$PSD_{in}(k) \propto \frac{1}{k^2 \Delta} \left(1 - \cos \frac{k\Delta}{2}\right)^2, \qquad (5)$$

where Δ is the bit interval. Other codes, for example Miller (sometimes known as Modified Frequency Modulation) coding, have very low average dc powers. In addition to the PSD, another measure of the long wavelength content of codes, which has recently been employed, is the digital sum variation.[12]

The limited short wavelength response of the recorder causes the unequalized output pulse $e(x)$ to be undesirably wide. If the reproduce gap loss is negligible, the 50% amplitude pulse width is given by $2\sqrt{(a+f)(a+f+d)}$; this usually exceeds $2\ \mu m$ and may well be much greater. At high transition densities, many adjacent pulses overlap and this is the basic cause of "peak-shift", "pulse-crowding" and other manifestations of intersymbol interference.[13] In peak-shift, the pulses behave as though they mutually repel each other; several pre-equalization schemes, for example time or phase compensations, have been used to mitigate the effect. Here we discuss only linear post-equalization, because it is both widely employed and simply implemented.

The correct post-equalizer filter transfer function is determined by the play-off of two conflicting requirements. In order to reduce intersymbol interference, by pulse compression, the short wavelengths must be amplified which reduces the signal-to-noise ratio (SNR); increasing the SNR, on the other hand, broadens the pulse and causes greater intersymbol interference. A logical solution to this dilemma is to regard the intersymbol interference as a noise and at very high densities, when tens of pulses overlap, this is statistically justifiable. The problem then reduces to the familiar case of detecting one pulse (or set of pulses) against the noise due to the other pulses and other noise sources.

The PSD of the intersymbol noise is approximately equal to the PSD of the recorder output (omitting the one subject pulse makes very little difference at high densities) and when linear superposition holds this is,

$$PSD_{out}(k) = PSD_{in}(k) \cdot |W(k)|^2. \qquad (6)$$

Moreover, for codes having a fundamental repeating character (or set of pulses) $g(x)$, with output spectrum $G(k)$, it follows that[3]

$$PSD_{out}(k) = \frac{1}{\Delta} |G(k)|^2. \qquad (7)$$

Other noise sources include the medium, the reproduce head and the preamplifier but, in well-designed recorders, the medium noise is dominant. The noise power spectrum (NSP) due solely to the particulate nature of the medium is known,[14] is purely additive and has the form,

$$NPS(k) \propto |k| (1-e^{-2|k|d})\, e^{-2|k|a}. \qquad (8)$$

Although this spectrum is modified by multiplicative noises due to physical and magnetic nonuniformities in the medium, the SNR computed on the basis of (1) and (8) is generally within 6 dB of that measured. The total noise voltage spectrum is, therefore,

$$N(k) = \sqrt{NPS(k) + \frac{|G(k)|^2}{\Delta}} \text{ (volts)}. \qquad (9)$$

against which the polarity of the code symbol $g(x) \leftrightarrow G(k)$ must be determined.

Communication theory prescribes an optimum filter structure for this problem,[15] which yields the lowest probability of error. It is simplest first to discuss this optimum post-equalizer as a cascade of two filters: the first may be termed a noise whitening filter (NWF) and the second a matched filter (MF) (see Fig. 2).

The transfer function of the NWF is just the reciprocal of (9)

$$NWF(k) = \left[NPS(k) + \frac{1}{\Delta}|G(k)|^2\right]^{-1/2} \qquad (10)$$

and it changes the code symbol g(x) to h(x) where H(k) = G(k) · NWF(k). The matched filter is such that its impulse response is h(-x) and it follows, therefore, that

$$MF(k) = G^*(k) \cdot NWF(k) , \qquad (11)$$

where * denotes the complex conjugate.

Now regarding the NWF and MF cascade as but a single entity, the post-equalization filter (PEF), we have

$$PEF(k) = G^*(k) \left[NPS(k) + \frac{1}{\Delta} \left| G(k) \right|^2 \right]^{-1} \qquad (12)$$

and the equalized code symbol i(x) is now such that

$$I(k) = \left|G(k)\right|^2 \left[NPS(k) + \frac{1}{\Delta} \left| G(k) \right|^2 \right]^{-1} \qquad (13)$$

It is instructive to examine (12) and (13) in the limiting condition, of greatest interest in high density recording, where the intersymbol interference noise dominates. Now the post-equalizer transfer function becomes simply $\Delta \left[G(k)\right]^{-1}$, which is a "flat" equalizer, and the output i(x) becomes an ideal impulse. Naturally, this condition cannot be realized in practice. It is only possible to achieve the "flat" equalization over a limited bandpass and the output i(x) is, therefore, oscillatory.

These oscillations suggest the possibility of further refinements in the post-equalizer. It has been shown[16] recently that, by judiciously adjusting the amplitude and/or phase response of a bandpass channel, the output pulse zero crossings may be forced to occur at equal intervals. When such a zero-forcing equalizer is adjusted to make this interval equal to the bit interval Δ, the intersymbol interference vanishes instantaneously during each bit interval and, at those instants, the detection process need only be limited by the recording medium noise.

The lineal and areal bit densities which may be achieved when the medium noise so dominates have been analyzed[17] recently. It has been shown that, for a given SNR, the lineal and areal densities are proportional to $[w]^{1/2}$ and $[w]^{-1/2}$ respectively, where w is the trackwidth. Thus, in seeking higher areal densities, it is always advantageous to use narrower tracks. For γ-Fe_2O_3 media it has been computed that a satisfactorily high SNR (about 20 dB) is available on 25 μm (1 mil) track width at an areal density of 3×10^6 bits/cm^2 (2×10^7 bits/in^2). Similar conclusions were reached in a separate study which included also the effects of preamplifier noise.[18]

CONCLUSIONS

The relationship of the major theories on high density recording has been discussed and it has been shown that no basic contradictions exist. The application of these theories leads to the conclusion that standard γ-Fe_2O_3 media can support areal densities greater than 3×10^6 bits/cm^2. Since this density is a factor of ten higher than occurs in current production machines, we may conclude that the development of systems with significantly higher densities is not precluded by consideration of signal and noise alone.

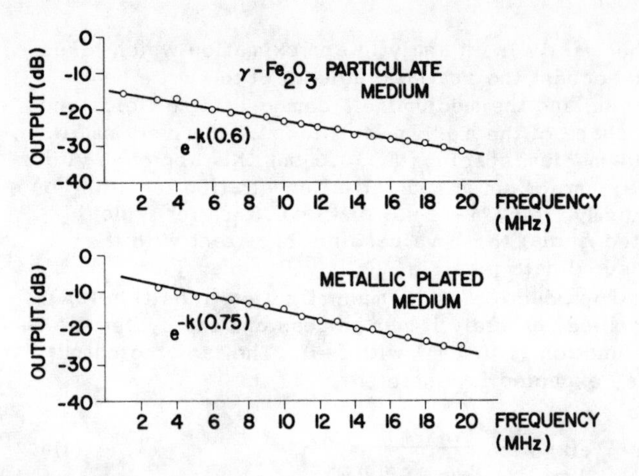

Fig. 1, Output in decibels versus frequency (MHz) for γ-Fe_2O_3 and plated media. The output has been adjusted for write and read head efficiencies, read head gap-length, thickness of the medium and the reproduce system response. The ferrite head-to-medium speed is 40 m/sec. The effective spacings are 0.6 and 0.75 μm respectively.

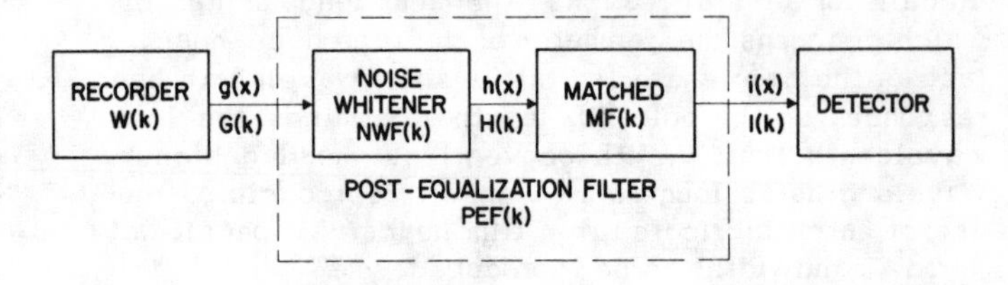

Fig. 2, An optimum filter structure showing how the post-equalizer may be regarded as having two functions, noise-whitening and matched filtering.

REFERENCES

1. J. Mallinson and V. Ragosine, "Bulk Storage Technology - Magnetic Recording", IEEE Trans. Magn., Vol. MAG-7, pp 598-600, Sept 1971.
2. J. Mallinson and C. Steele, "A Computer Simulation of Unbiased Sine Wave Recording", IEEE Trans. Magn., Vol. MAG-7, pp 249-254, June 1971.
3. J. Mallinson, "Applications of Fourier Transforms in Digital Magnetic Recording Theory", IEEE Trans. Magn., Vol. MAG-10, pp 69-77, March 1974.
4. R. Potter, "Digital Magnetic Recording Theory", IEEE Trans. Magn., Vol. MAG-10, pp 502-508, Sept. 1974.
5. J. Mallinson and C. Steele, "Theory of Linear Superposition in Tape Recording", IEEE Trans. Magn., Vol. MAG-5, pp 886-890, Dec. 1969.
6. D.L.A. Tjaden, "Some Notes on Superposition in Digital Magnetic Recording", IEEE Trans. Magn., Vol. MAG-9, pp 331-335, Sept. 1973.
7. D. W. Chapman, "Theoretical Limit on Digital Recording Density", Proc. IEEE, Vol. 51, pp 394-395, 1963.
8. R. Potter and R. Schmulian, "Self-Consistently Computed Magnetization Patterns in Thin Magnetic Recording Media", IEEE Trans. Magn., Vol. MAG-7, pp 873-880, Dec. 1971.

9. M. Williams and R. Comstock, "An Analytic Model of the Write Process in Digital Magnetic Recording", AIP Conf. Proc. Magnetism and Magnetic Materials, Vol. 5, pp 738-742, 1971.
10. J. Mallinson, "On Recording Head Field Theory", IEEE Trans. Magn., Vol. MAG-10, pp 773-775, Sept. 1974.
11. A. Knoll, "Spectrum Analysis of Digital Magnetic Recording Waveforms", IEEE Trans. Elec. Comps., Vol. EC-16, pp 732-743, Dec. 1967.
12. R. Kiwimagi, J. McDowell, M. Ottesen, "Channel Coding for Digital Recording", IEEE Trans. Magn., Vol. MAG-10, pp 515-518, Sept. 1974.
13. G. Jacoby, "Signal Equalization in Digital Magnetic Recording", IEEE Trans. Magn., Vol. MAG-4, pp 302-305, Sept. 1968.
14. J. Mallinson, "Maximum Signal-to-Noise Ratio of a Tape Recorder" , IEEE Trans. Magn., Vol. MAG-5, pp 182-186, Sept. 1969.
15. B. Lathi, "Communication Systems", J. Wiley, New York 1969, p. 400 et. seq.
16. R. A. Gibby and J. W. Smith, "Some Extensions of Nyquist's Telegraph Transmission Theory", Bell System Technical Journal, Vol. 44, pp 1487-1510, Sept. 1965.
17. J. Mallinson, "On Extremely High Density Digital Recording", IEEE Trans. Magn., Vol. MAG-10, pp 368-373, June 1974.
18. W. B. Philips and H. P. McDonough, "Maximizing the Areal Density of Magnetic Recording", IEEE COMPCON Proc. pp 101-103, 1974.

ACKNOWLEDGEMENT

The writer thanks J. Miller (Ampex), for permission to use his recording data shown in Fig. 1, and Prof. E. Wong (U. C. Berkeley), for many helpful suggestions.

Author Index (Reprint Papers)

Subject Index

T

V

W

Editor's Biography

Robert M. White (M'71) received the B.S. degree in physics from the Massachusetts Institute of Technology, Cambridge, in 1960, and the Ph.D. degree from Stanford University, Stanford, CA.

After a year at the University of California, Berkeley, he returned to Stanford as Assistant Professor of Physics. He spent 1970 as a National Science Foundation Senior Postdoctoral Fellow at the Cavendish Laboratory. In 1971 he joined Xerox Corporation's Palo Alto Research Center (PARC), where he was Manager of Solid State Research and Manager of Storage Technology. While at PARC, he was also Consulting Professor in Applied Physics at Stanford. He is currently Vice President of Engineering and Technology at Control Data, Minneapolis, MN. His research has focused on magnetic phenomena, in particular the optical properties of magnetic materials, the magnetic properties of amorphous materials, and the physics of magnetic recording. He is the author of 90 technical papers as well as *Quantum Theory of Magnetism* (New York: McGraw-Hill, 1970, 1st ed.; New York: Springer-Verlag, 1985, 2nd ed.) and *Long Range Order in Solids* (with T. Geballe; New York: Academic, 1979).

Dr. White is a Fellow of the American Physical Society, was a Distinguished Lecturer of the Magnetics Society in 1980, and received an Alexander von Humboldt Prize from West Germany in 1980. He recently chaired the National Materials Advisory Board Committee on Magnetic Materials and is Program Chairman for the 1985 International Conference on Magnetism.